Markus Maximilian Scherer

Turbulent coherent structures, Secondary currents and Sediment ridges

Dissertationsreihe am Institut für Hydromechanik
Karlsruher Institut für Technologie (KIT)

Heft 2023/2

Eine Übersicht aller bisher in dieser Schriftenreihe erschienenen Bände finden Sie am Ende des Buches.

Turbulent coherent structures, Secondary currents and Sediment ridges

by
Markus Maximilian Scherer

KIT Scientific Publishing

Karlsruher Institut für Technologie
Institut für Hydromechanik

Turbulent coherent structures, Secondary currents and Sediment ridges

Zur Erlangung des akademischen Grades eines Doktor-Ingenieurs von der KIT-Fakultät für Bauingenieur-, Geo- und Umweltwissenschaften des Karlsruher Instituts für Technologie (KIT) genehmigte Dissertation

von Markus Maximilian Scherer, M.Sc.

Tag der mündlichen Prüfung: 4. Februar 2022
Referent: Prof. Dr. Markus Uhlmann
Korreferent: Prof. Dr. Paolo Blondeaux

Impressum

KIT Scientific Publishing

Karlsruher Institut für Technologie (KIT)
KIT Scientific Publishing
Straße am Forum 2
D-76131 Karlsruhe

KIT Scientific Publishing is a registered trademark
of Karlsruhe Institute of Technology.
Reprint using the book cover is not allowed.

www.ksp.kit.edu

Print on Demand 2023 – Gedruckt auf FSC-zertifiziertem Papier

ISSN 1439-4111
ISBN 978-3-7315-1318-6
DOI 10.5445/KSP/1000161247

To the memory of Günter,
Gertrud and Heinz Scherer

Thank you for everything

Abstract

Sediment bedforms arise in many fluvial ecosystems including rivers, estuaries or the oceans due to a complex interaction between a turbulent shear flow and the erodible river or seabed. The resulting sedimentary patterns can take a variety of different shapes and occur at a wide range of scales, from small sediment ridges over medium sized sand dunes to massive marine sand banks. A fundamental understanding of the physical mechanisms that underlie this pattern formation processes is crucial for environmental and hydraulic engineers likewise, as the sedimentary bedforms have the potential to significantly modify the river's flow characteristics – possibly with severe consequences for the transport processes in the fluvial system and thus for its ecological status. In the engineering practice, it is therefore of great importance to have reliable tools to hand that allow accurate predictions of both the future river bed evolution and its impact on the natural ecosystem. Of special interest are in this regard the evolution of streamwise-aligned sediment ridges and accompanying large-scale secondary currents in the turbulent flow, as those latter are well-known to significantly enhance mixing processes of heat and momentum across the water body. Despite decades of intensive research, the processes which give rise to these sediment ridges and the related turbulent secondary flow cells are even nowadays not fully understood and accurate reduced-complexity models are still lacking.

In this thesis, our aim therefore is to contribute to the fundamental understanding of the mechanisms that lead to the formation of sediment ridges and secondary currents by means of high-fidelity direct numerical simulations of open channel and open duct flows. The performed computations reach bulk and friction Reynolds numbers of 9500 and 830, respectively, and include up to 1.4 million fully-resolved mobile particles. In order to resolve all flow scales including those smaller than the particle size, the applied numerical method incorporates an immersed boundary technique, which is then coupled with a soft-sphere discrete element model suitable to accurately describe the exchange of momentum during particle contacts. The resulting high-fidelity datasets allow a deep insight into the interplay between turbulent coherent structures and the erodible sediment bed and will hopefully help to improve existing engineering-type models in the long run.

Based on the newly-created database, we discuss the physical processes that trigger the evolution of sediment ridges on an initially macroscopically flat sediment bed and the corresponding development of secondary currents in the turbulent flow. In the classical literature, two conceivable formation mechanisms have been proposed (Nezu & Nakagawa, *IAHR Monograph*, 1993), but sound experimental or numerical evidence for their validity is still lacking to the best of our knowledge. In the first mechanism, an initial lateral sediment bed perturbation causes a spanwise disturbance of the bed shear stress, which, in turn, triggers the development of large-scale secondary currents. The second conceived mechanism is 'reversed' compared to the former in that the flow field exhibits a lateral variation that induces a laterally varying bed shear stress and thus a heterogeneous sediment erosion along the bed that ultimately leads to the development of local troughs and ridges.

In the current work, we analyse both conceived mechanisms concerning their relevance for the sediment ridge evolution under fully-turbulent conditions. First, we implement and reanalyse Colombini's theoretical model (*J. Fluid Mech.*, vol. 254, 1993, p. 701-719): In agreement with the first proposed mechanism, the linearised stationary model can be seen as a 'perturbation-response' system in which the flow is perturbed by a lateral sinusoidal bed variation of infinitesimal amplitude, while the most-amplified wavelength of this bed perturbation is determined by the strongest response of the linearised

Navier-Stokes operator in form of a maximised lateral shear stress. In this sense, the sediment bed takes the role of a stationary external forcing on the linear operator, as is shown by comparison with studies on the forced Orr-Sommerfeld problem and on transient growth in canonical channel flows.

In contrast to the linearised model system, the newly performed direct numerical simulations reveal that the impact of small bed perturbations on the turbulent flow is negligible and thus is of minor importance for the formation of sediment ridges in the fully-turbulent time-dependent case. Instead, laterally alternating turbulent large-scale velocity streaks that represent a finite-amplitude perturbation of the mean flow profile induce a laterally varying bed shear stress and erosion rate. That is, sediment ridges evolve in regions of relatively weaker erosion activity below large-scale low-speed streaks and troughs *vice versa* below the corresponding high-speed streaks. Such large-scale coherent velocity structures are a characteristic feature of wall-bounded shear flows over both smooth bottom walls and mobile sediment beds that typically exhibit a lateral spacing of between one and two mean fluid heights, explaining the regular spacing of sediment ridges at a comparable lateral wavelength. The interaction of large-scale velocity streaks with the erodible bed can be described as a 'top-down process', since the lateral organisation of the sediment ridges is observed to adapt to changes in the arrangement of large-scale flow structures with a time lag of several bulk time units, which is in fairly good agreement with the conceptual model on causality in canonical flows by Jiménez (*J. Fluid Mech.*, vol. 842, 2018, P1, § 5.6). The mean secondary currents, finally, are seen to represent the statistical footprint of the well-organised large-scale streaks and their associated Reynolds stress-carrying structures when averaged in time and the streamwise direction.

Sediment ridges are observed to form due to the outlined mechanism in the absence of nearby lateral sidewalls in the open channel and the core region of wide open ducts likewise, while a slightly different formation process takes place in the vicinity of sidewalls, where the sediment ridges interact with a 'pre-existing' sidewall-induced mean secondary flow. In this context, a similar 'top-down mechanism' is seen to initiate the formation of sediment ridges, but the relevant coherent velocity structures are located closer to the wall and are of smaller size than the afore described. The mobile sediment bed conversely alters the sidewall-induced mean secondary flow whose overall intensity increases compared to the smooth-wall case due to a strengthening of the outer secondary currents, while the mean secondary bottom vortices near the sediment bed-sidewall corners are seen to reduce in strength. The latter phenomenon is related to a rearrangement of instantaneous quasi-streamwise vortices in the vicinity of a permeable mobile sediment bed compared to an impermeable bottom wall. Agglomerations of such small vortices in form of larger-scale vortex clusters reveal a preferential organisation very similar to the mean secondary currents, which could imply that the mean secondary currents are the statistical footprint of these vortex clusters.

The analysis concludes with an outlook on the final stage of sediment ridge development in both laterally periodic channels and wide open ducts, during which transverse ripple-like patterns evolve and eventually dominate or entirely replace the faster-developing sediment ridges. The ridge-covered sediment bed is seen to gradually transform into a ripple-dominated bed by the evolution of small sediment seeds superimposed on the sediment ridges that propagate downstream while their amplitude is growing. Having reached a certain amplitude, a trough region develops in the lee side of these initial seeds that eventually leads to a streamwise disconnection of the former sediment ridges. In a final phase, the localised seeds synchronise to form higher-amplitude transverse bedforms that span over widths of up to 16 times the mean fluid height.

Zusammenfassung

Sedimenttransportkörper (oft auch als Sedimentformationen bezeichnet) entstehen in vielen natürlichen Ökosystemen durch die Wechselwirkung einer turbulenten Scherströmung mit einem erodierbaren Sedimentbett. Die Bandbreite von in der Natur auftretenden Sedimenttransportkörpern reicht dabei von kleinen Rippeln in schmalen Bachläufen bis zu Sanddünen von mehreren Kilometern Länge. Sogenannte "sediment ridges" (in deutscher Literatur etwa "Sedimentgrate", "Sedimentrücken") sind längliche Sedimenttransportkörper, die sich parallel zur Hauptströmungsrichtung ausbilden und eng mit Sekundärströmungen der zweiten Prandtl'schen Art zusammenhängen. Ein grundlegendes Verständnis der komplexen Wechselwirkungen zwischen turbulenter Scherströmung und den sich ausbildenden "sediment ridges" ist von großer Wichtigkeit sowohl für die Unterhaltung bestehender Wasserstraßen und hydraulischer Bauwerke als auch für den Schutz bestehender aquatischer Ökosysteme: Die sich entwickelnden Sedimentstrukturen und die damit einhergehenden Sekundärströmungen verändern nicht nur das Flussbett und die Ablagerung von Sediment entlang diesem, sondern üben auch einen starken Einfluss auf zentrale Misch– und Transportvorgänge im Gewässer aus. Dies betrifft zum einen die Ausbreitung von Wärme und gelösten Gasen wie Sauerstoff im Gewässer, zum anderen kann aber auch die Verbreitung umweltschädlicher Stoffe erhöht werden.

Ziel der vorliegenden Arbeit ist es, zu einem besseren Verständnis derjenigen Mechanismen beizutragen, die die Entwicklung dieser "sediment ridges" und der mit ihnen einhergehenden Sekundärströmungen beeinflussen. Insbesondere gilt es zu untersuchen,

(i) welche Mechanismen zur Entstehung von "sediment ridges" und Sekundärströmungen führen,

(ii) wie beide in Wechselwirkung mit kohärenten Strukturen in der turbulenten Strömung stehen,

(iii) welchen Einfluss eine durch Seitenwände induzierte Sekundärströmung auf die Entwicklung der Sedimentstrukturen ausübt und

(iv) wie sich schließlich größere Rippel aus den "sediment ridges" entwickeln.

Zu diesem Zweck werden hochaufgelöste direkte numerische Simulationen (DNS) von offenen Kanalströmungen mit und ohne Seitenwände(n) bis zu einer Reynoldszahl von 9500 (definiert basierend auf der mittleren Fluidhöhe und der mittleren Strömungsgeschwindigkeit) durchgeführt, was einer Reibungs-Reynoldszahl von ca. 830 entspricht. Bis zu 1,4 Millionen voll-aufgelöste, frei bewegliche Partikel bilden dabei das erodierbare Sedimentbett. Um alle relevanten Skalen der turbulenten Strömung inklusive der lokalen Umströmung einzelner Partikel darstellen zu können, wird eine "immersed boundary"–Methode verwendet, die wiederum mit einem "soft-sphere discrete element model" gekoppelt ist, um den Impulsaustausch beim Kontakt mehrerer Partikel zu bestimmen.

Klassischerweise werden zwei mögliche Mechanismen unterschieden, die zur Entwicklung von "sediment ridges" und der entsprechenden Sekundärströmung über einem anfangs flachen Bett führen können (Nezu & Nakagawa, *IAHR Monograph*, 1993). Jedoch ist bisher nicht geklärt, welcher der beiden tatsächlich für die Entwicklung der "sediment ridges" in einer voll-turbulenten Strömung verantwortlich ist. Im ersten der beiden Mechanismen regt eine Störung des ansonsten flachen Sedimentbettes in der Querrichtung eine laterale Variation der Bettschubspannung an, welche dann wiederum zur Enwicklung einer großskaligen Sekundärströmung führt. Der zweite Mechanismus wirkt quasi umgekehrt, indem eine anfängliche laterale Variation des Strömungsfeldes eine Oszillation der Bettschubspannung quer zur Hauptströmungsrichtung induziert, mit der eine ebenso lateral variierende

Erosionsrate einhergeht. Die inhomogene Sedimenterosion in der Querrrichtung führt schlußendlich zur Ausbildung lokaler Senken und Erhebungen auf dem Sedimentbett, wobei letztere die "sediment ridges" darstellen.

Lineare Instabilität einer turbulenten Strömung über "sediment ridges". In der vorliegenden Arbeit werden basierend auf neugewonnenen Datensätzen beide möglichen Mechanismen auf ihre Relevanz für den Sedimenttransport in voll-turbulenten Strömungen untersucht. Zu diesem Zweck wird zunächst das theoretische Modell von Colombini (*J. Fluid Mech.*, vol. 254, 1993, S. 701-719) zur Entwicklung von "sediment ridges" und großskaligen Sekundärströmungen implementiert und eingehend analysiert. Letzteres beschreibt, wie eine laterale sinusförmige Variation des Sedimentbettes mit infinitesimaler Amplitude in einer ebenen Kanalströmung entsprechend dem ersten Mechanismus die Entwicklung einer turbulente Sekundärströmung anregen kann. Die mittlere laterale Distanz zwischen zwei "sediment ridges" entspricht dabei derjenigen Störungswellenlänge, die den linearisierten Navier-Stokes-Operator zur maximalen lateralen Bettschubspannung anregt. In diesem Sinne übernimmt die laterale Variation der Sedimentbettoberfläche hier die Rolle eines externen Kraftfeldes, welches auf den Navier-Stokes-Operator wirkt. Diese Analogie wird durch einen Vergleich mit anderen Studien herausgearbeitet, die sich mit dem klassischen Orr-Sommerfeld-Problem unter externem Krafteinfluss sowie dem vorübergehenden Wachstum optimaler Eingangsbedingungen ("transient growth") für den Navier-Stokes-Operator in kanonischen Kanalströmungen beschäftigen.

Großskalige kohärente Strukturen und ihre Wechselwirkung mit "sediment ridges" in offenen Kanalströmungen ohne Seitenwände. Die im Rahmen der Arbeit durchgeführten direkten numerischen Simulationen zeigen im Gegensatz zu diesem linearen Modell, dass der erste Mechanismus einer durch das Bett induzierten Sekundärströmung im Falle einer voll-turbulenten, zeitabhängigen Strömung von untergeordneter Bedeutung ist. Stattdessen wird beobachtet, dass die laterale Variation von Bettschubspannung und Erosionsrate in diesem Fall durch großskalige kohärente Strukturen, sogenannte "large-scale velocity streaks" (großräumige zusammenhängende Regionen relativ höherer oder niedrigerer Geschwindigkeit), induziert wird. Im Gegensatz zum vorher diskutierten linearen Modell stellen diese Strukturen eine Störung des mittleren Geschwindigkeitsprofils mit finiter Amplitude dar. Durch die Interaktion der großskaligen "velocity streaks" mit dem Sedimentbett entwickeln sich Sedimentansammlungen in Form von "sediment ridges" in Regionen geringerer Strömungsgeschwindigkeit und somit schwächerer Erosionsaktivität, wohingegen sich lokale Senken dort ausbilden, wo die Erosionsrate aufgrund höherer Fließgeschwindigkeiten erhöht ist. Großskalige "velocity streaks" dieser Art sind ein charakteristischer Bestandteil wandnaher Scherströmungen und treten gleichermaßen über einer glatten Wand und über einer erodierbaren beweglichen Flusssohle auf. Typischerweise weisen diese gut organisierten Strukturen einen mittleren Abstand vom ein– bis zweifachen der mittleren Fluidhöhe in der Richtung quer zur Hauptströmungsrichtung auf, was die gleichmäßige Anordnung der "sediment ridges" auf dem Sedimentbett mit vergleichbarem lateralen Abstand erklärt. Der beschriebene Interaktionsprozess zwischen den großskaligen Geschwindigkeitsstrukturen und dem erodierbaren Sedimentbett ist in diesem Sinne ein "top-down"–Mechanismus, in Rahmen dessen sich die laterale Organisation der Sedimenttransportkörper mit einem gewissen Zeitverzug von mehreren "bulk time units" (Zeiteinheit skaliert mit der mittleren Strömungsgeschwindigkeit und der mittleren Fluidhöhe) an die Anordnung der großskalige Geschwindigkeitsstrukturen anpasst. Der beobachtete Mechanismus weist damit eine gute Übereinstimmung mit dem konzeptionellen Modell von Jiménez

(*J. Fluid Mech.*, vol. 842, 2018, P1, § 5.6) über die Kausalität in kanonischen turbulenten Strömungen auf. Die mittlere Sekundärströmung stellt schließlich den statistischen Fußabdruck der regelmäßig organisierten großskaligen Geschwindigkeitsstrukturen und der mit ihnen einhergehenden Strukturen starker Reynolds-Spannungen dar, wenn diese über die Hauptströmungsrichtung und die Zeit gemittelt werden.

Kohärente Strukturen und "sediment ridges" in offenen Kanalströmungen mit Seitenwänden. Der beschriebene Mechanismus führt zur Entwicklung von "sediment ridges" in ausreichender Entfernung zu lateralen Berandungen wie den Seitenwänden eines Kanals. In der Nähe solcher Seitenwände besteht dagegen eine Wechselwirkung zwischen den Sedimenttransportkörpern und der durch die Seitenwand induzierten Sekundärströmung, weswegen "sediment ridges" in diesen Regionen einen leicht veränderten Entwicklungsprozess durchlaufen. Zwar ist die Interaktion des Strömungsfeldes mit dem Sedimentbett auch hier ein "top-down"–Prozess, aber die dabei relevanten Geschwindigkeitsstrukturen befinden sich in kleinerer Entfernung zum Sedimentbett und sind von geringerer Größe. Umgekehrt verändert die Anwesenheit einer beweglichen Kanalsohle jedoch auch die durch die Seitenwand induzierte Sekundärströmung: Während deren mittlere Intensität durch eine Verstärkung der großskaligen äußeren Sekundärwirbel zunimmt, reduziert sich die Stärke und Größe der bodennahen Sekundärwirbel in den Ecken zwischen Bett und Seitenwand im Vergleich zum Fall einer einphasigen Strömung über einer glatten Wand. Letzteres Phänomen kann auf eine Umorganisation kohärenter Wirbelstrukturen mit Rotationsachse parallel zur Hauptströmungsrichtung zurückgeführt werden, welche sich über einem beweglichen Sedimentbett anders anordnen als über einer undurchlässigen glatten Wand. Dabei zeigt sich zudem, dass sich größere Gruppen solcher kleinen Wirbel in größeren "Wirbel-Clustern" zusammenschließen, deren bevorzugte Anordnung der Struktur der mittleren Sekundärströmung sehr ähnlich ist. Diese Beobachtung könnte darauf schließen lassen, dass die mittlere Sekundärströmung den statistische Fußabdruck dieser instantanen "Wirbel-Cluster" darstellt.

Der Übergang zu einem Rippel-dominierten Sedimentbett. Am Ende der Arbeit wird ein Ausblick auf die letzte Entwicklungsstufe der "sediment ridges" in breiten Kanalströmungen mit und ohne Seitenwände(n) gegeben. In dieser finalen Phase entwickeln sich auf dem Sedimentbett Rippel–ähnliche Transportkörper, deren Kamm quer zur Hauptströmungsrichtung verläuft und die mit der Zeit langsam stromabwärts wandern. Nach einer Übergangsphase in der beide Arten von Transportkörpern mit ähnlichen Abmessungen beobachtet werden können, beginnen die Rippel die Entwicklung des Sedimentbettes zu dominieren, wobei die zunächst beobachteten "sediment ridges" überlagert und zum Teil gänzlich verdrängt werden. Der Übergang eines von "sediment ridges" dominierten Sedimentbettes hin zu einer Rippel-dominierte Kanalsohle ist fließend: Zunächst entwickeln sich überlagert mit den "sediment ridges" kleine lokale Sedimentanhäufungen, welche mit der Zeit anwachsen und gleichzeitig stromabwärts wandern. Sobald deren Höhe eine bestimmte Schwelle überschreitet, entwickeln sich auf den Leeseiten lokale Senken aus, was schließlich zu einer Aufspaltung der früheren "sediment ridges" in mehrere einzelne Transportkörper führt. In der letzten Phase verbinden sich mehrere dieser lokalen Transportkörper zu größeren Rippeln, deren Kamm sich über die gesamte Kanalbreite vom 16–fachen der mittleren Fluidhöhe erstreckt.

Acknowledgements

First and foremost, I owe sincere gratitude to my advisor Prof. Markus Uhlmann. During the past three and a half years, Markus has been a true mentor to me who gave me the freedom to pursue my own ideas and to bring my fascination for applied mathematics and the physics of turbulence into this work, while making himself always available for me to offer his guidance whenever I felt I would need it. Looking back to the time when I commenced my doctoral studies makes me realise how much I have learned from Markus and benefitted from his tremendous physical and mathematical insights. Without his careful guidance and the freedom he provided me to drive this project in my way, the manuscript that lies now in front of me would not be the same. I thank you for your extraordinary commitment as my supervisor, it was a great pleasure for me to be your PhD student.

Also, I want to express my thanks to Prof. Paolo Blondeaux, University of Genoa, for his interest in my work, for providing me his expertise in several discussions throughout the course of this thesis and for his kind disposition as the second referee. I also thank Prof. Olivier Eiff, Prof. Willy Dörfler and Prof. Mário Franca, KIT, for their interest in my research and their availability as members of my PhD committee. I am further grateful to Prof. Genta Kawahara, Osaka University, for his support, the very insightful discussions on duct flow turbulence and for his willingness to join my PhD committee. I want to thank him and Prof. Susumu Goto as well as their groups at Osaka University for their kind hospitality during our research stay in Japan, which was generously supported by the German Academic Exchange Service (DAAD) and the Japan Society for the Promotion of Science (JSPS). The joint effort of Prof. Markus Uhlmann and Prof. Genta Kawahara to establish this collaboration is very much appreciated and I am thankful for the opportunity to join this research project which was very inspiring for me in both a scientific and a cultural aspect.

Next, I would like to thank all members of the Institute for Hydromechanics (IfH). Since I started to work as a student assistant at the institute more than five years ago, I have truly enjoyed the outstanding working environment full of mutual respect and helpfulness between the colleagues, among which I have found many good friends. The pandemic has shown more than ever how important it is to have a great team of colleagues around you and so I want to thank all current and former scientific members of the IfH team, including Lina, Yulia, Sina, Erni, Manu, Tiago, Swagat, Michele P., Michele T., Antoine, Victor, Luca, Christof and Yoshi. Especially, I want to thank Yoshi for sharing his dataset and discussing many aspects of his work on single-phase open duct flows with me. Having such detailed reference data at my disposal has helped me a lot to assess the impact of the sediment bed on the secondary flow features. Also, it is important to me to express my gratitude to Heidi, Angelika and Helmut for their continuous help throughout my time at the institute. They often worked behind the scenes of the institute's scientific work, but their support in solving the daily IT problems, in organising exams, student projects, tutor contracts and so many things more is essential for all of us. Eventually, I owe special thanks to Agathe, Aman and Michael for their scientific and emotional support right from my very first day at the institute and for their motivation especially in difficult situations. Thank you my friends!

Just as much, I want to thank my friends outside the university for many joyful hours during the past three and a half years which have been an important counterweight to my work on this thesis. Especially, I want to thank Lucas - our discussions and running sessions have helped me to take my mind off and to afterwards return to my computer with new energy.

Next, I want to express my deep gratitude to my family: first and foremost my parents, my brother Patrick, as well as Christine and Micha, Rita and Albert and especially my dear grandparents which we have lost during these last years. Thank you for your support and encouragements, for cheering me up whenever I was struggling and for believing in me during all this time – often more than I did myself. Finally, I want to thank Nadine for her enduring love and for always being by my side. Having you around me was the best "medicine" I can think of to protect me from panic and discouragement. With your empathy, you always noticed when I needed your encouragement, even if it was unspoken. Thank you for all this!

Markus Scherer
Hambach, February 2022

The preceding acknowledgments have been formulated before the defence of this PhD thesis – a day that I was privileged enough to celebrate together with my parents and my entire family, also with my friends and colleagues; the people that have accompanied me throughout the past three and a half years. Shortly after that day, however, we lost my dear father, unexpectedly and far too young. It is therefore my personal need to add this last paragraph that I dedicate to him: Throughout my life, he has supported me in my fascination for science and has encouraged me to pursue my way into research. A physical chemist by training, he had developed an impressively broad scientific interest in topics going far beyond his own field of expertise, not least in fluid mechanics. And so, whenever I was struggling with my research during these last years, it gave him great pleasure to discuss my work with me, seeking a solution together. Often enough, just explaining my problems to him helped me to see things from a different point of view and indeed gave me new ideas how to address them. His ability to think "outside the box", paired with a deep analytical understanding and a distinct helpfulness have been a true example for me and have left their mark on me; more than I had the chance to tell him.

Even though you will now never read these lines, I feel this is just the right place to thank you for everything you did for me throughout my life, I miss you!

Markus Scherer
Hambach, September 2022

The scientific work reported in this thesis was supported by the German Research Foundation (DFG) through grant UH242/12-1, which the author gratefully acknowledges. Part of the simulations were performed on the supercomputers BWUniCluster/ForHLR II/HoreKA at the Steinbuch Centre for Computing (SCC) of the Karlsruhe Institute of Technology (KIT), funded by the Ministry of Science, Research and the Arts Baden-Württemberg and by the Federal Ministry of Education and Research. The remaining simulations have been carried out on SuperMUC-NG at the Leibniz Supercomputing Centre at the Bavarian Academy of Science and Humanities (LRZ) and on Hazel Hen/Hawk at the High-Performance Computing Center at the University Stuttgart (HLRS). The computer resources, technical expertise and assistance provided by the staff at these computing centres are gratefully acknowledged.

Related publications

The content of this thesis is based in parts on the following references, published during the course of this PhD thesis. The contributions of the co-authors are highly acknowledged.

Peer-reviewed publications

(i) **M. Scherer**, A. G. Kidanemariam and M. Uhlmann. On the scaling of the instability of a flat sediment bed with respect to ripple-like patterns. *J. Fluid Mech.*, 900:A1, 2020.

Declaration: MS conducted the simulations, post-processed the data and prepared the manuscript supported by MU. MU contributed to the preparation of the discussion section. MU and AGK revised the manuscript.

(ii) **M. Scherer**, M. Uhlmann, A. G. Kidanemariam and M. Krayer. On the role of turbulent streaks in generating sediment ridges. *J. Fluid Mech.*, 930:A11, 2022.

Declaration: MS and MK conducted the simulations, MU provided reference simulation data. MS post-processed the data and prepared the manuscript supported by MU. MU, AGK and MK revised the manuscript.

(iii) A. G. Kidanemariam, **M. Scherer** and M. Uhlmann. Open channel flow over evolving subaqueous ripples. *J. Fluid Mech.*, 937:A26, 2022.

Declaration: AGK conducted the simulations, post-processed the data and prepared the manuscript supported by MU and MS. MU and MS revised the manuscript.

Conference contributions

(i) A. G. Kidanemariam, **M. Scherer**, M. Moriche, M. Garcia-Villalba and M. Uhlmann (2018). "Turbulent flow over two-dimensional subaqueous bedforms: A comparison between stationary and evolving dunes", *12th Euromech European Fluid Mechanics Conference*, Vienna (Austria).

(ii) **M. Scherer**, A. G. Kidanemariam and M. Uhlmann (2019). "Scaling parameters of subaqueous sediment bedforms in turbulent open channel flows", *17th European Turbulence Conference*, Torino (Italy).

(iii) **M. Scherer**, M. Uhlmann and A. G. Kidanemariam (2021). "Large-scale streaks and sediment ridges in turbulent open channel flows", *Euromech Colloquium 609: "Granular patterns in oscillatory flows"*, Genoa (Italy, remotely).

(iv) **M. Scherer** and M. Uhlmann (2022). "Large-scale coherent structures in sediment-laden turbulent channels" (Poster presentation), *Workshop: "Wall-bounded turbulence: beyond current boundaries"* in *"Mathematical Aspects of Turbulence: Where Do We Stand?"*, Isaac Newton Institute for Mathematical Sciences, Cambridge (UK).

(v) **M. Scherer** and M. Uhlmann (2022). "Causality in sedimentary turbulent channel flows", *Workshop: "Causality in turbulence and transition"*, Universidad Politécnica de Madrid (Spain).

"The universe has quite a bit of randomness in it,
but also quite a bit of structure."
— David Ruelle in *Chance and Chaos* (1991)

Contents

Nomenclature

Mathematical symbols

$\mathrm{d}(\bullet)/\mathrm{d}i$	total derivative in the ith dimension (including time) with $i \in \{x, y, z, t\}$
$\partial(\bullet)/\partial i$, $\partial_i(\bullet)$	partial derivative in the ith dimension (including time) with $i \in \{x, y, z, t\}$
$\partial^n(\bullet)/\partial i^n$, $\partial_i^n(\bullet)$	nth partial derivative in ith dimension (including time) with $i \in \{x, y, z, t\}$
$\bullet \cdot \bullet$	scalar (inner) product between two tensor fields
$\bullet \times \bullet$	cross product between two vector fields
$\bullet \otimes \bullet$	dyadic (outer) product between two tensor fields
$(\bullet)^T$	transpose of a tensor fields
$(\bullet)^{-1}$	inverse of a matrix
∇	Nabla operator, for Cartesian coordinates: $\nabla = (\partial_x, \partial_y, \partial_z)^T$
$\nabla_\perp$	cross-plane gradient operator, for Cartesian coordinates: $\nabla_\perp = (\partial_y, \partial_z)^T$
$\nabla \cdot (\bullet)$	Divergence operator
$\nabla^2(\bullet), \Delta(\bullet)$	Laplace operator, for Cartesian coordinates: $\nabla^2() = (\partial_x^2 + \partial_y^2 + \partial_z^2)()$
$\Delta_\perp(\bullet)$	cross-plane Laplace operator, for Cartesian coordinates: $\Delta_\perp() = (\partial_y^2 + \partial_z^2)()$
$\nabla \times (\bullet)$	curl operator, for Cartesian coordinates: $\varepsilon_{ijk} e_i \dfrac{\partial a_k}{\partial x_j}$
$\langle \bullet \rangle$	quantity averaged in all homogeneous spatial directions and time
$\langle \bullet \rangle_i$	quantity averaged in the ith dimension(s) with $i \in \{x, y, z, t\}$
$\langle \bullet \rangle_{[i_1, i_2]}$	quantity averaged in the ith dimension(s) with $i \in \{x, y, z, t\}$ restricted to the interval $[i_1, i_2]$
$\langle \bullet \rangle_{i,ens}$	quantity averaged in the ith dimension(s) with $i \in \{x, y, z, t\}$ and over an ensemble of several individual ridges
$\langle \bullet \rangle_{\mathcal{V}_f}$	quantity averaged over a subvolume $\mathcal{V}_f$ of the fluid domain Ω_f

$\bullet'$	fluctuation of a quantity w.r.t. to its mean value $\langle\bullet\rangle$ (in the LSA alternatively used to mark the total derivative w.r.t. the wall-normal direction)
$\bullet''$	fluctuation of a quantity w.r.t. to its streamwise average $\langle\bullet\rangle_x$
$\bullet'''$	fluctuation of a quantity w.r.t. its local time average $\langle\bullet\rangle_t$
$\bullet''''$	fluctuation of a time–averaged quantity $\langle\bullet\rangle_t$ w.r.t. its plane average $\langle\bullet\rangle_{xzt}$
$\bullet^+$	quantity scaled in wall units
$\bullet^{+_l}$	quantity scaled in local wall units
$\bar{\bullet}$	dimensional quantity in the context of the LSA
$\bullet_0$, $\bullet_1$	Base flow and perturbation profiles in the LSA
$\bullet_{lat}$	lateral component of a field variable tangential to the sediment bed in the cross-plane
$\tilde{\bullet}$	quantity filtered with a Gaussian cut-off filter
$\min(\bullet), \max(\bullet)$	minimum, maximum of a function
$\mathrm{argmax}(\bullet)$	argument for which a function attains its maximum
$\bullet_{rms}$	root mean square (r.m.s.) value of a field variable
$\Re(\bullet), \Im(\bullet)$	real and imaginary part of a complex number
$\bullet^*$	complex conjugate of a complex quantity
$\hat{\bullet}$	Fourier transform of quantity, in the context of the LSA alternatively used to denote Chebyshev coefficients
$\mathcal{F}(\bullet)$	Fourier transform operator
$\check{\bullet}$	reference locations for the two-point correlations
$\check{x}$	relative normalised position of a point w.r.t. to a conditionally-averaged Q2-Q4-pair
$\mathbb{R}$	real numbers
$\mathbb{N}$	natural numbers
$\mathbb{Z}$	integer numbers
$\mathbb{S}^n$	n-sphere
$\mathbb{T}^n$	n-torus
$\mathcal{O}(\bullet)$	Landau symbol, order of magnitude

Greek Symbols

$\partial\Omega_f, \partial\Omega_p$	fluid and particle domain boundary
$\partial\Omega$	full domain boundary
$\partial\Omega^D, \partial\Omega^N$	part of the domain boundary at which Dirichlet and Neumann boundary conditions are applied
$\partial\Omega_\perp^D, \partial\Omega_\perp^N$	part of the domain boundary of the cross-section at which Dirichlet and Neumann boundary conditions are applied
Δ_c	force range in the collision model
ΔH	reduction of the mean bed height
Δ_S	minimal Euclidean distance between two particles' surfaces
Δt	time step in the numerical scheme
Δt_{av}	effective time averaging interval
ΔU^+	roughness function, i.e. the offset of the velocity profile in virtue of wall roughness in the law of the wall
$\Delta x, \Delta y, \Delta z$	streamwise, wall-normal and spanwise grid widths
$\Delta x_{bin}, \Delta y_{bin}, \Delta z_{bin}$	streamwise, wall-normal and spanwise bin widths
$\Delta_x, \Delta_y, \Delta_z$	streamwise, wall-normal and spanwise cutoff-filter widths for the Gaussian filtering
$\widetilde{\Delta x}, \widetilde{\Delta y}, \widetilde{\Delta z}$	streamwise, wall-normal and spanwise dimensions of the bounding box enclosing a three-dimensional flow structure
$\mathbf{\Theta}_p^{(l)}$	current angular orientation of lth particle
Π	Time-dependent streamwise pressure gradient
$\mathcal{T}_f$	Cauchy stress tensor field
Φ	Particle erosion-deposition balance per unit length and time
$\Phi(\theta)$	Meyer-Peter and Müller (1948) particle flux formula
Ω_f, Ω_p	fluid and particle domain
$\Omega_\perp$	cross-sectional domain
Ω	full domain $\Omega = \Omega_f \cup \Omega_p$
α_b	local lateral bed inclination

α_s	α-shape threshold parameter
$\alpha_x, \alpha_z = \alpha$	streamwise and spanwise wavenumber in the LSA
γ	streamwise bottom inclination w.r.t. a horizontal reference plane
δ	angle between the particle flux vector and the streamwise direction
δ_{ij}	Kronecker Delta
δt	time separation in correlations
$\delta x, \delta z, \boldsymbol{\delta}$	streamwise/spanwise separation length and corresponding spatial separation vector in correlations
$\delta x_{min}, \delta z_{min}$	streamwise and spanwise separation at which a correlation attains its minimum
$\delta_c^{(i,j)}$	penetration length between particles i, j in the contact model
δ_ν	viscous length scale $\delta_\nu = \nu_f / u_\tau$
ε	dissipation
ε_d	'dry' coefficient of restitution
ε_{ijk}	Levi-Civita symbol
ϵ	infinitesimal perturbation amplitude in the LSA
θ	Shields parameter
θ_{bed}	bed Shields parameter in open duct simulations
θ_c	critical Shields parameter
θ_l	local Shields parameter computed based on the local bed shear stress
η_f	Kolmogorov length scale
η_0	reference level at which the logarithmic velocity profile reaches zero
κ	von Kármán's constant
λ_f	first Lamé constant
λ_2	second eigenvalue of the Hessian of the pressure
λ_{crit}	minimal streamwise wavelength of transverse bedforms
$\lambda_{h,z}$	mean lateral pattern wavelength
λ_x, λ_z	streamwise and spanwise wavelength

μ_b	dynamic Coulomb friction coefficient of a mobile particle on a sediment bed
μ_c	Coulomb friction coefficient in the contact model
μ_f	dynamic viscosity
ν_{eff}	effective viscosity $\nu_{eff} = \nu_f + \nu_t$
ν_f	kinematic viscosity
ν_t	turbulent viscosity
$\boldsymbol{\xi} = (\xi, \eta, \zeta)^T$	position vector in the non-orthogonal coordinate system used in the LSA
ρ_f, ρ_p	fluid and particle density
ρ_{hh}	two-point correlation coefficient of the sediment bed height fluctuation with itself
ρ_{uh_b}	instantaneous two—point correlation coefficient between the streamwise velocity and the sediment bed profile fluctuations, both streamwise-averaged
ρ^t_{ua}	two—point two-time correlation coefficient between the streamwise velocity fluctuation and that of an arbitrary quantity a, both streamwise-averaged
$\rho_{u\tau}$	instantaneous two—point correlation coefficient between the streamwise velocity and the bottom wall shear stress fluctuation
ρ_{uu}	instantaneous two—point correlation coefficient between the streamwise velocity and the centreline velocity fluctuations, both streamwise-averaged
σ	Sediment bed growth rate in the LSA
$\sigma_{h,2D}$	Two-dimensional root mean square of the non-averaged sediment bed height fluctuation
$\sigma_{h,x}, \sigma_{h,z}$	Root mean square of the streamwise- and spanwise-averaged sediment bed height fluctuation
$\hat{\sigma}_s$	r.m.s. of the local lateral meandering amplitude of a streamwise streak $\hat{A}_s$
$\boldsymbol{\tau}$	Reynolds stress tensor
τ_b	instantaneous wall shear stress at a smooth bottom wall
τ_{bed}	mean bed shear stress in open duct flows (ignoring sidewall-contributions)
τ_{disp}	dispersive stress
τ_η	Kolmogorov time scale
τ_f	mean fluid shear stress

$\boldsymbol{\tau}^L$	Linear approximation of the Reynolds stress tensor
$\boldsymbol{\tau}^{NL}$	Non-linear approximation of the Reynolds stress tensor
τ_{part}	stress due to particle-fluid interaction
τ_t	lateral bed shear stress component tangential to the bed contour
τ_{tot}	total shear stress
τ_{turb}	turbulent shear stress
τ_{visc}	viscous shear stress
τ_w	mean wall/bed shear stress
τ_w^i	instantaneous wall/bed shear stress
$\tau_{w,l}$	local wall/bed shear stress
ϕ	poloidal velocity potential
$\phi_{\alpha\alpha}$	energy spectra associated with velocity component $\alpha \in \{u, v, w\}$
$\hat{\varphi}$	inclination of the streak axis w.r.t. the x-direction
ϕ_s	global solid volume fraction
ϕ^*	shape factor in the empirical formula of Jones (1976)
ψ	toroidal velocity potential
ψ_{th}	threshold of the toroidal potential ψ for the extraction of two-dimensional streamwise rolls in the cross-plane
ψ_{vw}	cross–plane streamfunction
$\langle\psi\rangle_{xt}$	mean secondary flow streamfunction
$\boldsymbol{\omega}_f$	vorticity vector
$\boldsymbol{\omega}_p^{(l)}$	Lagrangian angular velocity of the lth particle
$\omega_{x,\mathcal{V}}$	streamwise vorticity averaged over a vortex cluster's volume

Roman Symbols

A, $\boldsymbol{B}$, $\boldsymbol{R}$	linearised stationary Navier-Stokes operator and right hand side vectors of the linearised system in the LSA
A_{bot}, A_{top}	v. Driest damping coefficient at the bottom wall and free surface

$\hat{A}^h$	single–sided amplitude spectra of the sediment bed height fluctuation
AR	open/closed duct cross-sectional aspect ratio
$\hat{A}_s$	local lateral meandering amplitude of a streamwise streak
$\mathcal{A}_s$	cross-sectional area of a two-dimensional flow structure in the cross-plane
B	bed height in the LSA
B_u	mean velocity profile coefficient
C_0, c_f	friction coefficients (C_0: in LSA, c_f: else)
C_d, C_e	parameters in the turbulence model of Speziale (1987)
$C_{u_i u_j}$	two-point velocity correlations between the ith and jth component
D	particle diameter
D^+	particle Reynolds number
$\mathcal{D}$	discriminant of the velocity gradient tensor $\nabla \boldsymbol{u}_f$
$\mathcal{D}_\perp$	discriminant of the cross-plane velocity gradient tensor $\nabla_\perp \boldsymbol{u}_f$
$\boldsymbol{D}$	rate of strain tensor
$\boldsymbol{D}_\eta$	Chebyshev collocation derivative matrix
D_{hy}	hydraulic diameter
$\widehat{D_{ij}}$	Oldroyd derivative of the rate of strain tensor
$E_\perp$	secondary flow kinetic energy in the xt-averaged flow field
E_{vw}	secondary flow kinetic energy
$\boldsymbol{F}$	External volume force field acting on the system in the LSA
$\boldsymbol{F}^{d(i,j)}$	normal damping collision force between particles i, j
$\boldsymbol{F}^{el(i,j)}$	normal elastic collision force between particles i, j
$\boldsymbol{F}^{H(l)}, \boldsymbol{F}^{C(l)}$	hydrodynamic and collision forces acting on the lth particle
$\boldsymbol{F}^{n(i,j)}$	total normal collision force between particles i, j
Fr	Froude number
$\boldsymbol{F}^{t(i,j)}$	tangential damping collision force between particles i, j

$\boldsymbol{F}^{tot(l)}$	total force acting on the lth particle
F_W	submerged weight of a single spherical particle
G	coefficient in the convolution associated with the Gaussian filtering
Ga	Galileo number
H	characteristic length scale in canonical flows (pipe radius, half channel height)
H_b	mean sediment bed height
H_f	mean fluid height
H_i	convective terms in the ith momentum equation
$H(x,z,t)$	location of the free surface in the LSA
I	imaginary unit $I = \sqrt{-1}$
$\mathcal{I}$	time interval
$I_c^{(i,j)}$	particle contact indicator function for particles i, j
Id	identity tensor
$I_{depos}^{(l)}$	deposition particle indicator function for particle l
$I_{erod}^{(l)}$	erosion particle indicator function for particle l
I_f	fluid indicator function
$I_{(i,k)}^{(l)}$	xz-bin indicator function for particle l
$I_{(j)}^{(l)}$	y-bin indicator function for particle l
$I_p^{(l)}$	particle indicator function for particle l
$I_{stat}^{(l)}$	stationary particle indicator function for particle l
$\mathcal{J}_p^{(l)}$	moment of inertia tensor of the lth particle
$\mathcal{L}$	characteristic length scale of the system
L_x, L_y, L_z	streamwise, wall-normal and spanwise extension of the computational domain
$N_{x,bin}$, $N_{y,bin}$, $N_{z,bin}$	number of averaging bins in the streamwise, wall-normal and spanwise direction
N_η	number of Chebyshev collocation points
N_p	total number of particles in the system

N_p^{depos}	number of deposited particles in the system
N_p^{erod}	number of eroded particles in the system
N_t	number of discrete time steps
$N_{t,sub}$	number of substeps for the integration of the Newton-Euler equations per time step of the fluid solver
N_x, N_y, N_z	number of grid points in the streamwise, wall-normal and spanwise direction
$\mathcal{P}_b$	set of bed particles
$\mathcal{P}_{depos}$	set of deposited particles
$\mathcal{P}_{erod}$	set of eroded particles
$\mathcal{P}_m$	set of mobile particles
Q	transverse sediment transport rate in the LSA
$\mathcal{Q}$	second invariant of the velocity gradient tensor $\nabla \boldsymbol{u}_f$
$\mathcal{R}$	third invariant of the velocity gradient tensor $\nabla \boldsymbol{u}_f$
Re	Reynolds number
Re_b	bulk Reynolds number
Re_τ	friction Reynolds number
Re_τ^i	instantaneous friction Reynolds number
$R_{u_i u_j}$	two-point velocity covariances between the ith and jth component
S	mean shear
S_b	streamwise channel slope in the LSA
$\mathcal{S}_p$	surface of a single particle
$\boldsymbol{T}$	transformation matrix between the Cartesian and the non-orthogonal coordinate system in the LSA
T	bed evolution time scale in the LSA
T_b	bulk time unit
$\boldsymbol{T}^{C(l)}$	torque due to collision forces acting on the lth particle
T_i	ith Chebyshev polynomial of the first kind

T_{obs}	total observation time
$\mathcal{U}$	characteristic velocity scale of the system
$\boldsymbol{u}_r^{(i,j)}$	relative particle velocity of particles i, j, with components $\boldsymbol{u}_{r,n}^{(i,j)}$ and $\boldsymbol{u}_{r,t}^{(i,j)}$ normal and tangential to the line connecting their centres of gravity, respectively
$\boldsymbol{u}_p^{(l)}$	Lagrangian translational velocity of the lth particle
$\mathcal{V}_D$	volume of an individual vortex cluster
$\mathcal{V}_\Omega, \mathcal{V}_{\Omega_f}, \mathcal{V}_{\Omega_p}$	volume of the entire domain Ω as well as subvolumes occupied by the fluid and solid phase
$\mathcal{V}_p$	volume of a single particle
$\mathcal{V}_u$	volume of an individual streamwise velocity streak
$\boldsymbol{X}^m$	mth Lagrangian force point
$\boldsymbol{x}_p^{(l)}$	Position of the lth particle's centre of gravity
c	gravitational parameter in the empirical formulae for the lateral particle flux
c_n, c_t	normal and tangential friction coefficient in the contact model
c_x, c_z, c_r, c_i	streamwise and spanwise travelling wave speeds as well as their real and imaginary parts
d	amplitude of the velocity dip at the duct bisector
$\boldsymbol{e}_i$	Cartesian basis vector in the ith direction
$\boldsymbol{e}_n^{(i,j)}, \boldsymbol{e}_t^{(i,j)}$	Unit vector parallel and tangential to the line connecting the centres of gravity of particles i, j
$\boldsymbol{f}$	(Immersed boundary) volume force acting on the fluid
$\boldsymbol{f}^H, \boldsymbol{f}^C, \boldsymbol{f}^{tot}$	surrogate Eulerian force fields associated with the hydrodynamic, collision and total particle forces
f_{ref}	reference force based on wall shear stress and projected particle area
$f(V)$	probability density function (p.d.f.) of a single random variable V
$\boldsymbol{g}$	gravitational acceleration with absolute value g
h_b	sediment bed height
h_f	fluid height

h_v, h_ω	RHS terms of the governing equations in v_f-$\omega_{f,x}$-form
$\boldsymbol{j}$, $\boldsymbol{k}$	unit vectors normal and tangential to the streamwise-averaged fluid-bed interface
k	turbulent kinetic energy
k_0	hydrodynamic roughness length
k_n	normal stiffness coefficient in the contact model
k_s	equivalent sand grain roughness
k_x, k_z	streamwise and spanwise wavenumber
l	mixing length
$\widetilde{l}_{yz}$	characteristic cross-sectional length scale of three-dimensional flow structures
l_s	characteristic cross-sectional length scale of two-dimensional flow structures in the cross-plane
$\boldsymbol{n}$	inward-/outward-pointing unit normal vector
$\boldsymbol{n}_p$	outward unit normal vector on the particles' surfaces
n_{ridges}	number of individual ridges in the cross-section
p_f	fluid pressure field
p_{bed}	bed porosity
q_f	volumetric flow rate of the fluid phase
q_i	inertial scale
$\boldsymbol{q}_p$	volumetric flow rate of the solid phase
q_{ref}	Wong and Parker (2006) reference flux
$\boldsymbol{r}$	position vector of points on a particle's surface w.r.t. its centre of gravity
r_b	ratio between particle lift and drag force
s	density ratio between solid and fluid phase
$s_{h,z}$	spanwise distance between a sidewall and the closest ridge crest
t	time
t_m	discrete instant in time at the mth time step
u_b, $U_{b,0}$	bulk velocity

u_η	Kolmogorov velocity scale
$\boldsymbol{u}_{f0}$	Initial condition of the velocity field
$\boldsymbol{u}_f, \boldsymbol{u}_p$	Eulerian fluid and particle velocity field with components $(u_f, v_f, w_f)^T$ and $(u_p, v_p, w_p)^T$, respectively
u_g	gravitational velocity scale
u_{ref}	duct core reference velocity
$u_\perp, U_\perp, U_{\perp,cum}$	instantaneous, time-averaged and cumulatively-averaged secondary flow intensity
$u_\tau, u_{\tau,0}$	friction velocity
$u_{\tau,l}$	local friction velocity
u_{th}	streamwise velocity threshold for the extraction of two-dimensional velocity streaks in the cross-plane
$u'_\mathcal{V}$	velocity fluctuation averaged over a streamwise streak's volume
$\boldsymbol{x} = (x, y, z)^T$	Eulerian coordinate vector in the fluid phase
$\boldsymbol{x}_p = (x_p, y_p, z_p)^T$	Eulerian coordinate vector in the particle phase
$\boldsymbol{x}_c$	centre of gravity of an individual three-dimensional flow structure
$\boldsymbol{x}_{c,2D}$	velocity-weighted centre point of an individual two-dimensional flow structure in the cross-plane
$\boldsymbol{x}_{ijk} = (x_i, y_j, z_k)^T$	discrete grid point position vector
$x_{i,min}, x_{i,max}$	minimal and maximal coordinate of a flow structure in the ith direction
$\hat{x}$	streamwise streak length in wall-parallel planes
$\tilde{y}$	shifted y-coordinate such that the origin is located at the mean fluid-bed interface
y_{cg}	y-coordinate of the centre of gravity of a velocity/vortex structure
y_s	shifted y-coordinate in the Cess (1958) viscosity profile
$\tilde{z}$	spanwise coordinate whose origin is shifted to the crest of the ensemble-averaged fluid-bed interface profile

Abbreviations

Q1	outward motion structures/events
Q2	ejection structures/events

Q3	inward motion structures/events
Q4	sweep structures/events
Q^-	'gradient supporting' Reynolds stress-carrying Q2 and Q4 structures
DAAD	Deutscher Akademischer Austauschdienst e.V.
DEM	discrete element model
DFG	Deutsche Forschungsgemeinschaft
DNS	direct numerical simulation
HLRS	Höchstleistungsrechenzentrum Stuttgart
HMP, LMP	high and low momentum pathways
IBM	immersed boundary method
joint p.d.f.	joint probability density function
JSPS	Japan Society for the Promotion of Science
LES	large eddy simulation
LHS	left hand side of an equation
LRZ	Leibniz-Rechenzentrum der Bayerischen Akademie der Wissenschaften, München
LSA	linear stability analysis
LSM	large-scale motion
p.d.f.	probability density function
r.m.s.	root mean square
RANS	Reynolds averaged Navier-Stokes equations
RHS	right hand side of an equation
SCC	Steinbuch Centre for Computing, KIT, Karlsruhe
VLSM	very large-scale motion

Chapter 1

Introduction

1.1 Motivation

Many natural fluvial ecosystems such as rivers or estuaries feature an erodible sediment bed composed of different cohesive and non-cohesive materials including clay, sand, gravel and possibly larger rocks. Under suitable conditions, the mutual interaction of individual sediment grains and a turbulent flow gives rise to a variety of sedimentary patterns or 'bedforms', ranging from small sediment ridges to massive marine sand banks and dunes. The presence of such bedforms in a river have a substantial impact on its flow behaviour, so that for the engineering practice, reliable tools to predict the sediment bed dynamics are of immense importance. This is especially true in light of the expected increased frequency of extreme flooding events due to global warming, which make accurate numerical simulations of both the river dynamics and the riverbed evolution more than ever necessary in order to estimate the future demands on flood protection and hydraulic infrastructure.

Incorporating the full dynamics of each individual sediment grain in such large-scale hydraulic simulations is however computationally unfeasible, instead significant simplifications in form of engineering-type models that estimate the sediment transport are typically considered. Further development of these models requires a deep understanding of the complex sediment-turbulence interactions, which are, however, even nowadays only partly understood. A central objective of the current thesis therefore is to provide high-fidelity datasets of sediment pattern evolution under the action of a turbulent shear flow obtained by fully-resolved direct numerical simulations (DNS) foregoing *a priori* simplifications of the two-phase system. Based on the newly generated datasets, we aim to contribute to a better understanding of the fundamental physical mechanisms that cause sediment patterns to emerge, which will provide opportunities to assess and improve existing simplified models by identifying the central processes that must be captured therein (Hsu, 2022).

In this context, we focus specifically on the formation of streamwise-elongated sediment ridges that appear in a variety of geophysical flows as a result of a complex interaction between the turbulent flow and a mobile sediment bed (Nezu and Nakagawa, 1993). Two such essentially parallel sediment ridges are shown in figure 1.1 for one of the open channel flow simulations performed in the course of this work. From an ecological point of view, these sediment patterns are of particular interest in virtue of their close relation to large-scale secondary flow cells of Prandtl's second kind, which – despite their comparably low amplitude of only a few percent of the mean flow velocity – significantly enhance the transport of heat and momentum in the water body (Demuren and Rodi, 1984). The secondary currents associated with the sediment ridges are clearly visible at the downstream end of the periodic simulation domain in figure 1.1, indicating that fluid is, on average, moved away from the bed over

Figure 1.1: Streamwise-elongated sediment ridges in a direct numerical simulation of turbulent open channel flow. Darker and brighter colours of the sediment bed particles indicate lower and higher regions of the sediment bed, respectively, flow is from left to right. The light blue three-dimensional structure represents an isolated large-scale low-velocity streak (isocontour of the streamwise velocity fluctuations), while the contours at the downstream end of the periodic domain indicate the instantaneous filtered and streamwise-averaged secondary flow field (red: clockwise, blue: anticlockwise rotation; isocontour of the secondary flow streamfunction). Shown are simulation results for case $CM850^{H3}$ ($Re_b = 9500$, $Re_\tau = 830$, $Ga = 57$, $\rho_p/\rho_f = 2.5$, $H_f/D \approx 25$, cf. section 6.1 for the definitions and further details).

the crests of the sediment ridges and is brought down to the sediment bed in local trough regions, indicating an increased mixing activity in these regions.

1.2 State of research

Even though the close relation between sediment ridges and secondary currents of Prandtl's second kind was expected already when sediment ridges were first scientifically investigated (Casey, 1935; Vanoni, 1946), the exact mechanism of how both interact is even nowadays not fully understood. Nezu and Nakagawa (1993) argue that there are basically two conceivable formation mechanisms owing to which sediment ridges and mean secondary vortices can arise. The first one is controlled by the bed in that an initial modulation of the sediment bed topography and/or the bed roughness leads to a perturbation of the bed shear stress, which in turn triggers the appearance of secondary circulations in the turbulent flow. The second mechanism works the other way round in that a lateral variation of the flow field induces a lateral variation of the bed shear stress, which then causes a laterally varying erosion rate that results in the formation of local troughs and ridges. Important to note is that both mechanisms are independent of sidewall-induced secondary currents, which had been claimed earlier to be the main driver of sediment ridge formation (Nezu and Nakagawa, 1984).

Which of the two mechanisms is of relevance in a turbulent flow is not entirely clear. While Colombini (1993) could show theoretically that a lateral variation of a turbulent base flow in form of depth-spanning secondary currents can in general be triggered by an initial sinusoidal bed perturbation, considerations based on experimental observations tend to support the second proposed mechanism (Shvidchenko and Pender, 2001; Nezu, 2005). According to the latter authors, instantaneous large-scale coherent structures whose size is of the order of the mean fluid height H_f might be responsible for the changing shear stress conditions along the bed that give rise to sediment ridges. As pointed out by Adrian and Marusic (2012), the large-scale velocity structures reported for hydraulic flows (Tamburrino and Gulliver, 1999; Shvidchenko and Pender, 2001; Tamburrino and Gulliver, 2007) bear a strong resemblance to similar flow patterns in canonical wall-bounded flows that are typically referred to as 'large-scale streaks' or 'large-scale motions' (LSM) (Smits et al., 2011; Jiménez, 2018). The instantaneous isolated low-speed streak, extracted from a simulation performed in this thesis and visualised in figure 1.1, represents one example for such a large-scale turbulent structure. Much is known about the organisation and dynamics of these flow structures in canonical wall-bounded flows thanks to a vigorous research activity predominantly during the past two decades, which was strongly stimulated by the tremendous increase in computing power and the thereby offered possibilities to perform highly-resolved direct numerical simulations of turbulent flows even at considerably high Reynolds numbers (cf. chapter 3 for an extended summary of the most important findings in this field).

In the context of sediment-laden turbulent flows, simulations of a similar degree of detail that resolve also the flow around the individual particles became feasible only recently. In particular, such simulations have proven capable of providing valuable insights into basic physical interactions between turbulent flow structures and mobile particles (Kidanemariam et al., 2013; Kidanemariam and Uhlmann, 2014a, 2017; Vowinckel et al., 2017a,b). Most importantly, simulations of this kind represent a powerful alternative to laboratory experiments, in which accurate simultaneous measurements of the secondary fluid motion and the sediment dynamics remain a challenging task up to the present day (Wang and Cheng, 2005).

1.3 Research objectives

The purpose of the current thesis is to contribute to a better understanding of the interaction between instantaneous coherent structures, sediment ridges and mean secondary currents. To this end, we have performed several series of direct numerical simulations of open channel (laterally periodic domains) and open duct flows (laterally bounded by sidewalls) over mobile sediment beds on the one hand and under single-phase smooth-wall conditions on the other hand. The thereby developed high-fidelity database allows a detailed analysis of the dynamics of both large-scale flow structures and sediment bedforms and offers the possibility to compare the coherent structures in canonical flows with those in hydraulic environments. We endeavour to scrutinise which of the above discussed conceivable formation mechanisms for sediment ridges and secondary currents are of relevance in the fully-turbulent state. To this end, we first revisit the linear model proposed by Colombini (1993), before moving on to analyse the simulation data and compare the therein observed formation mechanism with the concept of the linearised model. Even though secondary currents induced by lateral sidewalls have been shown to be no necessary condition for the evolution of sediment ridges, it remains unclear how such 'pre-existing' secondary currents affect the bed evolution and, *vice versa*, how the mean secondary flow changes in presence of a mobile sediment bed. These questions will be tackled in the remainder of this work based on the simulation results for open duct flows. In this context, we will specifically investigate how the presence of a mobile sediment bed affects the organisation of turbulent coherent structures in the cross-section of narrow open duct flows. In a final outlook, we will discuss the transformation of the ridge-covered sediment bed into a state dominated by transverse ripple-like patterns.

1.4 Outline of the thesis

The thesis is organised as follows: In chapter 2, we derive the governing equations for the fluid and the dispersed phase, which will form the basis for our following investigations. A detailed review of the state of research concerning both single-phase wall-bounded turbulence and sediment ridge formation in hydraulic flows is provided in chapter 3. In the subsequent chapter 4, the numerical techniques used to conduct the single- and multiphase simulations of open channel and duct flow are briefly presented. Thereafter, the linear model of Colombini (1993) that describes the instability of a turbulent base flow w.r.t. a bottom undulation is revisited, analysed and modified in chapter 5. Chapter 6 is devoted to the analysis of sediment ridge formation in doubly-periodic open channel flows and to the role that large-scale velocity streaks play in this process. A similar analysis is performed for laterally bounded open duct flows in chapter 7, with a special focus on the impact of sediment ridges on the already existing sidewall-induced mean secondary motion and the preferential spatial organisation of the turbulent coherent structures. Finally, a brief outlook on later stages of bedform evolution during which transverse ripple-like patterns become dominant over the streamwise-aligned sediment ridges is provided in chapter 8. A possible mechanism in virtue of which individual sediment ridges gradually transform into transverse bedforms is discussed. We close with a summary of the most important findings of this work in chapter 9, highlighting potentially interesting points of departure for future studies.

Chapter 2

Mathematical framework

"Nonlinearity means that the act of playing the game has a way of changing the rules. (...) Analyzing the behavior of a nonlinear equation like the Navier-Stokes equation is like walking through a maze whose walls rearrange themselves with each step you take. (...) The world would be a different place – and science would not need chaos – if only the Navier-Stokes equation did not contain the demon of nonlinearity."
— James Gleick in *Chaos: Making a new science* (1987)

The subsequently presented derivation of the governing equations of the carrying fluid phase partly follows the lecture notes of Uhlmann (2014) and Dörfler (2019) as well as the classical textbooks by Landau and Lifshitz (1959), Pope (2000), Chorin and Marsden (2000) and Majda and Bertozzi (2002). In the second section which focuses on the dynamics of the dispersed particulate phase, the derivation of the Newton-Euler equations partially follows the textbooks of Clift et al. (1978) and Jackson (2000) and the notations used in Uhlmann (2005) and Henn (2016), respectively.

2.1 Governing equations of the fluid phase

2.1.1 Conservation of mass and momentum

Consider a domain $\Omega_f \subset \mathbb{R}^d$ ($d = 2, 3$) occupied by a fluid and a time interval $\mathcal{I} = [0, T] \subset \mathbb{R}$ with $T > 0$. Let $\mathcal{V}(t) \subset \Omega_f$ be an arbitrary sub-volume of the domain at time $t \in \mathcal{I}$, then 'Reynolds' transport theorem' for a sufficiently smooth function $\varphi(\bullet, t) : \Omega_f \times \mathcal{I} \to \mathbb{R}$ reads

$$\frac{\mathrm{d}}{\mathrm{d}t} \int_{\mathcal{V}(t)} \varphi(\boldsymbol{x}, t) \mathrm{d}\boldsymbol{x} = \int_{\mathcal{V}(t)} \left\{ \partial_t\, \varphi(\boldsymbol{x}, t) + \nabla \cdot \left(\varphi(\boldsymbol{x}, t) \boldsymbol{u}_f(\boldsymbol{x}, t) \right) \right\} \mathrm{d}\boldsymbol{x} \qquad \forall\, \mathcal{V}(t) \subset \Omega_f. \tag{2.1}$$

Therein, $\mathrm{d}()/\mathrm{d}t = (\partial_t + \boldsymbol{u}_f(\boldsymbol{x}, t) \cdot \nabla)()$ is the total derivative w.r.t. time t and $\partial_i^n() = \partial^n()/\partial i^n$ is the nth partial derivative w.r.t. to the ith Cartesian coordinate direction or time. The divergence operator when applied to an arbitrary vector field $\boldsymbol{a}(\boldsymbol{x}, t)$ reads

$$\nabla \cdot \boldsymbol{a} = \sum_{j=1}^{d} \partial_j \boldsymbol{a}_j = (a_j)_{,j}.$$

In the last definition, we have used the Einstein summation convention that implies summation over repeatedly appearing indices, the index after the comma indicating derivation w.r.t. j. In this context, the nabla operator is introduced as $\nabla = (\partial_{x_1}, \dots, \partial_{x_d})^T$, where $(\bullet)^T$ indicates transposition of the respective vector or tensor field. Equation (2.1) further includes the fluid velocity field $\boldsymbol{u}_f(\boldsymbol{x}, t) : \Omega_f \times \mathcal{I} \to \mathbb{R}^d$. An associated pressure and density field are introduced as $p_f(\boldsymbol{x}, t) : \Omega_f \times \mathcal{I} \to \mathbb{R}$ and $\rho_f(\boldsymbol{x}, t) : \Omega_f \times \mathcal{I} \to \mathbb{R}$, respectively. Hereafter, variables that feature a subindex f are related to the fluid phase, while physical quantities associated with the dispersed particulate phase introduced in the subsequent section are indicated by a subscript p.

The total fluid mass in an arbitrary volume $\mathcal{V}(t)$ can be computed as

$$M_{\mathcal{V}(t)} = \int_{\mathcal{V}(t)} \rho_f(\boldsymbol{x}, t) \mathrm{d}\boldsymbol{x}. \tag{2.2}$$

Supposing that no mass is generated or lost in the domain, the total mass is conserved for each volume $\mathcal{V}(t)$, that is, $\mathrm{d}M_{\mathcal{V}(t)}/\mathrm{d}t = 0$. Making use of Reynolds' transport theorem (2.1) and recalling that $\mathcal{V}(t)$ was chosen to be arbitrary, it can be shown that for sufficiently regular functions $\boldsymbol{u}_f$ and p_f the conservation of mass holds also pointwise, viz.

$$\partial_t \rho_f(\boldsymbol{x}, t) + \nabla \cdot (\rho_f(\boldsymbol{x}, t) \boldsymbol{u}_f(\boldsymbol{x}, t)) = 0 \quad \forall \boldsymbol{x} \in \Omega_f(t), t \in \mathcal{I}. \tag{2.3}$$

Hereafter, we consider exclusively incompressible fluids in which the density field is constant $\rho_f(\boldsymbol{x}, t) = \rho_f$, in which case equation (2.3) reduces to the continuity equation

$$\nabla \cdot \boldsymbol{u}_f(\boldsymbol{x}, t) = 0 = (u_{f,i})_{,i}, \quad \forall \boldsymbol{x} \in \Omega_f(t), t \in \mathcal{I}, \tag{2.4}$$

which represents an extra condition on the flow field to be solenoidal or divergence-free.

In a similar way, considering Newton's second law, the momentum balance can be formulated in an integral sense as

$$\frac{\mathrm{d}}{\mathrm{d}t} \left[\int_{\mathcal{V}(t)} \rho_f \boldsymbol{u}_f(\boldsymbol{x}, t) \mathrm{d}\boldsymbol{x} \right] = \int_{\partial\mathcal{V}(t)} \boldsymbol{b}(\boldsymbol{x}, t) \mathrm{d}S + \int_{\mathcal{V}(t)} \boldsymbol{f}(\boldsymbol{x}, t) \mathrm{d}\boldsymbol{x}, \tag{2.5}$$

where $\rho_f(\boldsymbol{x}, t)\boldsymbol{u}_f(\boldsymbol{x}, t)$ takes the role of a momentum density. The left hand side (LHS) of equation (2.5) represents the rate of change of the total momentum, while the right hand sight (RHS) indicates the sum of all forces acting on the fluid in volume $\mathcal{V}(t)$, respectively. This includes both surface forces $\boldsymbol{b}(\boldsymbol{x}, t) : \Omega_f \times \mathcal{I} \to \mathbb{R}^d$ acting on the domain boundary $\partial\mathcal{V}(t)$ and volume or body forces $\boldsymbol{f}(\boldsymbol{x}, t) : \Omega_f \times \mathcal{I} \to \mathbb{R}^d$ acting on the entire volume $\mathcal{V}(t)$, respectively. We introduce Cauchy's stress law, according to which

$$\boldsymbol{b}(\boldsymbol{x}, t) = \boldsymbol{\mathcal{T}}_f(\boldsymbol{x}, t) \boldsymbol{n}(\boldsymbol{x}, t), \tag{2.6}$$

where $\boldsymbol{n}(\boldsymbol{x}, t)$ is an outward pointing normal vector and $\boldsymbol{\mathcal{T}}_f(\boldsymbol{x}, t) : \Omega_f \times \mathcal{I} \to \mathbb{R}^{d \times d}$ is the second order Cauchy stress tensor field. Then, making use of the Reynolds transport theorem as well as the divergence theorem and recalling that $\mathcal{V}(t)$ is chosen arbitrarily in equation (2.5), the local conservation of momentum for sufficiently smooth functions follows

$$\partial_t (\rho_f \boldsymbol{u}_f(\boldsymbol{x}, t)) + \nabla \cdot (\rho_f \boldsymbol{u}_f(\boldsymbol{x}, t) \otimes \boldsymbol{u}_f(\boldsymbol{x}, t)) = \boldsymbol{f}(\boldsymbol{x}, t) + \nabla \cdot \boldsymbol{\mathcal{T}}_f(\boldsymbol{x}, t) \quad \forall \boldsymbol{x} \in \Omega_f(t), t \in \mathcal{I}, \tag{2.7}$$

where $\otimes : \mathbb{R}^m \times \mathbb{R}^n \to \mathbb{R}^{m\times n}$, $[\boldsymbol{a} \otimes \boldsymbol{b}]_{ij} = (a_i b_j)\ \forall \boldsymbol{a} \in \mathbb{R}^m, \boldsymbol{b} \in \mathbb{R}^n$ is the outer product between two vector fields $\boldsymbol{a}$ and $\boldsymbol{b}$. The Cauchy stress tensor field $\mathcal{T}_f(\boldsymbol{x},t)$, which can be shown to be symmetric when evaluating the angular momentum balance, is a material-specific quantity and is typically determined by additional constitutive relations. Hereafter, we restrict ourselves to the investigation of constant-property Newtonian fluids, for which the stress law reads

$$\mathcal{T}_f(\boldsymbol{x},t) = \left[-p_f(\boldsymbol{x},t) + \lambda_f \nabla \cdot \boldsymbol{u}_f(\boldsymbol{x},t)\right] \mathbf{Id} + 2\mu_f \boldsymbol{D}(\boldsymbol{x},t), \tag{2.8}$$

where λ_f is the first Lamé constant, μ_f is the dynamic viscosity of the fluid, $\mathbf{Id}$ is the identity tensor and

$$\boldsymbol{D}(\boldsymbol{x},t) = \frac{1}{2}\left(\nabla \boldsymbol{u}_f(\boldsymbol{x},t) + (\nabla \boldsymbol{u}_f(\boldsymbol{x},t))^T\right) \tag{2.9}$$

is the rate of strain tensor $\boldsymbol{D}(\boldsymbol{x},t) : \Omega_f \times \mathcal{I} \to \mathbb{R}^{d\times d}$ which represents the symmetric part of the velocity gradient tensor field $\nabla \boldsymbol{u}_f(\boldsymbol{x},t)$. Assuming constant material parameters ρ_f and μ_f, the divergence of the Cauchy stress tensor field simplifies to

$$\nabla \cdot \mathcal{T}_f(\boldsymbol{x},t) = -\nabla p_f(\boldsymbol{x},t) + \mu_f \Delta \boldsymbol{u}_f(\boldsymbol{x},t) = -\nabla p_f(\boldsymbol{x},t) + \nu_f \rho_f \Delta \boldsymbol{u}_f(\boldsymbol{x},t), \tag{2.10}$$

where $\nu_f = \mu_f/\rho_f$ is the kinematic fluid viscosity and the Laplace operator reads

$$\Delta() = \nabla^2() = \nabla \cdot (\nabla()) = \sum_{j=1}^{d} \frac{\partial^2()}{\partial x_j^2}.$$

Finally, introducing expression (2.10) into the local momentum balance (2.7) and expanding the divergence of the convective term $\rho_f \boldsymbol{u}_f \otimes \boldsymbol{u}_f$ subject to the divergence-free constraint, the Navier-Stokes problem for an incompressible Newtonian fluid follows:

We seek a velocity vector field $\boldsymbol{u}_f(\boldsymbol{x},t) : \Omega_f \times \mathcal{I} \to \mathbb{R}^d$ and an associated scalar pressure field $p_f(\boldsymbol{x},t) : \Omega_f \times \mathcal{I} \to \mathbb{R}$ with material parameters $\rho_f, \nu_f > 0$ under the action of a body force field $\boldsymbol{f}(\boldsymbol{x},t) : \Omega_f \times \mathcal{I} \to \mathbb{R}^d$, s.t. for a sufficiently smooth initial condition $\boldsymbol{u}_{f_0}(\boldsymbol{x}) : \Omega_f \to \mathbb{R}^d$ and appropriate boundary conditions (cf. discussion below)

$$\begin{cases} \partial_t \boldsymbol{u}_f - \nu_f \Delta \boldsymbol{u}_f + (\boldsymbol{u}_f \cdot \nabla)\boldsymbol{u}_f + \dfrac{1}{\rho_f}\nabla p_f = \boldsymbol{f} & \text{in } \Omega_f \times (0,T] \\ \nabla \cdot \boldsymbol{u}_f = 0 & \text{in } \Omega_f \times (0,T] \\ \boldsymbol{u}_f = g_D & \text{on } \partial\Omega_f^D \times (0,T] \\ \dfrac{\partial \boldsymbol{u}_f}{\partial n} = g_N & \text{on } \partial\Omega_f^N \times (0,T] \\ \boldsymbol{u}_f(\bullet,0) = \boldsymbol{u}_{f_0} & \text{in } \Omega_f \end{cases} . \tag{2.11}$$

Note that in order to simplify the notation, we have absorbed the fluid density in the force term without introducing a new sign for the force density. A special type of body force is the gravitational acceleration field $\boldsymbol{g}$, which can be expressed as the gradient of a gravitational potential (Pope, 2000). This allows to subsume the gravitational potential and the dynamic fluid pressure into a modified total pressure field. Since the pressure field $p_f(\boldsymbol{x},t)$ appears in form of a gradient in the Navier-Stokes equations, it is determined only up to a constant which can be freely chosen. Typically, the constant is determined by either setting the pressure at a single point of the domain to zero, or by requesting that the mean value of $p_f(x,t)$ over the entire domain $\Omega_f(t)$ vanishes identically.

It might be worth mentioning that the role of the pressure field $p_f(x,t)$ in an incompressible fluid is different from that in a compressible material, where $p_f(x,t)$ is a thermodynamical quantity that is related to the local density and temperature by some relation of state. In an isothermal fluid of constant density, on the other hand, the pressure field is by definition independent of the fluid density and temperature, but it turns out that it is closely related to the divergence-freedom of the velocity field $u_f(x,t)$. In particular, one can derive the following Poisson equation for the pressure field which represents a necessary and sufficient condition for the velocity field to be solenoidal:

We seek a velocity vector field $u_f(x,t) : \Omega_f \times \mathcal{I} \to \mathbb{R}^d$ and an associated scalar pressure field $p_f(x,t) : \Omega_f \times \mathcal{I} \to \mathbb{R}$, s.t.

$$\begin{cases} \Delta p_f = \rho_f \nabla \cdot [(u_f \cdot \nabla) u_f] & \text{in } \Omega_f \times \mathcal{I} \\ p_f = p_D & \text{on } \partial\Omega_f^D \times \mathcal{I} \\ \dfrac{\partial p_f}{\partial n} = p_N & \text{on } \partial\Omega_f^N \times \mathcal{I} \end{cases}, \tag{2.12}$$

where $p_D(x,t) : \partial\Omega_f^D \times \mathcal{I} \to \mathbb{R}$ and $p_N(x,t) : \partial\Omega_f^N \times \mathcal{I} \to \mathbb{R}$ are appropriate functions that represent the Dirichlet and Neumann boundary values.

2.1.2 Boundary conditions

Note that in the Navier-Stokes problem (2.11), we have subdivided the boundary of the fluid domain $\partial\Omega_f(t)$ into two sets $\partial\Omega_f^D$ and $\partial\Omega_f^N$, which denote the parts of the domain boundary at which homogeneous Dirichlet g_D and Neumann boundary conditions g_N are applied, respectively, such that $\partial\Omega_f = \partial\Omega_f^D \cup \partial\Omega_f^N$. The two sets are disjoint for each individual velocity component such that the boundary condition is well-posed (i.e. at each position only one condition is imposed), whereas for the boundary condition of the full vector field in general $\partial\Omega_f^D \cap \partial\Omega_f^N \neq \emptyset$. For instance, we will see below for the free-surface boundary conditions that Dirichlet and Neumann boundary conditions can be applied to different components of the velocity field at one and the same spatial position.

Dirichlet boundary conditions with $g_D = \mathbf{0}$ will be applied in the remainder of this work to describe the behaviour of a fluid at a stationary solid impermeable wall $\mathcal{S}_{wall} \subset \partial\Omega_f$. That is, the fluid velocity tangential to the solid surface vanishes ('no-slip condition') and so does the wall-normal velocity component, the latter ensuring that there is no flux across the impermeable wall ('impermeability condition'), viz.

$$\left(u_f - \left(n \cdot u_f\right) n\right) = \mathbf{0} \quad \text{on } \mathcal{S}_{wall} \times (0,T] \tag{2.13a}$$

$$n \cdot u_f = 0 \quad \text{on } \mathcal{S}_{wall} \times (0,T], \tag{2.13b}$$

which directly implies that $u_f = \mathbf{0}$ at the solid wall. At the free surface of open channels and ducts $\mathcal{S}_{surf} \subset \partial\Omega_f$, on the other hand, the no-slip condition is replaced by a 'free-slip condition'. That is, a symmetry condition is imposed for the surface-parallel velocity components in form of a homogeneous

Neumann boundary condition while maintaining the impermeability condition for the surface-normal component, viz.

$$\frac{\partial}{\partial n}\left(u_f - \left(n \cdot u_f\right) n\right) = \mathbf{0} \quad \text{on } \mathcal{S}_{surf} \times (0, T] \tag{2.14a}$$

$$n \cdot u_f = 0 \quad \text{on } \mathcal{S}_{surf} \times (0, T]. \tag{2.14b}$$

For the pressure field, typically homogeneous Neumann boundary conditions are imposed along the free surface.

In both numerical simulations and theoretical considerations, it is moreover common practice to apply periodic boundary conditions to the Navier-Stokes system. We will follow this approach and impose periodic boundary conditions in the homogeneous directions of the later considered open channel and duct flows, that are, the streamwise and spanwise direction in the open channel and the streamwise direction in the open duct case. Consider, for instance, a single-phase flow in an infinitely long rectangular open duct with cross-sectional dimensions L_y and L_z, respectively, such that the entire domain is $\mathbb{R} \times [0, L_y] \times [0, L_z]$. Instead of full $\mathbb{R}$, we consider in practice the interval $x \in [0, L_x)$ and assume streamwise periodicity with fundamental period L_x, such that $u_f(x - hL, t) = u_f(x, t) \; \forall \; h \in \mathbb{Z}$ and $L = (L_x, 0, 0)^T$. For the open channel case, analogous boundary conditions are applied with $L = (L_x, 0, L_z)^T$ and $h \in \mathbb{Z}^3$. Consequently, the open duct case can be topologically associated with the one torus $\mathbb{T}^1$ and thus the one sphere $\mathbb{S}^1$, while doubly-periodic channels are associated with the two torus $\mathbb{T}^2 = \mathbb{S}^1 \times \mathbb{S}^1$ (Marqués, 1990; Majda and Bertozzi, 2002).

2.2 Governing equations of the dispersed phase

In the following, we consider the case of a number of N_p solid rigid particles that are immersed in the fluid domain. For this purpose, we consider a domain $\Omega \subset \mathbb{R}^d$ with disjoint subsets $\Omega_f(t), \Omega_p(t) \subseteq \Omega$ s.t. $\Omega_f(t) \cup \Omega_p(t) = \Omega$, from which the former denotes the subdomain occupied by the fluid phase at time t and

$$\Omega_p(t) = \dot{\bigcup}_{1 \le j \le N_p} \Omega_p^{(j)}(t)$$

is the particle domain composed of N_p pairwise disjoint non-overlapping subdomains $\Omega_p^{(j)}(t)$. The boundary of the fluid phase $\partial\Omega_f(t)$ now comprises also the phase boundary between fluid and solid phase

$$\partial\Omega_p(t) = \dot{\bigcup}_{1 \le j \le N_p} \partial\Omega_p^{(j)}(t) \; \subset \; \partial\Omega_f(t).$$

The coupling between fluid and particulate phase along the surface $\partial\Omega_p^{(j)}(t)$ of the jth particle is described in form of a similar set of no-slip and impermeability conditions as along a stationary solid wall (cf. equation 2.13), with the difference that points $x_p \in \Omega_p^{(j)}(t)$ move at a velocity

$$u_p^{(j)}(x_p, t) = U_p^{(j)}(t) + \omega_p^{(j)}(t) \times \left(x_p - X_p^{(l)}(t)\right) \quad \text{in } \Omega_p^{(j)}(t) \times \mathcal{I} \tag{2.15}$$

as a consequence of the rigid-body motion of the particle. Here, $u_p^{(j)}(x_p, t) : \Omega_p^{(j)} \times \mathcal{I} \to \mathbb{R}^d$ is the local particle velocity within a rigid body that originates in a superposition of the translational motion at a Lagrangian velocity $U_p^{(j)}(t) : \mathcal{I} \to \mathbb{R}^d$ and the rotational motion with an angular velocity $\omega_p^{(j)}(t) : \mathcal{I} \to$

$\mathbb{R}^d$, respectively. The current position of the centre of gravity of the jth particle is $\boldsymbol{X}_p^{(j)}(t) : \mathcal{I} \to \mathbb{R}^d$. The no-slip and impermeability boundary conditions are then formulated for all points $\boldsymbol{x} = \boldsymbol{x}_p \in \partial\Omega_p^{(j)}(t)$ on the surface of all particles $1 \leq j \leq N_p$ as

$$\left(\boldsymbol{u}_f - \boldsymbol{u}_p^{(j)}\right) - \boldsymbol{n}_p^{(j)} \cdot \left(\boldsymbol{u}_f - \boldsymbol{u}_p^{(j)}\right) \boldsymbol{n}_p^{(j)} = \boldsymbol{0} \quad \text{on } \partial\Omega_p^{(j)}(t) \times (0,T] \quad \forall\, 1 \leq j \leq N_p, \tag{2.16a}$$

$$\boldsymbol{n}_p^{(j)} \cdot \left(\boldsymbol{u}_f - \boldsymbol{u}_p^{(j)}\right) = 0 \quad \text{on } \partial\Omega_p^{(j)}(t) \times (0,T] \quad \forall\, 1 \leq j \leq N_p. \tag{2.16b}$$

Here, $\boldsymbol{n}_p^{(j)}$ is an outward pointing normal vector at the surface of particle j. The resulting condition is that fluid and particle velocity have to be identical at the surface of each particle, viz. $\boldsymbol{u}_f(\boldsymbol{x}_p, t) = \boldsymbol{u}_p(\boldsymbol{x}_p, t)$ on $\partial\Omega_p^{(j)}(t) \times (0, T]$.

The dynamics of the individual particles are governed by the Newton-Euler equations, which combine Newton's equations of linear and angular momentum for a rigid body immersed in a fluid. In the former case, Newton's second law implies that the rate of change of the linear momentum of particle j is balanced by a set of forces acting on the sphere. In the current case, three forces are exerted on the individual particles, including contributions due to gravity, buoyancy and contacts with nearby particles or solid walls. Newton's equations of linear momentum for a spherical rigid body j immersed in a Newtonian fluid are then formulated as (Jackson, 2000)

$$\begin{cases} \rho_p \mathcal{V}_p{}^{(j)} \dfrac{\mathrm{d}\boldsymbol{U}_p^{(j)}(t)}{\mathrm{d}t} = \displaystyle\oint_{\partial\Omega_p^{(j)}(t)} \boldsymbol{\mathcal{T}}_f(\boldsymbol{x}_p, t) \cdot \boldsymbol{n}_p^{(j)}\, \mathrm{d}S + \left(\rho_p - \rho_f\right) \mathcal{V}_p \boldsymbol{g} + \boldsymbol{F}^{C(j)} & \text{in } \mathcal{I} \\ \dfrac{\mathrm{d}\boldsymbol{X}_p^{(j)}(t)}{\mathrm{d}t} = \boldsymbol{U}_p^{(j)}(t) & \text{in } \mathcal{I} \end{cases}, \tag{2.17}$$

where $\mathcal{V}_p{}^{(j)}$ is the volume of the jth particle. The third term on the RHS of equation (2.17), $\boldsymbol{F}^{C^{(j)}}(t) : \mathcal{I} \to \mathbb{R}^d$, subsumes all force contributions due to contacts of the jth particle with nearby particles or solid walls. The first term on the RHS includes the fluid stress tensor and thus represents with the fluid traction force the coupling between the fluid and the solid phase equations, whereas the second term on the RHS represents the net force due to gravity and buoyancy.

Similarly, the conservation of angular momentum is formulated as a balance between the rate of change of angular momentum and torque acting on the jth particle, leading to Newton's equation of angular momentum, viz. (Jackson, 2000)

$$\begin{cases} \dfrac{\mathrm{d}(\boldsymbol{\mathcal{J}}_p^{(j)}(t)\, \boldsymbol{\omega}_p^{(j)}(t))}{\mathrm{d}t} = \displaystyle\oint_{\partial\Omega_p^{(j)}(t)} \boldsymbol{r}^{(j)} \times (\boldsymbol{\mathcal{T}}_f(\boldsymbol{x}_p, t) \cdot \boldsymbol{n}_p^{(j)})\mathrm{d}S + \boldsymbol{r}^{(j)} \times \boldsymbol{F}^{C(j)} & \text{in } \mathcal{I} \\ \dfrac{\mathrm{d}\boldsymbol{\Theta}_p^{(j)}(t)}{\mathrm{d}t} = \boldsymbol{\omega}_p^{(j)}(t) & \text{in } \mathcal{I} \end{cases}, \tag{2.18}$$

where $\boldsymbol{r}^{(j)}(t) = \boldsymbol{x}_p - \boldsymbol{X}_p^{(l)}(t)\ \forall\ \boldsymbol{x}_p \in \partial\Omega_p^{(j)}(t)$ is the position vector of all points on the particle surface w.r.t. the particle's centre of gravity. Accordingly, $|\boldsymbol{r}^{(j)}| = D^{(j)}/2$ and $D^{(j)}$ are the particle radius and diameter in case of a rigid spherical particle, respectively. Furthermore, $\boldsymbol{\Theta}_p^{(j)}(t) : \mathcal{I} \to \mathbb{R}^d$ is the current angular orientation of the jth particle and $\boldsymbol{\mathcal{J}}_p^{(j)}(t) : \mathcal{I} \to \mathbb{R}^{d \times d}$ is the moment of inertia tensor, which reduces for the special case of a spherical rigid body to the constant scalar moment of inertia $\mathcal{J}_p^{(j)} = 2/5 \rho_p \mathcal{V}_p{}^{(j)} (D^{(j)}/2)^2$. In analogy to the translational linear momentum in equation (2.17), the rate of change of angular momentum is caused by a balance of the hydraulic torque and the torque induced by contact forces during particle-particle and particle-wall contacts, respectively, which form the first and second term on the RHS of equation (2.18).

In equation (2.18), the cross-product between two vector fields $\boldsymbol{a}(x,t)$ and $\boldsymbol{b}(x,t)$ is introduced as

$$\boldsymbol{a} \times \boldsymbol{b} = \varepsilon_{ijk}\boldsymbol{e}_i a_j b_k, \tag{2.19}$$

where ε_{ijk} is the Levi-Civita symbol and $\boldsymbol{e}_i$ is the ith unit normal vector of the Cartesian basis. In analogy, the curl of a three-dimensional vector field $\boldsymbol{a}(x,t)$ is given by

$$\nabla \times \boldsymbol{a} = \varepsilon_{ijk}\boldsymbol{e}_i \frac{\partial a_k}{\partial x_j}. \tag{2.20}$$

In this context, let us introduce for later use the fluid vorticity $\boldsymbol{\omega}_f(x,t) : \Omega_f \times \mathcal{I} \rightarrow \mathbb{R}^d$ as the curl of the velocity vector field $\boldsymbol{\omega}_f = \nabla \times \boldsymbol{u}_f$, with components $\boldsymbol{\omega}_f = (\omega_{f,x}, \omega_{f,y}, \omega_{f,z})^T$ w.r.t. the Cartesian basis.

Chapter 3

Literature review

"Where chaos begins, classical science stops."
— James Gleick in *Chaos: Making a new science* (1987)

3.1 Historical remarks

Turbulence is one of the most complicated phenomena in classical physics and in contrast to other fields in this discipline, we are still far from a full theory on how turbulence works nor is there even a unique definition of what turbulence actually is. Instead, turbulence is typically described phenomenologically by its fundamental properties, which are immanent to all its forms and thus represent a minimal consensus of what turbulence is. A rather comprehensive list of properties is given by Tennekes and Lumley (1972), who describe turbulent flows as a form of irregular, three-dimensional and multi-scale fluid motion, further characterised by its tendency to vortical motion, its strong ability of mixing and its typically high momentum and heat transfer intensity. But turbulence is also a dissipative system, meaning that kinetic energy is at some point dissipated and transferred to heat. We will come back to this specific point below.

The historical development of classical mechanics and fluid dynamics in particular is well summarised in the textbook by Dugas (2012), whereas Davidson et al. (2011) illustrate the history of turbulence research from the 1880s to the present day based on the biographies of a number of outstanding researcher personalities, including O. Reynolds, L. Prandtl, G. Bachelor and A. A. Townsend. Even though turbulence has been probably observed by humans for thousands of years, the fundament that allowed its scientific investigation was laid not before the 19th century, when Navier and Stokes provided the theoretical basis through the fundamental laws of fluid dynamics that bear their names up to the present day. Only a few years after the formulation of the fundamental equations of fluid motion, Hagen (1854) and Darcy (1857) studied independently from each other the pressure drop along straight ducts and observed that only part of the observed decay depended on the fluid's viscosity, while a second contribution was apparently independent of viscous effects and scaled quadratically with the fluid velocity. The latter non-linear effect was reported to come along with a 'disordered motion' and it was claimed that the observed increase in drag was due to the necessity of additional energy for the creation of these complicated eddies (Jiménez, 2004b). These first ideas of two distinct effects that contribute to the overall drag were further developed in the seminal work of Boussinesq (1877), in which, for the first time, a clear differentiation in a laminar and a turbulent regime was proposed, even though still using different terms. Apart from this pioneering classification, the study of Boussinesq contained some fundamental concepts that have influenced and inspired turbulence research up to the present day. Just to mention one example, Boussinesq developed the eddy viscosity hypothesis, in which he described the effect of turbulence on the flow as an increase of the molecular viscosity by an additional turbulent contribution, exploiting an analogy to the classical relation between stress and the rate-of-strain for Newtonian fluids.

The perhaps most important achievements for todays turbulence research, however, date back to Osborne Reynolds. Based on his experiments on pipe flow instability, Reynolds (1883) presented a criterion for the transition between the regular, ordered 'laminar' motion and the chaotic, irregular 'turbulent' motion as a critical value for the ratio of inertial to viscous forces, viz.

$$Re = \frac{\mathcal{U}\mathcal{L}}{\nu_f}, \tag{3.1}$$

which has later been termed 'Reynolds number' to his honour. Here, we have used the most general formulation of the Reynolds number, including a characteristic velocity $\mathcal{U}$ and length scale $\mathcal{L}$ of the flow of interest, respectively, which gives us the freedom to define different Reynolds numbers depending on our needs in the remainder of this work. Considering again Reynolds' original experiment of the flow in an elongated pipe, a small perturbation of the flow will be damped and eventually dissipated into heat due to the viscous effects if the Reynolds number is sufficiently low. Is the Reynolds number on the other hand sufficiently high, viscous damping is much weaker than the inertial forces and the perturbation cannot be completely dissipated into heat. The flow therefore compensates the additional energy input by giving rise to disordered fluid motion that will be further amplified causing a transition to a fully-turbulent state in which viscous damping is much weaker than the inertial forces.

The second inestimable contribution of Reynolds to turbulence research is the 'Reynolds averaging concept' first published in Reynolds (1894), in which he proposed the fundamental idea to decompose all physical quantities into an ensemble-averaged mean contribution and a fluctuation around this mean. Hereafter, the ensemble-averaging operator will be denoted by $\langle \bullet \rangle$ and is defined as (Pope, 2000)

$$\langle \varphi \rangle = \int_{-\infty}^{\infty} V f(V) \mathrm{d}V, \tag{3.2}$$

where $f(V)$ is the probability density function (p.d.f.) and V is the sample space parameter, representing the set of all values that φ can possibly attain. If not otherwise stated, $\langle \bullet \rangle$ refers to averaging over all statistically homogeneous directions and time. The fluctuation of a field variable $\varphi(x,t)$ w.r.t. its mean $\langle \varphi \rangle$ is typically denoted by a superscripted prime, viz. $\varphi'(x,t)$, such that

$$\varphi(x,t) = \langle \varphi(x,t) \rangle + \varphi'(x,t). \tag{3.3}$$

For the sake of brevity, we forgo a detailed presentation of the mathematical properties of the Reynolds averaging-operator and the temporal fluctuation of the signal, but the interested reader is encouraged to check the detailed explanations on the statistical treatment of random variables in chapters 3 and 4 of Pope (2000).

Upon substituting the decomposition for the velocity and pressure field into the Navier-Stokes and continuity equations (2.11), Reynolds derived a system of equations for the mean flow field that would later become the 'Reynolds-averaged Navier Stokes (RANS) equations'. The RANS equations for an incompressible, Newtonian fluid with constant viscosity in the absence of external force fields then read for the mean velocity field $\langle \boldsymbol{u}_f \rangle(\boldsymbol{x},t) : \Omega_f \times \mathcal{I} \to \mathbb{R}^d$, the mean pressure field $\langle p_f \rangle(\boldsymbol{x},t) : \Omega_f \times \mathcal{I} \to \mathbb{R}$, and a fluctuation velocity field $\langle \boldsymbol{u}'_f \rangle(\boldsymbol{x},t) : \Omega_f \times \mathcal{I} \to \mathbb{R}^d$

$$\nabla \cdot \langle \boldsymbol{u}_f \rangle = 0 \tag{3.4a}$$

$$\partial_t \langle \boldsymbol{u}_f \rangle + \left(\langle \boldsymbol{u}_f \rangle \cdot \nabla \right) \langle \boldsymbol{u}_f \rangle = -\frac{1}{\rho_f} \nabla \langle p_f \rangle + \nu_f \Delta \langle \boldsymbol{u}_f \rangle - \nabla \cdot \langle \boldsymbol{u}'_f \otimes \boldsymbol{u}'_f \rangle. \tag{3.4b}$$

Note that in this context, both the mean and the fluctuating velocity field are exactly solenoidal due to the linearity of the divergence operator. The averaging, however, comes not for free, and the price we have to pay is a 'closure problem', that is, the new system of equations features an unknown symmetric tensor field $\langle \boldsymbol{u}'_f \otimes \boldsymbol{u}'_f \rangle : \Omega_f \times \mathcal{I} \to \mathbb{R}^{d \times d}$ that represents velocity fluctuation (co)variances. The fact that the tensor (multiplied by the fluid density ρ_f) is commonly termed 'Reynolds stress tensor' is motivated by the fact that it appears in the role of a stress term in the above equations. Note that the appearance of unknown statistical moments is inherent to the application of Reynolds operators and further averaging of unclosed terms as done to obtain Reynolds-stress transport equations leads to even higher-order unknown moments (Pope, 2000).

In the first half of the 20th century, the dissipative and multi-scale nature of turbulence came into the focus of turbulence research, in the context of investigations on the energy transfer in the flow. The concept of an 'energy cascade' was first posited by Richardson (1922) and later extended by Kolmogorov (1941a,b) to a mathematically more rigorous theory. The basic idea therein was that in a fully-developed turbulent flow at sufficiently high Reynolds number, the overwhelming fraction of the kinetic energy is injected in a number of large-scale eddies of size comparable to the largest length scales of the system, while it is dissipated exclusively by the smallest eddies of the system. The intrinsic Reynolds numbers of these small scales are of order $Re = \mathcal{O}(1)$, such that inner-system energy-conservation is broken and energy is dissipated to heat. Kolmogorov determined the minimal length scale of the flow at which energy is dissipated based on dimensional considerations to be $\eta_f = (\nu_f^3/\varepsilon)^{1/4}$, the Kolmogorov length, wherein ε indicates the dissipation rate. A velocity scale $u_\eta = (\nu_f \varepsilon)^{1/4}$ and a time scale $\tau_\eta = (\nu_f/\varepsilon)^{1/2}$ can be derived accordingly. In between the energy-containing large eddies and the dissipative range of small-scale eddies, an essentially inviscid cascade of successively decreasing scales is acting to transfer energy from the largest to the smallest flow scales, with the astonishing consequence that the dissipation in the smallest scales is directly linked to energy production in the large scales to satisfy the conservation of energy outside the dissipative regime, i.e. along the cascade itself.

Note that even though the general concepts of the energy cascades conceived by Richardson (1922) and Kolmogorov (1941a,b) resemble each other, strictly speaking, the underlying ideas are not the same. The Kolmogorov cascade is a pure scale-space concept that does not further specify how exactly the energy is transferred in physical space. Richardson and after him Obukhov (1941), on the other hand, imagined a cascade local in physical space, in which large eddies become unstable and break, giving rise to smaller eddies and so on and so forth down to the smallest eddies of dimensions η_f, which are eventually dissipated into heat. The isotropic theory of Kolmogorov represents a milestone in the understanding of turbulence, but nevertheless it has its limitations and shortcomings. Ever since its original formulation, it has been extended and modified (see, for instance, Batchelor and Townsend (1949), Barenblatt and Chorin (1998) and the review in section 6.7 of Pope (2000)).

3.2 Wall-bounded turbulence

3.2.1 The classical theory

Turbulent shear flows that are surrounded by solid boundaries form one of the most important subclasses of turbulence due to their indisputable importance for our every-day life. Popular examples include the flows around airfoils, the oil or gas transport through industrial pipelines or the dynamics of the atmospheric boundary layer, to list only a few. It is therefore less surprising that the investigation of wall-bounded turbulence has attracted immense interest among engineers, physicists and applied mathematicians for more than a century, beginning with the pioneering discovery of the near-wall boundary layer by Prandtl (1904). In the following, we will first give a brief overview of what we will hereafter call the classical or mean flow theory of wall-bounded turbulence, following the classical textbooks of Hinze (1959), Tennekes and Lumley (1972), Townsend (1976), McComb (1990) and Pope (2000), while our nowadays view on it is summarised in a number of more recent review articles in the subsequent section (Jiménez, 2000; Marusic et al., 2010b; Smits et al., 2011; Jiménez, 2013a, 2018). The classical theory bases predominantly on the study of flows in some 'canonical' flow situations which feature comparably simple properties and geometries, making them ideal candidates for the investigation of fundamental physical processes. The list of canonical flows usually includes the equilibrium flow between two infinite horizontal smooth plates (henceforth referred to as 'closed channel') or through straight pipes of circular cross-section, both driven by a mild streamwise pressure gradient. These 'internal' flows are completed with 'external' boundary layers developing over a flat smooth bottom plate without an top bound. In order to introduce a general notation for all three configurations, we shall use as a characteristic outer length scale H either the half channel height, the pipe radius or the boundary layer thickness, respectively. For the following considerations, let us focus on statistically stationary fully-developed closed channel flows at non-marginal Reynolds numbers between two horizontal plates, separated by a wall-normal distance $2H$ and featuring infinite extensions in the two wall-parallel directions. As a Cartesian basis, we impose a coordinate system with unit vectors parallel to the streamwise x-, wall-normal y- and spanwise z-direction centred at the lower wall $y = 0$. It follows that the position and fluid velocity vectors can be expressed w.r.t. the aforementioned basis as $\boldsymbol{x} = (x, y, z)^T$ and $\boldsymbol{u}_f = (u_f, v_f, w_f)^T$, respectively. Under the given assumptions, the flow is statistically stationary and one-dimensional, that is, the statistics of the flow depend solely on the wall-normal position y. We further choose the bulk velocity

$$u_b = \frac{1}{2H}\int_0^{2H} \langle u_f \rangle \mathrm{d}y \tag{3.5}$$

as characteristic velocity scale of the system of interest, based on which a bulk Reynolds number representative for the outer bulk flow can be formally defined as $Re_b = u_b H/\nu_f$. Averaging the streamwise momentum equations for the case of a fully-developed channel flow, it can be easily shown that the momentum equations reduce to a simple balance between the axial pressure gradient and the resisting wall-normal gradient of the fluid shear stress, viz.

$$\frac{\mathrm{d}\tau_f}{\mathrm{d}y} = \frac{\mathrm{d}\langle p_f \rangle}{\mathrm{d}x}, \tag{3.6}$$

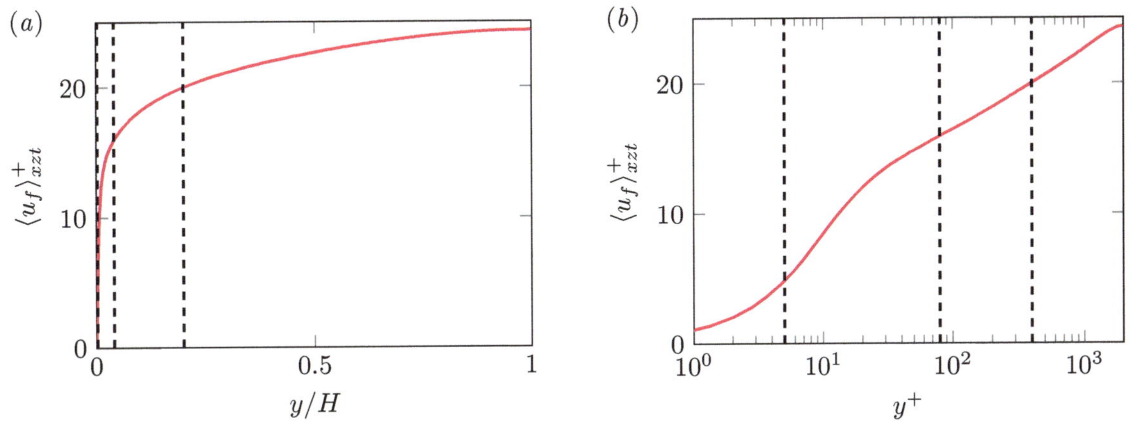

Figure 3.1: Wall-normal profile of the inner-scaled mean velocity in a closed channel at $Re_\tau \approx 2000$ scaled in (*a*) outer and (*b*) inner coordinates. Vertical dashed lines indicate conventional upper limits of the viscous sublayer ($y^+ = 5$), the buffer layer ($y^+ = 80$) and the log-layer ($y/H = 0.2$), respectively. Data is from Hoyas and Jiménez (2006).

wherein the total mean shear stress $\tau_f(y)$ comprises a viscous and a turbulent contribution, viz.

$$\tau_f(y) = \rho_f \nu_f \frac{\mathrm{d}\langle u_f \rangle}{\mathrm{d}y} - \rho_f \langle u'_f v'_f \rangle. \tag{3.7}$$

It follows directly from equation (3.6) that the wall-normal gradient of τ_f and the streamwise gradient of $\langle p_f \rangle$ both have to be constant, and hence that the wall-normal profile of τ_f follows a simple linear relation

$$\tau_f = \tau_w \left(1 - \frac{y}{H}\right), \tag{3.8}$$

where τ_w is the shear stress at the solid wall, viz. $\tau_w = \tau_f(y)|_{y=0}$.

Formally, it is possible to define a second set of velocity and length scales from the wall shear stress τ_w that turn out to be representative for the flow structures in proximity to the wall, viz.

$$u_\tau = \sqrt{\tau_w / \rho_f}, \tag{3.9a}$$

$$\delta_\nu = \nu_f / u_\tau, \tag{3.9b}$$

In the remainder, we will refer to quantities normalised with u_τ and ν_f interchangeably as scaled in 'inner units' or 'wall units' and mark them by a superscripted plus, viz. $\bullet^+$. In particular, we can define with this set of characteristic scales the friction Reynolds number

$$Re_\tau = \frac{u_\tau H}{\nu_f} = \frac{H}{\delta_\nu} = H^+, \tag{3.10}$$

which serves as a measure for the scale separation between the outer and inner characteristic length scale and consequently grows as the Reynolds number increases.

The presence of solid domain boundaries drastically alters the dynamics of and the energy distribution within the flow compared, for instance, with the situation in homogeneous isotropic turbulence in the absence of walls. The no-slip and impermeability conditions which manifest the influence of the wall on the flow enforce a mean shear $S = \mathrm{d}\langle u_f \rangle / \mathrm{d}y$ whose intensity increases when approaching the solid wall and which acts as the main source of energy for turbulence. Obviously, the presence of the wall introduces a strong inhomogeneity and anisotropy of turbulence, and it seems therefore natural to study turbulence in dependence of its wall-normal distance. In the classical theory, a classification into different wall-parallel layers has proved itself, which are visualised in figure 3.1 together with a characteristic mean velocity profile, exemplary for a turbulent closed channel flow at $Re_\tau \approx 2003$ (Hoyas and Jiménez, 2006).

In the near wall or 'inner region', the flow is assumed to be essentially unaffected by the outer scales H and u_b such that the flow dynamics therein should exclusively depend on inner scales. The subregion closest to the wall in a region $y^+ \in [0, 5]$ is typically called 'viscous sublayer', as the proximity of the wall sufficiently dampens the local velocity fluctuations such that the Reynolds stresses become negligible and viscous stresses in fact dominate the fluid shear stress (Pope, 2000). Under the assumption that the velocity profile in this region depends on y^+ only, Prandtl (1925) derived by fundamental dimensional arguments the 'law of the wall' $\langle u_f \rangle = f(y^+)$. For values of y^+ as small as in the viscous sublayer, truncation of a Taylor series expansion after the linear contribution gives a good approximation for the velocity profile as

$$\langle u_f \rangle^+ = y^+. \tag{3.11}$$

As can be expected, this argumentation loses its validity if we turn our attention to those regions of the flow that are still in the inner region, but sufficiently far away from the wall that we cannot expect viscous effects to play a significant role. As pointed out by Jiménez (2018), there is thus no applicable characteristic length scale available in this intermediate region and structures therein feature a characteristic aspect ratio rather than a characteristic size. That is, the largest structures grow linearly with increasing distance to the wall such that the only relevant length scale can be their local distance to the wall (Townsend, 1976). Note that this 'logarithmic layer' (referring to the typical velocity profile in this range) is in that sense another incarnation of a self-similar cascade as that of Kolmogorov turbulence, but this time in physical space rather than in scale space (Jiménez, 2013b, and references therein) and thus more resembles the concepts of Richardson (1922) and Obukhov (1941), respectively. A more detailed discussion on individual flow structures in the logarithmic layer will be given in the subsequent section of this literature review.

Again, one way to derive a formulation for the mean velocity profile in this regime is by considering dimensional arguments including the independence of the mean velocity (gradient) of both y^+ and y/H. The resulting approximation is the classical 'logarithmic' or 'log law'

$$\langle u_f \rangle^+ = \frac{1}{\kappa} \ln(y^+) + B_u. \tag{3.12}$$

Two separate ways to arrive at the log law can be found in Hinze (1959), which we shall not repeat at this point for the sake of brevity. For the v. Kármán constant κ and the additive constant B_u, slightly varying values can be found in the literature, we here exemplarily refer to the values $\kappa = 0.41$ and $B_u = 5.2$ reported by Pope (2000). Also, there exists no unique definition of where the logarithmic layer starts and where it ends. Commonly applied values are, for instance, $30\delta_\nu \lesssim y \lesssim 0.3H$ (Pope, 2000) or $80\delta_\nu \lesssim y \lesssim 0.2H$ (Jiménez, 2018). It is directly clear from the proposed limits and our earlier assumption according to which the log-layer is essentially independent of inner and outer length scales that a well-developed logarithmic layer requires a certain minimal scale separation, that is, a minimal friction Reynolds number that is at least of the order of $Re_\tau = \mathcal{O}(10^3)$.

The above definitions of the viscous sublayer and the logarithmic layer leave some gap in between, where one would expect both viscous and turbulent effects to be of relevance. Such a transitional regime between the viscous and the turbulence-dominated regime can indeed be found in form of the 'buffer layer', which we will discuss in detail in the following.

3.2.2 The structural view on wall-bounded turbulence

Looking back, the statistical treatment of turbulence that has its roots in the seminal works of Reynolds paved the way for our today's understanding of turbulence and its role in the development of the just reviewed mean flow theory is beyond doubt. In fact, it appears conclusive that a phenomenon as irregular and unpredictable as turbulence is treated statistically.

However, chaos and irregularity should not be erroneously equated with 'randomness'. In the following, we change our perspective and think of turbulence for the moment as a purely mathematical phenomenon and a necessary consequence of the peculiar non-linear nature of the Navier-Stokes equations. From this viewpoint, we have to admit that the governing equations are strictly deterministic rather than incorporating any stochastic contribution. That is, identical initial conditions to the equations will always lead to the exactly same evolution of the flow and there is thus no way of how turbulence could feature any degree of 'randomness'. Nonetheless, also deterministic systems can loose their predictability leading to what is called 'deterministic chaos' (Lorenz, 1963). In such chaotic systems, slight variations of the initial conditions can cause the system to follow a completely different trajectory in phase space, and that is exactly what the 'irregular motion' of turbulence has to be considered as.

Wall-bounded turbulent flows are chaotic dynamical systems, in which the fully-turbulent state represents a strange attractor in the phase space (Ruelle and Takens, 1971; Lanford, 1982; Holmes et al., 1996) to which nearby trajectories are 'drawn' (cf. Guckenheimer and Holmes, 1983; Brin and Stuck, 2002, for a rigorous definition). Even though typically of high dimension, the turbulent attractor can be shown to possess a finite number of degrees of freedom which scale with some global Reynolds number as $Re^{9/4}$ (Landau and Lifshitz, 1959). The temporal evolution of an appropriate initial state of the velocity field $\boldsymbol{q}_0 \in \mathbb{R}^n$ in state space can accordingly be expressed by a high- but finite-dimensional (autonomous) system of ordinary differential equations of the form (Kawahara et al., 2012b)

$$\frac{\mathrm{d}\boldsymbol{q}(t)}{\mathrm{d}t} = \boldsymbol{f}(\boldsymbol{q}(t), p). \tag{3.13}$$

Here, $\boldsymbol{q}(t) : \mathcal{I} \to \mathbb{R}^n$ (and accordingly $\boldsymbol{q}_0$) are vectors of n unknown independent physical quantities which can be, for instance, Fourier or Chebyshev coefficients of the flow field expansions that together represent a full instantaneous state of the flow field (a viewpoint on turbulence that was strongly coined by Hopf, 1948). The right hand side of equation (3.13) represents a smooth vector field $\boldsymbol{f}(\boldsymbol{q}(t), p) : U \times \mathbb{R}^m \to \mathbb{R}^n$, where $U \subseteq \mathbb{R}^n$ is a subset of the state space and $p \in \mathbb{R}^m$ is a set of m additional parameters such as the Reynolds number. The reason for the finite-dimensionality of the turbulent attractor can be physically interpreted as a consequence of viscosity that smoothes the smaller scales (Kawahara et al., 2012b, and references therein), such that no scales smaller than the Kolmogorov length can occur.

Even though the system is inherently chaotic, trajectories can spend a considerable amount of time in the neighbourhood of low-dimensional manifolds in the turbulent attractor. Such low-dimensional state-space patterns represent 'coherent structures' in the flow (Jiménez, 1987), which are local, recurrent and spatiotemporally organised flow patterns typically carrying a relevant portion of energy or contribute to the dissipation of the latter. The investigation of turbulence on the basis of its coherent structures is a significantly younger discipline than the statistical approach and, consequently, a fast changing field. We refer in the following on some recent monographs of the group around Javier Jiménez in Madrid (Jiménez, 2013a; Jiménez and Kawahara, 2013; Jiménez, 2018) that naturally possess a strong affinity to results obtained by direct numerical simulations, while very high Reynolds number results obtained from experiments and measurements are reviewed, for instance, in the recent works of the group around Ivan Marusic in Melbourne (Marusic et al., 2010b; Smits et al., 2011), which together give a comprehensive overview of the state of the art at the time this thesis is completed.

Historically, the search for coherent motions in wall-bounded turbulence first focused on the structures inherent to the buffer layer (Kline et al., 1967; Kim et al., 1971) as summarised in the early review of Robinson (1991). Thanks to a number of seminal theoretical and numerical studies in the 1990s when the first direct numerical simulations of turbulent channel flows were performed (Kim et al., 1987; Jiménez and Moin, 1991; Hamilton et al., 1995), the processes that are dominating this regime are nowadays relatively well understood (Jiménez and Kawahara, 2013). The buffer layer accommodates in fact a few types of coherent structures that are associated with different physical quantities and partly interact with each other. Of largest importance are high- and low-speed streaks of the streamwise velocity and quasi-streamwise vortices, the latter representing structures of locally intense streamwise vorticity. Kline et al. (1967) first observed high- and low-speed streaks in their hydrogen bubble visualisations by detecting streamwise-elongated regions of spanwise alternating positive and negative velocity fluctuations w.r.t. the mean velocity at that height. The characteristic spacing between two streaks of the same sign is around $z^+ = 100$, whereas their average streamwise length is typically a few hundred up to one thousand wall units. Quasi-streamwise vortices are shorter than the velocity streaks (Jeong et al., 1997), such that several quasi-streamwise vortices are associated with a single velocity streak and flank this latter alternately (in a statistical sense) on both sides with a spacing of roughly $x^+ \approx 300$ (Jiménez, 2013a). The special arrangement of the quasi-streamwise vortices causes and amplifies sinusoidal bending of the streaks that eventually leads to their destruction with simultaneous generation of vorticity, giving birth to a next generation of quasi-streamwise vortices. These latter lead, in turn, to the creation of a new set of streamwise streaks by modulation of the local velocity profile. The validity of this self-sustaining regeneration cycle has been shown in a series of numerical and analytical studies that were focused on the exact underlying instability process which causes streaks to break up (Hamilton et al., 1995; Waleffe, 1997; Schoppa and Hussain, 2002; Kawahara et al., 2003).

Jiménez and Pinelli (1999) could moreover show by artificially suppressing all fluctuations above $y^+ > 60$ that the near-wall cycle is indeed 'autonomous' in that is survives in the absence of outer influences. The autonomy of the regeneration cycle is, however, not equivalent to the absence of outer influences in the near wall region, it only states that these are not a necessary condition for the cycle to be maintained. In fact, it has been clearly revealed that near-wall buffer layer structures actively interact with larger structures centring further away from the wall (Toh and Itano, 2005; Marusic et al., 2010a; Jiménez, 2012), even though the exact processes are still only partly understood (Jiménez, 2018).

A substantial fraction of our knowledge about buffer layer dynamics also originates from the insights gained in simplified systems such as invariant solutions to the Navier-Stokes equations. Such 'exact coherent structures' (Waleffe, 2003) are typically simple solutions of the dynamical system (3.13), including travelling waves or simple equilibria for which $\mathrm{d}q/\mathrm{d}t = 0$ holds in a suitable inertial frame of reference. Similarly simple but time-dependent solutions are (relative) periodic orbits, for which the trajectory of the solution repeats after a time T as $q(t+T,p) = q(t,p)$ in a (possibly moving) inertial frame of reference, where the smallest T that fulfils this condition is called the 'fundamental period'.

The first simple three-dimensional invariant solutions related to turbulence were found around the same time at which the first DNS studies became feasible by Nagata (1990) in Couette flow, but their relevance for the understanding of many processes such as the laminar-turbulence transition or the self-sustained near wall cycle were not directly accepted (see van Veen, 2019, on the historical evolution). The limitations of the current review do not allow to give this topic the space it would deserve, and the interested reader is therefore referred to the extensive reviews of Kawahara et al. (2012b) and Graham and Floryan (2021) concerning solutions relevant to fully-turbulent wall-bounded flows and to those by Kerswell (2005) and Eckhardt et al. (2008) for their relevance in the transition from the laminar to the turbulent regime. To list only a few important solutions that strongly resemble the buffer-layer dynamics in different aspects and that impressively well represent the typical low-order turbulence statistics, we want to draw the readers' attention to the traveling wave solutions of Nagata (1990), Clever and Busse (1997) and Waleffe (2001, 2003) as well as the (relative) periodic solutions detected by Kawahara and Kida (2001) and Viswanath (2007, 2009) in plane Couette, Poiseuille and pipe flow, respectively. The latter set of (relative) periodic orbits is in this context of particular interest, as their periodic cycle closely resembles the different phases of the self-sustained regeneration cycle, including the banding and eventual 'bursting' of the streaks.

While individual invariant solutions can thus give some insight into specific turbulent processes, periodic orbit theory suggests that the statistics of a dynamical system's evolution can be represented by an appropriately weighted average over the periodic orbits around which the turbulent trajectory of the system is circulating (Auerbach et al., 1987; Cvitanović, 2013; Cvitanović et al., 2022), since the periodic orbits lie dense in the chaotic attractor. An attempt to use this kind of theory in turbulence has recently been made by Chandler and Kerswell (2013) for two-dimensional doubly-periodic Kolmogorov flows, for which the authors found about 50 individual periodic orbits and combined them in order to comment on the statistics of the entire flow. Note that in order to perform such analysis, an easy and fast detection of a large number of (relative) periodic orbits is crucial. In practice, typically time series of flow field realisations are scanned for near-recurrences in state space, which then serve as initial guesses for a glorified Newton-Raphson search (typically coupled with a Krylov-subspace method), with the aim to converge the nearly recurrent trajectory to a truly periodic orbit (Viswanath, 2009; Chandler and Kerswell, 2013).

However, finding suitable initial guesses for the Newton search itself is a delicate task, such that, recently, much effort has been put in developing more efficient methods to search for periodic orbits, including data-driven search methods based on, for instance, Koopman theory and dynamic mode decomposition (Page and Kerswell, 2018, 2019, 2020) or machine learning techniques (Page et al., 2021). At higher Reynolds numbers, identifying such near-recurrences generally becomes complicated as the chaotic turbulent trajectories reside for shorter times nearby a periodic orbit, often shadowing only part of the full period. Together with the notorious sensitivity of the Newton method's convergence on the quality of the initial guesses, this has recently motivated the development of structurally different approaches, including the stabilisation of exact recurrent states by external 'non-intrusive' force fields added to the RHS of the dynamical system (3.13) (Lucas and Yasuda, 2022) or the application of adjoint-based variational methods (Farano et al., 2019; Azimi et al., 2022; Parker and Schneider, 2022).

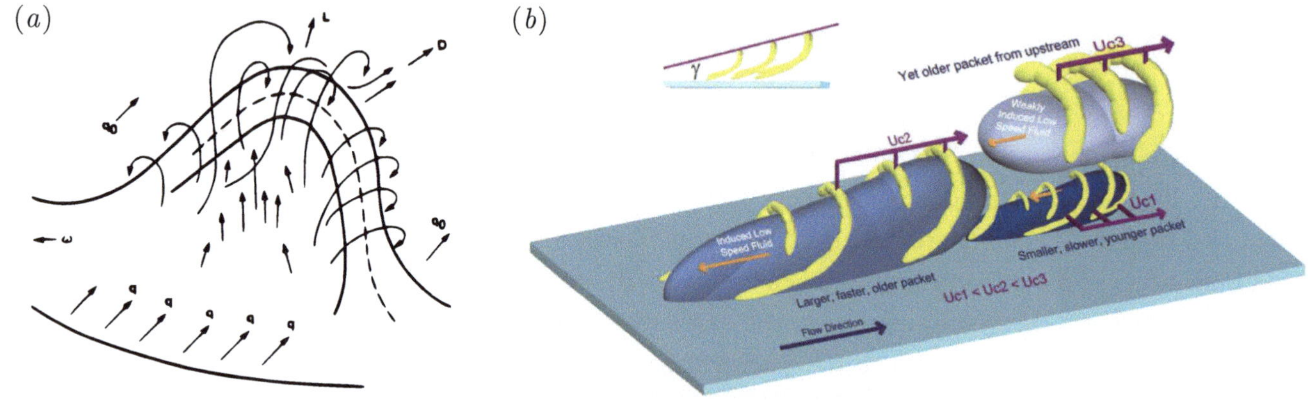

Figure 3.2: (*a*) Conceptual sketch of an individual horseshoe-/Λ-vortex as imagined by Theodorsen (1952). (*b*) Proposed mechanism of a low-speed region generation due to the collective effect of individual hairpin vortices, grouped in form of hairpin packets, as proposed by Adrian et al. (2000) and Adrian (2007). The visualisations are from Adrian (2007) (copyright 2007, reprinted with permission of AIP Publishing).

Direct numerical simulations of turbulence in highly spatially constrained domains, on the other hand, allow to study fully-chaotic but isolated coherent structures and their dynamics. In the seminal work of Jiménez and Moin (1991), the simulation domain was just large enough to accommodate a single streamwise streak flanked by, on average, two staggered counter-rotating quasi-streamwise vortices. A major achievement of the latter study was to quantify a 'minimal flow unit' necessary to sustain the turbulent regeneration cycle in a sequence of simulations with successively reduced box dimensions. It was observed that if the box dimensions fall below a streamwise and spanwise period of $L_x^+ \approx 250$-350 and $L_z^+ \approx 100$, respectively, the flow effectively relaminarises. This, in turn, implied that the observed regeneration cycle is indeed the smallest turbulent process and essential for turbulence to be maintained. Even though conclusions for spatially chaotic turbulent flows on the base of strongly reduced systems such as minimal box simulations and invariant solutions have to be drawn with caution, it has been shown that the predictions obtained from such simulations agree well with statistics in subdomains of large and thus less constrained fully-turbulent simulations (Jiménez et al., 2005).

In contrast to the buffer layer dynamics, the processes that drive the logarithmic layer are less well understood for several reasons: First, direct numerical simulations with a sufficiently high scale separation $Re_\tau \approx 4000$-$10\,000$ that allow for a clear log-layer to be detectable have become feasible only recently (Bernardini et al., 2014; Lozano-Durán and Jiménez, 2014a; Lee and Moser, 2015; Yamamoto and Tsuji, 2018; Hoyas et al., 2022). Second, structures in the logarithmic layer are neither smooth nor are they single-scale objects as their counterparts in the buffer layer, but they are themselves non-smooth multi-scale turbulent objects with a characteristic Reynolds number far from unity (Jiménez, 2013a). Also, the scales of the energy- and enstrophy-carrying structures, that are, velocity streaks and vortices separate outside the buffer layer: While the organisation of velocity in streaks seems to be a fundamental and robust concept of shear flows as they are observed at practically all scales from the near-wall to the outer region (Jiménez, 1998, 2018), vorticity is observed to be much more isotropically distributed outside the buffer layer and organises in structures of scales that are only slightly larger than those near the wall (Jiménez, 2013a). Interpreting the dominant wavelengths of the enstrophy spectrum as the characteristic scale of dissipation, this is in conceptual agreement with Kolmogorov turbulence applied locally to each wall-normal layer in that with increasing scale of the high-kinetic energy structures, the distance between the largest and smallest scales (i.e. the inertial range) increases as a function of the wall-distance.

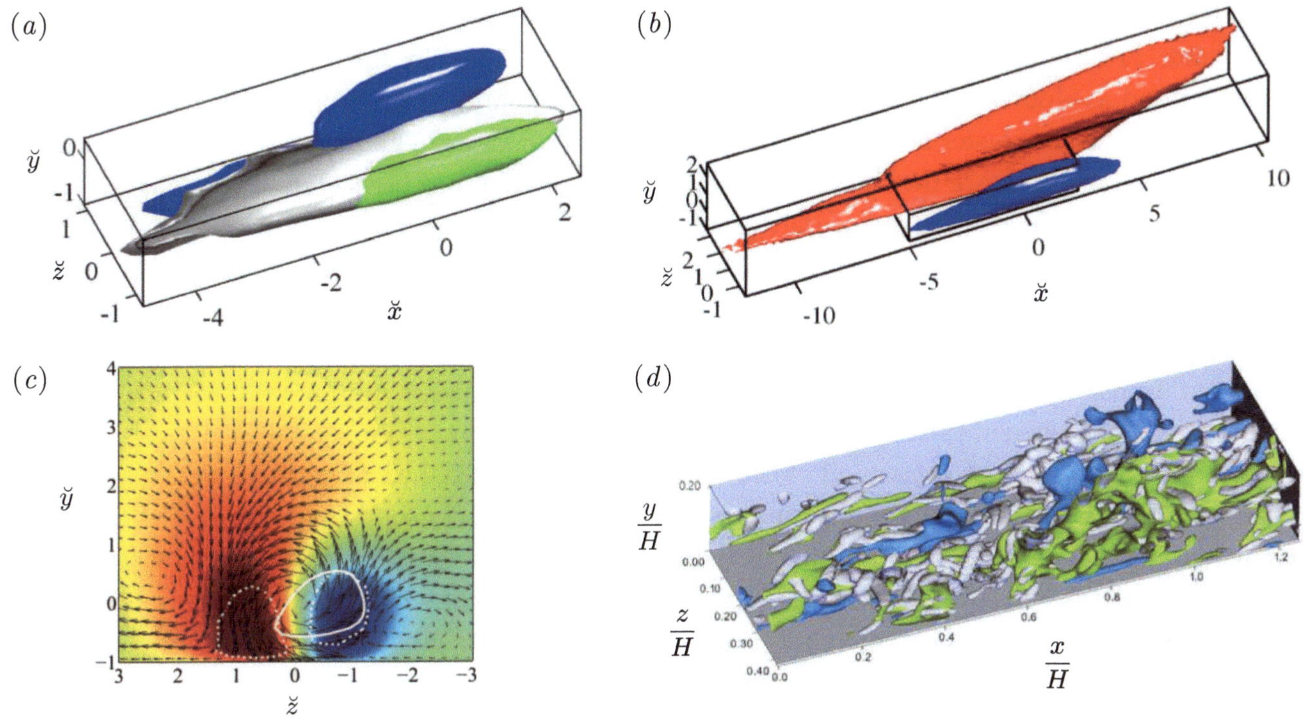

Figure 3.3: (*a-c*) Conditionally-averaged coherent flow structures in the log-layer of a turbulent channel flow at $Re_\tau \approx 2000$ (Hoyas and Jiménez, 2006; Lozano-Durán et al., 2012): (*a*) P.d.f.s of points belonging to ejection structures (Q2, green), sweep structures (Q4, blue) and vortex clusters (grey); (*b*) the same for low- (blue) and high-speed streaks (red), respectively. The smaller parallelepiped in (*b*) indicates the bounding box of the objects shown in (*a*). Length scales are defined as $\breve{x}_i = (x_i - x_{c,i})/y_c$, with $x_{c,i}$ being the ith coordinate of the centre of gravity of the Q2-Q4-pair. (*c*) Cross-section of the conditional velocity field at $\breve{x} = 0$, corresponding to the objects shown in (*b*). The arrows indicate the cross-flow velocity field $(v_f, w_f)^T$, while the background colourmap expresses the local amplitude of the conditional streamwise velocity fluctuations. The white dotted lines indicate the cross-sectional locations of the Q2 and Q4 objects in (*a*), while the solid line marks the location of the corresponding vortex cluster. (*d*) Exemplary instantaneous state of the flow field from the same dataset, with colour code as in (*a*). Flow is from bottom left to top right in all panels. The visualisations in (*a,b,d*) are from Lozano-Durán (2015) and that in (*c*) is from Jiménez (2013a) (copyright 2013, reprinted with permission of AIP Publishing); the notation has been adapted accordingly.

Today, mainly two different concepts exist that aim to describe the coherent structures present in the logarithmic layer. The first bases on 'horseshoe', 'hairpin' or 'Λ-vortices' first studied by Theodorsen (1952), that consist (in their simplest configurations) of two essentially horizontal legs and a connecting arch or head that is inclined w.r.t. the wall at which it is born, as can be seen in the original conceptual sketch in figure 3.2(*a*). Ever since, several conceptual models were build on hairpins of different size, scale and shape as a fundamental building block of turbulence (Perry and Chong, 1982; Robinson, 1991). The most recent attempt are the well-organised 'hairpin packets' described by Adrian (2007): As can be seen in the conceptual sketch in figure 3.2(*b*), the idea of this theory is that a number of individual hairpins arranges in a comparably regular pattern in the streamwise direction to create larger-scale structures, here a large low-speed streak. Hairpin vortex packets can grow in time and thus, with increasing size, spread over successively larger wall-normal distances (Adrian et al., 2000; Ganapathisubramani et al., 2003; Adrian, 2007), resulting in a set of self-similar families of hairpins and vortex packets at different scales spanning over different wall-parallel layers (Kim and Adrian, 1999; Marusic and Monty, 2019).

A different concept is described by the group of Javier Jiménez in Madrid by investigating the log-layer in terms of coherent structures related to different physical quantities such as velocity streaks (Sillero,

2014), Reynolds stress-carrying 'Q'-objects (Lozano-Durán et al., 2012; Lozano-Durán and Jiménez, 2014b) and vortex clusters (Del Álamo et al., 2006; Jiménez, 2013a). In this regard, vortex clusters can be understood as clouds of intense vorticity that resides in a large number of aggregated small-scale vortices which fill the cluster interior without clear spatial structure. The Reynolds stress carrying objects, on the other hand, are the three-dimensional analogues to the classical quadrant analysis (Wallace et al., 1972) in which velocity covariances $|u'_f v'_f|$ are classified depending on the quadrant in which they are located in the u'_f-v'_f parameter plane (Lozano-Durán et al., 2012; Lozano-Durán and Jiménez, 2014b). Of particular interest are those contributions that act in favour of the mean Reynolds stress ($u'_f v'_f < 0$), i.e. 'ejections' ($Q2$-event, transport of low momentum away from the wall) and 'sweeps' ($Q4$-event, transport of high momentum towards the wall). Summarising the individual findings as in Jiménez (2018), a general structural view of the relation between the individual structures emerges. Generally, structures can be classified as either attached to or detached from the wall, following the concept of the attached-eddy hypothesis posited by Townsend (Townsend, 1961, 1976). The basic idea of this latter theory is that the log-layer of turbulent shear flows can be represented by families of self-similar superposed eddies of different size. The largest of these structures at a given distance y to the wall spread until they reach the bottom wall and thus their height is $\approx 2y$. Assuming a random superposition of such eddies, Townsend (1976) showed that such a composition of structures directly implies the logarithmic shape of the velocity profile in this regime as well as a logarithmic scaling of some of the velocity fluctuations. Note that the attached-eddy hypothesis is nowadays widely accepted to hold (Marusic and Monty, 2019), and it is in accordance with both the hairpin-based model and the structural model presented in this paragraph.

Lozano-Durán et al. (2012) showed that, in an ensemble-averaged sense, an ejection and a sweep structure appear as pairs. As a consequence of continuity, the are connected by a quasi-streamwise roller, which is oriented in such a way that it leaves the ejection object at the top and the sweep at the near-wall side, as can be seen in figure 3.3(*a*,*c*). Vortex clusters are, in turn, associated with at least one ejection object and are typically found in-between the ejection and sweep of such pairs (Del Álamo et al., 2006; Lozano-Durán and Jiménez, 2014b). The conditional rollers take the place of the quasi-streamwise vortices in the buffer layer cycle and are probably (at least partly) driven by some local anisotropy of the small vortices associated to the vortex clusters (Jiménez, 2018). By definition, ejections and sweeps are located in low- and high-speed streaks, respectively, such that the sweep-ejection pairs are usually found at the transition between the streaks in figure 3.3(*b*). Figure 3.3(*c*) highlights visualises how the quasi-streamwise conditional rollers consequently transport low momentum fluid away from the wall in the low-speed streaks and *vice versa* in the high-speed streaks (cf. figure 3.3(*c*).

It has been stated earlier that the logarithmic layer itself is a self-similar cascade, and so are the families of streaks, ejections/sweeps and vortex clusters in that they repeat at varying scales and wall-distances in a similar way (Hwang and Cossu, 2010a, 2011; Cossu and Hwang, 2017; Jiménez, 2018). While the smallest structures in this context are those in the buffer layer, the largest streaks or 'large-scale motions' (LSM) at the upper end of the log-layer and in the outer layer reach dimensions of $\mathcal{O}(H)$, featuring a characteristic lateral spacing of 1-2H (Smits et al., 2011; Jiménez, 2013a). Important to recall is that the described mutual organisation of the different structures is valid for the ensemble-average, while the instantaneous structures are surely less regularly arranged, as becomes clear when comparing the instantaneous flow field shown in figure 3.3(*d*) with the conditionally-averaged patterns in figure 3.3(*a*-*c*). Nonetheless, the findings give a conclusive idea of the structures' interaction in the logarithmic layer and the ideas have been recently supported by the conceptual model and ensemble-averaged statistics of Kevin et al. (2019b) as well as by the analysis of low-pass filtered velocity fields in boundary layer and canonical channel flows by Motoori and Goto (2019, 2021).

Minimal box simulations have also been performed for the logarithmic layer, with the difference that a specific box is minimal w.r.t. a specific distance to the wall due the self-similar scaling of the log-layer structures (Flores and Jiménez, 2010). The minimal flow unit of Jiménez and Moin (1991) thus is the smallest possible of a cascade of minimal boxes just as the buffer layer is the lower end of the logarithmic layer cascade. Flores and Jiménez (2010) observed that the interval of wall-normal coordinates for which a box is minimal collapses with the height at which the spanwise domain period L_z is just wide enough to accommodate the vertical velocity energy spectra, and boxes are minimal for a wall-normal coordinate $y \approx L_z/3$. They furthermore observed that the isolated streaks surrounded by a single sweep and ejection intermittently burst quite similar to their counterparts in the buffer layer at time scales that vary with the local shear or equivalently with the wall-distance.

The two types of structures, i.e. hairpin packets and the structures described by Jiménez (2018) and his group, might share some similarities such as their support of the attached eddy hypothesis of Townsend (1961, 1976), but they fundamentally differ in their concept on where coherent structures are generated and how information and momentum is transferred across the different layers in wall-bounded turbulence. Hairpin-like structures are born at the wall and undergo a continuous growth, during which they penetrate layers further away from the wall and thus control the flow from the near-wall region. The conceptual model of Jiménez and co-workers, on the other hand, includes the possibility that structures are generated at each height independently as a consequence of the local shear, while the main role of the solid wall is in this context to provide the necessary shear to feed the turbulent field with energy (see, for instance, the wake-like process illustrated in Del Álamo et al., 2006 and the discussion about causality in section 5.6 of Jiménez, 2018). In the latter work, it was also mentioned that there is some evidence from the results obtained by Flores and Jiménez (2010) and Lozano-Durán and Jiménez (2014b) that the dominant direction of information propagation might be from outer regions towards the wall rather than the other way round, and that the characteristic velocity with which wall-normal momentum transfer occurs is comparable to the friction velocity u_τ. Recent studies provide further support for such an asymmetric interaction between the near-wall region and the outer flow (Zhou et al., 2022).

3.3 Turbulence in rectangular duct flows

3.3.1 The classical theory

While canonical closed channel flows have been extensively studied during the past decades, less is known about turbulent ducts which are basically pipe flows with rectangular cross-section. In that sense, closed channels represent the asymptotic state of a rectangular duct whose aspect ratio AR tends to infinity. In the remainder, we will refer to the aspect ratio AR of closed rectangular ducts as the quotient of the duct half-span $L_z/2$ over its half-height $L_y/2$, viz. $AR = L_z/L_y$ (Sakai, 2016). In experiments, it has been shown that substantial aspect ratios of $AR \gtrsim 25$ are required to sufficiently reduce the influence of the lateral sidewalls such that the core of the duct can be assumed to be essentially laterally homogeneous (Vinuesa et al., 2014a). If the aspect ratio is close to unity, on the other hand, the flow features four boundary layers that interact in the corner regions such that all three mean velocity components are non-zero, even though their statistics are independent of the streamwise direction.

In this regard, the most striking phenomenon unquestionably is the mean secondary flow pattern that is induced by the presence of the four solid sidewalls. First quantitative studies on secondary flows in non-circular pipes date back to Prandtl and Nikuradse (Nikuradse, 1926; Prandtl, 1926) and it was also Prandtl who coined today's standard classification of secondary flows into two categories: Secondary flows of Prandtl's first kind are due to the skewness of the mean flow axis as in meandering rivers or curved pipes, while secondary flows of the second kind are a pure turbulent phenomenon and originate in anisotropy and non-homogeneity of the Reynolds stresses across the flow domain (Bradshaw, 1987). In the remainder of this work, we will exclusively investigate secondary currents of the second kind and so we will refer to this type when using the term secondary motion hereafter. The relation of the turbulence-induced secondary motion with the Reynolds stresses is readily shown when considering the transport equation for the streamwise component of the mean streamwise vorticity field $\langle\omega_{f,x}\rangle_{xt}(y,z) = \nabla \times \langle \boldsymbol{u}_f\rangle_{xt}(y,z)$ in a straight and fully-developed duct. The transport equation for $\langle\omega_{f,x}\rangle_{xt}$ can be derived either by applying the curl operator to the RANS equations (3.4), that is, by cross-differentiation and substraction of these latter (Bradshaw, 1987; Gavrilakis, 1992), leading to

$$\begin{aligned}\langle v_f\rangle_{xt}\frac{\partial\langle\omega_{f,x}\rangle_{xt}}{\partial y} + \langle w_f\rangle_{xt}\frac{\partial\langle\omega_{f,x}\rangle_{xt}}{\partial z} &= \nu_f\Delta\langle\omega_{f,x}\rangle_{xt}\\ &+ \frac{\partial^2}{\partial y\partial z}\left(\langle v'_f v'_f\rangle_{xt} - \langle w'_f w'_f\rangle_{xt}\right)\\ &- \left(\frac{\partial^2}{\partial y^2} - \frac{\partial^2}{\partial z^2}\right)\langle v'_f w'_f\rangle_{xt}.\end{aligned} \tag{3.14}$$

The absence of the streamwise velocity component in equation (3.14) highlights that the system is effectively decoupled from the streamwise mean velocity field, such that interactions among the different velocity components can occur only in an indirect way via the Reynolds stresses. Particularly, the right hand side of equation (3.14) underlines that except for regions close to the wall where the viscous term is of relevance, secondary flows are driven by a balance between gradients of the normal Reynolds stress difference $\langle v'_f v'_f\rangle_{xt} - \langle w'_f w'_f\rangle_{xt}$ on the one hand and the gradients of the cross-stream Reynolds shear stress $\langle v'_f w'_f\rangle_{xt}$ on the other hand. In fact, it can be shown that a sufficient condition for the occurrence of secondary flows is that the cross-stream normal Reynolds stresses are not equal, i.e. $\langle v'_f v'_f\rangle_{xt} \neq \langle w'_f w'_f\rangle_{xt}$ (Speziale, 1982).

Two major reasons have complicated investigations of turbulent duct flow throughout the time since Prandtl and are probably responsible for the fact that in spite of its high relevance for many technical applications, the flow through rectangular ducts at low AR is much less frequently studied than that in the canonical doubly-periodic channel. First, the secondary flow and the transverse Reynolds stresses are weak in amplitude, the former usually reported as a few percent of the bulk velocity (Nezu and Nakagawa, 1993), such that accurate measurements have been and are still hard to perform in laboratory experiments (Wang and Cheng, 2005). Some important contributions from experimental studies are the works by Brundrett and Baines (1964), Gessner (1973) and Melling and Whitelaw (1976). In these early studies, the characteristic eight-vortex state of the secondary flow field in square ducts (cf. discussion of figure 3.4(*c*) below) was investigated in detail and studies of the different Reynolds stresses were performed. Also, the contributions of the individual terms in the vorticity budget (3.14) were determined in the experiments. Knight and Patel (1985), in turn, focused on the relation between secondary motions and the boundary shear stress along the duct perimeter for duct flows with rectangular cross-section.

Numerical simulations, on the other hand, often failed to predict secondary flows since most simple eddy-viscosity models are incapable of capturing the anisotropy of the normal Reynolds stresses $\langle v'_f v'_f \rangle_{xt} \neq \langle w'_f w'_f \rangle_{xt}$ (Speziale, 1982). Consequently, it requires more complex closure techniques for RANS-type models such as the non-linear eddy-viscosity model of Speziale (1987) or the algebraic stress models of Naot and Rodi (1982) and Demuren and Rodi (1984) to account for the secondary currents. Even though these models were able to reproduce a secondary motion in the cross-plane, the intensity that was obtained with these models clearly underpredicted that measured in experiments. The first accurate quantification of the secondary currents was provided by Gavrilakis (1992) and Huser and Biringen (1993) based on direct numerical simulations for the flow in a square duct, only a few years after the pioneering DNS study of Kim et al. (1987) in low Reynolds number channel flow. Several experimental and numerical studies of the mean secondary flow patterns and the streamwise vorticity budget in square and rectangular duct flows have been performed ever since (Zhang et al., 2015; Gavrilakis, 2019). For instance, the group at KTH in Stockholm investigated in a number of experimental and DNS-based studies the dependence of the secondary flow pattern on the aspect ratio in a range $1 \lesssim AR \lesssim 48$, the decay of the secondary flow intensity towards the duct centre and the velocity statistics in the limit $AR \gg 1$ for moderate Reynolds numbers (Vinuesa et al., 2014a,b, 2015, 2018). Recently, Pirozzoli et al. (2018) and Modesti et al. (2018) extended the Reynolds number range of direct numerical simulations of square duct flow to values $Re_\tau \geq 1000$ and studied how the mean secondary flow and the vorticity organisation change while increasing the Reynolds number. Also, numerical experiments were performed in which the secondary motion in the cross-section was artificially suppressed. It was concluded that the role of secondary flows in the duct cross-section is to serve for an equilibration between corner and bulk regions to compensate for the momentum deficit in the former region. For the absence of the secondary flow, it was concluded that the standard law of the wall was less well recovered along the duct walls.

3.3.2 The structural view on duct turbulence

Did earlier studies mainly focus on the relevance of the Reynolds stresses and the turbulent kinetic energy budget, the development of computing power and experimental techniques in the past decades also allowed to identify and study the dynamics of individual coherent structures similar to those in canonical flows.

One of the first works that followed this path was the study of Kawahara and Kamada (2000), in which the authors studied, based on experimental data, a quasi-streamwise vortex in the corner of a square duct whose centre was located roughly on the corner bisector, reporting the close similarity in size and structure to the well-known quasi-streamwise vortices in canonical flows. In the following years, a group around G. Kawahara, M. Uhlmann and A. Pinelli exploited a variety of different analytical and numerical techniques to study the transition to turbulence and the fully-turbulent state, mainly focussing on square duct flow. Uhlmann and Nagata (2006) investigated the stability of an internally heated rectangular duct by means of linear stability analysis, extending the work of Tatsumi and Yoshimura (1990) and Theofilis et al. (2004) in the isothermal case. While in the isothermal case, the lateral sidewalls stabilise the flow w.r.t. infinitesimal perturbations for all aspect ratios $AR \lesssim 3.2$, the additional thermal effects can under certain conditions destabilise the flow such that it becomes linearly unstable even for lower aspect ratios.

In analogy to the minimal flow unit of Jiménez and Moin (1991), Uhlmann et al. (2007) determined the minimal requirements to maintain turbulence in a square duct at a critical Reynolds number of $Re_b = 1077$ ($Re_\tau = 77$) or, equivalently, for a minimum inner-scaled duct span of $L_z^+ = 154$. In this context, the bulk and friction Reynolds numbers are defined based on the duct half height and Re_τ is determined based on the mean wall-shear stress averaged over the entire perimeter of the duct. It was shown by means of a coherent structure eduction study that the mean streamwise vorticity distribution as well as the mean secondary flow pattern are directly related to the preferential location of the streaks and their adjacent quasi-streamwise vortices. Interestingly, the long-time averaged secondary flow pattern induced at this marginal conditions consists of a fourfold symmetric eight vortex pattern that is relatively similar to the secondary flow patterns in fully-developed square duct turbulence. Shorter time averages, however, reveal that due to the marginal Reynolds numbers and thus the marginal relative wall lengths $L_z^+ = L_y^+$, not all four walls can accommodate a full turbulent regeneration cycle simultaneously. The consequence is an intermittent four-vortex behaviour in which always two facing walls feature a single streak-vortex group each.

To further clarify the origin of the four- and eight-vortex states, Sekimoto (2011) performed simulations at marginal Reynolds numbers $Re_b = 1100$ ($Re_\tau = 78$) in extremely long domains up to $L_x = 160\pi H$ ($H = L_y/2 = L_z/2$ indicating the half height of the square duct), compared to the longest domain in Uhlmann et al. (2007) that measured $L_x = 4\pi H$. Sekimoto (2011) could show that in an essentially spatially unconstrained domain (regarding the streamwise direction), the flow was not only intermittent in time, but also in space. More specifically, regions of locally enhanced wall shear and dissipation alternate with more quiescent regions of significantly lower wall shear stress, leading the author to the conclusion that the flow in a rectangular duct reveals comparable localised turbulent 'puffs' as have been observed in other canonical flows such as pipe flows under marginal conditions (Kerswell, 2005; Mullin, 2011; Barkley, 2016). Sekimoto (2011) in that sense generalised the conclusions of Uhlmann et al. (2007) to the spatiotemporal context, as he reported that within the localised turbulent puffs, the mean flow field exhibits the turbulent eight-vortex secondary flow state, whereas outside (at a locally smaller Reynolds number) again only one pair of opposite walls can accommodate a streak-vortex family and thus the cross-flow is characterised by one or the other four-vortex state. While the geometry of the square duct poses relatively strong spatial constraints on the localised turbulent puffs, large aspect ratio do less so: Takeishi et al. (2015) observed that sufficiently high aspect ratios allow the turbulent patterns to localise not only in the streamwise, but also in the spanwise direction, leading to the appearance of turbulent 'spots' analogous to those in doubly-periodic channels (Lemoult et al., 2013).

Pinelli et al. (2010) later extended the database studied by Uhlmann et al. (2007) to the non-marginal regime up to Reynolds numbers $Re_b = 3500$ ($Re_\tau = 230$). In particular, the authors concluded that with increasing relative length of the sidewalls L_z^+, the near-wall regions of the duct are able to accommodate several streamwise buffer-layer streaks and quasi-streamwise vortices, whose preferential positions can be inferred from the profile of the mean wall shear stress. The simultaneously increasing scale separation between the inner and outer length scale revealed moreover a different scaling behaviour of the mean streamwise vorticity and the secondary flow streamfunction: The streamwise vorticity that is associated with the preferential organisation of small-scale quasi-streamwise vortices scales in inner units, while the secondary flow streamfunction is most likely related to outer-scaling flow structures

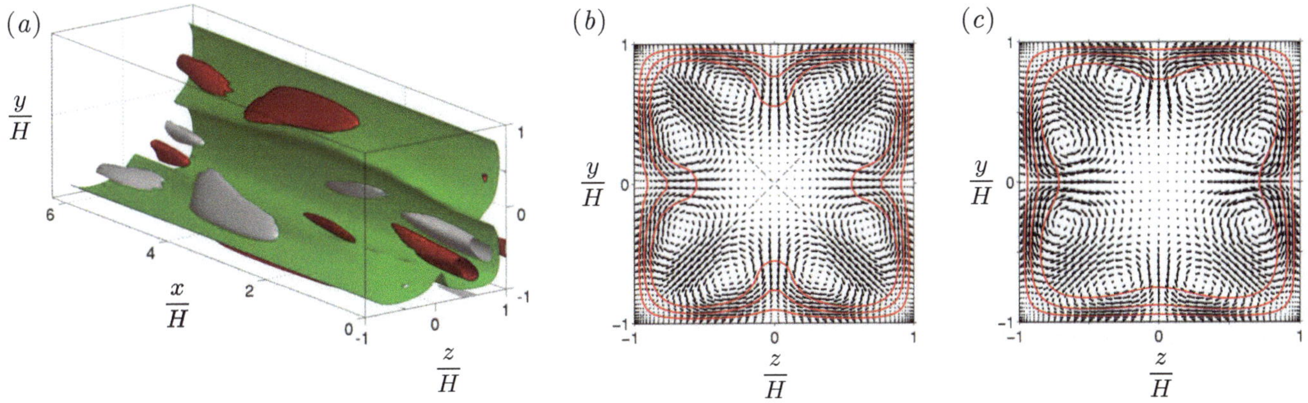

Figure 3.4: (*a*) Upper-branch travelling-wave solution of turbulent square duct flow found by Uhlmann et al. (2010) at $Re_b = u_b H/\nu_f = 1404.1$ (H: half duct height). Green: isosurface of the streamwise velocity fluctuation, red/gray: regions of intense positive/negative vorticity. Flow is from bottom right to top left. (*b*) Corresponding mean streamwise (red isolines) and mean secondary flow field (arrows) in the duct cross-plane, computed for the travelling-wave solution shown in (*a*). (*c*) Same quantities as in (*b*), but for a full DNS of square duct flow from Pinelli et al. (2010) at quasi identical Reynolds number $Re_b = 1400$. Visualisations are from Uhlmann et al. (2010) with adapted notation (copyright 2010, reprinted with permission of AIP Publishing).

and is thus less affected by the increasing Reynolds number. Sakai (2016) confirmed these observations for higher Reynolds numbers up to $Re_b = 7000$ ($Re_\tau = 411$), and Zhang et al. (2015) concluded based on their simulations up to $Re_\tau = 1200$ in a however relatively short box with $L_x = 2\pi H$ that low-Reynolds number effects remain observable for $Re_\tau \lesssim 600$. Sekimoto (2011) reached Reynolds numbers up to $Re_b = 8300$ ($Re_\tau = 480.7$) in the same box and determined the mean cross-sectional dimensions of the large-scale streamwise velocity structures to $1H$-$1.2H$ by evaluation of the two-point velocity correlations. Recently, Pirozzoli et al. (2018) commented on the relation between secondary streamfunction and mean vorticity based on simulations up to $Re_b = 20\,000$ ($Re_\tau = 1055$), highlighting the strong resemblance of their high Reynolds number DNS results to analytically obtained eigenfunctions of the Laplace operator.

The role of coherent structures in the generation of the secondary flow patterns was further underlined by a family of traveling-wave solutions to the Navier-Stokes equations found by Uhlmann et al. (2010) using the homotopy method proposed by Waleffe (2003) adapted to the duct configuration. From the detected traveling waves, in particular the upper branch solution whose individual vortices are depicted in figure 3.4(*a*) generates a characteristic eight-vortex secondary flow pattern when averaged over the streamwise direction, cf. figure 3.4(*b*). The similarity to the time- and streamwise-averaged mean field in the fully-turbulent flow shown in figure 3.4(*c*) based on the DNS data of Pinelli et al. (2010) is striking, being caused by the similar arrangement of the individual vortices in the invariant solution and the mean secondary vortices in the secondary flow field (Kawahara et al., 2012b). Other invariant solutions which share some similarities with the four-vortex state in marginal turbulence described by Uhlmann et al. (2007) were found by Okino et al. (2010) and Okino and Nagata (2012), using a homotopy approach based on the instability observed by Uhlmann and Nagata (2006). Sekimoto (2011), on the other hand, detected an eight-vortex state travelling wave by using the solution found by Uhlmann et al. (2010) as initial condition for his Newton-Krylov search. Arguably, the exact coherent structure analysed by Sekimoto (2011) represents an edge state in duct turbulence, that is, a local saddle on the boundary between the laminar and turbulent basin of attraction that acts as attractor for trajectories approaching the saddle along the edge manifold (Schneider and Eckhardt, 2006).

Recently, Atzori et al. (2018, 2021) applied the structural concept of the Q-events and the related percolation analysis originally introduced in canonical closed channel flow by Lozano-Durán et al. (2012) (cf. the discussion in the previous section 3.2) to the coherent structures in turbulent duct flows. It was concluded that the structures located in the vicinity of the duct corners differ from those in canonical channel flow concerning their geometry, while those outside the corners feature a relatively similar shape and aspect ratio.

3.4 Free-surface effects in wall-bounded turbulence

3.4.1 Open channel flows

While closed channel and closed rectangular ducts have important applications in many industrial processes, open channel and duct flows in which the upper solid wall is replaced by a flat free-slip plane are common models for hydraulic flows in natural rivers or man-made canals. Note that the representation of the free surface as a strictly flat free-slip plane is a common simplification in numerical simulations and excludes the evolution of surface waves. There are other models that are able to capture such phenomena, but a review of these methods is out of the scope of the current work. In this section, we will restrict ourselves to the study of single-phase smooth-wall open channel and duct flows, while we postpone a detailed review of those studies that simultaneously investigate effects of roughness and sediment transport to section 3.6 below. In the following, we use as characteristic outer scale the full fluid height H_f to allow for a just comparison between open channels or ducts and their closed counterparts. Also, in case of the open duct, we define the mean wall shear stress as the average of the wall shear stress along the wetted perimeter, that is, the three solid walls excluding the free surface.

Free surface flows share many similarities with their closed counterparts especially in the near-wall region and it is usually assumed that regions away from the surface do not feel its presence. In the vicinity of the free-slip boundary, on the other hand, the organisation of the turbulent flow field significantly differs from that in closed channels or ducts at the same distance to the bottom wall, where this height represents the channel/duct centreline. When approaching the free surface, the wall-normal velocity is successively damped as a consequence of the impermeability boundary condition and energy is transferred to the two horizontal velocity components as a consequence of continuity (Nezu and Nakagawa, 1993). The redistribution of kinetic energy has been observed to be in favour of the spanwise velocity rather than of the streamwise component and the main driving seems to originate in the pressure-strain term (Handler et al., 1993). The peculiar situation that only one velocity component is damped while the remaining two are unaffected or even enhanced leads to a highly anisotropic turbulent layer in which, in contrast to the near-wall region, essentially no turbulent kinetic energy is gained as the mean shear tends to zero at the surface. The dominance of the two horizontal over the wall-normal velocity components causes special vortex structures to arise, as has been outlined by Nagaosa (1999). This includes, on the one hand, surface-attached vortices that are connected to the free surface with their rotation axes roughly perpendicular to the free surface, such that they are characterised by a locally enhanced level of wall-normal vorticity $\omega_{f,y}$. On the other hand, elongated quasi-streamwise vortices form more or less parallel to the free surface, where they are of particular relevance for the momentum exchange between the free surface and the underlying fluid layers.

Possible interactions of structures born near the wall that are advected to the free surface were investigated by Nagaosa and Handler (2003) and their relevance for heat transfer in the vicinity of the free surface were discussed. More recently, Nagaosa and Handler (2012) and Pinelli et al. (2022) investigated the relevance of these near-surface vortices for the interfacial gas transfer. Bauer et al. (2022), in turn, showed that the largest scales in free-surface flows equivalent to the 'very large-scale motions' (VLSM) or 'global modes' in canonical flows (Kim and Adrian, 1999; Del Álamo and Jiménez, 2003; Hutchins and Marusic, 2007a; Monty et al., 2009) appear at lower Reynolds numbers than in closed channel flows and are about twice as long and twice as wide as those in the closed channel case. Also, these structures penetrate deep into the near-wall region, indicating that there might be in fact no region that is truly unaffected by the presence of the free surface. In general, however, the role of large- and very large-scale motions in free surface flows and their interaction with mean secondary currents is not yet fully understood (Adrian and Marusic, 2012), and the topic has only recently attracted some attention in the community (Cameron et al., 2017; Zhang et al., 2019; Cameron et al., 2020; Duan et al., 2020; Peruzzi et al., 2020; Camporeale et al., 2021). The similarity between coherent large-scale structures in canonical and free surface flows will be further discussed in the section on ridges in hydraulic flows below.

3.4.2 Open duct flows

In open duct flows, the free surface and the lateral sidewalls form a 'mixed-boundary corner' with different boundary conditions on the two adjacent sides of the corner. High-fidelity data from direct numerical simulations of turbulent open duct flows to compare experiments with is rare, and the study of Sakai (2016) is to the best of our knowledge the most extensive database of fully-resolved open duct flow simulations. Only few other fully-resolved numerical simulations of open duct flow such as those of Joung and Choi (2010) are known, which however feature a relatively short computational box ($L_x = 2\pi H_f$) that can hardly be expected to be sufficient to capture all relevant structures and length scales. Indeed, Sakai (2016) showed the necessity of domain lengths of at least $L_x = 8\pi H_f$ to allow for a decoupling of the two-point correlations in the streamwise direction, whereas in closed duct flows domains of streamwise length $L_x = 4\pi H_f$ were shown to be sufficient to do so. The longest coherent structures in presence of a free surface are thus apparently much longer than those in closed ducts, which agrees with the respective observations in open and closed channel flows above.

Sakai (2016) further observed that the minimal Reynolds number required to sustain turbulence in an open duct is somewhat higher than for the closed square duct at matching aspect ratio $AR = 1$ (using $AR = L_z/(2L_y)$ in the open duct case), in that a minimal Reynolds number $Re_b = 1450$ ($Re_\tau \approx 102$) is required to keep the system from relaminarisation. Interestingly, this critical Reynolds number turns out to be a monotonically decreasing function of the aspect ratio for both open and closed duct flow.

For non-marginal conditions, the open duct flow develops a characteristic secondary flow pattern that differs from the closed counterpart in several aspects. Figure 3.5 shows such a secondary flow field together with the mean streamwise velocity exemplary for case $DL400^{H2}_{smooth}$ ($Re_b \approx 7000$, $Re_\tau = 416$), which was originally presented by Sakai (2016) and will be later in this work used as reference data for the multiphase simulations. Three characteristic secondary flow patterns are clearly distinguishable in figure 3.5: First, we observe the 'inner secondary vortex' in the mixed boundary between free surface and sidewalls that has been first studied by Grega et al. (1995),Hsu et al. (2000) and Grega et al. (2002). This local mean swirling motion features a sense of rotation that is oriented in such a way that

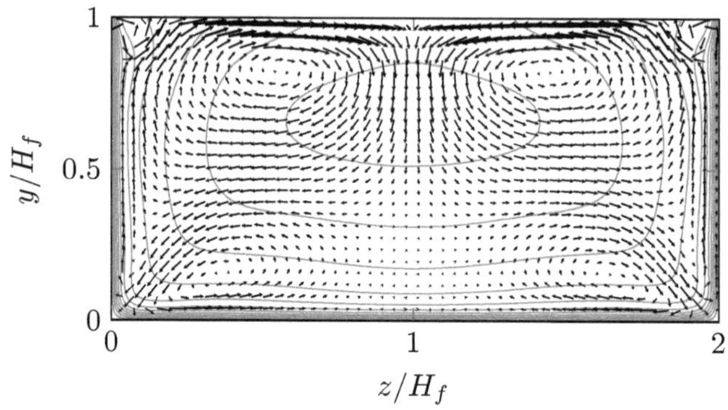

Figure 3.5: Organisation of the mean streamwise velocity $\langle u_f \rangle_{xt}/u_b$ and secondary motion $(\langle v_f \rangle_{xt}/u_b, \langle w_f \rangle_{xt}/u_b)^T$ in open duct simulations. Grey lines are isocontours of $\langle u_f \rangle_{xt}/u_b$, shown for values 0.1(0.1)1.3. Intensity and orientation of the mean cross-stream secondary flow field are indicated as vector plot in the background. The shown data is from case $DL400^{H2}_{smooth}$ (cf. section 7.1).

fluid is transported along the free surface towards the wall and downwards along the solid boundary. According to Sakai (2016), the existence of this inner-scaling mean vortex is closely linked to a preferential agglomeration of quasi-streamwise surface-parallel vortices with a common sense of rotation in the mixed corners. Vortices with a reverse sense of rotation move away from the sidewall along the surface towards the duct bisector. Sakai (2016) illustrates this proposed 'vortex-sorting mechanism' by a simple two-dimensional vortex dipole consisting of the actual vortex and its virtual mirror on the other side of the free surface which, in this context, acts as a symmetry plane (Orlandi, 1990). Depending on the orientation of the 'real' vortex, the dipole will either induce an advection velocity towards the corners where the vortices stick and appear in the mean field as inner vortex, or they move out of the corner towards the duct centre. The fact that the mean inner vortex scales in inner units further supports this concept. The large mean secondary flow cell which occupies most of the cross-section in figure 3.5, in turn, is typically termed 'outer' or 'free-surface vortex'. In contrast to the inner vortex, this latter is weakly affected by changes of the Reynolds number as it scales in outer units similar to the large-scale secondary vortices in the closed square duct. A third, medium-sized mean 'bottom vortex' is finally visible in the lower left and right half of the cross-section shown in figure 3.5, which is reminiscent of the eight-vortex state in the closed duct counterpart.

Besides the mean secondary flow field, figure 3.5 also shows the characteristic pattern of the mean streamwise flow field $\langle u_f \rangle_{xt}(y,z)$. We observe what is typically called the 'velocity dip phenomenon' ([first observed more than a century ago by Stearns, 1883): Along the duct bisector at $z = L_z/2$, the streamwise velocity attains its maximum not at the free surface, but the highest velocity is found somewhat below it. Tominaga et al. (1989) quantified the wall-normal location of the streamwise velocity peak at an offset of 0.2-0.3H_f from the free surface and Nezu and Rodi (1985) reported that the velocity dip is only visible for ducts with low aspect ratio $AR \lesssim 2.5$. In the classical literature, its appearance is reported to be a consequence of the concentrated downward motion in between two secondary currents, where high momentum fluid is brought downwards from the free surface, with the consequence that the maximum of the streamwise velocity is shifted downwards (Nezu and Nakagawa, 1993). Sakai (2016) reports a quite similar value of $0.33H_f$ for the offset of the velocity maximum from the free surface in his direct numerical simulations, underlining that the scaling in outer units indicates its close relation to outer-scaling large-scale flow structures. Also, in order to investigate the origin of the velocity dip, he determined the preferential location of low-speed streaks in the cross-section and concluded that the dip phenomenon is related to the transport of low velocity streaks from the sidewalls towards the duct centre as a consequence of the mean secondary flow, in accordance with the explanation given by Nezu and Nakagawa (1993).

With increasing aspect ratio, the influence of the sidewalls on the duct core is seen to considerably reduce (Nezu and Rodi, 1985) and for sufficiently large aspect ratios, no mean secondary flow is observed in the duct core far away from the sidewalls. Tominaga et al. (1989) described that in such situations, the outer secondary flow cell extends roughly $2H_f$ from the sidewall towards the duct core, while the bottom vortex is stronger confined to the near-sidewall region with maximum spanwise extensions of $z < H_f$. These observations were recently confirmed by the DNS results of Sakai (2016) for open duct flows. For high aspect ratio closed duct flows, similar conclusions seem to hold true as suggested by the experiments of Nezu and Rodi (1985) and the numerical work of Vinuesa et al. (2015).

While the existence of intense and apparently outer-scaling mean secondary vortices in the near-sidewall area of wide ducts can be interpreted as the statistical footprint of instantaneous large-scale structures, the absence of secondary currents in their centre does not allow a direct conclusion on the instantaneous structures in this area. On the contrary, there is experimental evidence that 'instantaneous secondary motions' are also present away from the lateral domain boundary (see Tamburrino and Gulliver, 1999; Nezu, 2005, and references therein). Due to their freedom to laterally meander and propagate in the absence of lateral confinements, however, their footprints do not appear in the long-time average. We will discuss this phenomenon in more detail in section 7.2.

Interestingly, a significant bottom roughness seems to influence this effect by reducing the mobility of instantaneous large-scale velocity structures. Rodríguez and García (2008) report based on measurements in ducts with medium aspect ratios $AR = [3.15, 4.25]$ that the mean secondary flow intensity does not significantly decay from the sidewall region towards the duct core over a fully-rough bed, as it is observed in smooth-wall duct flows, for instance, by Nezu and Rodi (1985). In fact, the entire cross-section is more or less covered with mean secondary flow cells, in agreement with earlier experimental observations (Tominaga et al., 1989). Rodríguez and García (2008) argue that the difference in roughness between the side- and the bottom-wall are the main cause for the persistent mean flow patterns, claiming that the instantaneous large-scale structures were less strongly meandering over a rough bed than over a smooth wall, such that their statistical footprint remains visible even in the long-time statistics.

3.5 Secondary currents in the absence of sidewalls

While analysing the transport equation of the mean streamwise vorticity (3.14), it was pointed out that anisotropy and heterogeneity of the Reynolds stress tensor field are necessary conditions for the evolution of secondary currents. So far, we have focussed on sidewall-induced secondary currents exclusively, but it turns out that a variety of lateral inhomogeneities can cause such an anisotropy of the Reynolds stress tensor (Anderson et al., 2015). Hinze (1967, 1973) observed that secondary flow cells evolved over a bottom wall covered with alternating rough and smooth stripes, specifically centred over the roughness transition. He argued that the secondary momentum transport is closely related to the local production-dissipation balance in that secondary currents transport weakly turbulent fluid into the regions of intense turbulence production, while they simultaneously carry fluid with high turbulent kinetic energy out of this region. In particular during the past two decades, the interest in secondary motions over such lateral heterogeneities has significantly increased, partly motivated by the observation of secondary flows over complex three-dimensional industrial surfaces such as the replicated surface of a damaged turbine blade that was scrutinised by Mejia-Alvarez and Christensen (2010) and Barros and Christensen (2014). Interestingly, secondary currents seem to be a ubiquitous phenomenon arising over a variety of different bottom heterogeneities including, amongst others, the flow over straight (Goldstein and Tuan, 1998) and converging/diverging riblets (Kevin et al., 2017, 2019a), spanwise alternating roughness stripes (McLelland et al., 1999; Wang and Cheng, 2005; Willingham et al., 2014; Wangsawijaya et al., 2020), transverse alternating non-/hydrophobic roughness areas (Türk et al., 2014; Stroh et al., 2016), streamwise aligned artificial ridges of different cross-sectional shape (Wang and Cheng, 2006; Vanderwel and Ganapathisubramani, 2015; Vanderwel et al., 2019; Medjnoun et al., 2020; Stroh et al., 2020a; Zampiron et al., 2020a) or combinations of the aforementioned, as in Stroh et al. (2020b). Most of the listed studies mainly focus on the size, location and intensity of the mean secondary currents, their sense of rotation (Yang and Anderson, 2018; Anderson, 2019) or the impact of the lateral spacing s of the inhomogeneities on the secondary flow (Chung et al., 2018; Wangsawijaya et al., 2020).

The last point is of central importance for engineering-type applications as it allows to control the appearance, size and strength of the cross-stream circulation. For most applications, three different regimes can be distinguished (Vanderwel and Ganapathisubramani, 2015; Vanderwel et al., 2019; Wangsawijaya et al., 2020; Zampiron et al., 2020b): Is s much larger than the characteristic outer length scale H, depth-spanning secondary currents are localised and appear solely close to the roughness transition. Is s on the other hand clearly smaller than H, secondary currents are relatively small and restricted in the vicinity of the bottom wall, while the outer flow is rather weakly affected by the topography and roughness of the heterogeneous bottom wall. For most cases, an 'optimal' spacing that maximises the intensity of the secondary currents is reached for the typical lateral spacing of the large-scale velocity structures $1 \lesssim s/H \lesssim 2$ (Jiménez, 2013a, 2018): At these spacings, the secondary currents become space-filling, which means that the entire cross-section of the domain is occupied by counter-rotating, depth-spanning secondary currents.

The relation between mean secondary flows and large-scale instantaneous coherent structures over heterogeneous bottom walls has attracted much less interest. Recently, Kevin et al. (2019b) claimed that the secondary currents could be the statistical footprint of the conditional quasi-streamwise large-scale rollers observed in canonical channel flows (Lozano-Durán et al., 2012) that are 'locked' in their lateral position by the bottom heterogeneity. Locking means in this context that the average spanwise position of the instantaneous large-scale structures changes little in time, while the flow structures still exhibit a lateral meandering around this relatively persistent mean position (Kevin et al., 2017, 2019a). The reduced lateral mobility of large-scale structures over heterogeneous bottom walls makes them visible in the long-time average, while their counterparts over spanwise homogeneous walls do not underly such constraints and can form at all spanwise positions with the same probability. As a consequence, contributions of instantaneous rotating motions with opposite sign will cancel out for sufficiently long averaging time intervals, such that these flows do not reveal a mean secondary motion.

The lateral meandering behaviour that is observed for the flow over both homogeneous and heterogeneous bottom walls (Wangsawijaya et al., 2020; Zampiron et al., 2020a) is assumed to be part of a large-scale self-similar regeneration cycle in analogy to the well-known buffer-layer process (Hamilton et al., 1995; Jeong et al., 1997; Schoppa and Hussain, 2002), now including however large-scale velocity streaks and shorter quasi-streamwise large-scale rollers (Kevin et al., 2019b). Note that, in this context, large-scale quasi-streamwise rotating motions are typically assumed to be the collective effect of a large amount of individual globally-organised vortices or vortex clusters, rather than single large-scale vortices that do not exist in such form in a fully-turbulent flow (Del Álamo et al., 2006; Jiménez, 2018). As in the buffer layer, it is expected that these quasi-streamwise rollers laterally flank a large-scale streak in a staggered way, such that the streak gets unstable and eventually breaks. Similar mechanisms have been proposed earlier by Flores and Jiménez (2010) and Hwang and Cossu (2011) for the log-layer region and by Hwang and Cossu (2010a) for the outer layer large-scale streaks over homogeneous bottom walls (Cossu and Hwang, 2017).

3.6 Secondary flows and subaqueous sediment ridges

In hydraulic engineering, sediment ridges and secondary currents usually evolve side by side at very similar lateral wavelengths $1H_f$-$3H_f$ as the 'optimal' spacing of the bottom wall heterogeneities discussed in the previous section, which is in turn comparable to the typical lateral wavelength of the large-scale streaks in canonical wall-bounded turbulence (Smits et al., 2011).

The existence of secondary currents in hydraulic flows were posited even before the pioneering studies of Nikuradse and Prandtl on the flow through pipes with non-circular cross-sections, for instance, by Stearns (1883), Möller (1883) and Gibson (1909). Later, Casey (1935) and Vanoni (1946) observed straight elongated sediment ridges to evolve on the mobile sediment bed in their experimental flumes and claimed that these features should be related to secondary currents of Prandtl's second kind. Vanoni (1946) argued based on observed lateral oscillations of the sediment concentration in the flume that the proposed mechanism of interactions between secondary flow cells and the mobile bed could similarly occur in open ducts with significantly larger aspect ratios than those used in their study, for instance, in natural rivers. Wolman and Brush (1961) observed ridges to appear unbroken over considerable streamwise distances of $\mathcal{O}(100H_f)$ at lateral spacings in a range 1.74-2.36H_f and maximum amplitudes of several diameters D at a relative submergence $H_f/D \approx$ 12-24. On the other hand, much larger sediment ridges than those in the laboratory flumes were detected in field observations of dried river beds of wide fluvial channels (Culbertson, 1967; Karcz, 1967), indicating that a similar interaction of sediment ridges and secondary currents might exist also on a much larger scale. Around the same time, Kinoshita (1967) took aerial photographies of several Japanese rivers shortly after typhoon events and visually estimated the size of large-scale secondary currents based on spanwise alternating regions of strong and weak boiling with a spacing of roughly $2H_f$ on the river surface.

Even though there was a consensus on the mutual interaction of secondary currents and sediment ridges in these early studies, there was no agreement on the actual origin of the process. Furthermore, it was known from experiments that sediment ridges and their associated secondary flow cells often covered the entire cross-section of the domain rather than only the region close to lateral domain boundaries. One attempt to explain the process was that of Nezu and Nakagawa (1984) who imagined a cascade-like process that has its origin in the secondary flows induced by the sidewalls of their experimental flume: From earlier studies it was known that the bottom wall shear stress in ducts oscillates laterally with distinct peaks in the near-sidewall region. Nezu and Nakagawa (1984) thus argued that the near-sidewall peak of the wall-shear stress would give rise to a first ridge-trough pair in the region closest to the sidewalls due to the locally varying erosion and deposition in consequence of the varying wall shear stress. As soon as the ridge and trough would have reached a certain height difference, they were able to trigger a new secondary current next to them and so on and so furth, so the argument. In a subsequent study, though, Nezu and Nakagawa (1989) investigated the flow over a mobile sediment bed in a wide open duct and observed that ridges evolved essentially simultaneously over the entire span of the duct, which indicated that the formation mechanism could hardly be predominantly controlled by the lateral sidewalls and their associated secondary currents.

Ikeda (1981) had a different mechanism in mind and conjectured that the appearance of sediment ridges was an instability problem between an erodible bed and a turbulent shear flow that does not require the presence of lateral sidewalls. Nezu and Nakagawa (1993) later argued that such an instability could be due to two conceivable mechanisms: In the first, the instability is initiated by a lateral perturbation of the bed topography or the bottom roughness, which causes a lateral variation of the bed shear stress that, in turn, triggers the formation of secondary currents in the turbulent flow. The second conceivable process acts the other way round in that a spanwise varying flow induces a laterally varying bed shear stress, which initiates the formation of sediment ridges due to a heterogeneous erosion activity along the bed.

Colombini (1993) considered the first mechanism and could show based on a linearised model of the Navier-Stokes equations coupled with the sediment bed continuity equation that a sinusoidal modulation of the sediment bed can indeed trigger a lateral variation of the flow field in form of large-scale depth-spanning secondary currents. The induced secondary motion, in turn, induces a bed shear stress that supports a further growth of the bottom perturbation. For the set of investigated parameters, a most amplified wavelength of $\lambda_{h,z} = 1.3H_f$ was obtained, which closely resembles the values in the previously reported experiments. Here, we forego a deeper discussion of the model details, as those will be analysed in great detail in chapter 5.

Observations in experiments, on the other hand, suggest that the second formation process initiated by a lateral variation of the flow in form of large-scale turbulent structures might be more relevant to the fully-turbulent flow (Onitsuka and Nezu, 2001; Shvidchenko and Pender, 2001; Nezu, 2005; Adrian and Marusic, 2012), but clear evidence for this claim is still lacking. Historically, the investigation of large-scale structures with streamwise extensions of $\mathcal{O}(1H_f\text{-}10H_f)$ in hydraulic flows developed parallel to but more or less independent from the study of coherent structures in the wall-bounded turbulence community (Adrian and Marusic, 2012). Extensive reviews that discuss such similarities of structures observed in smooth-wall and hydraulic flows and the possible interactions with sediment can be found in Adrian and Marusic (2012) or in the more general work of Venditti et al. (2013) on coherence in geophysical flows, while the review of Nikora and Roy (2012) mainly focusses on their role in the context of secondary flows in rivers. A special focus on open duct turbulence and the importance of the observed processes for fluvial systems is given in the classical textbook of Nezu and Nakagawa (1993) and the more recent review of Nezu (2005).

Coherent large-scale motions that closely resemble those known from smooth-wall turbulence are, amongst others, the 'boils of the second kind' described by Nezu and Nakagawa (1993), the 'large longitudinal eddies' of Imamoto and Ishigaki (1986) or the 'large vortex structures' described by Tamburrino and Gulliver (1999, 2007), Shvidchenko and Pender (2001) and Rodríguez and García (2008). All these flow structures reveal very similar properties and appear in form of elongated streaks with a streamwise extent of several multiples of the mean fluid height H_f and a lateral width of about 1-2H_f, such that, most likely, they describe one and the same phenomenon (Adrian and Marusic, 2012). Also, immanent to these instantaneous structures seems to be a direct relation to large-scale rotating motions

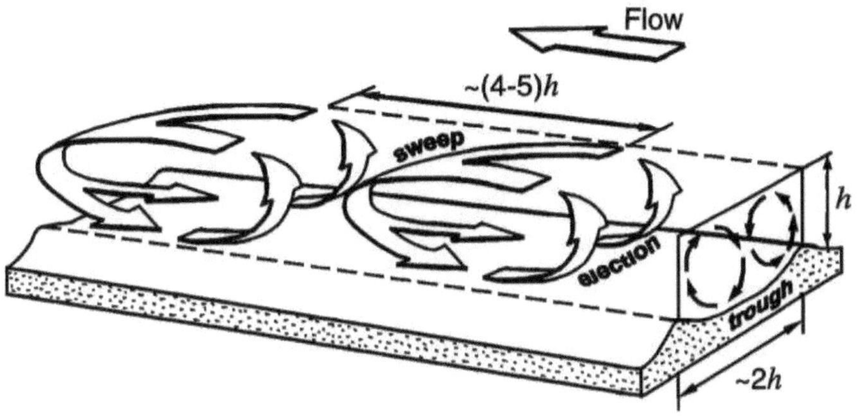

Figure 3.6: Conceptual sketch of the interaction between sediment ridges and large-scale turbulent structures in open channel flows as proposed by Shvidchenko and Pender (2001) (copyright 2001, reprinted with permission of John Wiley and Sons).

that are regularly described as 'instantaneous secondary currents' (Nezu, 2005), in analogy to the mean secondary currents of Prandtl's second kind (Nikora and Roy, 2012). Such large-scale instantaneous rotating motions were reported, for instance, by Nezu and Rodi (1986) and Onitsuka and Nezu (2001) to occur even in those regions of the flow where no mean secondary flow is visible in the long-time average (cf. the discussion on large-scale meandering coherent structures in sufficient distance to sidewalls in the previous sections).

Nezu (2005) claimed a mutual interaction between these instantaneous large-scale flow structures and evolving sediment ridges, in the context of which the instantaneous large-scale rotations cause a laterally varying erosion of sediment that ultimately leads to the evolution of sediment ridges. Once developed, these latter are assumed to stabilise the large-scale flow structures by reducing their lateral mobility and meandering tendency, making them visible also in the long-time average. A more detailed model was presented by Shvidchenko and Pender (2001), who argued that the large-scale structures and their associated ejection and sweep events lead to a regular distribution of zones with high and low particle erosion and transport. The process is illustrated in figure 3.6, where it is seen that erosion is dominant in regions where 'macroturbulent structures' cause laterally varying regions of strong and weak sediment erosion and sediment transport inside and outside their travelling path, respectively. Local troughs and ridges form accordingly in the regions of strong and weak erosion, respectively, while in a streamwise-averaged framework, the laterally organised large-scale structures give rise to a mean secondary motion in the cross-plane that is visible at the upstream end of the sketched domain in figure 3.6. The transverse variation of the sediment transport intensity is at least partly related to the preferential organisation of Reynolds-stress carrying structures, as sediment erosion is predominantly driven through high-speed sweeps that reach the sediment bed, rather than by the action of ejection events which could lift particles up in the low-speed regions (Gyr and Schmid, 1997; Cameron et al., 2020). Bagherimiyab and Lemmin (2018) recently showed that such sweeps, when reaching the sediment bed in open channel flows, indeed induce a local increase of the bed shear stress that enhances the local erosion rate.

The global hydraulic and morphologic conditions under which sediment ridges form are not fully clarified, but it is worth to mention that in experimental works, ridges are predominantly observed at low sediment transport rate (Wolman and Brush, 1961). This generally agrees with the observations of

Kleinhans et al. (2002), who observed in experiments with moderate to strong sediment-supply limitations that ridges are the dominant - if not the only - developing sediment patterns. Such a situation of highly constrained sediment availability occurs, for instance, in the case of fine sediment transport over armoured beds of coarse grains, in which case sediment ridges of the fine material are found to evolve on top of the non-erodible coarser material (Bertin and Friedrich, 2019; Venditti et al., 2019). Kleinhans et al. (2002) also postulates that with increasing sediment transport rate, there is a continuous transition from sediment ridges first to larger three-dimensional sediment patterns such as barchans and eventually to transverse ripple-like bedforms.

Only recently, DNS studies of sediment transport with fully-resolved particles became feasible that further strengthen these results. So, Kidanemariam and Uhlmann (2014a, 2017) and Scherer et al. (2020) studied the evolution of bedforms on an initially flat sediment bed and concluded that ridges appeared right after the particle release, when only few particles were in motion, while transverse patterns with a higher amplitude grew at a significantly lower rate until they eventually dominated the small-amplitude ridges. That already a comparably small amount of mobile sediment grains is indeed sufficient to interact with a relatively robust secondary flows was already discussed by Vanoni (1946), and recently Vowinckel et al. (2017b) confirmed this observation. In their DNS-based study of sediment transport for a relatively low number of fully-resolved particles in motion over a uniformly roughened bottom wall, secondary flow cells and streamwise aligned 'chains' of sediment grains were clearly detectable.

3.7 Knowledge gaps

In the preceding sections, we have reviewed the relevant literature that contains valuable information for our investigations on the interplay between sediment ridges, secondary currents and turbulent coherent structures. Specific care has been taken to clearly differentiate in each field between the established classical view and the current state of the art. It has become evident that coherent structures play an important role in the energy and momentum transfer of wall-bounded turbulent flows. Most of today's knowledge on coherent structures originates from rather simple flow configurations such as smooth-wall channels for which fully-resolved direct numerical simulations and high-quality measurements are available. Turbulence in many hydraulic flows such as rivers or estuaries is significantly more complex, for instance, due to the continuous interaction with a mobile sediment bed that gives rise to different types of sediment bedforms.

It has been highlighted that some conceptual ideas exist of how coherent structures and mean secondary currents could take part in the formation cycle that leads to the evolution of streamwise-aligned sediment ridges. But to the best of the author's knowledge, clear evidence for the outlined mechanisms is still lacking. In particular, it has been discussed that different contradicting ideas exist on whether the development of sediment ridges and the associated secondary currents are controlled from the sediment bed, triggered by an initial perturbation of the latter or whether the sediment bed evolution is initiated and dominated by the dynamics of the turbulent flow.

Also, a clear understanding of how sediment ridges interact with a 'pre-existing' sidewall-induced secondary flow field as it occurs in rectangular open ducts such as man-made canals or even laboratory flumes lacks up to the present day. In virtue of the challenging measurements necessary to study both secondary motion and sediment transport at the same time, often such investigations have been performed with fixed artificial sediment ridges, which did not allow clear conclusions about the formation of sediment ridges in the vicinity of such sidewalls (Nezu and Nakagawa, 1984). However, both from a practical and from a scientific point of view, the impact of sidewalls on the formation of sediment ridges and the converse influence of a mobile sediment bed on the typical secondary flow patterns in such open ducts is of significant relevance. In particular, it is crucial to assess to which degree experimental measurements conducted in narrow to moderately wide ducts are comparable with simulations in infinitely wide or laterally periodic channels.

In both cases, a profound understanding of the interaction between turbulent coherent structures and sediment ridges is of fundamental importance. As was stated above, however, rigorous analyses of the analogy between such structures in canonical wall-bounded flows and their counterparts in hydraulic free-surface flows over mobile sediment beds have just started (Adrian and Marusic, 2012). One of the central objectives of this study therefore is to discuss the role and characteristics of turbulent coherent structures in both situations and to highlight their fundamental similarities. In this context, we will greatly benefit from our today's knowledge about the dynamics of coherent structures in the simpler canonical flows when interpreting the role of such structures in sediment transport.

Chapter 4

Numerical methodology

4.1 Direct numerical simulations

In the previous chapter, we have established a picture of turbulence as a chaotic, but deterministic multi-scale dynamical system that is appropriately described by the non-linear Navier-Stokes equations. The multi-scale character manifests itself in form of a range of flow scales between the largest flow structures that are limited by the characteristic length scale of the domain and the smallest ones that are of the order of the Kolmogorov length. As stated earlier, the range of scales naturally increases with the Reynolds number and the number of degrees of freedom is typically estimated as $\sim Re^{9/4}$ (Landau and Lifshitz, 1959).

Needless to say, the direct numerical simulation (DNS) of a turbulent flow at Reynolds numbers of considerable size that resolves all scales from the smallest dissipative to the largest energetic one is therefore a costly undertaking from a computational point of view. The pioneers that established the method in the second half of the past century (Orszag, 1971a; Kim et al., 1987) were accordingly bounded in the achievable Reynolds numbers by the available computing power at that time. But even today, several decades and a tremendous growth of computing speed later, accurate fully-resolving direct numerical simulations of many natural and technical flows are still beyond what is computationally feasible. Examples for such situations include the simulation of hydraulic flows over entire sections of rivers or simulations in the atmospheric boundary layer that are performed in the context of weather forecasting and climate prediction. In both examples, the intrinsic Reynolds number of the flow and the immense scale of the domain make it even nowadays impossible to resolve all scales at reasonable computational expenses.

This is where modelling comes into play. In this context, the fundamental difference in terminology between simulation and modelling should be underlined: Turbulence simulation denotes the accurate numerical treatment of the Navier-Stokes equations in their original form, where deviations compared to the continuous analytical form arise solely in form of numerical errors due to the discretisation of the problem. Modelling turbulence, on the other hand, means to not resolve specific dynamics of turbulence and to account for the 'ignored' by empirical relations (Pope, 2000). Classically, simulations that do not resolve all scales of turbulence are broadly classified into two families, that are, large-eddy simulations (LES) and Reynolds-averaged Navier-Stokes (RANS) models. As the names suggest, the former approach resolves, i.e. simulates, all flow scales larger than a given filter width and so describes the dynamics of a low-pass filtered turbulent field, whereas the dynamics of all scales smaller than this low-pass filter width are modelled. RANS models follow a different approach: Instead of the actual Navier-Stokes equations, these kind of models solve the RANS equations (3.4) and thus describe the dynamics of the mean velocity and pressure field only. As indicated in the previous chapter, these equations are not closed in that they contain the unknown Reynolds stress tensor and so every RANS model has to provide an appropriate modelling approach for this term. A wide range from very simple to highly complex turbulence closures has been developed during the past decades for different applications and flow configurations, a detailed overview of which is given in the textbook of Pope (2000).

In the current work, we are interested in the fundamental physical interactions between individual turbulent coherent structures of different scale and size and sediment particles that eventually lead to the development of sediment bedforms. We therefore forego any modelling of turbulence and exclusively analyse data obtained by means of direct numerical simulations in order to study these processes 'from first principles', avoiding any influence of sub-grid models or turbulence closures.

4.2 Multiphase simulations

The interaction of turbulent eddies with a dispersed particulate phase crucially depends, among others, on the ratio between the characteristic size of the flow structures and that of the particles (Balachandar and Eaton, 2010). In the context of the previous chapter, it was discussed that the smallest dissipative length scale in a turbulent flow is of the order of the Kolmogorov length η_f, whereby follows that particles whose characteristic length scale D is much smaller than η_f can hardly affect the structure of turbulence, assuming that the particle Reynolds number based on D and the relative particle velocity is similarly small. Is the characteristic size of the particles on the other hand comparable to or larger than η_f and the particle Reynolds number of considerable size, the presence of a solid obstacle locally alters the flow and possibly leads to the evolution of a particle wake (Clift et al., 1978). In wall-bounded turbulent flows, the characteristic length scale of the near-wall fluid motion is $\delta_\nu = \nu_f / u_\tau$. In such flows, the influence of individual particles on the turbulent structures can thus be assumed to be weak if the particle Reynolds number defined as $D^+ = D/\delta_\nu$ is sufficiently small.

Numerical methods for the simulation of particulate flows can be correspondingly classified into two families, those which fully resolve the particle geometry and the flow around each individual particle and those which do not. The first class of schemes is naturally suitable for particles whose size is not negligible compared to the characteristic flow scales, while the other family which includes Eulerian-Eulerian and point-particle approaches is typically applied for particles clearly smaller than η_f (Balachandar and Eaton, 2010; Maxey, 2017). In the context of direct numerical simulations of the turbulent carrying phase, Balachandar and Eaton (2010) propose a critical ratio of $D/\eta_f \sim \mathcal{O}(0.1)$ above which fully-resolved methods are typically chosen. Nonetheless, point-particle approaches in which the individual particles are represented as point masses are also popular for the simulation of larger particles under appropriate extensions of the governing equations (Maxey and Riley, 1983), mainly due to their simpler implementation and their much lower computational cost compared to fully-resolved methods.

While individual particles of considerable size influence the turbulent flow locally by creation of a particle wake possibly including vortex shedding, the question whether or not the presence of mobile particles affect the turbulent flow as a whole strongly depends on the the global solid volume fraction

$$\phi_s = \frac{1}{\mathcal{V}_\Omega} \sum_{i=1}^{N_p} \mathcal{V}_p{}^{(i)} = \frac{\mathcal{V}_{\Omega_p}}{\mathcal{V}_\Omega}, \tag{4.1}$$

where $\mathcal{V}_\Omega$ is the volume of the domain $\Omega \subset \mathbb{R}^d$, N_p denotes the number of particles contained in the system under consideration and $\mathcal{V}_p{}^{(i)}$ is the volume of the ith particle. At very low values of $\phi_s \lesssim \mathcal{O}(10^{-6})$ (Elghobashi, 1994), few particles are distributed over the domain Ω such that their cumulated volume $\mathcal{V}_{\Omega_p}$ is small compared to that occupied by the carrying phase $\mathcal{V}_{\Omega_f}$. Under such circumstances, the interaction between particles and the surrounding fluid is often simplified as 'one-way coupled', that is, the fluid influences the dynamics of the individual particles via hydrodynamic forces and torque that act on the particles' surfaces, whereas the reversed momentum transfer from particles to the turbulent flow field is neglected. For higher values of the solid volume fraction $\mathcal{O}(10^{-6}) \lesssim \phi_s \lesssim \mathcal{O}(10^{-3})$ (Elghobashi, 1994), this simplified point of view is no longer justified as the impact of the particulate phase on the flow cannot be neglected any more. In such situations, the system is considered to be 'two-way coupled', i.e. the feedback of the particle motion on the fluid phase has to be taken into account in numerical simulations. Some authors further classify numerical techniques that deal with dense particulate systems at even higher solid volume fractions $\mathcal{O}(10^{-3}) \lesssim \phi_s$ as 'four-way coupled' if they additionally take care of the momentum exchange during particle-particle contacts or collisions (Crowe et al., 1996).

Bedload-dominated sediment transport in a turbulent carrying fluid that will be considered in the remainder of this work is in this sense a particularly formidable challenge, as it represents a dense particulate system with both frequent particle contacts and particle diameters typically larger than the smallest length scales of the flow. A numerical approach that reduces modelling to an absolute minimum thus needs to fully resolve the flow around individual particles, while it also has to accurately describe particle-fluid and particle-particle interactions in a four-way coupled way.

4.2.1 Immersed boundary method

Two main classes of simulation techniques that are capable to resolve the geometry of individual rigid obstacles and the flow around them have been developed in the past decades. The most accurate, but also computationally most expensive way to treat such problems is to choose a spatial discretisation scheme that adapts to the particles' surface curvatures such as body-fitted meshes in the context of, for instance, spectral element simulations (Ghidersa and Dušek, 2000; Jenny and Dušek, 2004) or finite element simulations based on arbitrary Lagrangian Eulerian formulations (cf. Maxey, 2017, and references therein). While the former technique is particularly useful for systems with stationary particles, the latter allows at least simulations including between a few hundred and a few thousand mobile particles (Maxey, 2017). On the other hand, methods that rely on a body-fitted discretisation scheme are limited in case of dense particulate systems with a large number of individual mobile particles, since they require continuous adaptions of the numerical mesh at each time step to take care of the changing fluid domain Ω_f. In the context of sediment transport simulations, this would lead to unacceptably high simulation times and costs (Wachs, 2019).

A popular alternative to the accurate representation of the particles' surfaces by body-fitted meshes are so-called 'fictitious domain methods'. In this approach, the Navier-Stokes equations are solved numerically on a regular grid in the entire simulation domain Ω comprising the carrying and the dispersed phase. Appropriate forcing terms added to the momentum equations ensure that the no-slip condition at the interface between solid and fluid phase is fulfilled and that the fluid motion inside the particle domain Ω_p is that of a rigid body (Maxey, 2017). A variety of different approaches exist that differ in the way the force field is formulated, in the spatial locations at which it acts and in the spatial discretisation schemes used to solve the governing equations. In the following we focus predominantly on the class of 'direct forcing immersed boundary methods (IBM)' to which the here adopted numerical technique belongs. Alternative remarkable concepts include, for instance, the finite-element distributed Lagrange-multiplier method introduced by Glowinski et al. (1999), wherein the Lagrange-multiplier in the Lagrangian function of the optimisation problem takes the role of the body force.

The original immersed boundary formulation dates back to the work of Peskin (1972) who conceived the method to describe haemodynamic flows in the human heart (Peskin, 2002). The basic concept of the technique is to apply a suitable singular force field at a finite set of N discrete Lagrangian force points X^m ($1 \leq m \leq N$) along the phase boundary. In the original formulation of Peskin (1972), the immersed structure is a flexible valve of the heart and so the forcing points advect with the fluid velocity. This leads to internal stresses in the flexible material, which in turn cause a feed-back force on the surrounding fluid. In theory, the method can be directly extended to rigid bodies by modelling the specific material properties with a spring-like coupling of nearby forcing points at high stiffness, which however leads to the introduction of additional model-dependent parameters and other undesirable numerical effects (Uhlmann, 2005). Therefore, a direct forcing approach is nowadays preferred in which the appropriate forcing required to obtain the desired velocity at the phase boundary is directly applied at the discrete Lagrangian force points X^m (Mohd-Yusof, 1997; Fadlun et al., 2000), thus avoiding the feed-back mechanism. Ever since its original formulation, a multitude of different variants of the IBM have been proposed (cf., for instance, the review by Mittal and Iaccarino, 2005), which differ, amongst others, in the way the desired force is determined (Wachs, 2019).

The immersed boundary technique used in the present work was proposed by Uhlmann (2005) specifically for the use in direct numerical simulations of systems that accommodate a large number of individual immersed particles. In the algorithm outlined by Uhlmann (2005), first a preliminary Eulerian velocity field is determined by solving the governing equations without considering the immersed boundary force. The obtained velocity information is subsequently transferred to the discrete Lagrangian marker points that are equally distributed over the phase boundary, such that their current positions do, in general, not coincide with those of the Eulerian velocity grid. The desired forcing term is then determined as the force required to enforce the no-slip condition at each Lagrangian marker point X^m, that is, the velocities of both phases have to identically match at each of these discrete locations (cf. equations (2.16) in section 2.2). In a final step, the obtained forces are spread back from the Lagrangian marker points to the surrounding nodes of the Eulerian fluid grid. In analogy to the original method of Peskin (1972), Uhlmann (2005) uses a set of regularised Dirac delta functions as kernels in the transfer functions that communicate velocity and force information between the Eulerian fluid grid nodes and the Lagrangian force points X^m. The support of these discretised distributions typically includes three or four grid nodes of the fluid grid (Peskin, 2002). An important advantage of this procedure is that the immersed boundary force and torque are enforced exactly at the particle surface, such that the hydrodynamic force and torque are straightforward obtained from the individual force contributions at the Lagrangian marker points (Wachs, 2019), which we will make use of later in this work.

The immersed boundary formulation is implemented in the context of a standard fractional step method, in which the Navier-Stokes equations are first solved disregarding the divergence-free constraint in form of the continuity equation. In a second step, the preliminary velocity field is subsequently projected onto the space of solenoidal velocity fields. The governing equations are evaluated on a uniformly-spaced staggered finite difference grid with grid width $\Delta x = \Delta y = \Delta z$ using second order accurate approximations of the spatial derivatives. Time integration is performed in a mixed explicit-implicit framework that consists of a three-step low-storage Runge-Kutta scheme for the non-linear convective terms and a Crank-Nicholson scheme for the viscous terms (Verzicco and Orlandi, 1996). A constant discrete time step Δt is chosen small enough to ensure that the non-dimensional time step is with $(\Delta t \max_{x}(\boldsymbol{u}_f))/\Delta x \lesssim 0.6$ clearly below the Courant-Friedrichs-Lewy limit (Courant et al., 1928). In the past, the numerical method has proven its accuracy and versatility in a number of physical problems for the simulation of single spherical (Uhlmann and Dušek, 2014) and non-spherical particles (Moriche et al., 2021) as well as for large numbers of particles from the rather dilute (Uhlmann and Doychev, 2014; Uhlmann and Chouippe, 2017) to the very dense regime (Kidanemariam and Uhlmann, 2017; Scherer et al., 2022), including partly up to $\mathcal{O}(10^6)$ individual particles.

4.2.2 Particle dynamics and contact modelling

The particle dynamics are determined by integration of the discretised Newton-Euler equations (2.17) and (2.18) forward in time. The hydrodynamic force and torque therein are directly obtained from the immersed boundary method, while the terms related to contact forces and torque are still unclosed.

In order to quantify the exchange of linear and angular momentum during particle-particle and particle-wall collisions, the immersed boundary technique is coupled with a discrete or distinct element model (DEM) first introduced by Cundall and Strack (1979). Models of this kind are typically used to describe the motion of large numbers of particles as well as their mutual interaction during contact sequences. How exactly the contact phase is treated numerically depends on the specific model, but all DEMs can be generally divided into hard- and soft-sphere approaches. Hard-sphere models describe the momentum exchange during binary contact events, where it is assumed that the two particles in contact are perfectly rigid such that these models do not allow for any deformation of the two contact partners. The contact phase itself is not resolved in time in hard-sphere models, instead a number of time-integral impulse-momentum equations are considered. With the help of the coefficient of restitution and the coefficient of Coulomb friction which can be determined experimentally, the post-collision velocities and the 'impulsive force', i.e. the contact force integrated over the contact phase, can be determined. A detailed overview of the method can be found in Crowe et al. (1998).

While relatively straightforward to implement and less expensive in terms of computing time, the main disadvantage of hard-sphere models clearly is that they are, by definition, unable to resolve the exact particle behaviour during the contact phase in its temporal evolution. Soft-sphere models, on the other hand, are capable to describe the dynamics of all collision partners during the entire contact phase by solving the full Newton-Euler equations for each of them. The unknown contact forces are modelled in analogy to classical mechanical model systems such as springs and dampers. Particles are allowed to slightly overlap each other (Cundall and Strack, 1979), thereby inducing repulsive forces that are proportional to the overlap length and that eventually lead to the separation of the collision partners. The strength of the repulsive force and the post-collision velocities are controlled by the parameters of the analogous mechanical model system (Kidanemariam, 2016), including spring stiffnesses and damping parameters.

The numerical method used in this work follows the soft-sphere approach applying a model that was proposed, implemented and validated by Kidanemariam and Uhlmann (2014b) and Kidanemariam (2016). In the following, we will review the main elements of the model, while the interested reader is referred to the original works for a more comprehensive overview of the method's details. In the current approach, the total collision force acting on the ith particle at time t due to contacts with its nearby particles

$$\boldsymbol{F}^{C^{(i)}}(t) = \sum_{\substack{j=1 \\ j\neq i}}^{N_p} \left(\boldsymbol{F}^{el^{(i,j)}}(t) + \boldsymbol{F}^{d^{(i,j)}}(t) + \boldsymbol{F}^{t^{(i,j)}}(t) \right) I_c^{(i,j)}(t) \tag{4.2}$$

comprises three different force contributions, including a normal elastic force $\boldsymbol{F}^{el^{(i,j)}}$, a normal damping force $\boldsymbol{F}^{d^{(i,j)}}$ and a tangential force $\boldsymbol{F}^{t^{(i,j)}}$, respectively. The superscript $\bullet^{(i,j)}$ indicates that the force emanates from the contact between the ith and jth particle. The information whether two particles i and j are in contact is contained in the contact identifier function $I_c^{(i,j)}(t)$, which attains a value of unity if particles i and j are in contact and zero if not. Two particles are considered as being in contact if the minimal Euclidean distance between their surfaces Δ_S falls below a given force range Δ_c. Only then, particles 'feel' the contact with the neighbouring obstacle in form of a non-zero finite collision force $\boldsymbol{F}^{C^{(i)}}$.

For the following definitions, we introduce a unit normal vector $\boldsymbol{e}_n^{(i,j)}(t)$ that points along the line connecting the centre of gravity of the ith particle with that of the jth particle. The relative velocity $\boldsymbol{U}_r^{(i,j)}(t)$ between both particles can be thus decomposed into one component along this line and one tangential to it, viz.

$$\boldsymbol{U}_r^{(i,j)}(t) = \boldsymbol{U}_{r,n}^{(i,j)}(t) + \boldsymbol{U}_{r,t}^{(i,j)}(t), \tag{4.3}$$

where $\boldsymbol{e}_t^{(i,j)}(t) = \boldsymbol{U}_{r,t}^{(i,j)}(t)/|\boldsymbol{U}_{r,t}^{(i,j)}(t)|$ uniquely defines an associated tangential normal vector for non-zero values of $\boldsymbol{U}_{r,t}^{(i,j)}(t)$. The individual force components are then modelled in analogy to the classical mass-spring-damper system as

$$\boldsymbol{F}^{el^{(i,j)}}(t) = -k_n \delta_c^{(i,j)}(t)\boldsymbol{e}_n^{(i,j)}(t), \tag{4.4a}$$

$$\boldsymbol{F}^{d^{(i,j)}}(t) = -c_n \boldsymbol{U}_{r,n}^{(i,j)}(t), \tag{4.4b}$$

$$\boldsymbol{F}^{n^{(i,j)}}(t) = \boldsymbol{F}^{el^{(i,j)}}(t) + \boldsymbol{F}^{d^{(i,j)}}(t), \tag{4.4c}$$

$$\boldsymbol{F}^{t^{(i,j)}}(t) = \begin{cases} -\left[\min\left(\mu_c|\boldsymbol{F}^{n^{(i,j)}}(t)|, c_t|\boldsymbol{U}_{r,t}^{(i,j)}(t)|\right)\right] \boldsymbol{e}_t^{(i,j)}(t) & \text{if } |\boldsymbol{U}_{r,t}^{(i,j)}(t)| \neq 0 \\ 0 & \text{else} \end{cases} \tag{4.4d}$$

where $\boldsymbol{F}^{n^{(i,j)}}(t)$ is the total normal contact force and a penetration length has been introduced as

$$\delta_c^{(i,j)}(t) = \frac{1}{2}\left(D^{(i)} + D^{(j)}\right) - |\boldsymbol{X}_p^{(j)}(t) - \boldsymbol{X}_p^{(i)}(t)| + \Delta_c. \tag{4.5}$$

Therein, $\boldsymbol{X}_p^{(i)}$ and $D^{(i)}$ are the centre of gravity and diameter of the ith particle, k_n is the normal stiffness coefficient and c_n and c_t are the normal and tangential friction coefficients, respectively. For the tangential component, it is seen in equation (4.4) that the force amplitude is limited by the Coulomb traction limit, with μ_c being the Coulomb friction coefficient. Since the tangential force acts in an off-centric way on the particles, it creates an additional torque on the latter as

$$\boldsymbol{T}^{C^{(i)}}(t) = \frac{D^{(i)}}{2} \sum_{\substack{j=1 \\ j \neq i}}^{N_p} \boldsymbol{e}_n^{(i,j)}(t) \times \boldsymbol{F}^{t^{(i,j)}}(t)\; I_c^{(i,j)}(t). \tag{4.6}$$

The entire model thus features a set of five free parameters $(k_n, c_n, c_t, \mu_c, \Delta_c)$ that have to be specified. Following Crowe et al. (1998), a functional relationship between k_n and c_n can be derived by introduction of the dry restitution coefficient ε_d that represents the ratio between the relative normal velocity $\boldsymbol{U}_{r,n}$ in the post- and in the pre-collision phase. The dry restitution coefficient ε_d can thus be used to eliminate the normal friction coefficient from the set of unknown parameters. Further choosing the tangential friction coefficient c_t to be identical to its normal counterpart c_n, the yet to choose model parameters reduce to $(k_n, \varepsilon_d, \mu_c, \Delta_c)$.

This alternative set of free model parameters is advantageous in that ε_d and μ_c are material properties and can be obtained from accurate experiments. In practice, the Coulomb friction coefficient was set to a typical range $\mu_c = 0.4$-0.5, while the normal stiffness coefficient was varied in a range $k_n = 17\,000$-$40\,000$ times the submerged weight of a single particle divided by the particle diameter D. The latter parameter was adapted in such a way that the particle overlap was limited to a few percent of the force range Δ_c. The dry restitution coefficient was set to $\varepsilon_d = 0.3$ in most cases, except for the high Reynolds number open channel cases $CS850$ and $CM850^{H3}$ where a higher value of $\varepsilon_d = 0.9$ was adopted. The latter value is comparable to that measured for the contact of a glass sphere with a thick smooth glass wall, $\varepsilon_d = 0.97$ (Gondret et al., 2002), while the former is somewhat smaller and thus implies a stronger damping of the velocities during the contact.

Note that a parameter sensitivity study performed in the course of Scherer et al. (2020) underlines that a variation of ε_d in this range has a rather negligible effect on the development of the subaqueous bedforms in the here considered bedload-dominated physical systems. In the latter study, also the sensitivity of the model to some of the remaining parameters such as the force range which is chosen here equal to the width of the finite difference grid was scrutinised for the case of sediment bedform evolution, while a detailed validation for the case of a single sphere colliding with a solid wall is provided in Kidanemariam (2016).

Finally, it might be worth mentioning that the characteristic time scale of the here considered collision events is typically clearly smaller than the characteristic time scale of the turbulent dynamics, such that for an accurate temporal resolution of the collision phase the Newton-Euler equations are integrated in the framework of a sub-stepping algorithm. That is, within each time-step Δt of the fluid solver, a number of $N_{t,sub}$ time integration steps of the Newton-Euler equations are performed while the flow field is frozen in its latest state. As was approximated by Kidanemariam (2016), the typically chosen number of $N_{t,sub} = \mathcal{O}(100)$ particle sub-steps is more than enough to capture all relevant phases of the collision event.

4.3 Single-phase simulations

Single-phase reference simulations in open channel and duct flows have been performed using two different high-order pseudo-spectral in-house codes whose main concepts will be briefly presented in the following.

4.3.1 Open channel flow

The numerical technique applied to perform simulations of single-phase open channel flow follows the pseudo-spectral method first presented by Kim et al. (1987) for the simulation of turbulent closed channel flow. Here and in the following section, a Cartesian basis is adopted with unit normal basis vectors pointing in the streamwise (e_x), wall-normal (e_y) and spanwise directions (e_z), respectively.

The current implementation with appropriate free-slip boundary conditions at the top of the domain was recently used to investigate turbulent single-phase open channel flows in Bauer (2015) and Bauer et al. (2022), respectively. Here we briefly present the main concept of the methodology, while the interested reader is referred to the latter works as well as the original study of Kim et al. (1987) for the numerical details. The original Navier-Stokes system is transformed into a fourth-order transport equation for the wall-normal velocity v_f and a second-order one for the wall-normal vorticity $\omega_{f,y}$ alongside the continuity equation, viz.

$$\frac{\partial}{\partial t}\nabla^2 v_f = h_v + \frac{1}{Re_\tau}\nabla^4 v_f, \tag{4.7a}$$

$$\frac{\partial}{\partial t}\omega_{f,y} = h_\omega + \frac{1}{Re_\tau}\nabla^2 \omega_{f,y}, \tag{4.7b}$$

$$0 = \left(\frac{\partial u_f}{\partial x} + \frac{\partial w_f}{\partial z}\right) + \frac{\partial v_f}{\partial y}. \tag{4.7c}$$

The transformation into the v_f-$\omega_{f,y}$-system (4.7) leads to an elimination of the pressure term in the governing equations, whose right hand side now features merely viscous and convective terms, viz.

$$h_v = -\frac{\partial}{\partial y}\left(\frac{\partial H_1}{\partial x} + \frac{\partial H_3}{\partial z}\right) + \left(\frac{\partial^2}{\partial x^2} + \frac{\partial^2}{\partial z^2}\right) H_2, \tag{4.8a}$$

$$h_\omega = \frac{\partial H_1}{\partial z} - \frac{\partial H_3}{\partial x}, \tag{4.8b}$$

$$H_i = -u_j \frac{\partial u_i}{\partial x_j}. \tag{4.8c}$$

All variables are thereafter expanded in terms of truncated Fourier series in the two wall-parallel directions and Chebyshev polynomials in the wall-normal direction, respectively. In the framework of the Fourier expansion, N_x and N_z points are equally distributed in the streamwise and spanwise direction, respectively, whereas in the wall-normal direction the governing equations are enforced in the context of a Chebyshev collocation method at a set of non-uniformly spaced Gauss-Lobatto points (cf. section 5.2 for the definition). The method is pseudo-spectral in that the non-linear terms are evaluated in physical space, while the viscous terms are treated numerically in Fourier space. Back and forth transformation between physical and Fourier space along the two homogeneous wall-parallel directions can be efficiently performed via fast Fourier transformations, in which it is taken care of aliasing errors following the 2/3-rule of Orszag (1971b) (cf. also Uhlmann, 2000; Canuto et al., 2006). Time integration of the governing equations is performed using an implicit Euler scheme for the viscous terms, while a third order low-storage Runge-Kutta scheme is adopted to treat the advective terms. As in the finite difference case, the flow is driven by a time-dependent pressure gradient that ensures a constant mass flow rate throughout the simulation. Free-slip and no-slip boundary conditions are directly implemented in the discretised system (Chebyshev tau-method), rather than by modification of the Chebyshev basis functions.

4.3.2 Open duct flow

The numerical technique considered to perform simulations of open duct flow is, in its basic idea, similar to that used for open channel flow simulations, but the exact numerical treatment of the governing equations is fairly different. The implementation presented in the remainder of this section was first used in Uhlmann et al. (2007) and has been outlined and validated by Pinelli et al. (2010), both for closed duct flows. Later, the solver was adapted to the time integration of free-surface open duct flows in the context of Sakai (2016), wherein the numerical details of the method are explained in a much more comprehensive way than in the current work.

The necessity of a different computational method arises from the non-periodicity of the spanwise direction in the presence of lateral sidewalls which do not admit the application of Fourier methods in this direction. Instead, the flow is therefore expanded in terms of Chebyshev polynomials in both cross-stream directions, discretised in form of Gauss-Lobatto points which naturally provide a refinement of the resulting grid near the domain boundaries. Note that all computations are conducted on the standard cross-sectional square $[-1,1] \times [-1,1]$, for which purpose the distributed collocation points are projected and scaled from their actual positions to this area in case of non-square duct cross-sections.

Time integration is again performed in a semi-implicit way, with a Crank-Nicholson scheme applied to the linear viscous terms and an explicit three-step low-storage Runge–Kutta method for the non-linear convective terms. In contrast to the numerical method used for open channel flows, however, the primitive-variable formulation in form of the Navier-Stokes equations is considered, rather than a velocity-vorticity formulation. The discretised equations are therefore solved in the framework of an incremental-pressure projection method which splits the governing equations into two fractional steps, identical to the fractional step method used in the finite-difference IBM discussed above. The splitting leads to a Helmholtz equation for the provisional, non-solenoidal velocity field and a Poisson problem for the pseudo-pressure, both of which are solved for each streamwise wavenumber separately using a fast-diagonalisation technique (Haldenwang et al., 1984; Sekimoto, 2011). As in the open channel case, the numerical method follows a pseudo-spectral approach: The fast-diagonalisation procedure for the Helmholtz and the Poisson equation is executed for each wavenumber separately after having transformed the governing equations to Fourier space along the x-direction, while the non-linear terms therein are evaluated in physical space. Again, transformations of the expanded quantities between physical and Fourier space are executed by using fast Fourier methods, following the de-aliasing 2/3-rule in the homogeneous streamwise direction. Finally, also here a constant mass flow rate is ensured by imposing a time-dependent pressure gradient to the zero streamwise Fourier mode at each time step (Pinelli et al., 2010; Sakai, 2016).

CHAPTER 5

LINEAR INSTABILITY OF A TURBULENT FLOW OVER INITIAL SEDIMENT RIDGES

According to Nezu and Nakagawa (1993), two basic mechanisms of how sediment ridges and secondary currents can evolve in a natural river are conceivable. In the first, an initial deformation of the sediment bed generates a laterally varying bed shear stress, which in turn gives rise to the necessary turbulence anisotropy that results in the development of depth-spanning mean secondary currents. In the second mechanism, *vice versa,* intermittent rotating motions are part of the flow field and induce a laterally varying bed shear stress and erosion rate along the bed that ultimately lead to the rise of sediment ridges. The first mechanism was investigated theoretically by means of a linear stability analysis in the work of Colombini (1993), wherein it was shown that a turbulent base flow profile can be lastingly perturbed by a sinusoidal modulation of the lower domain boundary, inducing large-scale secondary currents if the wavenumber is in the appropriate range.

The goal of the following chapter is to analyse the original model formulation of Colombini (1993), focussing predominantly on how the most-amplified wavenumber is chosen by the linearised system and which role the bed deformation plays in this process. We are striving to show that the key contribution comes in this context from the linearised Navier-Stokes operator, while the sediment bed acts as a stationary outer forcing required to maintain the lateral perturbation of the flow and thus the secondary flow.

To this end, we will derive the linearised equations following the approach presented in Colombini (1993) and discuss the different assumptions that are necessary to arrive at the linearised model. Subsequently, we will give a brief overview of the numerical implementation in terms of a Chebyshev collocation method that is considered to solve the linear algebraic system arising from the discretised equations. This is followed by a discussion of the model results in the context of recent investigations on secondary currents induced by laterally inhomogeneous bottom walls. In the final section, the original model is modified to describe the flow in a smooth flat open channel at finite Reynolds number, in which the lateral perturbation is alternatively maintained by an external force field. It will be shown that essentially the same wavenumber as in the original model is amplified for very different forcing schemes, underlining that the most unstable wavelength is effectively chosen by the linearised Navier-Stokes operator rather than by the sediment bed itself. Similarities and differences with recent works on transient growth and the externally-forced Orr-Sommerfeld problem are eventually discussed.

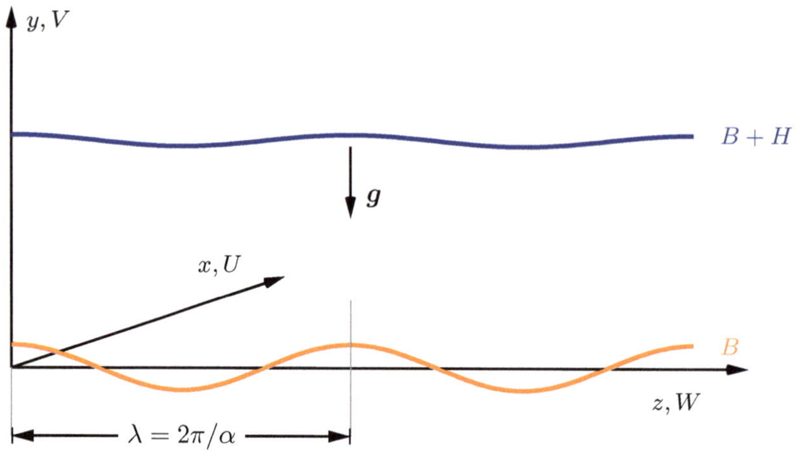

Figure 5.1: Sketch of the idealised physical system considered in the linear stability analysis. The contour of the sediment bed B is shown as orange line, the location of the free surface $B + H$ is indicated as blue solid line (with H denoting the local fluid depth) and g refers to the gravitational acceleration. λ and α are the lateral wavelength and wavenumber of the sinusoidal bed undulation. The sketch is created in analogy to figure 1 in Colombini (1993).

5.1 Mathematical formulation

5.1.1 Governing equations

We consider the flow of an incompressible fluid in a straight and infinitely wide open channel, bounded on the bottom by a deformable sediment bed and at the top by a deformable free surface. In the remainder of this chapter, all quantities are expressed w.r.t. a Cartesian basis, with the associated coordinate system placed at the bottom of the domain such that the three unit normal vectors $\boldsymbol{e}_x$, $\boldsymbol{e}_y$ and $\boldsymbol{e}_z$ point in the streamwise, wall-normal and spanwise directions, respectively. An arbitrary position vector can then be decomposed as $\bar{\boldsymbol{x}} = (\bar{x}, \bar{y}, \bar{z})^T$. Note that $\bar{\bullet}$ here denotes dimensional quantities, while quantities without bars are non-dimensionalised using the characteristic scales that will be outlined below. The flow is driven by the gravitational field in form of the gravitational acceleration vector $\bar{\boldsymbol{g}}$ with $g = |\bar{\boldsymbol{g}}|$, since the mean flow direction of the channel $\boldsymbol{e}_x$ is tilted at a constant angle γ w.r.t. the horizontal reference plane. The streamwise slope of the channel, S_b, is assumed to be small in size such that the approximations $\sin(\gamma) \approx \tan(\gamma) = S_b$ and $\cos(\gamma) \approx 1$ are justified. In the adopted frame of reference, we denote the location of the lower domain boundary as $B(\bar{x}, \bar{z}, t)$ and the local fluid height as $H(\bar{x}, \bar{z}, t)$, such that the location of the free surface is $B + H$. The physical system and the orientation of the coordinate axes are sketched in figure 5.1.

In the following, the Reynolds-averaged Navier-Stokes equations (3.4) as well as the continuity equation are considered to describe mass and momentum balances in the system, subject to the additional assumption that all averaged quantities are effectively independent of the homogeneous streamwise direction, i.e. $\partial(\langle\,\rangle)\partial\bar{x} = 0$. Further neglecting the effect of the viscous terms in equations (3.4), the reduced set of equations reads

$$\begin{aligned}
\frac{\partial\langle\bar{\boldsymbol{u}}_f\rangle}{\partial\bar{t}} + \langle\bar{v}_f\rangle\frac{\partial\langle\bar{\boldsymbol{u}}_f\rangle}{\partial\bar{y}} + \langle\bar{w}_f\rangle\frac{\partial\langle\bar{\boldsymbol{u}}_f\rangle}{\partial\bar{z}} &= \begin{pmatrix}\bar{g}S_b\\ -\bar{g}\\ 0\end{pmatrix} - \frac{1}{\bar{\rho}_f}\begin{pmatrix}0\\ \partial_{\bar{y}}\langle\bar{p}_f\rangle\\ \partial_{\bar{z}}\langle\bar{p}_f\rangle\end{pmatrix} + \bar{\nabla}\cdot\bar{\boldsymbol{\tau}} \\
\frac{\partial\langle\bar{v}_f\rangle}{\partial\bar{y}} + \frac{\partial\langle\bar{w}_f\rangle}{\partial\bar{z}} &= 0,
\end{aligned} \tag{5.1}$$

where we have introduced the Reynolds stress tensor $\bar{\tau} = \langle \bar{u}'_f \otimes \bar{u}'_f \rangle$ or, in index notation, $\bar{\tau}_{ij} = \langle \bar{u}'_{fi} \bar{u}'_{fj} \rangle$. To bring the governing equations (5.1) in a non-dimensional form, we introduce an equivalent set of quantities normalised by the fluid density $\bar{\rho}_f$, the bulk velocity $\bar{U}_{b,0}$ and the mean fluid height $\bar{H}_0$ in the base flow (cf. equation (5.14) below), viz.

$$
\begin{aligned}
\boldsymbol{x} = (x, y, z)^T &= \bar{\boldsymbol{x}}/\bar{H}_0, & t &= \bar{t}/(\bar{H}_0/\bar{U}_{b,0}), \\
\boldsymbol{U} = (U, V, W)^T &= \langle \bar{\boldsymbol{u}}_f \rangle / \bar{U}_{b,0}, & & \\
P &= \bar{p}_f/(\bar{\rho}_f \bar{U}_{b,0}^2), & \boldsymbol{\tau} &= \bar{\boldsymbol{\tau}}/(\bar{\rho}_f \bar{U}_{b,0}^2). \\
B &= \bar{B}/\bar{H}_0, & H &= \bar{H}/\bar{H}_0.
\end{aligned}
$$

In order to keep notation short, we have omitted averaging brackets and the subindices relating to the fluid phase for the non-dimensional quantities and we will continue to do so in the remainder of this chapter. Substituting the non-dimensional quantities into the RANS-equations (5.1), we obtain the non-dimensionalised equations as

$$\frac{\partial U}{\partial t} + V\frac{\partial U}{\partial y} + W\frac{\partial U}{\partial z} = \frac{S_b}{Fr^2} + \frac{\partial \tau_{xy}}{\partial y} + \frac{\partial \tau_{xz}}{\partial z} \tag{5.2a}$$

$$\frac{\partial V}{\partial t} + V\frac{\partial V}{\partial y} + W\frac{\partial V}{\partial z} = -\frac{\partial P}{\partial y} + \frac{\partial \tau_{yy}}{\partial y} + \frac{\partial \tau_{yz}}{\partial z} \tag{5.2b}$$

$$\frac{\partial W}{\partial t} + V\frac{\partial W}{\partial y} + W\frac{\partial W}{\partial z} = -\frac{\partial P}{\partial z} + \frac{\partial \tau_{yz}}{\partial y} + \frac{\partial \tau_{zz}}{\partial z} \tag{5.2c}$$

$$\frac{\partial V}{\partial y} + \frac{\partial W}{\partial z} = 0. \tag{5.2d}$$

Note that by neglecting the viscous terms, we implicitly assume that the system is in the limit $Re \to \infty$ such that the non-dimensional system (5.2) features, apart from the amplitude of the streamwise slope S_b that indicates the strength of the driving force, only a single non-dimensional control parameter in form of the Froude number $Fr = \bar{U}_{b,0}/\sqrt{\bar{g}\bar{H}_0}$.

The governing equations of fluid motion are coupled with the sediment bed continuity equation (Exner, 1925; Seminara, 2010) that captures the time evolution of the sediment bed in dependence of the particle flow rate. In dimensional form assuming again independence of the streamwise coordinate $\bar{x}$, the sediment bed continuity equation reads

$$\frac{\partial \bar{B}}{\partial \bar{t}} = -\frac{1}{1 - p_{bed}} \frac{\partial \bar{Q}}{\partial \bar{z}}, \tag{5.3}$$

where p_{bed} and $\bar{Q}$ denote the porosity of the sediment bed and the sediment flow rate in the spanwise direction, respectively. With the same reference scales as above, equation (5.3) can be rewritten in non-dimensional form as

$$\frac{\partial B}{\partial t} = -Q_{ref}\frac{\partial Q}{\partial z} = -\frac{\sqrt{(s-1)\bar{g}\bar{D}}\bar{D}}{(1 - p_{bed})\,\bar{H}_0 \bar{U}_{b,0}} \frac{\partial Q}{\partial z} \tag{5.4}$$

where $\bar{D}$ and $s = \bar{\rho}_p/\bar{\rho}_f$ are the sediment diameter and the density ratio, respectively. As outlined by Colombini (1993), the newly introduced non-dimensional variable Q_{ref} basically represents the ratio between the inertial particle flux scale $\bar{q}_i = \bar{u}_g \bar{D}$ and the mean fluid mass flow rate $\bar{q}_f = \bar{H}_0 \bar{U}_{b,0}$, with the gravitational velocity $\bar{u}_g = \sqrt{(s-1)\bar{g}\bar{D}}$. In Colombini (1993), Q_{ref} is used to derive a new time scale for the bed evolution, viz. $T = Q_{ref}t$. Assuming that $Q_{ref} \ll 1$, it follows that the sediment bed evolves at a much slower time scale than the fluid. In other words, the flow adapts quasi-instantaneously to a change of the sediment bed and, *vice versa*, the sediment bed 'sees' a quasi-steady flow. Indeed, for the simulation data we are going to present in the subsequent two chapters, Q_{ref} is typically of order $\mathcal{O}(10^{-3})$, such that the assumption is also justified for the here investigated parameter points. As a consequence of the differing time scales, terms including a derivative w.r.t. the sediment bed time scale within the governing fluid equations can be assumed to be of negligible size such that $\partial()/\partial T = 0$, which somewhat simplifies the resulting system that to solve will be the task in the following sections.

The sediment bed continuity equation (5.4) in its current form is unclosed, as a functional expression for the non-dimensional spanwise sediment flux Q is lacking. Colombini (1993) makes use of the following ansatz, which includes concepts from Ikeda (1982) and Parker (1984):

$$Q = \Phi(\theta)\left(\frac{\tau_t}{\tau_0} - \frac{c}{\theta^{1/2}}\frac{\partial B}{\partial z}\right) \tag{5.5a}$$

$$\Phi(\theta) = 8(\theta - 0.047)^{2/3}. \tag{5.5b}$$

Here, c is an empirical parameter, $\theta = (\bar{u}_{\tau,0}/\bar{u}_g)^2$ is the non-dimensional Shields number, τ_0 is the amplitude of the bottom shear stress in the base flow and τ_t is the component of the bed shear stress in the direction tangential to the bed itself and perpendicular to the streamwise direction. The function $\Phi(\theta)$ is the original version of the empirical formula of Meyer-Peter and Müller (1948), based on which the transverse component of the sediment flux vector $\boldsymbol{Q}$ is computed as

$$Q = \sin(\delta)\,\boldsymbol{Q} = \sin(\delta)\,\Phi(\theta). \tag{5.6}$$

Here, δ denotes the angle at which the orientation of $\boldsymbol{Q}$ deviates from the streamwise direction. The above formulation is found when approximating $\sin(\delta)$ under the assumption of a straight channel and δ being small as (Engelund, 1981; Blondeaux and Seminara, 1985; Colombini et al., 1987)

$$\sin(\delta) \approx \left(\frac{\tau_t}{\tau} - \frac{c}{\theta^{1/2}}\frac{\partial B}{\partial z}\right). \tag{5.7}$$

The exact value of c remains a matter of debate: While originally proposed to lie in a range $c \approx 0.5$-0.6 (Engelund, 1981), smaller values of $c \approx 0.3$ (Olesen, 1983; Blondeaux and Seminara, 1985; Colombini et al., 1987) or $c \approx 0.1$ (Colombini and Stocchino, 2012) were adopted in later studies. A more detailed discussion of the empirical sediment flux formulae and a comparison with data from our fully-resolved open channel flow simulations will be provided in section 6.3.2 in the following chapter. Therein, it will be shown that choosing a value of $c \approx 0.09$ leads to the best match between simulation data and the corresponding empirical relations.

In order to approximate the Reynolds stresses τ_{ij} in equation (5.2), an adequate turbulence model is required. Simple linear models based on the Boussinesq hypothesis naturally fail to correctly describe turbulent secondary motion of Prandtl's second kind, as they are not able to reproduce the anisotropy of the normal Reynolds stresses that is a necessary condition for the occurrence of secondary currents (Speziale, 1982; Demuren and Rodi, 1984), as was discussed in section 3.3. More elaborate turbulence models are therefore required to capture the mean secondary flow. Possible candidates are, among others, non-linear extensions of the Boussinesq approximation (Speziale, 1987; Craft et al., 1996), algebraic stress (Naot and Rodi, 1982; Demuren and Rodi, 1984) or full Reynolds-stress models (see Pope, 2000, chapter 11 for a detailed overview of all three approaches). We here follow Colombini (1993) and choose the first approach in form of the quadratic eddy-viscosity model of Speziale (1987). This model extends the typical linear ansatz for the turbulent Reynolds stresses of Boussinesq-type models, viz.

$$\tau_{ij}^{L} = -2\nu_t D_{ij} + \frac{2}{3}k\delta_{ij}, \tag{5.8}$$

by second order terms which are quadratic in the mean velocity gradients. In the above equation, k is the turbulent kinetic energy, ν_t is the turbulent viscosity, $D_{ij} = 1/2(\partial_j U_i + \partial_i U_j)$ is the mean rate of strain tensor and δ_{ij} is the Kronecker-Delta. The formulation of Speziale (1987) obeys some basic mathematical properties of the Navier-Stokes equations, including general coordinate and dimension invariance, positiveness of k and material frame-indifference in the limit of 2D pseudo-turbulence. The next higher order equation for the turbulent Reynolds stresses that fulfils these requirements is quadratic in ∇U and reads

$$\tau_{ij}^{L} + \tau_{ij}^{NL} = \tau_{ij}^{L} + C_D l^2 \left(D_{im} D_{mj} - \frac{1}{3} D_{mn} D_{mn} \delta_{ij} \right) + C_E l^2 \left(\widehat{D_{ij}} - \frac{1}{3} \widehat{D_{mm}\delta_{ij}} \right). \tag{5.9}$$

As before, the Einstein summation convention over repeated indices holds. In equation (5.9), $\widehat{D_{ij}}$ denotes the 'Oldroyd derivative' (Oldroyd and Wilson, 1950) of the mean rate of strain tensor D_{ij}, viz.

$$\widehat{D_{ij}} = \frac{\partial D_{ij}}{\partial t} + (\boldsymbol{U} \cdot \nabla) D_{ij} - \frac{\partial U_i}{\partial x_k} D_{kj} - \frac{\partial U_j}{\partial x_k} D_{ki}, \tag{5.10}$$

which ensures the frame-indifference of the derivative. For the two model constants C_D and C_E, values of $C_D = C_E = 1.68$ are proposed in Speziale (1987), but it is stated that this choice bases on a comparison with only a single experimental data point. In Colombini (1993), on the other hand, higher values for both parameters are chosen, i.e. $C_D = C_E = 3.4$, which were seen to provide the best approximation to the Reynolds stresses measured by Nezu and Nakagawa (1984) over fixed long ridges of trapezoidal cross-section. Note that in contrast to the original model of Speziale (1987), the turbulent viscosity ν_t and the mixing length l are obtained in Colombini (1993) by simple algebraic expressions instead of full transport equations, which the author justifies by the "absence of separation phenomena due to the peculiar structure of sand ridges".

5.1.2 Ansatz functions

Classical hydrodynamic linear stability theory analyses the influence of an initial perturbation of infinitesimal amplitude on the stability of a given base flow (Schmid, 2007). When the base flow is linearly unstable w.r.t. such perturbations, the initial disturbance will grow exponentially as $t \to \infty$, otherwise it will decay (be stable) or at least it will not further grow (be neutrally stable). For this purpose, in case of plane Poiseuille flow conventionally a travelling-wave (or normal-mode) ansatz of the form (Drazin and Reid, 1981)

$$U(y) = U_0(y) + u_1(y) \exp[I(\alpha_x x + \alpha_z z - \alpha_x c_x t)] \tag{5.11}$$

is chosen for the velocity perturbations, where $I = \sqrt{-1}$ is the imaginary unit, $u_1(y)$ is the wall-normal profile of the perturbation, α_x and α_z are the streamwise and spanwise wavenumber and $c_x = c_r + Ic_i$

is the wave speed in the streamwise direction, with real and imaginary part $c_r = \Re\{c_x\}$ and $c_i = \Im\{c_x\}$, respectively. To obtain information on physical quantities, it is understood that the real part of (5.11) has to be considered. Also note that we have here introduce the convention to denote base flow quantities in uppercase letters with subindex zero and perturbation profiles in lowercase letters with subindex one. Inserting the travelling wave ansatz into the linearised equations typically leads to a (generalised) eigenvalue problem, such that the stability of the base flow can be analysed in terms of the spectrum of the linear operator (Schmid, 2007).

The situation in the current case differs from this concepts in several aspects: First, the current system is independent of the streamwise direction, such that all modes can be considered as being of infinite streamwise wavelength (i.e. $\alpha_x = 0$) and hence no growth or propagation of the wave in the streamwise direction occurs. Instead, we are interested in spanwise modes that grow in time but do not propagate in space, as the sediment bed perturbations are typically assumed to remain at their initial lateral positions. Thus, we consider in the following a conceptually similar approach as the travelling wave ansatz above, but based on the spanwise wavenumber $\alpha = \alpha_z$ (with a wavelength $\lambda = 2\pi/\alpha$) and a purely imaginary spanwise wave speed $c_z = Ic_i$, viz.

$$U(y) = U_0(y) + u_1(y)\exp[I\alpha z + \alpha c_i t]. \tag{5.12}$$

Taking the real part of the ansatz (5.12), choosing as characteristic time scale the sediment evolution time scale T and defining the growth rate as $\sigma = \alpha c_i$, it is readily shown that a perturbation of the base flow profile $U_0(y)$ with infinitesimal amplitude ϵ has the form

$$U(y) = U_0(y) + \exp(\sigma T)\, \epsilon u_1(y)\, \cos(\alpha z). \tag{5.13}$$

In Colombini (1993), corresponding perturbation ansatz functions are proposed for all quantities of interest, viz.

$$\left\{\begin{array}{lcll} U & = & U_0(y) & + \exp(\sigma T)\, \epsilon u_1(y)\, \cos(\alpha z) \\ V & = & 0 & + \exp(\sigma T)\, \epsilon v_1(y)\, \cos(\alpha z) \\ W & = & 0 & - \exp(\sigma T)\, \epsilon \alpha w_1(y)\, \sin(\alpha z) \\ P & = & P_0(y) & + \exp(\sigma T)\, \epsilon p_1(y)\, \cos(\alpha z) \\ & & & \\ B & = & 0 & + \exp(\sigma T)\, \epsilon\, \cos(\alpha z) \\ H & = & 1 & + \exp(\sigma T)\, \epsilon h_1\, \cos(\alpha z) \\ & & & \\ Q & = & 0 & - \exp(\sigma T)\, \epsilon \alpha q_1\, \sin(\alpha z) \\ & & & \\ \nu_t & = & \nu_{t0}(y) & + \exp(\sigma T)\, \epsilon \nu_{t1}(y)\, \cos(\alpha z) \\ l & = & l_0(y) & + \exp(\sigma T)\, \epsilon l_1(y)\, \cos(\alpha z). \end{array}\right. \tag{5.14}$$

Note that not all perturbation functions in (5.14) follow exactly an ansatz function analogous to equation (5.13). The reason is that the resulting physical quantities have to fulfil a set of symmetry conditions, and here the perturbation ansatz functions are directly chosen to take care of these conditions. In particular, all variables except for W and Q are in phase with the sediment bed contour B and, as such, possess an even symmetry, whereas W and Q feature an odd symmetry w.r.t. to the bedform crest since they are instead in phase with the first transverse derivative of the bed contour, viz. $\partial_z B \sim -\alpha \sin(\alpha z)$.

5.1.3 Non-orthogonal coordinate transformation

In order to take into account the variability of the flow domain as a consequence of the sediment bed deformation, the cross-section of the flow is mapped onto a rectangular domain using the following non-orthogonal coordinate transformation

$$\begin{aligned} \eta &= \frac{y - B(z)}{H(z)} \iff y = \eta H(z) + B(z) \\ \zeta &= z. \end{aligned} \tag{5.15}$$

The coordinate transformation leads to additional terms in the governing equations. Let ψ be an arbitrary scalar, vector or tensor component w.r.t. the standard Cartesian basis $\boldsymbol{x} = y\boldsymbol{e}_y + z\boldsymbol{e}_z$ where we ignore the streamwise direction that is of no importance here. Alternatively, ψ can be expanded in the new non-orthogonal basis $\boldsymbol{\xi} = \eta\boldsymbol{e}_\eta + \zeta\boldsymbol{e}_\zeta$. Application of the chain rule allows to transform associate derivatives from one coordinate system to the other, viz.

$$\begin{aligned} \frac{\partial\psi}{\partial x_i} &= \frac{\partial\xi_j}{\partial x_i}\frac{\partial\psi}{\partial\xi_j} \quad \text{or} \quad \frac{\partial\psi}{\partial\xi_i} = \frac{\partial x_j}{\partial\xi_i}\frac{\partial\psi}{\partial x_j} \\ \nabla_x\psi &= \boldsymbol{T}^{-1}\nabla_\xi\psi \quad \text{or} \quad \nabla_\xi\psi = \boldsymbol{T}\nabla_x\psi, \end{aligned} \tag{5.16}$$

where ∇_x and ∇_ξ are the nabla operators w.r.t. the Cartesian and the non-orthogonal coordinate basis, respectively. The transformation matrix and its inverse are denoted by $\boldsymbol{T}$ and $\boldsymbol{T}^{-1}$, respectively, and are defined as

$$\boldsymbol{T} = \left[\frac{\partial x_j}{\partial\xi_i}\right] = \begin{bmatrix} \frac{\partial y}{\partial\eta} & \frac{\partial z}{\partial\eta} \\ \frac{\partial y}{\partial\zeta} & \frac{\partial z}{\partial\zeta} \end{bmatrix} = \begin{bmatrix} H(z) & 0 \\ \eta\frac{\mathrm{d}H(z)}{\mathrm{d}z} + \frac{\mathrm{d}B(z)}{\mathrm{d}z} & 1 \end{bmatrix}, \tag{5.17a}$$

$$\boldsymbol{T}^{-1} = \left[\frac{\partial\xi_j}{\partial x_i}\right] = \begin{bmatrix} \frac{\partial\eta}{\partial y} & \frac{\partial\zeta}{\partial y} \\ \frac{\partial\eta}{\partial z} & \frac{\partial\zeta}{\partial z} \end{bmatrix} = \begin{bmatrix} \frac{1}{H(z)} & 0 \\ -\frac{1}{H(z)}\left[\eta\frac{\mathrm{d}H(z)}{\mathrm{d}z} + \frac{\mathrm{d}B(z)}{\mathrm{d}z}\right] & 1 \end{bmatrix}. \tag{5.17b}$$

Inserting these expressions into equation (5.16), the transformation rules for the partial derivatives of ψ are obtained as

$$\frac{\partial\psi}{\partial y} = \frac{\partial\eta}{\partial y}\frac{\partial\psi}{\partial\eta} + \frac{\partial\zeta}{\partial y}\frac{\partial\psi}{\partial\zeta} = \frac{1}{H(\zeta)}\frac{\partial\psi}{\partial\eta} \tag{5.18a}$$

$$\frac{\partial\psi}{\partial z} = \frac{\partial\eta}{\partial z}\frac{\partial\psi}{\partial\eta} + \frac{\partial\zeta}{\partial z}\frac{\partial\psi}{\partial\zeta} = -\frac{1}{H(\zeta)}\left[\eta\frac{\mathrm{d}H(\zeta)}{\mathrm{d}\zeta} + \frac{\mathrm{d}B(\zeta)}{\mathrm{d}\zeta}\right]\frac{\partial\psi}{\partial\eta} + \frac{\partial\psi}{\partial\zeta} \tag{5.18b}$$

Substituting the transformed derivatives into the reduced non-dimensional system (5.2), we obtain the transformed set of equations (A.1) that is presented in appendix A.1.1 for the sake of clarity.

5.1.4 Base flow equations

In the following, we first collect terms of order $\mathcal{O}(\epsilon^0)$ that describe the configuration of the base flow. In Colombini (1993), a strictly logarithmic velocity profile is assumed for the turbulent uni-directional base flow. To this end, a friction coefficient $C_0 = (\bar{u}_{\tau,0}/\bar{U}_{b,0})^2 = S/Fr^2$ is introduced that represents the squared ratio between the friction velocity $\bar{u}_{\tau,0}$ and the bulk velocity $\bar{U}_{b,0}$ in the base flow, respectively. For the sake of completeness, let us point out that C_0 differs by factor two from the skin friction factor c_f found in many textbooks (e.g. Pope, 2000, p. 267) and used later in this work.

In the asymptotic case of a fully-rough lower boundary, the turbulent velocity profile reads (Einstein, 1950; Pope, 2000)

$$\frac{U_0(\eta)}{u_{\tau,0}} = 8.5 + \frac{1}{\kappa} \ln\left(\frac{\eta}{k_s}\right), \tag{5.19}$$

where k_s is the non-dimensional equivalent sand grain roughness (Jiménez, 2004a) which is usually approximated as $k_s \approx 2.5D$ (Engelund and Hansen, 1967; Colombini et al., 1987) and a value of $\kappa = 0.4$ is chosen for von Kármán's constant. Integration of the velocity profile (5.19) over the wall-normal direction $\eta \in (0,1]$ using l'Hôpital's rule in the lower limit leads to the following expression for the friction coefficient C_0:

$$\begin{aligned} \frac{U_{b,0}}{u_{\tau,0}} = \frac{1}{\sqrt{C_0}} &= [8.5\eta + 2.5\,(\eta\,\ln(\eta) - \eta) - 2.5\ln(2.5D)\;\eta]_0^1 \\ &= 8.5 - 2.5 - 2.5\ln(2.5D) + \lim_{\eta\to 0+}(2.5\,(\eta\,\ln(\eta) - \eta)) \\ &= 6 - 2.5\,\ln(2.5D). \end{aligned} \tag{5.20}$$

Equation (5.19) can be alternatively written in a more compact form as

$$\frac{U_0(\eta)}{u_{\tau,0}} = 2.5\ln\left(\frac{\eta}{\eta_0}\right), \tag{5.21}$$

where η_0 is the non-dimensional reference level at which the logarithmic velocity profile reaches zero, which in turn marks the wall-normal location where the no-slip boundary conditions are imposed. Comparison of equation (5.21) with equation (5.19) then directly implies that $\eta_0 \approx k_s/30 \approx D/12$ in the fully-rough case. A more general relation between the friction coefficient and the reference level which is valid for all roughness regimes is then found by integration of equation (5.21) over the channel height as before (Colombini, 2004):

$$\begin{aligned} \frac{U_{b,0}}{u_{\tau,0}} = \frac{1}{\sqrt{C_0}} &= 2.5\,[(\eta\,\ln(\eta) - \eta)]_0^1 - \ln(\eta_0) \\ &= 2.5\,[-1 - \ln(\eta_0))] \\ \Longleftrightarrow \eta_0 &= \exp\left(-\kappa/\sqrt{C_0} - 1\right). \end{aligned} \tag{5.22}$$

The turbulence parameters, namely the turbulent viscosity ν_t and the mixing length l, are determined following a formulation of the mixing length hypothesis that was earlier used by De Vriend (1977). After linearising the terms, the associated expressions for ν_t and l in the base flow read

$$l_0 = \kappa\eta\,\sqrt{1-\eta} \quad \text{and} \quad \nu_{t0} = l_0^2\,U_0'. \tag{5.23}$$

Note that in order to keep notation short, we will use in the remainder of this work n primes to indicate the nth derivative w.r.t. the wall-normal direction, $\mathrm{d}^n()/\mathrm{d}\eta^n$, which should not be confused with its meaning in the context of fluctuating quantities elsewhere in this work.

In appendix A.1, it is shown in more detail how the system (A.1) is linearised around $\epsilon = 0$ by neglecting all terms that are of order $\mathcal{O}(\epsilon^2)$ or higher. To this end, first the standing wave-like ansatz functions (5.14) are substituted into the governing equations (A.1) in the new coordinate system, in which the individual Reynolds stresses are approximated by the non-linear expressions for $\tau_{ij} = \tau_{ij}^L + \tau_{ij}^{NL}$ introduced in equations (5.8) and (5.9). Note that the exponential growth in the ansatz functions (5.14) was formulated w.r.t. to the much slower time scale T that is associated with the bedform evolution. When inserted in equations (A.1), the temporal derivatives transform under application of the chain rule into

$$\frac{\partial}{\partial t}() = \frac{\partial T}{\partial t}\frac{\partial}{\partial T}() = Q_{ref}\frac{\partial}{\partial T}().$$

Recalling that $Q_{ref} \ll 1$, all temporal derivatives in equations (A.1) become negligibly small and are thus not further considered in the following.

In the resulting linearised system, all remaining contributions that are associated with the non-linear convective terms of equations (A.1) are at least of first order in ϵ and thus do not contribute to the base flow. While the continuity equation exclusively features terms of $\mathcal{O}(\epsilon^1)$, the momentum equations for the base flow accordingly reduce to a balance between the turbulent Reynolds stresses and the driving forcing term S_b/Fr^2 on the one hand and the wall-normal pressure gradient on the other hand, viz.

$$-C_0 = \nu_{t0}'U_0' + \nu_{t0}U_0'' = -\frac{S_b}{Fr^2} \tag{5.24a}$$

$$P_0' = \left(\frac{1}{6}C_D + \frac{2}{3}C_E\right)\left(l_0 l_0'(U_0')^2 + l_0^2 U_0' U_0''\right). \tag{5.24b}$$

Making use of the relation $\nu_{t0} = l_0^2 U_0'$ (cf. equations (5.23)), the system above can be further simplified to obtain

$$-C_0 = \nu_{t0}'U_0' + \nu_{t0}U_0'' = 2\left(l_0 l_0'(U_0')^2 + l_0^2 U_0' U_0''\right) = -\frac{S}{Fr^2} \tag{5.25a}$$

$$P_0' = -\left(\frac{1}{6}C_D + \frac{2}{3}C_E\right)\frac{C_0}{2}. \tag{5.25b}$$

In fact, the problem has thus been reduced to a single ordinary differential equation for the base flow $\boldsymbol{U}_0 = (U_0(\eta), 0, 0)^T$, while the linear pressure profile P_0 over the channel depth is described by an algebraic expression which is especially independent of the base flow itself. Interestingly, the wall-normal pressure gradient still depends on a pre-factor which originates in the non-linear turbulent Reynolds stresses.

The assumption of a quasi-stationary flow reduces the original initial boundary value problem to a pure boundary value problem, thus it requires only the definition of boundary conditions at the bottom of the domain (i.e. at a wall-normal location η_0 above the bed) and at the free surface $\eta = 1$. For the base flow, the no-slip condition at the bottom and the stress-free condition at the free surface read

$$\begin{aligned} U_0(\eta = \eta_0) &= 0 && \text{(no-slip)},\\ \tau_{xy}(\eta = 1) &= \nu_{t0} U_0' = (l_0 U_0')^2 = 0, \quad P_0(\eta = 1) = 0 && \text{(stress-free)}. \end{aligned}$$

Note that the free-surface condition for the shear stress is here fulfilled by design of the mixing length approach, as both l_0 and ν_{t0} vanish at the free surface. Taking this into account, an analytic solution for U_0 and ν_{t0} can be obtained by first integrating the first equation in (5.25) subject to the upper boundary condition for the turbulent shear stress:

$$\begin{aligned} C_0(1-\eta) &= \nu_{t0} U_0' \\ \overset{\eta \geq \eta_0}{\Longrightarrow} \quad \nu_{t0} &= l_0^2 U_0' = l_0^2 \frac{C_0(1-\eta)}{\nu_{t0}} \\ \Longleftrightarrow \quad \nu_{t0} &= l_0 \sqrt{C_0}\sqrt{1-\eta} \overset{(5.23)}{=} \sqrt{C_0}\,\kappa\,\eta\,(1-\eta). \end{aligned} \tag{5.26}$$

As stated in equation (5.21) above, the analytical solution for the vertical base velocity profile is then strictly logarithmic over the entire channel height, fulfilling the no-slip condition at a wall-normal distance η_0 above the sediment bed:

$$U_0(\eta) = \frac{\sqrt{C_0}}{\kappa} \ln\left(\frac{\eta}{\eta_0}\right). \tag{5.27}$$

5.1.5 Normal-mode analysis

In the framework of a normal-mode analysis, perturbations of different wavenumbers are studied in the following separately, seeking for those that cause a maximisation of the growth rate σ. In the classical normal-mode approach, linearisation of the governing equations together with a travelling wave ansatz of the form that we have introduced above typically results in a (generalised) eigenvalue problem. In this context, the analysis of the spectrum of the linear operator is indicative of stable and unstable subspaces of the parameter space. In the current case, on the other hand, collecting all terms of order $\mathcal{O}(\epsilon^1)$ results in a simple linear system of the form

$$A\boldsymbol{q} = h_1 \boldsymbol{B} + \boldsymbol{R}, \tag{5.28}$$

in which we assume that we can write the unknown functions in a single vector

$$\boldsymbol{q}(\eta) = (u_1(\eta), v_1(\eta), w_1(\eta), p_1(\eta))^T \tag{5.29}$$

of possibly infinite dimension, e.g. in terms of Fourier or Chebyshev coefficients. The individual entries of the linear operator A and the vectors $\boldsymbol{B}$ and $\boldsymbol{R}$ can be found in equations (A.5) and (A.6) of appendix A.1, respectively. Also, the exact form of the boundary conditions for the perturbed functions will be discussed therein.

As a consequence of the neglected time dependence, the linear system (5.28) has become independent of the growth rate of the sediment bed σ and is instead a function of the wavenumber α and the governing non-dimensional parameters only. As a consequence, σ can only be determined by solving the sediment continuity equation (5.4), whose linearised form eventually leads to the following dispersion relation for the growth rate σ (see appendix A.1 for the detailed derivation)

$$\frac{\sigma}{\Phi\left(\theta_0\right)\alpha^2} = \left[\frac{w_1'}{U_0'}\right]_{\eta_0} - \frac{c}{\theta_0^{1/2}}, \tag{5.30}$$

where a parameter $c = 0.3$ is chosen as in Colombini (1993). The angular brackets denote that the respective term is evaluated at the non-dimensional reference level $\eta = \eta_0$, and $\theta_0 = Fr^2\ C_0/[(s-1)D]$ is the Shields number in the base flow. So, eventually, the problem formulation requires three non-dimensional (physical) input parameters to be uniquely defined, that are: the relative particle diameter $D = \bar{D}/\bar{H}_0$ (inverse of the relative submergence), the Froude number Fr and the density ratio s. The dispersion relation (5.30) can be physically interpreted in that the sediment bed growth rate σ arises from the balance between the bed-destabilising lateral bed shear stress and a counteracting and thus stabilising gravitational term that depends only on the constant parameter c and the Shields number. For a given set of physical non-dimensional parameters, σ is thus a function of the spanwise wavenumber α only. We will discuss the exact consequences of this dependencies in section 5.4 below.

5.2 Numerical implementation

5.2.1 Chebyshev collocation method

In the following, we choose a standard Chebyshev collocation method to solve the linearised system, in the context of which the governing equations are enforced at a set of discrete collocation points in physical space (Boyd, 2001). To this end, all quantities are approximated as truncated series of Chebyshev polynomials, viz.

$$\hat{\phi}(\eta) = \sum_{i=0}^{N_\eta} \bar{\phi}_i T_i(\eta), \tag{5.31}$$

where $T_i(\eta)$ is the ith Chebyshev polynomial of the first kind, $\bar{\phi}_i$ is the ith coefficient of the series and N_η is the highest considered polynomial degree. For the collocation points at which the equations of the linear system will be enforced, we choose a set of $N_\eta + 1$ Gauss-Lobatto grid points, defined as the locations at which the N_ηth Chebyshev polynomial of the first kind T_{N_η} attains its extrema, viz.

$$\eta_k = \cos\left(\frac{\pi k}{N_\eta}\right), \quad k = 0, \ldots, N_\eta. \tag{5.32}$$

We obtain $N_\eta + 1$ degrees of freedom for each of the unknown physical variables u_1, v_1, w_1 and p_1, respectively, that can be collected in analogy to $\boldsymbol{q}$ in a single vector $\boldsymbol{q}_N \in \mathbb{R}^{4(N_\eta+1)}$

$$\begin{aligned} \boldsymbol{q}_N &= \left(u_N, v_N, w_N, p_N\right)^T \\ &= \left(\hat{u}_1(\eta_0), \ldots, \hat{u}_1(\eta_{N_\eta}), \hat{v}_1(\eta_0), \ldots, \hat{p}_1(\eta_{N_\eta})\right)^T, \end{aligned} \tag{5.33}$$

with the discrete Chebyshev coefficient vectors $u_N, v_N, w_N, p_N \in \mathbb{R}^{(N_\eta+1)}$. Variables with a superscript $\hat{\bullet}$ denote the individual coefficients associated with the respective collocation points. Together with the discretised matrix $A_N \in \mathbb{R}^{4(N_\eta+1)\times 4(N_\eta+1)}$ and vectors $\boldsymbol{B}_N, \boldsymbol{R}_N \in \mathbb{R}^{4(N_\eta+1)}$, we obtain the following inhomogeneous linear system:

$$A_N \boldsymbol{q}_N = \boldsymbol{B}_N h_1 + \boldsymbol{R}_N. \tag{5.34}$$

The individual entries of A_N, $\boldsymbol{B}_N$ and $\boldsymbol{R}_N$ will be provided in appendix A.2. The appropriate discretised boundary conditions are directly imposed into the linear system (Chebyshev tau-method) by modification of the respective entries in the matrix and the RHS vectors (cf. appendix A.2.1 for a detailed explanation of how the boundary conditions are applied to the discrete system).

Note that the matrix A_N represents a discretised linear differential operator that is build on individual collocation derivative matrices $\boldsymbol{D}_\eta$ of the form (cf. Canuto et al. (2006, p.89) or Boyd (2001, p.570))

$$(\boldsymbol{D}_\eta)_{jl} = \begin{cases} \dfrac{\bar{c}_j}{\bar{c}_l} \dfrac{(-1)^{j+l}}{\eta_j - \eta_l} & j \neq l \\ -\dfrac{\eta_l}{2\left(1-\eta_l^2\right)} & 1 \leq j = l \leq N_\eta - 1 \\ \dfrac{2N_\eta^2+1}{6} & j = l = 0 \\ -\dfrac{2N_\eta^2+1}{6} & j = l = N_\eta \end{cases}, \tag{5.35}$$

where the coefficients $\bar{c}_j$ are defined as

$$\bar{c}_j = \begin{cases} 2, & j = 0, N_\eta \\ 1, & j = 1, \ldots, N_\eta - 1 \end{cases}.$$

Higher derivatives are readily obtained as repeated matrix products of $\boldsymbol{D}_\eta$ with itself, i.e. $\boldsymbol{D}_{\eta\eta} = \boldsymbol{D}_\eta \boldsymbol{D}_\eta$ and $\boldsymbol{D}_{\eta\eta\eta} = \boldsymbol{D}_\eta \boldsymbol{D}_\eta \boldsymbol{D}_\eta$, respectively.

5.2.2 Validation against the original results

In order to validate the numerical method, we first recompute the results of Colombini (1993) with the afore outlined Chebyshev collocation method. As can be seen in figure 5.2, both the destabilising term $[w_1'/U_0']_{\eta_0}$ in the dispersion relation (A.13) and the variation of the growth rate σ as a function of the wavenumber α show a quasi perfect match with the original model implementation of Colombini (1993). Note that the shown datasets are the only results provided in Colombini (1993) for the problem of interest, such that a more detailed validation against the original model is not possible, but the very good match of the destabilising term and the growth rate convincingly shows that the current method is able to accurately reproduce the results of the reference study. The results shown in figure 5.2 have been obtained by evaluating the governing equations at 64 Chebyshev-Gauss-Lobatto points distributed over the channel depth. Reference computations with 128 and 256 Chebyshev-Gauss-Lobatto points

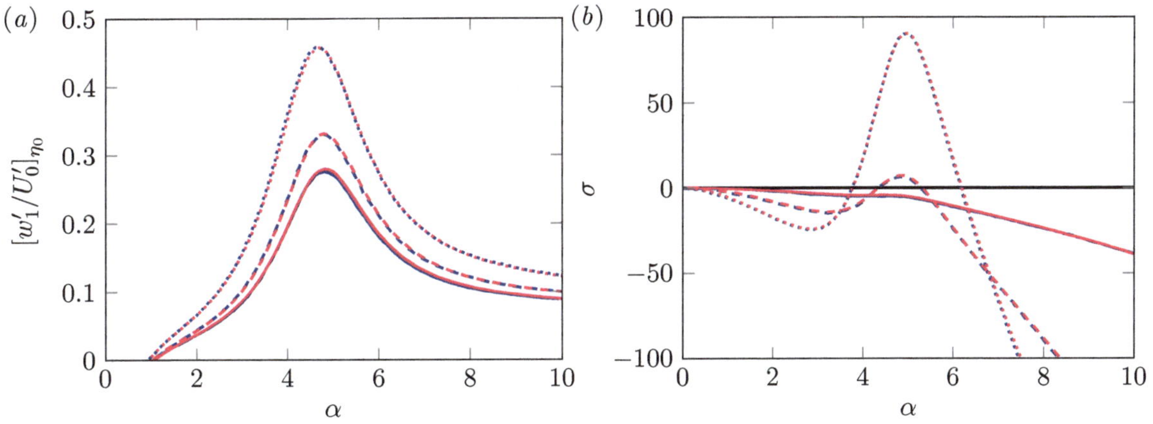

Figure 5.2: Validation of the current numerical linear stability analysis against the original results in Colombini (1993) in terms of (*a*) the destabilising shear stress term $[w_1'/U_0']_{\eta_0}$ and (*b*) the growth rate σ as a function of the wavenumber α. In (*a*), different lines indicate different values of the relative roughness: (—) $D = 0.005$; (---) $D = 0.01$; (···) $D = 0.05$. In (*b*), lines of different style refer to different Froude numbers for a constant relative roughness $D = 0.01$ and density ratio $s = 2.65$: (—) $Fr = 1$; (--- $Fr = 2$); (···) $Fr = 3$. (—) Data extracted from figures 9 and 10 in Colombini (1993); (—) results obtained with the current numerical method.

show no visible differences to the results obtained with 64 collocation points, verifying that the chosen production grid is sufficiently fine for the current needs.

For the sake of consistency with our direct numerical simulations discussed later in this work, we have set the value of the fluid depth perturbation to $h_1 = -1$, which is equivalent to enforcing that the free surface remains flat. In the original study, it is not entirely clear if the free surface was allowed to deform with h_1 being a variable parameter that is determined alongside the solution of the linear system or whether the top of the domain was also kept flat for the here shown results. The very good match between the original data and our results seen in figure 5.2 however implies that h_1 is in any case close to -1, that is, the bed undulations have a relatively weak influence on the free surface.

For the stabilising gravitational term, on the other hand, the Froude number is of significant importance, as it is directly related to the Shields number in the base flow. In Colombini (1993), it is argued that a parametrisation of the problem in terms of the Shields number rather than the Froude number would be more suitable. As can be seen in figure 5.2(*b*), a Froude number of $Fr = 1$ is under the current model assumptions not sufficient to allow for any mode to grow in time, but for all wavenumbers the fluid-driven destabilising term is too weak to overcome the stabilising gravity term. A critical Froude number can be defined as the lowest value of Fr for which at least a single mode is amplified rather than damped in time. It turns out that the Shields number associated with this critical Froude number attains in this case an unrealistically high value of $\mathcal{O}(1)$, as pointed out by Colombini (1994). As mentioned in the latter publication, a possible reason for the overestimation of the critical Shields and Froude number could be that the adopted turbulence model underpredicts the actual lateral bed shear stress. Not considered is that, possibly in addition to the former effect, that the gravitational term might be overpredicted. In fact, we will see in the following chapter (cf. section 6.3.2) that for the simulations considered therein, a gravitational factor of $c \approx 0.09$ leads to a better match between the simulation results and the theoretical prediction following equation (5.7) than the earlier adopted value of $c = 0.3$, which agrees well with the findings of Colombini and Stocchino (2012). Replacing $c = 0.3$ in the original model by $c = 0.1$ indeed leads to an increase of the growth rate as the gravitational term reduces

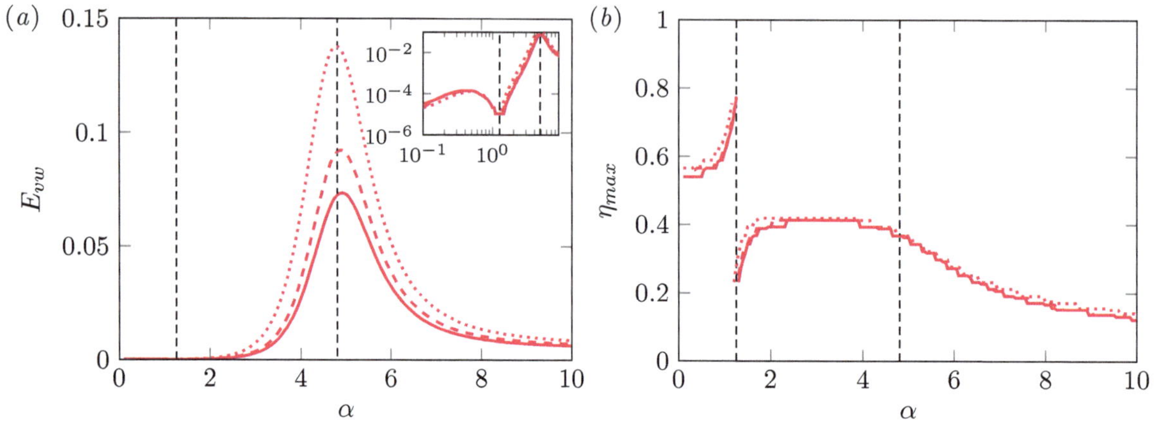

Figure 5.3: (*a*) Secondary flow kinetic energy $E_{vw}(\alpha)$ as a function of the wavenumber α. The inset shows the same data in doubly-logarithmic scaling. (*b*) Wall-normal location η_{max} at which the amplitude of the wall-normal perturbation velocity v_1 attains its maximum as a function of the wavenumber α. $D = 0.005$ (—); $D = 0.01$ (- - -); $D = 0.05$ (· · ·). The vertical dashed lines indicate the two wavenumbers $\alpha = \{1.2548, 4.8\}$. In (*b*), η_{max} is defined as $\eta_{max} = \underset{\eta}{\operatorname{argmax}}(|v_1(\eta)|)$.

(plots omitted), but the critical Shields and Froude number still remain much higher than those observed in experiments and numerical simulations, underlining the need to critically assess the chosen turbulence closure and its ability to correctly reproduce the strength of the lateral bed shear stress.

5.3 Analysis of the original model

In figure 5.3(*a*), it is seen that not only the lateral bed shear stress is maximised around a wavenumber of $\alpha = 4.8$ ($\lambda = 1.3$), but also the secondary flow kinetic energy

$$E_{vw}(\alpha) = \frac{1}{\lambda}\int_0^{\lambda}\int_0^1 \left[(v_1\cos(\alpha\zeta))^2 + (-w_1\alpha\sin(\alpha\zeta))^2\right] \mathrm{d}\eta\mathrm{d}\zeta \tag{5.36}$$

attains its maximum peak at this wavenumber, which highlights the fact that the secondary currents themselves are indeed most amplified for a characteristic bed perturbation wavelength of $\lambda = 1.3$. For both very small and very large wavenumbers, the secondary flow kinetic energy strongly decreases, but only for $\alpha \to 0$ it tends to negligible values. In the large wavenumber regime, on the other hand, E_{vw} is seen to settle at a small but finite value that is essentially constant even for much higher values up to $\alpha = 50$ (plot not shown). The inset in figure 5.3(*a*) moreover uncovers an additional, very weak local minimum at a smaller wavenumber $\alpha \approx 1.2548$ that is interestingly not identical to the value $\alpha \approx 1$ at which the sign of the bed shear stress changes in figure 5.2(*a*). Instead, it is seen in figure 5.3(*b*) that at the same wavenumber $\alpha \approx 1.2548$, the wall-normal location of the maximum amplitude of $|v_1(\eta)|$ that represents the position of the secondary currents' centres changes abruptly from near the free surface ($\eta_{max} \gtrsim 0.75$) to near the curved bottom contour ($\eta_{max} \lesssim 0.25$). At the most amplified wavenumber, $|v_1(\eta)|$ attains its maximum somewhat below the channel centreline at $\eta_{max} \approx 0.38$. As the wavenumber is further increased, the maximum of $|v_1(\eta)|$ is seen to successively approach the lower domain boundary.

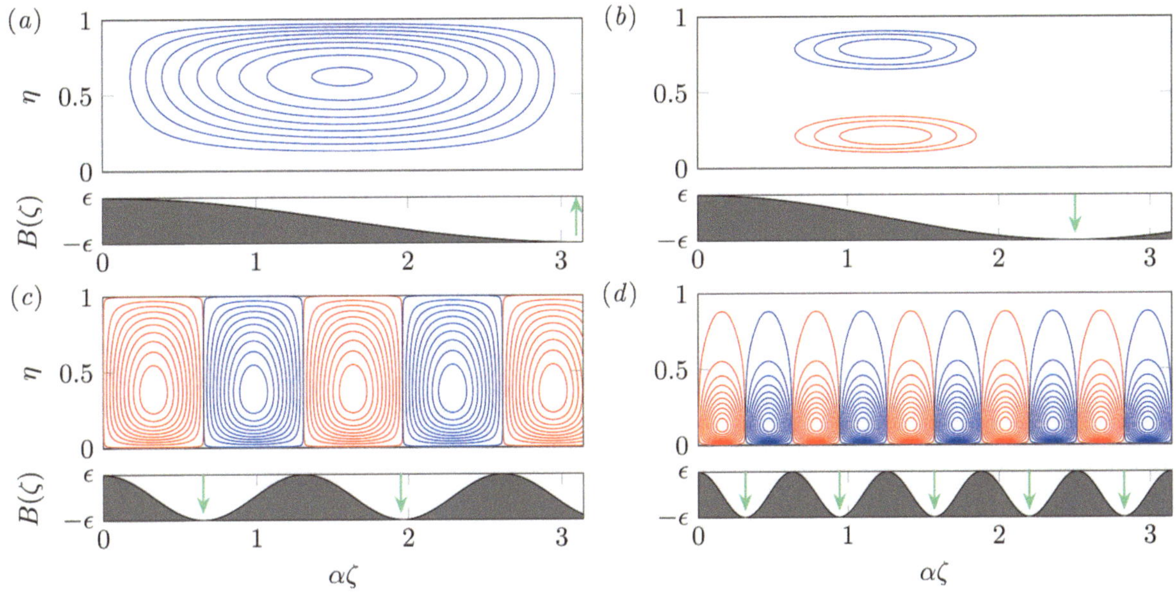

Figure 5.4: Secondary flow organisation in terms of the secondary flow streamfunction ψ_{vw} for $Fr = 2$, $D = 0.01$, $s = 2.65$ and different wavenumbers: (a) $\alpha = 1$ ($\lambda = 6.28$); (b) $\alpha = 1.2548$ ($\lambda = 5.01$); (c) $\alpha = 4.8$ ($\lambda = 1.30$); (d) $\alpha = 10$ ($\lambda = 0.63$). The bed contours B for the respective wavenumbers are shown below each panel, with green arrows in the trough regions indicating whether fluid is moving upward or downward in the respective area. Clockwise (counterclockwise) secondary flow rotation is indicated by red (blue) isolines. The shown contours indicate 20 equally spaced subintervals of the interval $[-\max_{y,z}|\psi_{vw}|, \max_{y,z}|\psi_{vw}|]$.

Figure 5.4 shows the perturbed secondary flow field for selected wavenumbers $\alpha \in \{1, 1.2548, 4.8, 10\}$ or, equivalently, for wavelengths $\lambda \in \{6.28, 5.01, 1.30, 0.63\}$. The secondary fluid motion is therein indicated by isocontours of the cross-plane streamfunction

$$\nabla_{\perp}\psi_{vw}(\eta,\zeta) = (\partial_\eta \psi_{vw}, \partial_\zeta \psi_{vw})^T = (-(-w_1 \alpha \sin(\alpha\zeta)), (v_1 \cos(\alpha\zeta)))^T, \tag{5.37}$$

that is obtained by integration of the velocity perturbations.

In agreement with the observation that the lateral bed shear stress changes its sign around a wavenumber of $\alpha \approx 1$ in figure 5.2(a), the large secondary vortex with a wavelength $\lambda = 2\pi \approx 6.28$ shown in figure 5.4(a) induces a 'damping' motion near the bed. That is, fluid is moving towards the free surface over the initial trough and *vice versa* over the initial crest of the bed profile such that the fluid motion counteracts a further growth of the bed perturbation. The opposite is true for the flows with perturbation wavenumbers $\alpha \gtrsim 1$ shown in figures 5.4(b-d), for which the secondary currents induce a lateral bed shear stress that is able to enhance sediment transport from the troughs towards the crests and thus to support the growth of the bed undulations. In this context, figure 5.4(b) is of particular interest as it provides the secondary flow patterns for the wavenumber $\alpha = 1.2548$ at which the secondary flow kinetic energy is minimal. At this specific wavelength, the system is seen to be in a transient phase between two states in which the channel depth accommodates pairs of stacked localised counter-rotating secondary vortices that are of identical size and amplitude and their centres feature essentially the same distance to the closest domain boundary. The abrupt change of η_{max} at this wavenumber that was detected in figure 5.3(b) marks the transition from the dominance of the upper 'damping' vortex to the dominance of the lower vortex that supports the growth of the sediment bed perturbation. Note that this specific secondary flow organisation at $\alpha = 1.2548$ is of no particular relevance for the actual bed formation as the amplitudes of both the secondary flow kinetic energy and the lateral bed shear

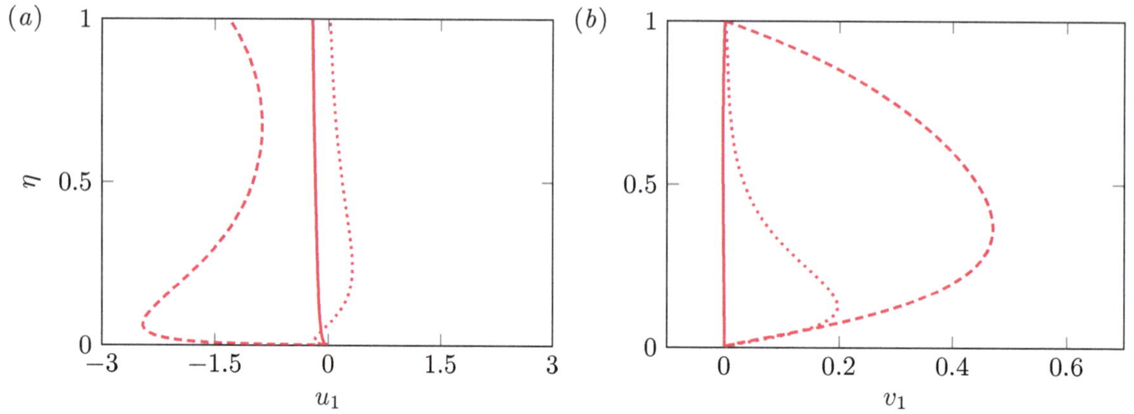

Figure 5.5: Wall-normal perturbation profiles of (*a*) the streamwise velocity, $u_1(\eta)$, and (*b*) the wall-normal velocity, $v_1(\eta)$, for $Fr = 2$, $D = 0.01$ and $s = 2.65$. Different lines indicate different wavenumbers: $\alpha = 1$ ($\lambda = 6.28$) (—), $\alpha = 4.8$ ($\lambda = 1.30$) (---) and $\alpha = 10$ ($\lambda = 0.63$) (· · ·).

stress are of negligible size at this wavenumber. But the shown data reveals that the system does not suddenly change its structure. Rather, the system is seen to gradually transform from a bed stabilising to a bed destabilising state.

With further growing wavenumber, the system reaches in figure 5.4(*c*) with a wavenumber of $\alpha = 4.8$ its most amplified state. The secondary currents become 'space-filling', that is, the secondary flow kinetic energy is of non-negligible size everywhere in the channel cross-section. In agreement with figure 5.3(*b*), the regions of intense secondary motion become more and more confined to the near bed region in the large-wavenumber regime $\alpha \to \infty$, while the secondary flow intensity in the outer flow decreases successively (cf. figure 5.4(*d*)).

The corresponding individual wall-normal profiles of the streamwise and wall-normal velocity perturbations are shown in figure 5.5. As expected, figure 5.5(*a*) shows that the total streamwise velocity U is generally reduced over the bedform crests compared to the laterally flat channel if depth-spanning secondary rollers such as those for $\alpha = 4.8$ lead to a mean upward motion of low-momentum fluid over the crest regions (cf. figure 5.5(*b*)). For the higher wavenumber $\alpha = 10$, on the other hand, the mean upflow over the bedform crests is markedly weaker and more confined to the near-bed region as was observed above. The streamwise velocity U, in turn, is reduced only in the vicinity of the bedform crest, but increases over the rest of the channel depth compared to the base flow velocity profile U_0. In contrast to the high wavenumber regime, the effect of bed undulations on the base flow at lower wavenumber $\alpha = 1$ and thus larger wavelength $\lambda = 2\pi$ is negligible compared to the remaining two cases, where the perturbation amplitudes are at least one order of magnitude higher.

These model predictions agree quite well with a series of recent experimental and numerical studies on the structure of secondary currents over non-homogeneous bottom surfaces (cf. section 3.5 for a more comprehensive overview of the current state of the art). A main outcome of these studies is that secondary currents over different types of bottom-wall inhomogeneities are most intense and fill the entire flow cross-section if the mean lateral spacing of these inhomogeneities is in a range of approximately (1-1.5) times the outer fluid length scale (Chung et al., 2018; Wangsawijaya et al., 2020; Zampiron et al., 2020a) – a range that is strikingly similar to the most amplified wavelength $\lambda = 1.3$ determined in the current analysis based on the model of Colombini (1993). In the same studies, it is moreover reported that while reducing the transverse spacing of the inhomogeneities, the secondary currents reduce in size and become more and more confined to the vicinity of the bottom wall. The outer flow, on the other hand, becomes less affected by the bottom wall heterogeneities as the wavenumber tends to infinity, which is not unexpected as for sufficiently small wavelengths (compared to the mean size of the large-scale turbulent structures) the flow 'sees' the bottom perturbations only as a laterally homogeneous bottom roughness. Both observations are in accordance with the results obtained above with the here considered linear model.

In a recent study, Zampino et al. (2022) investigated the steady flow in a closed channel at finite Reynolds numbers over individual sinusoidal and rectangular roughness elements with variable spacing by means of a linearised model that is structurally quite similar to the one discussed in this work. The authors adopt therein a more recent non-linear eddy-viscosity turbulence closure proposed by Spalart (2000) and include a separate transport equation for the eddy-viscosity which is coupled to the convective terms of the RANS equations. In complete agreement with the results obtained by the model of Colombini (1993), the linear modelling approach of Zampino et al. (2022) predicts most amplified secondary currents as a response to a sinusoidal bottom perturbation at wavelengths of $\lambda \approx 1.5$, whereas a reversal of the secondary currents' sense of rotation was found to occur at $\lambda \approx 6$ (compared to $\lambda \approx 6.28 = 2\pi$ in the current case).

5.4 Modified model for stationary viscous open channel flow

In the following, we are going to analyse the linearised system that has to be solved in the model of Colombini (1993) in more detail. The linear system in equation (5.28) that is in the following repeated for convenience, viz.

$$Aq = h_1\boldsymbol{B} + \boldsymbol{R},$$

basically represents a linear operator A acting on a velocity and pressure field expressed by $\boldsymbol{q}$ to map it onto a given right hand side. To be more precise, A is the linearised Navier-Stokes operator reduced according to the above listed assumptions including stationarity, streamwise independence of all quantities as well as an infinitely large Reynolds number that allows us to neglect viscous effects. The linearised Navier-Stokes operator is as such independent of the bed perturbations, whereas the right hand side $h_1\boldsymbol{B} + \boldsymbol{R}$ comprises merely source terms that originate in the lateral sinusoidal bottom undulations. In that sense, the sinusoidal perturbation of the bottom wall can be interpreted as an exterior stationary forcing that is exerted on the flow system under consideration (Zampino et al., 2022). If the bottom domain boundary remains flat, the external forcing disappears and the RHS of the system vanishes such that the 'trivial solution' of an unperturbed velocity and pressure field fulfils the above equation.

The following discussion of the linearised system and the modifications we apply to it requires some knowledge of the development in the field of hydrodynamic stability in the past decades. In the subsequent short excursus, we make the reader familiar with some of the main findings in this field, which is surely far from being complete. For a much more detailed overview of the topic, the interested reader is referred to the textbooks of Drazin and Reid (1981) and Schmid and Henningson (2001) and the more recent review articles of Schmid (2007), McKeon (2017) and Jiménez (2018), as well as that of Kerswell (2018) who discusses stability in a broader sense towards a fully non-linear dynamical systems description of the problem. As our main interest concerns the flow in an open channel, we restrict ourselves in the following to the discussion of the stability of plane Poiseuille flows, for which it was first assumed that the turbulent mean velocity profile could be marginally stable w.r.t. to a linear instability mechanism (Malkus, 1956). A decade later, however, it was shown by Reynolds and Tiederman (1967) that this is in fact not the case, but that the turbulent mean velocity profile is stable w.r.t. to infinitesimal perturbations in that all such perturbations eventually decay as $t \to \infty$, rather than following an exponential growth. It then took another two decades until the instability of Poiseuille flows received again more attention as in the course of a number of studies the non-normality of the linearised Navier-Stokes/the Orr-Sommerfeld operator was taken into account (Butler and Farrell, 1992, 1993). Non-normality in this context refers to the fact that not all eigenfunctions of the linear Orr-Sommerfeld operator are orthogonal to each other (Reddy et al., 1993), and so a normal-mode analysis that investigates solely the spectrum of the linear operator ignores the possibility of an interaction of individual modes in finite time intervals (Schmid, 2007). Indeed, it could be shown that for 'optimal' initial conditions of the flow field significant energy amplifications can be achieved over a finite time horizon in a process that is typically termed 'transient growth' (Jiménez, 2018). Butler and Farrell (1993) considered the classical Orr-Sommerfeld problem (Drazin and Reid, 1981) and concluded that such an optimal initial state consists of a set of infinitely long streamwise rolls that trigger the formation of streamwise-elongated streamwise velocity streaks. While an outer-scaling peak at a lateral wavelength $\lambda_z \approx 3$ related to large-scale velocity streaks was detected irrespective of the investigated time window, an optimal state that led to the evolution of inner-scaling buffer layer streaks with $\lambda_z^+ \approx 100$ was clearly seen only if the time window under consideration was restricted to times of the order of the buffer layer streaks' lifetime. Del Álamo and Jiménez (2006) and Pujals et al. (2009) later considered the same problem, but instead of restricting the time period during which the system is allowed to transiently grow, these authors supplemented the molecular fluid viscosity ν by a height-dependent turbulent viscosity ν_t as in classical linear eddy-viscosity models. In agreement with the results of Butler and Farrell (1993), they observed two peaks in the energy amplification, one local maximum at $\lambda_z^+ \approx 100$ that is indicative of the buffer-layer vortices and streaks and a second, global one at about $\lambda_z^+ \approx 3$ that refers to the large-scale streaks.

So far, we have implicitly assumed that the linearised Navier-Stokes system is free of external forces. If this is not the case and the governing equations are subject to a possibly time-dependent external force field, then both the initial condition and the specific force field affect the solution at a given time t (Schmid, 2007). In this vein, Hwang and Cossu (2010b) scrutinised how the non-normal linearised Orr-Sommerfeld operator reacts to external harmonic (deterministic) and stochastic forcing of small amplitude and sought for the optimal external forcing to achieve a maximum growth of energy in the perturbed system, rather than an optimal initial condition. The resulting forcing profiles possess a striking similarity with the optimal initial conditions detected by Del Álamo and Jiménez (2006) and Pujals et al. (2009). In particular, for both inner- and outer-scaling streamwise streaks the optimal 'source' forcing again takes the form of counterrotating streamwise rolls of appropriate size. As in the previous works on transient growth of 'optimal' initial conditions, the strongest energy amplification was observed for infinitely long modes with streamwise wavenumber $\alpha_x = 0$, while the optimal temporal forcing frequency was found to be zero, that is, the most 'critical' forcing is a permanent one.

In the following, we will discuss the properties of the here considered model of Colombini (1993) in comparison with the afore-mentioned time-dependent models. By assumption, all considered modes in the model of Colombini (1993) are of infinite streamwise wavelength ($\alpha_x = 0$), which agrees with the above discussed 'criticality' of these modes. In contrast to the above studies in which the temporal evolution of the flow field subject to either a particular initial state or an external forcing was investigated, the linear operator considered in the current model is stationary, that is, the flow reacts immediately to a change of the forcing. In order to highlight the similarities of the model of Colombini (1993) with those typically adopted in the stability analysis of plane channel flows, we adapt the original problem formulation in several aspects: Instead of a system that is perturbed by a laterally varying bottom wall, we consider in the following a plane open channel flow that is forced by a laterally varying external force field, in analogy to Hwang and Cossu (2010b). On the one hand, this will allow us to show in the remainder of this section that the choice of the most amplified wavelength in the original model is indeed exclusively determined by the fluid equations, rather than by the coupling with the bed. On the other hand, it greatly simplifies the computations as we can remain in the standard orthogonal Cartesian coordinate system with basis $(\boldsymbol{e}_y, \boldsymbol{e}_z)$, avoiding the non-orthogonal transformation to the coordinate system spanned by $\boldsymbol{e}_\eta$ and $\boldsymbol{e}_\zeta$. This is particularly attractive as we relax in the following the original assumption of an infinitely large Reynolds number by reintroducing the viscous terms in the governing equations and we can thus avoid to transform the Laplace operator to the non-orthogonal coordinate system. Note that it is absolutely possible to add viscous effects also to the original model, but it would require a significantly larger amount of algebra without providing any further insights.

For the new forced viscous channel flow, the set of non-dimensional governing equations then reads

$$V\frac{\partial U}{\partial y} + W\frac{\partial U}{\partial z} = \frac{S_b}{Fr^2} + \frac{\partial \tau_{xy}}{\partial y} + \frac{\partial \tau_{xz}}{\partial z} + \frac{1}{Re_b}\left(\frac{\partial^2 U}{\partial y^2} + \frac{\partial^2 U}{\partial z^2}\right) + F_x \tag{5.38a}$$

$$V\frac{\partial V}{\partial y} + W\frac{\partial V}{\partial z} = -\frac{\partial P}{\partial y} + \frac{\partial \tau_{yy}}{\partial y} + \frac{\partial \tau_{yz}}{\partial z} + \frac{1}{Re_b}\left(\frac{\partial^2 V}{\partial y^2} + \frac{\partial^2 V}{\partial z^2}\right) + F_y \tag{5.38b}$$

$$V\frac{\partial W}{\partial y} + W\frac{\partial W}{\partial z} = -\frac{\partial P}{\partial z} + \frac{\partial \tau_{yz}}{\partial y} + \frac{\partial \tau_{zz}}{\partial z} + \frac{1}{Re_b}\left(\frac{\partial^2 W}{\partial y^2} + \frac{\partial^2 W}{\partial z^2}\right) + F_z \tag{5.38c}$$

$$\frac{\partial V}{\partial y} + \frac{\partial W}{\partial z} = 0. \tag{5.38d}$$

where $Re_b = \bar{U}_{b,0}\bar{H}_0/\bar{\nu}$ and $Re_\tau = Re_b\sqrt{C_0} = \bar{u}_{\tau,0}\bar{H}_0/\bar{\nu}$ are the bulk and friction Reynolds numbers in the base flow and $\boldsymbol{F} = (F_x, F_y, F_z)^T$ is an arbitrary external force field. To allow for a comparison with the remaining models, we consider in the following the flow in a smooth open channel, for which (Pope, 2000)

$$Re_\tau \approx 0.166 Re_b^{0.88}. \tag{5.39}$$

Thus, choosing a fixed value for either of the two Reynolds numbers Re_τ or Re_b uniquely determines the values of both the remaining Reynolds number as well as the friction coefficient C_0.

Removing the bottom curvature results in $B = 0$ and $H = H_0 = 1$, respectively. The coordinate transformation consequently reduces to the identity transformation, such that no additional terms arise from this side. As the coupling with the sediment bed continuity equation is for now not of main interest and the equations do not contain any dependence on the time scale T, we can simplify our ansatz functions by setting $\sigma = 0$ and thus $\exp(\sigma T) = 1$. The ansatz function for the streamwise velocity then reads

$$U(y) = U_0(y) + \epsilon\, u_1(y)\cos(\alpha\zeta). \tag{5.40}$$

In analogy, appropriate ansatz functions of the laterally oscillating external force field $\boldsymbol{F} = (F_x, F_y, F_z)^T$ with wall-normal perturbation profiles $f_{x_i,1}(y)$ are introduced as

$$\begin{cases} F_x(y) &= 0 + \epsilon\, f_{x,1}(y)\cos(\alpha\zeta) \\ F_y(y) &= 0 + \epsilon\, f_{y,1}(y)\cos(\alpha\zeta) \\ F_z(y) &= 0 - \epsilon\, \alpha f_{z,1}(y)\sin(\alpha\zeta) \end{cases} . \tag{5.41}$$

Inserting the modified ansatz functions into the governing equations, the newly appearing linear diffusive terms contribute the following terms

$$\begin{aligned} \frac{1}{Re_b}\left[\partial_{yy}U + \partial_{zz}U\right] &= \frac{1}{Re_b}\left[U_0''(y) + (u_1''(y) - \alpha^2 u_1(y))\cos(\alpha\zeta)\right] \\ \frac{1}{Re_b}\left[\partial_{yy}V + \partial_{zz}V\right] &= \frac{1}{Re_b}\left[0 + (v_1''(y) - \alpha^2 v_1(y))\cos(\alpha\zeta)\right] \\ \frac{1}{Re_b}\left[\partial_{yy}W + \partial_{zz}W\right] &= \frac{1}{Re_b}\left[0 - \alpha(w_1''(y) - \alpha^2 w_1(y))\sin(\alpha\zeta)\right] \end{aligned} . \tag{5.42}$$

Only in the streamwise momentum equation, viscous stresses appear in the base flow equations, such that relation (5.24a) extends to

$$-\frac{S_b}{Fr^2} = \nu_{t0}'U_0' + \nu_{t0}U_0'' + \frac{1}{Re_b}U_0'' = \nu_{t0}'U_0' + \left(\nu_{t0} + \frac{1}{Re_b}\right)U_0'' = \left[\left(\nu_{t0} + \frac{1}{Re_b}\right)U_0'\right]' . \tag{5.43}$$

While in equation (5.24) the driving force term S_b/Fr^2 was balanced by turbulent stresses only, the total shear stress on the RHS of equation (5.43) now comprises viscous and turbulent contributions. The turbulent viscosity $\nu_{t0}(y)$ in the streamwise momentum balance is accordingly replaced by the non-dimensional effective viscosity $\nu_{eff} = (\nu_{t0}(y) + 1/Re_b)$ comprising both the constant molecular and the height-dependent turbulent viscosity. From now on, $\bullet^+$ denotes normalisation in inner units using $\tilde{\nu} = \tilde{\nu}_f$ and $\tilde{u}_{\tau,0}$ as characteristic scales. External forcing terms as well as the remaining contributions of the viscous terms are of order $\mathcal{O}(\epsilon^1)$ and thus appear solely in the perturbed system. The viscous terms contribute to the 'main diagonal' of the operator A, that are, the terms a_{11}, a_{22} and a_{33} in (A.5), respectively. At last, the boundary conditions at the free surface are adapted to the changed conditions (cf. appendix A.3).

In regard of the modification of the flow configuration to a smooth open channel flow at finite Reynolds number, we replace the fully-logarithmic base velocity profile and the associated profiles of the turbulent viscosity ν_{t0} and mixing length l_0 by an analytic expression that takes into account the variation of the Reynolds number. A popular choice is the analytic expression for the turbulent viscosity in closed channels originally proposed by Cess (1958) and later adopted by Reynolds and Tiederman (1967), which reads

$$\nu_{t0}^+(y) = \frac{1}{2}\left\{1 + \left[\frac{\kappa^2 Re_\tau^2}{9}\left(1 - y_s^2\right)^2\left(1 + 2y_s^2\right)^2\left(1 - \exp\left(\left(|y_s| - 1\right)\frac{Re_\tau}{A_{bot}}\right)\right)^2\right]\right\}^{1/2} - \frac{1}{2}, \tag{5.44}$$

where a shifted wall-normal coordinate $y_s = y - 1$ is used such that $y_s \in [-1, 0]$. The van Driest damping coefficient and the von Kármán constant are chosen as $A_{bot} = 25.4$ and $\kappa = 0.426$, respectively, as proposed by Del Álamo and Jiménez (2006). For open channel flow, however, expression (5.44) is not suitable in its original form as ν_{t0} attains finite values at $y_s = 0$ (which represents the centreline

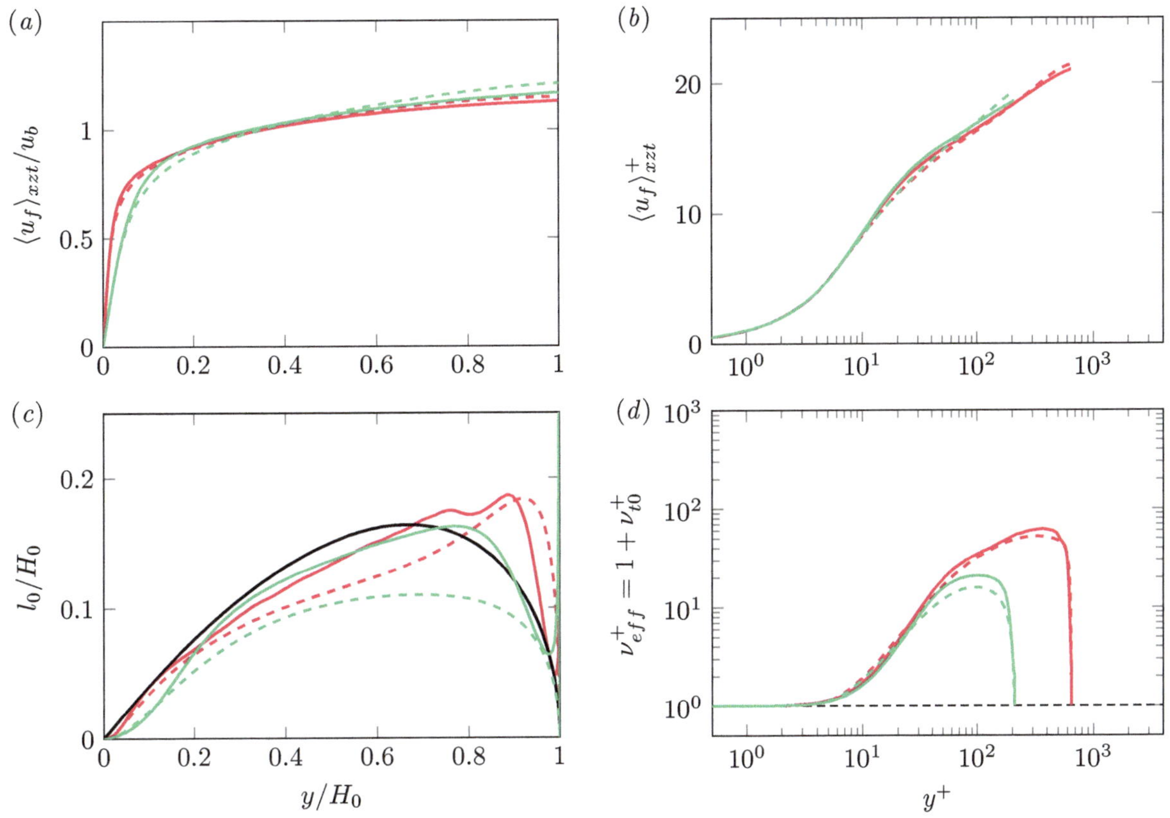

Figure 5.6: Wall-normal profiles of the mean velocity, mixing length and effective viscosity from smooth-wall open channel flow simulations as reference data for the linear stability analysis. *(a,b)* Mean velocity profile *(a)* in outer scaling, $\langle u_f \rangle_{xzt}/u_b$, and *(b)* in inner scaling, $\langle u_f \rangle^+_{xzt}$. *(c)* Mixing length normalised with the mean fluid height, l_0/H_0. The black line indicates the relation $l_0 = \kappa y\sqrt{1-y}$ used in De Vriend (1977). *(d)* Effective viscosity normalised with the molecular viscosity $\nu^+_{eff} = (\nu + \nu_{t0})/\nu$. Solid lines mark data from cases $CL250^{H16}_{smooth}$ (—, $Re_\tau = 210$) and $CM650^{H3}_{smooth}$ (—, $Re_\tau = 649$) that will be analysed in detail in the subsequent chapter 6. The respective physical and numerical parameters can be found in tables 6.1 and 6.2, respectively. Dashed lines indicate the corresponding profiles obtained by the modified analytic expressions of Cess (1958) and Reynolds and Tiederman (1967) for the turbulent viscosity following expression (5.46).

in a corresponding closed channel flow), while in open channels ν_{t0} should tend to zero at the free surface (Nezu and Rodi, 1986), as can be also seen in figure 5.6. Therein, the plane-averaged wall-normal profiles of the mean velocity $\langle u_f \rangle_{xzt}$, the turbulent viscosity ν_{t0} and the mixing length l_0 are presented for two direct numerical simulations of smooth wall open channel flow at Reynolds numbers $Re_\tau \in \{210, 650\}$, which will be investigated in detail in the subsequent chapter 6. The turbulence parameters ν_{t0} and l_0 have been recomputed from the simulation statistics as (Pope, 2000; Pirozzoli, 2014)

$$\nu_{t0} = \frac{-\langle u'_f v'_f \rangle_{xzt}}{\mathrm{d}\langle u_f \rangle_{xzt}/\mathrm{d}y}, \quad l_0 = \frac{\left(-\langle u'_f v'_f \rangle_{xzt}\right)^{1/2}}{|\mathrm{d}\langle u_f \rangle_{xzt}/\mathrm{d}y|}. \tag{5.45}$$

Note that these definitions hold everywhere but at the free surface itself, where $|\mathrm{d}\langle u_f \rangle_{xzt}/\mathrm{d}y|$ vanishes due to the applied free-slip boundary condition in the simulations. Whilst ν_{t0} tends indeed to zero when approaching the free surface, l_0 is seen to increase strongly in the direct vicinity of the free surface. This, in turn, suggests that the mean shear $\mathrm{d}\langle u_f \rangle_{xzt}/\mathrm{d}y$ decreases to zero at a slower rate than the turbulent stresses $-\langle u'_f v'_f \rangle_{xzt}$, but at a faster rate than their square root. Note that from a physical point of view, l_0 should nevertheless tend to zero when approaching the free surface, as the turbulent structures themselves are restricted in their size in the vicinity of the upper domain boundary such that

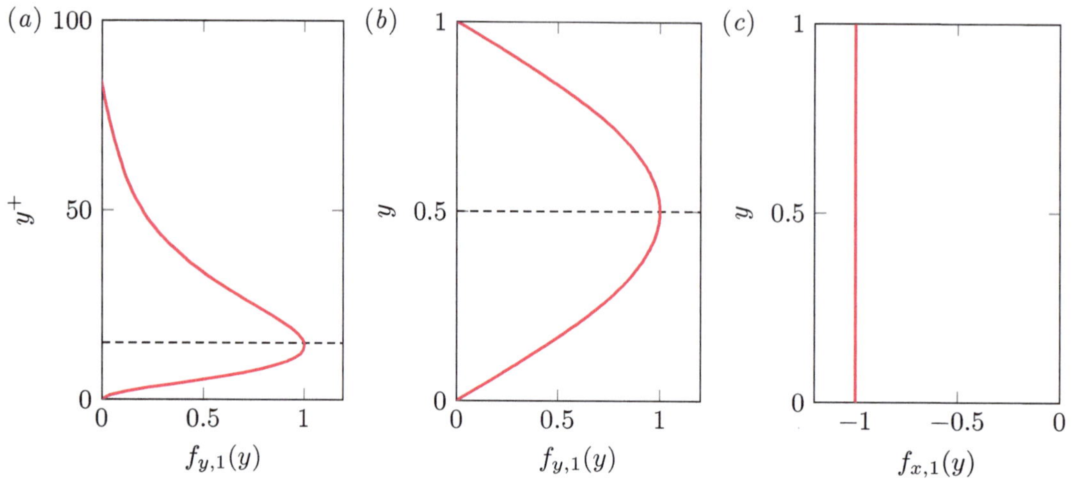

Figure 5.7: Wall-normal profiles of the external stationary forcing schemes in the modified linearised model of plane smooth open channel flow. (*a*) Near-optimal cross-sectional forcing for buffer layer streaks $\boldsymbol{F} = (0, F_y, F_z)^T$. Shown is the profile of the wall-normal forcing component, $f_{y,1}(y)$, which attains its peak value at approximately $y^+ = 15$ (dashed line). (*b*) Cross-sectional forcing in the outer layer $\boldsymbol{F} = (0, F_y, F_z)^T$, shown is the profile of the wall-normal force component with $f_{y,1} = \sin(\pi y)$. (*c*) Streamwise constant forcing $\boldsymbol{F} = (F_x, 0, 0)^T$, for which $f_{x,1}(y) = -1$ is presented.

a reasonable characteristic length scale should not grow without bounds in this region. We therefore underline that the definition (5.45) for l_0, even though mathematically defined everywhere but at the singularity $y = 1$, is appropriate only in some distance to the free surface.

For the sake of comparison, figure 5.6 provides the same quantities computed based on a slightly modified version of the Cess-Reynolds-Tiederman formula (5.44), viz.

$$\nu_{t0}^{+}(y) = \frac{1}{2}\left\{1 + \left[\frac{\kappa^2 Re_\tau^2}{9}\left(1-y_s^2\right)^2\left(1+2y_s^2\right)^2\left(1-\exp\left(\left(|y_s|-1\right)\frac{Re_\tau}{A_{bot}}\right)\right)^2 \left(1-\exp\left(y_s\frac{Re_\tau}{A_{top}}\right)\right)^2\right]^{1/2}\right\} - \frac{1}{2}. \tag{5.46}$$

Comparing relations (5.44) and (5.46) with each other, it is seen that we have added an additional van Driest damping function that ensures that ν_{t0} tends to zero at the free surface. For this purpose, a second damping coefficient $A_{top} = 40$ is chosen that is seen to reasonably well describe the decay of ν_{t0} when approaching the free surface, at least for the higher Reynolds number simulation. As we are interested in higher Reynolds numbers in the remainder of this study, a fitting of the parameter A_{top} to the simulation at higher Reynolds number appears reasonable. For the sake of completeness, we have performed the subsequent analysis also with the DNS data as base flow profiles, which lead to qualitatively similar results as those obtained with the analytic expression in (5.46). Surely, more elaborate analytic expressions than relation (5.46) are conceivable for ν_{t0} that will match the simulation data even better, but for a sound investigation similar to the study of Pirozzoli (2014) in closed channel flows, our current database would have to be extended to cover a wider range of Reynolds numbers. In the context of hydraulic turbulence models, however, such a study could be indeed of high interest and could be performed in a future project.

In the remainder of this section, we analyse how the large-scale secondary flow cells and their intensity depends on the forcing. To this end, three different force fields $\boldsymbol{F} = (F_x, F_y, F_z)^T$ are tested on their impact on the flow organisation. The first two force fields are of the form $\boldsymbol{F} = (0, F_y, F_z)^T$ and represent infinitely long streamwise vortex rolls of different size that enforce accordingly-arranged high- and low-speed streaks. Figure 5.7(*a*) shows a wall-normal forcing profile $f_{y,1}$ that has been constructed to approximate the optimal forcing for the buffer layer streaks found by Hwang and Cossu (2010b). The peak forcing is attained roughly 15 wall units above the bottom wall, while it rapidly tends to zero outside the buffer layer. Figure 5.7(*b*), on the other hand, represents a large-scale forcing of the form $f_{y,1} = \sin(\pi y)$ that is of finite size over the entire channel depth and peaks in the channel half height. In both cases, we are interested in solenoidal force fields, such that $f_{z,1}$ can be recomputed from the respective force profile of the wall-normal component as $f'_{y,1}(y) = \alpha^2 f_{z,1}(y)$, in analogy to the continuity equation for the streamwise independent solenoidal velocity field. In contrast to these vortex-like forcing approaches, the last considered force field $\boldsymbol{F} = (F_x, 0, 0)^T$ is unidirectional in the streamwise direction and its profile is constant over the entire channel depth with amplitude $f_{x,1} = -1$ (cf. figure 5.7(*c*)). That is, a global low-speed region is forced at position $z = 0$.

Note that comparable forcing schemes in form of infinitely long streamwise rolls are frequently used when seeking for invariant solutions of the full non-linear Navier-Stokes operator. Waleffe (2003) proposed a homotopy method in analogy to the buffer-layer self-sustaining process (Hamilton et al., 1995), in which infinitely long streamwise rolls are added to a base flow under the action of a continuous forcing that prevents them from decaying due to viscosity. The forced rolls trigger the formation of a streamwise velocity streak, that eventually becomes unstable and feeds back into the vortex rolls. In the actual homotopy step, the external forcing is then successively reduced until the original unforced system is recovered. While originally used by Waleffe (2003) in plane Couette and Poiseuille flow, it was later successfully applied also in other flow configurations, including pipe (Kerswell, 2005) and square duct flows (Uhlmann et al., 2010).

Figure 5.8 shows the maximum amplitude of the wall-normal velocity $v_1(y)$ in inner and outer scaling as a function of the lateral wavelength λ for all three forcing types, evaluated for Reynolds numbers $Re_\tau = \{500, 1000, 2000\}$. Notably, the maximum wall-normal velocity amplitude attains for all three forcing schemes a global peak at essentially the same wavelength $\lambda = 1.3$ ($\alpha = 4.8$) as in the original model of Colombini (1993), irrespective of the fact that we use here entirely different forcing schemes compared to that induced by the sinusoidal bed deformation. It is moreover astonishing that the peak persists even if the forcing is entirely restricted to the buffer layer (cf. figure 5.8(*b*)) and a relevant forcing amplitude is only attained in the lowest 100 wall units. In the highest Reynolds number case $Re_\tau = 2000$, for instance, the forcing peak is attained at a wall-normal position $y = 0.008$, but the forcing still suffices to trigger the formation of depth-spanning large-scale secondary currents when applied at a wavelength $\lambda = 1.3$. The peak value itself is seen to be essentially independent of the Reynolds number for each individual forcing when scaled in terms of the bulk velocity, whereas the amplitude varies between the different forcing schemes. This is not unexpected since, even though the maximum values of all three forcing profiles are of same size, the total force acting on the domain that is obtained by integration of the force density $\boldsymbol{F}$ over the cross-plane markedly varies. Note that the maximum velocity amplitude is partly two orders of magnitude larger than that of the associated forcing, which agrees with a comment of Hwang and Cossu (2010b) that the ratio between energy output and input in such forced systems is typically quite high.

A clear inner-scaling peak of the maximum wall-normal velocity amplitude is seen only for the quasi-optimal buffer layer forcing at a lateral wavelength $\lambda^+ = 81$ in figure 5.8(*a*), although also in the remaining cases there is partly a change of slope visible around this wavelength. In contrast to the global

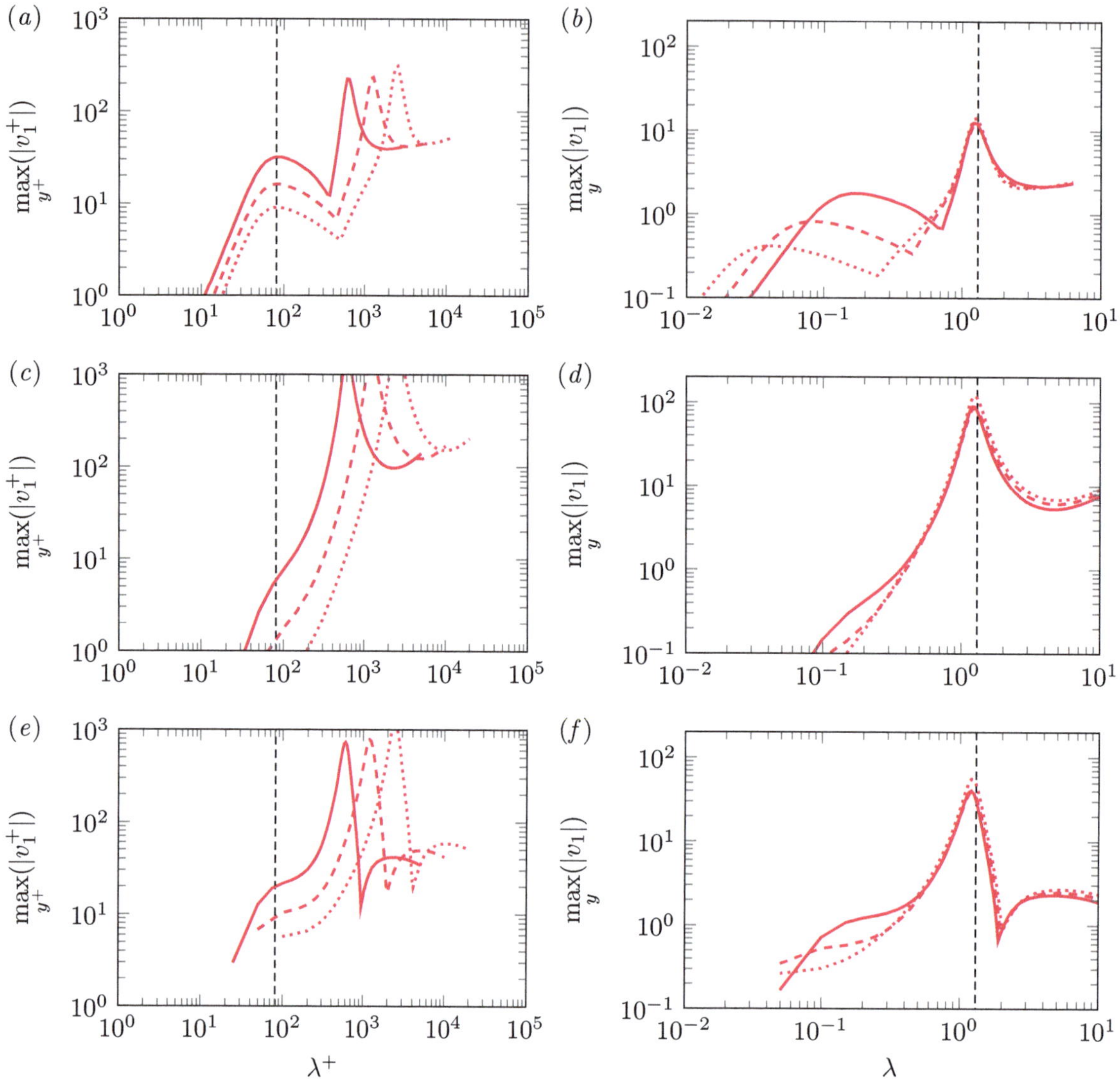

Figure 5.8: Maximum of the wall-normal velocity amplitude $|v_1(y)|$ as a function of the wavelength for different external stationary forcing schemes and Reynolds numbers in the modified linearised model of plane smooth open channel flow: *(a,b)* Near-optimal cross-sectional forcing for buffer layer streaks; *(c,d)* cross-sectional forcing in the outer layer; *(e,f)* streamwise constant forcing. Data is shown in *(a,c,e)* inner and *(b,d,f)* outer scaling with vertical dashed lines indicating $\lambda^+ = 81$ ($\alpha^+ = \alpha\nu/u_\tau = 0.078$) and $\lambda = 1.3$ ($\alpha = 4.8$), respectively. $Re_\tau = 500$ (—), $Re_\tau = 1000$ (- - -) and $Re_\tau = 2000$ (· · ·).

outer peak, the inner-scaling peak is seen to reduce in amplitude with increasing Reynolds number, irrespective of whether it is scaled with the outer or inner velocity scales $\bar{U}_{b,0}$ or $\bar{u}_{\tau,0}$, respectively. Note that at first glance, the appearance of an inner-scaling peak at $\lambda^+ = 81$ seems less surprising as Hwang and Cossu (2010b) extracted the optimal homogeneous force profile on which our buffer-layer forcing is based for essentially this wavelength. However, it should be underlined that it was not *a priori* clear that the current stationary linearised operator reacts in the same way as the full time-dependent Orr-Sommerfeld operator of Hwang and Cossu (2010b) to an outer forcing and whether the former would be able to enforce the characteristic streaky streamwise velocity structures for the appropriate wavelength.

Figure 5.9 shows the perturbed streamwise velocity field together with the perturbed cross-flow $(V, W)^T$ and the cross-flow components of the force field $(F_y, F_z)^T$ at the inner and outer peak wavelengths

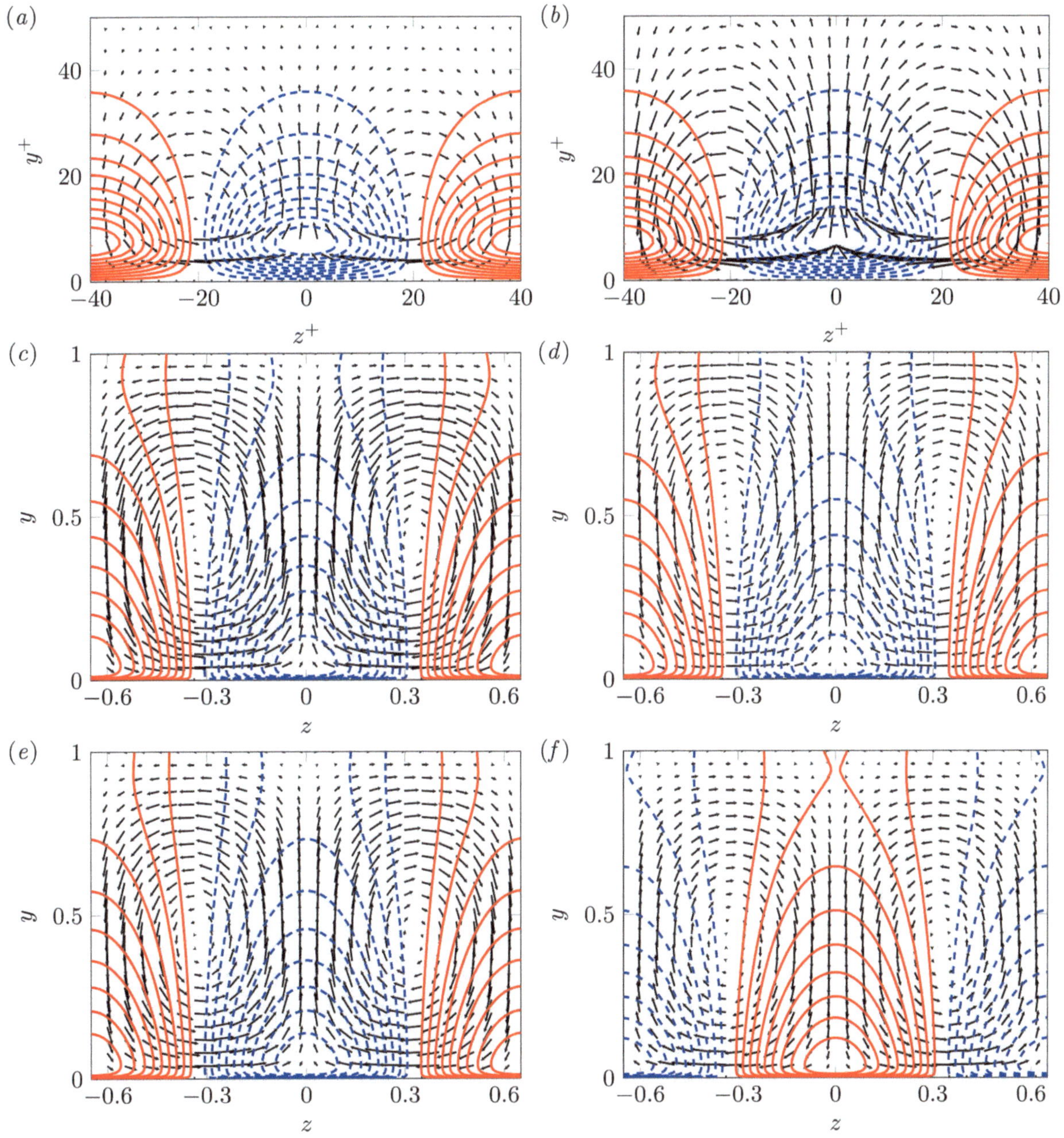

Figure 5.9: Cross-section of the perturbed velocity field driven by different external stationary forcing schemes in the modified linearised model of plane smooth open channel flow at $Re_\tau = 2000$. The vector field visualisations in (*a,c,e,f*) show intensity and direction of the cross-stream flow field $(V, W)^T$, whereas those in (*b,d*) represent the stationary cross-stream body force field $(F_y, F_z)^T$. (*a,b*) Near-optimal cross-sectional forcing for buffer layer streaks ($\lambda^+ \approx 81$); (*c,d*) cross-sectional forcing in the outer layer ($\lambda = 1.3$); (*e*) Streamwise constant forcing ($\lambda = 1.3$); (*f*) Same buffer-layer forcing as in (*a,b*), but this time for a given outer wavelength $\lambda = 1.3$. Red solid and blue dashed isocontours represent positive and negative values of the streamwise velocity perturbation $u_1(y)\cos(\alpha\zeta)$, respectively, in a range $[-\max_{y,z}|u_1(y)\cos(\alpha\zeta)|, \max_{y,z}|u_1(y)\cos(\alpha\zeta)|]$ with an increment of 1/20 of the interval.

$\lambda^+ = 81$ and $\lambda = 1.3$, respectively. In figures 5.9(*a*,*b*), the buffer-layer forcing as well as the resulting flow field are presented for a given wavelength $\lambda^+ = 81$. Indeed, the roll-like forcing induces streamwise-elongated vortex rolls that, in turn, enforce the streamwise velocity field to organise in laterally alternating high- and low-speed streaks.

The flow patterns match those obtained by Hwang and Cossu (2010b) for a similar forcing and those formed by transient growth from an 'optimal' initial condition (Del Álamo and Jiménez, 2006) quite well, further supporting our earlier observation that the linearised operator in the current stationary and streamwise-independent model captures nonetheless the basic linear physical processes of vortex-streak interactions (Hamilton et al., 1995; Waleffe, 1997; Schoppa and Hussain, 2002; Kawahara et al., 2003; Jiménez, 2018).

In a conceptually similar way, we observe for a given wavelength $\lambda = 1.3$ the formation of large-scale infinitely long vortex rolls which are equivalent to the large-sale secondary currents in the original model of Colombini (1993). Interestingly, there is essentially no visible difference between the flow organisation in the case where the forcing takes the form of two large-scale rollers $\boldsymbol{F} = (0, F_y, F_z)^T$ (cf. figure 5.9(*c*,*d*)) and that where we apply a constant streamwise forcing $\boldsymbol{F} = (F_x, 0, 0)^T$ (cf. figure 5.9(*e*)). In other words, in the first case forced rollers induce a streamwise velocity streak while in the second case, an externally forced low-speed streak causes large-scale secondary rolls to emerge.

Notably, essentially the same patterns of primary and secondary motions are obtained for a given wavelength $\lambda = 1.3$ even in the case of an exclusive buffer-layer forcing, as is seen in figure 5.9(*f*), even though the signs of the rolls and streaks are reversed. The latter effect is a consequence of the special shape of the forced buffer layer rolls: While the height of the forcing rolls is less than 100 wall units, pre-defined by the shape of the wall-normal forcing profile, their spanwise extent is given by half the forcing wavelength $\lambda/2 = 0.65$. As a consequence, the near-wall region features groups of counterrotating streamwise vortices which are thin but very wide. The outer flow 'sees' only the upper part of these thin rolls in which the fluid moves in the opposite direction as directly above the wall, such that the large-scale rolls are forced to counterrotate their thin near-wall counterparts.

The very similar shape of the large-scale secondary currents under different forcing schemes implies that the large-scale flow organisation in the here tested model is not very sensitive to the exact type and shape of the external force field. Even a forcing only in the direct vicinity of the bottom wall triggers the formation of large-scale secondary vortices. Remarkably, in agreement with the findings of Hwang and Cossu (2010b), it is seen that the maximum streamwise velocity perturbation occurs, even for the large-scale streaky structures, in the vicinity of the wall, irrespective of where the forcing profile attains its maximum.

In contrast to the near-wall region, the flow organisation at the outer-scaling peak in the current open channel flow differs in several aspects from that found by previous authors in the context of closed channel flows, focussing either on the transient growth of optimal initial conditions or on external deterministic and statistical forcing of the governing linearised equations (Butler and Farrell, 1993; Del Álamo and Jiménez, 2006; Pujals et al., 2009; Hwang and Cossu, 2010b). In both open and closed channel, the secondary flow in the cross-plane manifests itself in form of large-scale quasi-streamwise vortices that are of comparable aspect ratio $H/(\lambda/2)$, where H is the full channel height, i.e. $H = H_0$ in the open channel case and twice the half channel height in the closed channel. The streamwise velocity streaks, on the other hand, have different aspect ratios that differ by roughly factor two: In the open channel case, single streaks of either sign can spread over the entire channel depth, while their counterparts in the closed channel span only to the channel centreline and a 'mirror streak' of opposite sign forms in the other half of the channel (cf., for instance, fig. 3 in Hwang and Cossu (2010b)). This is in line with the orientation of the large-scale secondary rolls: The same large-scale circulation creates an upflow associated with a low-speed streak on one wall while it induces a downflow with associated high-speed region at the other wall, at the same lateral position. In the open channel, on the other hand, the free-slip/shear-free boundary condition at the free surface admits also solutions that are not symmetric w.r.t. the half height of the open channel $H_0/2$, but they enforce the wall-normal velocity v to vanish at the top of the domain. When scaled in terms of the typically adopted length scales, i.e. the full channel height in the open channel and the half channel height in the closed counterpart, the 'peak wavelengths' in the original and the modified open channel model are with $\lambda = 1.3$ smaller by a factor of roughly two compared to the values $\lambda \approx$ 3-4 detected by the above listed authors for closed channels. This difference has been recently discussed by Camporeale et al. (2021).

5.5 Discussion, summary and conclusion

In this chapter, we have analysed the linear response of a one-dimensional turbulent base flow in an infinitely wide open channel w.r.t. to a sinusoidal perturbation of the bottom wall with infinitesimal amplitude. In this context, we have revisited and modified the classical linear model of Colombini (1993) to better understand the functioning of the therein considered linear system and the physical processes it corresponds to. Subsequently, the results of both the original and the modified model were discussed in the context of recent studies on flow over laterally heterogeneous bottom walls on the one hand and compared with works on linear processes in turbulent channel flows on the other hand, including the investigation of externally forced dynamics and transient growth processes.

The original model formulation of Colombini (1993) is derived from the Reynolds-averaged Navier-Stokes equations, linearised around a fully-logarithmic turbulent open channel flow profile. The governing equations are simplified by neglecting viscous effects as well as by focussing on a two-dimensional problem in the cross-plane of the channel, assuming independence of the streamwise direction for all variables of interest. In order to take care of the deformation of the fluid domain in consequence of the bottom wall modulation, the governing equations are mapped onto an appropriate non-orthogonal coordinate system.

The governing equations of the fluid motion are coupled with the sediment bed evolution via the lateral particle flux in the sediment bed continuity equation, which in turn is formulated as a function of the lateral bed shear stress. Under the assumption that the characteristic time scale of the turbulent dynamics is much shorter than that of the comparably slow bedform evolution, it is in the following assumed that the flow field adapts quasi-immediately to a variation of the bottom wall curvature. As a consequence, the linearised system is one-way coupled in that first the governing, effectively stationary fluid equations are solved for all wavenumbers separately. The growth rate of the bedform amplitude is thereafter determined using a dispersion relation that represents the linearised sediment bed continuity equation. This latter equation is in fact a simple balance between the typically destabilising lateral bed shear stress and a counteracting stabilising gravitational term, from which the latter depends solely on a constant gravitational parameter and the Shields number in the base flow, i.e. $\sim \theta_0^{-1/2}$. In order to capture the secondary flow that arises as a consequence of the bottom wall-induced perturbations, the

Reynolds stress tensor is modelled using the non-linear eddy-viscosity model of Speziale (1987) which allows for the necessary anisotropy of the turbulent Reynolds stresses.

In the current work, the model equations were numerically solved in the framework of a standard Chebyshev tau-collocation method, in the context of which the governing equations are enforced at a set of Gauss-Lobatto points that are unevenly spaced over the wall-normal direction of the channel. A validation of the obtained results showed a quasi perfect agreement with the data in the original publication of Colombini (1993).

The data obtained with the original model formulation was in the following investigated in more detail than it was possible in the original study, in particular in the context of recent experimental and numerical studies on secondary flow over inhomogeneous bottom walls (Chung et al., 2018; Wangsawijaya et al., 2020; Zampiron et al., 2020a). It turned out that the linear model predicts the transverse wavelength $\lambda = 1.3$ at which the secondary flow kinetic energy is maximised and the secondary currents are 'space-filling' reasonably well. Also, it captures the trend of secondary currents being more restricted to the direct vicinity of the bottom wall as the lateral wavenumber α tends to infinity. In addition, the model of Colombini (1993) was compared to a recently published theoretical study of Zampino et al. (2022), which is based on similar model assumptions but uses a different turbulence model. The qualitatively very good agreement of the results in both models underlines that results discussed up to this point are no effect of the specific chosen turbulence closure, but that the observed phenomena are indeed of physical nature.

In the remainder of the chapter, we focused in our analysis on the linearised system to which the governing equations of fluid motion reduce under the given assumptions and the linearisation. The linearised Navier-Stokes operator therein is effectively independent of the sediment bed modulation, which takes the role of a stationary external forcing to the linear system. In this regard, comparable large-scale secondary currents could likewise originate from a laterally varying body force, for instance, a laterally varying driving pressure gradient or regions of blowing and suction. To underline the phenomenological similarity between the role of the sediment bed and a stationary forcing on the flow, we replaced the curved lower boundary by a flat smooth bottom wall, while the flow instability was instead driven by an external sinusoidal body force field. Also, diffusive terms were reintroduced into the governing equations to study flow organisation at finite Reynolds numbers both in the near-wall and outer-layer region. In this context, alternative analytical expressions for the base flow, the turbulent viscosity and the mixing length were introduced in order to take care of variable Reynolds numbers. These expressions were shown to reasonably well describe wall-normal reference profiles of the same quantities in direct numerical simulations of open channel flows.

It was then tested how the flow field organises under the influence of different external forcing schemes. First, an external forcing in form of infinitely long streamwise vortices in the buffer layer was applied, similar to the optimal forcing profile found by Hwang and Cossu (2010b) at $\lambda^+ = 81$. Despite the streamwise independence and the stationarity of the current model, the characteristic shape of the streaks that evolve due to the external forcing are astonishingly similar to those observed by Hwang and Cossu (2010b), including the wavelength at which they appear. In addition to the former forcing that was concentrated to the buffer layer, we furthermore investigated a similar roll-like forcing but for the outer layer that spanned over the entire channel height as well as a constant unidirectional streamwise force field. Interestingly, for all three forcing schemes the maximum secondary flow kinetic energy and thus the most intense secondary motion was found at essentially the same wavelength $\lambda = 1.3$ as in the original model, where the instability was triggered by the curved bottom wall.

That way, these investigations show that the most amplified wavelength in the model of Colombini (1993) which should represent the characteristic lateral spacing of initial sediment ridges is in fact entirely controlled by the linearised Navier-Stokes operator. The initial perturbation of the sediment bed is required to provide and maintain a stationary forcing that acts on the system, as the turbulent base flow is otherwise linearly stable (Reynolds and Tiederman, 1967). The 'most critical' bed perturbation wavelength that leads to the strongest response of the flow in form of space-filling large-scale secondary currents is, however, entirely determined by the governing fluid equations. The analogy of the modified model to the forced equations of Hwang and Cossu (2010b) and the transient growth analysis in closed channels (Butler and Farrell, 1993; Del Álamo and Jiménez, 2006; Pujals et al., 2009), on the other hand, underlines that the origin of the large-scale flow structures that arise in the model of Colombini (1993) is closely related to the linear processes that are assumed to play a major role in the formation and interaction processes of large-scale streamwise velocity streaks and quasi-streamwise rotating motions in canonical wall-bounded shear flows (Hwang and Cossu, 2011; Jiménez, 2018); rather than being a specific feature of the flow over deformable bottom walls. Conversely, based on the here discussed results, it is expected that a detailed transient growth analysis in open channel flows based on the full three-dimensional time-dependent Orr-Sommerfeld problem will lead to a maximum energy growth at a similar wavelength $\lambda \approx 1.3$. To the best of the author's knowledge, such analysis has not yet been performed, and the only work that goes into a similar direction is a recent study by Camporeale et al. (2021) on the linear (secondary) instability of the mean secondary flow in open channels with deformable free surface that could give rise to large-scale flow structures. While their problem is formulated considering the depth-averaged RANS equations with some extra terms, a full transient growth analysis would be of high interest to investigate how the free-slip boundary condition affects the transient growth of suitable initial conditions and whether or not the predictions obtained with the here discussed simplified two-dimensional and stationary model can be reproduced.

Chapter 6

Turbulent large-scale streaks and sediment ridges in open channel flow *

The theoretical investigations in the previous chapter have shown that a lateral undulation of the lower domain boundary is indeed capable of inducing large-scale secondary currents in an initially one-dimensional turbulent channel base flow. Clearly, an instantaneous fully-turbulent channel flow field is far from being one-dimensional and laterally alternating large-scale high- and low-speed streaks are well known to modulate the mean profile in wall-bounded shear flows (Marusic et al., 2010b; Smits et al., 2011; Jiménez, 2018).

The following chapter is dedicated to the analysis of such large-scale streaks over mobile sediment beds and of their relevance for the formation of sediment ridges. To this end, a series of direct numerical simulations of doubly-periodic turbulent open channel flow over a mobile sediment bed composed of fully-resolved spherical particles at various Reynolds numbers and in domains of different sizes has been performed. The dataset is supplemented by reference simulations in single-phase smooth-wall open channel flows. The analysis we are going to present provides evidence that in fully-turbulent flows, the organisation of the flow in form of large-scale high- and low-speed streaks is the main driver of sediment ridge formation by inducing a laterally varying bed shear stress and erosion rate along the sediment bed. The mechanism is thus similar to the second formation process conceived by Nezu and Nakagawa (1993), that is, a lateral variation of the flow field induces a lateral variation of the bed surface. Indeed, we will show that the large-scale structures located in the bulk of the channel interact in a kind of 'top-down mechanism' with the sediment bed, in accordance with the conceptual model of Jiménez (2018, § 5.6 and references therein) on the generation and organisation of coherent structures in canonical wall-bounded flows.

In the next section, we will first introduce the physical system under consideration, define the relevant physical parameters required to describe the system and elaborate how the initial states for the simulations are prepared. In a second section, we will outline a procedure used to detect the instantaneous interface between the sediment bed and the fluid-dominated region of the domain and define appropriate measures to quantify its temporal evolution. These definitions form the basis for the analysis of the simulations in the remainder of this chapter, which is structured as follows: First, the temporal evolution of the sediment bed from an initially macroscopically flat state to a ridge-covered bed will be investigated and a number of mean particle and flow field statistics will be discussed. In what follows,

* Part of the contents of this chapter have been published in *J. Fluid Mech.*, volume 930, pages A11, under the title *On the role of turbulent large-scale streaks in generating sediment ridges*, with M. Uhlmann, A. G. Kidanemariam and M. Krayer as coauthors. The publication was selected for a detailed discussion in the 'Focus on Fluids' article by Hsu (2022), underlining the novelty of our work and its relevance in particular for reduced-order modelling.

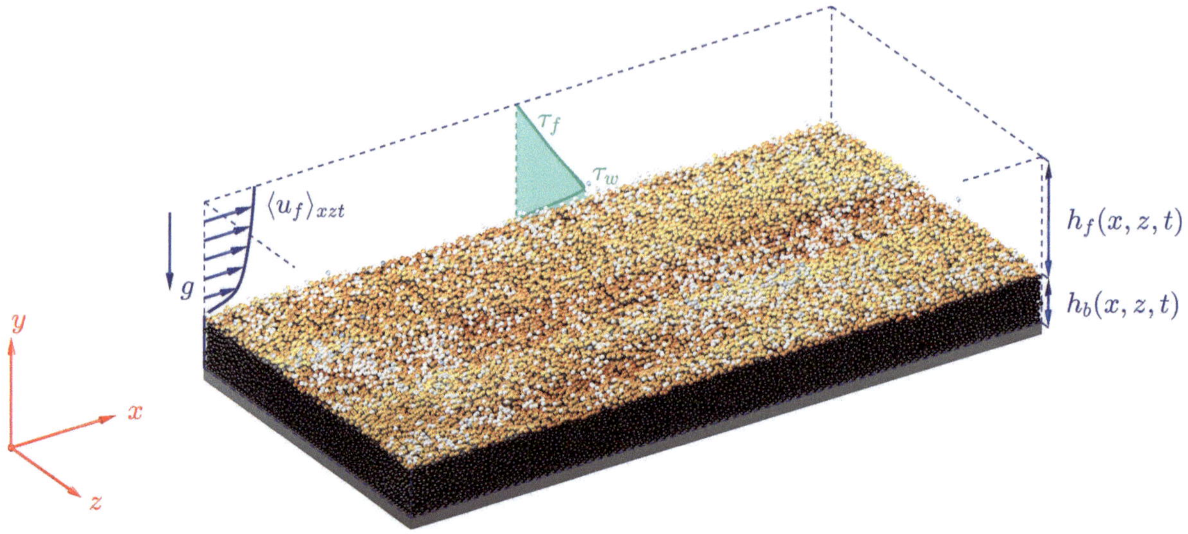

Figure 6.1: Sketch of the physical system analysed in the multiphase open channel simulations. Mean flow and gravity are pointing in positive x- and negative y-direction, respectively. The mean flow profile $\langle \boldsymbol{u}_f \rangle_{xzt} = (\langle u_f \rangle_{xzt}(y), 0, 0)^T$ is shown in blue, while the green curve represents the wall-normal variation of the mean fluid shear stress $\tau_f(y)$ and h_f and h_b denote the local instantaneous fluid and bed height, respectively. Particles are coloured depending on their location: Bed particles are coloured in black, interface particles in orange to yellow with increasing wall distance and transported particles are indicated by white colour (cf. section 6.2.1).

we will discuss the striking similarity between large-scale streaks in single- and multiphase simulations concerning both their lateral organisation and their dynamics. Correlations between the dynamics of these flow structures and the temporal evolution of the sediment ridges are then computed which reveal the causal connection between both structures. Finally, the loop back to the mean secondary flow is closed by showing that the depth-spanning secondary currents are basically the statistical footprint of the spatially well-organised large-sale streaks and their associated Reynolds stress-carrying structures. The analysis closes with a scrutinisation of the sediment ridges in later stages of the sediment bed evolution in additionally performed simulations featuring strongly constrained streamwise domain extensions to hinder the rise of transverse sediment bedforms. The observed 'top-down mechanism' and the dynamics of the large-scale structures over mobile sediment beds are finally discussed in comparison with the conceptual model of Jiménez (2018) for canonical smooth-wall flows on the one hand and the linear model proposed by Colombini (1993) on the other hand, which was the matter of the previous chapter.

6.1 Flow configuration and computational setup

6.1.1 Flow configuration

In the course of this study, we have performed a total number of 16 simulations of turbulent open channel flow over a mobile sediment bed. The database is supplemented with two reference simulations of single-phase smooth-wall open channel flow kindly provided by Prof. M. Uhlmann, which have been performed using the high-order pseudo-spectral method summarised in section 4.3.1 based on Fourier and Chebyshev expansions in the periodic and wall-normal directions, respectively (Kim et al., 1987).

The physical system under consideration is sketched in figure 6.1. In the remainder of this chapter, we will refer to a Cartesian coordinate system whose origin is positioned on the bottom wall of the channel, such that the coordinates of an arbitrary spatial position $\boldsymbol{x} = (x, y, z)^T$ are measured along the streamwise x-, wall-normal y- and spanwise z-direction, respectively. The components of the fluid velocity vector field at a spatial position $\boldsymbol{x}$ w.r.t. the Cartesian basis thus read $\boldsymbol{u}_f(\boldsymbol{x}, t) = (u_f, v_f, w_f)^T$. The fluctuating velocity field $\boldsymbol{u}'_f$ is defined w.r.t. the average in the two statistically homogeneous spatial directions and time, viz.

$$\boldsymbol{u}'_f(\boldsymbol{x}, t) = \boldsymbol{u}_f(\boldsymbol{x}, t) - \langle \boldsymbol{u}_f \rangle_{xzt}(y), \tag{6.1}$$

where angular brackets $\langle \bullet \rangle_i$ represent the averaging operator in the homogeneous directions and/or time indicated by the respective indices $i \in \{x, y, z, t\}$. A rigorous definition of the averaging operator in the current multiphase system is provided in appendix B. Note that the here introduced averaging scheme is shown in an exemplary manner for the three-dimensional velocity field, but is analogously applicable to lower-dimensional fields such as the two-dimensional fluid-bed interface which will be defined below. In addition to $\boldsymbol{u}'_f$, let us furthermore introduce the fluctuation of the streamwise-averaged field w.r.t. the instantaneous plane average, $\boldsymbol{u}''_f$, as

$$\boldsymbol{u}''_f(y, z, t) = \langle \boldsymbol{u}_f \rangle_x(y, z, t) - \langle \boldsymbol{u}_f \rangle_{xz}(y, t). \tag{6.2}$$

In both single- and multiphase simulations, the flow is driven by a time-dependent pressure gradient $\Pi(t)$ that ensures a constant fluid mass flow rate per unit width, q_f, in the streamwise direction. The gravitational acceleration vector $\boldsymbol{g}$ is pointing in the negative y-direction, with amplitude $g = |\boldsymbol{g}|$. The simulation domain is periodically repeated in the wall-parallel x- and z-directions with fundamental periods L_x and L_z, respectively, whereas in the wall-normal direction, it is bounded by a solid wall at the bottom (no-slip boundary condition) and a flat impermeable and shear-free boundary at the top that mimics the free surface of the open channel. The wall-normal extent of the simulation domain is denoted by L_y. In the single-phase simulations, L_y is identical to the mean fluid height H_f, whereas in the sediment-laden flows, the domain consists of two distinct regions, that are, the particle-dominated subdomain of mean height H_b, henceforth denoted as 'the sediment bed', and the upper fluid-dominated region of mean height $H_f = L_y - H_b$. A rigorous definition of the two distinct sub-domains and their mean heights will be given in the next section 6.2. In the sediment-laden simulations, the interface separating the sediment bed and the fluid dominated region takes the role of a virtual wall (Kidanemariam and Uhlmann, 2014a). For the sake of comparability with the smooth-wall single-flow cases, we will frequently refer to a shifted wall-normal coordinate $\tilde{y} = y - H_b$ that reduces to $\tilde{y} = y$ in the single-flow case, where the virtual wall collapses with the physical one. The wall-normal location of the virtual wall $\tilde{y} = 0$ is chosen as the position at which the mean wall shear stress is evaluated. We follow Chan-Braun et al. (2011) and determine the mean wall shear stress τ_w by extrapolating the pure-fluid stress

$$\tau_f(\tilde{y}) = \rho_f \nu_f \frac{\mathrm{d}\langle u_f \rangle_{xzt}}{\mathrm{d}\tilde{y}} - \rho_f \langle u'_f v'_f \rangle_{xzt} \tag{6.3}$$

from the region in the channel bulk where it varies linearly down to the location of the virtual wall $\tilde{y} = 0$. The last term on the RHS of equation (6.3) represents the contribution of the Reynolds stresses to τ_f and can be further decomposed into a turbulent and a dispersive contribution (Nikora et al., 2007). The origin of these two contributions in the current simulations will be analysed in section 6.3.9. Let us remark that for the smooth-wall case in which no bed exists, the Reynolds stresses identically vanish at the bottom wall and the standard definition of the wall-shear stress $\tau_w = \tau_f(\tilde{y} = 0) = \rho_f \nu_f \mathrm{d}\langle u_f \rangle_{xzt} / \mathrm{d}\tilde{y}|_{\tilde{y}=0}$ is recovered.

The physical system under consideration can be classified by three characteristic length scales in the featureless case, that are, the outer length scale H_f, the particle diameter D and the viscous length $\delta_\nu = \nu_f / u_\tau$, where $u_\tau = \sqrt{\tau_w/\rho_f}$ is the friction velocity. Following the general conventions, a superscript $\bullet^+$ shall indicate non-dimensionalisation in terms of u_τ and/or ν_f in the remainder of this work and we refer to quantities that way normalised as scaled in inner or wall units. Based on the three characteristic length scales H_f, D and δ_ν, we define the friction Reynolds number $Re_\tau = H_f/\delta_\nu = H_f^+$, the particle Reynolds number $D^+ = D/\delta_\nu$ and the relative submergence H_f/D, respectively. In addition, we introduce the bulk Reynolds number as

$$Re_b = \frac{q_f}{\nu_f} = \frac{u_b H_f}{\nu_f}, \tag{6.4}$$

where $u_b = q_f/H_f$ is the bulk velocity. Note that in virtue of the increased parameter space that comes with the additional degrees of freedom of the mobile particles, dimensional considerations following the methodology of Buckingham (1914) imply that two additional non-dimensional numbers are necessary to describe all relevant parameters of the multiphase system. Here, we choose the Galileo number $Ga = u_g D/\nu_f$ and the density ratio between the particle and fluid phase, $\rho_p/\rho_f = 2.5$, where a value of 2.5 is adopted which is close to the density ratio of glass beads or sand grains in water. The Galileo number Ga is a specific particle-related Reynolds number in which the gravitational velocity $u_g = \sqrt{(\rho_p/\rho_f - 1)|g|D}$ and the particle diameter D are chosen as characteristic velocity and length scales, respectively. The squared ratio of the particle Reynolds number and the Galileo number is known as Shields number

$$\theta = \left(\frac{D^+}{Ga}\right)^2 = \left(\frac{u_\tau}{u_g}\right)^2, \tag{6.5}$$

which represents the ratio between the bottom shear stress and the apparent weight of a single particle and, as such, is a measure for the ability of the flow to erode sediment grains of a certain size (Shields, 1936). Alternatively, one might interpret the Shields number as a relative turbulence intensity. A critical Shields number θ_c that marks the onset of sediment erosion in a turbulent flow is typically estimated as $\theta_c = 0.03$-0.05, with a slight dependence on the Galileo number (Soulsby et al., 1997; Wong and Parker, 2006; Franklin and Charru, 2011). In the current simulations, the Galileo number is chosen for each simulation individually such that for the expected particle Reynolds number D^+, the Shields number θ is sufficiently larger than θ_c such that particle motion is guaranteed.

6.1.2 Preparation of the simulations

The initial state of each simulation is prepared following the procedure described by Kidanemariam and Uhlmann (2014a,b): We first create a macroscopically flat pseudo-randomly arranged sediment bed by letting settle a certain amount of particles under the action of gravity in a dry-granular simulation – ignoring for the moment hydrodynamic effects – until a quasi-stationary bed at the bottom of the domain is established. As a controlling parameter, we monitor the particle velocity over the settling time interval, classifying a bed as being in a quasi-stationary state if the magnitude of all velocity components attains negligible values of $|U_{i,p}/u_b| \leq \mathcal{O}(10^{-3})\ \forall\, i = 1,2,3$. The number of particles included per simulation is adjusted to obtain the desired value of the relative submergence H_f/D, which varies in the current cases between $H_f/D \approx 26$ and $H_f/D \approx 52$. In a next step, the quasi-stationary sediment bed is exposed to a quiescent fluid to allow for possible relaxation effects under the influence of the

	Case	Re_b	Re_τ	ρ_p/ρ_f	Ga	D^+	H_f/D	H_b/D	H_f/H_b	θ
—	$CS250^{H3}$	3011	259.43	2.5	28.37	9.81	26.43	11.97	2.21	0.12
—	$CM250^{H3}$	3011	247.86	2.5	28.37	9.30	26.66	11.74	2.27	0.11
—	$CM650^{H3}_{smooth}$	12100	648.69	-	-	-	-	-	-	-
—	$CM850^{H3}$	9483	827.81	2.5	56.68	29.12	28.43	7.13	3.99	0.26
—	$CL250^{H16}_{smooth}$	3350	209.83	-	-	-	-	-	-	-
—	$CL250^{H16}$	3012	249.31	2.5	28.37	9.42	26.46	11.94	2.22	0.11

Table 6.1: Physical parameters of the short-time open channel flow simulations. Re_b, Re_τ and D^+ are the bulk, friction and particle Reynolds numbers, respectively. The density ratio ρ_p/ρ_f and the Galileo number Ga are imposed in each simulation, whereas the relative submergence H_f/D, the relative sediment bed height H_b/D, the fluid to bed height ratio H_f/H_b and the Shields number θ are computed a posteriori (cf. table 6.2).

	Case	$[L_x \times L_z]/H_f$	$[L_x \times L_z]/D$	$D/\Delta x$	$\min(\Delta y^+)$	N_p	T_{obs}/T_b
—	$CS250^{H3}$	1.94×2.91	51.2×76.8	10	0.98	43 730	678
—	$CM250^{H3}$	5.76×2.88	153.6×76.8	10	0.93	127 070	94
—	$CM650^{H3}_{smooth}$	5.33×2.67	-	-	0.05	0	397
—	$CM850^{H3}$	5.00×2.50	142.2×71.1	36	0.81	92 292	59
—	$CL250^{H16}_{smooth}$	12.00×16.00	-	-	0.06	0	431
—	$CL250^{H16}$	11.61×15.48	307.2×409.6	10	0.94	1 406 983	84

Table 6.2: Numerical parameters of the short-time open channel flow simulations. The computational domain has dimensions L_i in the ith spatial direction and is discretised using a uniform finite difference grid with mesh width $\Delta x = \Delta y = \Delta z$ for the multiphase simulations, while the smooth-wall single-phase simulations were performed using a spectral method with Fourier and Chebyshev expansions in the periodic and non-periodic directions, respectively, featuring a non-uniform distribution of the grid/collocation points in the three spatial directions. N_p is the total number of particles in the respective case and T_{obs} is the total observation time of each simulation, starting from the release of the mobile particles at $t = 0$. Time dependent physical and numerical parameters in tables 6.1 and 6.2 (Re_τ, D^+, H_f, H_b, θ, Δy^+) are computed as an average over the entire simulation period.

hydrodynamic forces and torque. After the completion of the bed preparation, all particles are fixed in their position and a fully-turbulent open channel flow at the desired Reynolds number is developed over the stationary sediment bed until it reaches a statistically stationary state. Note that in order to limit the computational effort in this preparation phase, the process of developing a statistically stationary flow has been partly performed on a coarser mesh. Once the desired state was reached in those cases, the obtained flow field was linearly interpolated to the fine 'production' grid for which $\Delta y^+ < 1$ and was further integrated in time for a time period $\mathcal{O}(10T_b)$ such that the smallest flow scales can evolve. Here, $T_b = H_f/u_b$ denote outer-scaled bulk time units. Eventually, the particles are released again starting from a time which we arbitrarily define as $t = 0$. Note that this does not include a small fraction of sediment grains close to the bottom wall that is kept fixed even after $t = 0$ in order to ensure a minimal amount of bottom roughness.

6.1.3 Simulation parameter values

A collection of the most relevant physical parameters in the current simulations can be found in table 6.1, while additional numerical information is provided in table 6.2. Note that these two tables include only those simulations which we will consider for our investigation of the initial ridge evolution. Additional simulations in shorter domains which we have performed in order to analyse the long-time development of sediment ridges belong to a similar parameter point as case $CS250^{H3}$ and their physical and numerical parameters will be provided in a later section 6.3.10 (cf. also the discussion on the domain length on sediment pattern development below). In the remainder of this thesis, each simulation is indicated by a unique name that comprises information about the flow configuration (C: open channel, D: open duct), the size of the computational domain (S: small, M: medium, L: large) and the friction Reynolds number. For smooth-wall single-phase cases, the name is extended by a respective subscript. The streamwise and lateral domain sizes have been varied in the range $L_x/H_f \in [2, 12]$ and $L_z/H_f \in [2.5, 16]$, respectively, reaching friction Reynolds numbers Re_τ between 210 and 830. The particle Reynolds number attains values up to $D^+ \approx 29$, leading (together with the accordingly chosen Galileo number Ga) to Shields numbers $\theta > 0.11$. For the smooth-wall reference simulations, the friction Reynolds numbers $Re_\tau = 210$ and $Re_\tau = 650$ were chosen to match those in the multiphase simulations $CL250^{H16}$ and $CM850^{H3}$ in the time period during which particles are still in rest and the bed is macroscopically flat. Note that the Reynolds number in the particle-laden cases later increase for both cases to $Re_\tau = 250$ and $Re_\tau = 830$ in the phase after the particle release (cf. table 6.1) as a consequence of the increased friction caused by the transport of mobile sediment, the development of sediment bedforms and the decrease of the bed height (cf. the discussion about figure 6.11 in the following section). Let us furthermore remark that case $CM250^{H3}$ is found at the same parameter point as case $H6$ of Kidanemariam and Uhlmann (2017), but the currently investigated data represents a new independent simulation conducted in the context of this study. Due to technical reasons, data in the first roughly $5T_b$ of the simulation are not available for the study which, however, does not affect the validity of the obtained results in later stages of the simulation.

For the here considered range of Shields numbers, initial sediment ridges are seen to rapidly evolve from the initially flat sediment bed within the first few bulk time units. Simultaneously, also transverse sediment patterns start to arise even though at a markedly slower rate. Nevertheless, these latter become the dominant sediment bedforms after between 100 and 200 bulk time units, in the sense that their amplitude is at least twice as large as those of the sediment ridges (Kidanemariam and Uhlmann, 2017). For the investigation of the initial formation of sediment ridges and its connection to large-scale streaks, the evolution of transverse bedforms does not pose any restrictions as our study focuses merely on the initial simulation time interval of $t/T_b < 100$, during which transverse bedforms are still small enough to be of minor importance. The total observation time intervals T_{obs} that quantify the time period from the onset of particle motion until the end of the simulation are accordingly comparably short. For the purpose of investigating also the flow over developed sediment ridges and longer time intervals of $\mathcal{O}(100T_b)$, on the other hand, we have performed a second simulation series in which we intentionally suppress the evolution of transverse bedforms by choosing strongly restricted streamwise domain sizes $L_x/D < 70$. These short domains are not able to accommodate the shortest unstable wavelength that leads to the formation of initial ripple-like patterns, whose length was recently found to be larger than $80D$ at the here considered parameter point (Scherer et al., 2020). As we shall see, indeed none of the cases in this second series of simulations allows for the growth of transverse bedforms even for time intervals of about $T_{obs} \approx 1000T_b$. In the first part of our analysis focussing on the initial evolution of sediment ridges, we include only a single case from this second simulation campaign, namely case $CS250^{H3}$, while the remaining cases together with their physical and numerical parameter will be presented in more detail in section 6.3.10.

The idea to study turbulent flows in strongly constrained domains in order to isolate individual features of the velocity field is quite common in the turbulence community. The most prominent example is surely the minimal flow unit in the pioneering work of Jiménez and Moin (1991) which represents the smallest possible domain that can still accommodate all elements of the self-sustained regeneration buffer layer cycle (Hamilton et al., 1995), in their case a single velocity streak flanked by two quasi-streamwise vortices (cf. the more detailed discussion in section 3.2). The here considered cases feature, on accord of their varying box dimensions and Reynolds numbers, a wide range of inner-scaled horizontal box dimensions L_x^+ and L_z^+ from small domain simulations whose extensions are of the order of the Jiménez and Moin (1991) minimal flow unit ($L_x^+ \approx$ 250-350, $L_z^+ \approx 100$) to the medium and large domain simulations, in which the boxes have streamwise and spanwise periods of $\mathcal{O}(1000\delta_v)$ (cf. cases $CM850^{H3}$ and $CL250^{H16}$). Note that due to the multi-scale nature of the logarithmic layer, minimal domains of different size can be found for structures at each distance to the bottom wall, as was shown by Flores and Jiménez (2010) for closed channels. The result is a self-similar cascade of minimal boxes, from which the smallest one is the minimal flow unit of Jiménez and Moin (1991) and a further reduction in size directly leads to the relaminarisation of the flow. The largest minimal domain, on the other hand, is that related to the largest flow structures at the centreline of the channel, for which Flores and Jiménez (2010) estimated a minimal box width of $L_z/H_f \approx 3$. A comparison of the outer-scaled box width L_z/H_f in our simulation shows that all simulations in table 6.1 are approximately wide enough to host at least a single full regeneration cycle of the largest log-layer streaks. Surely, the comparison to our open channel is limited in the vicinity of the free surface, but it should give a good estimate for the rest of the domain (Bauer et al., 2022). The here discussed small domain simulations, in turn, resemble the 'streamwise-minimal' channel simulations of Toh and Itano (2005) and Abe et al. (2018), whose intention was to study the interaction of small- and large-scale structures. To this end, they limited only the streamwise domain length to a value for which L_x^+ is just long enough to maintain the buffer layer process while L_z^+ was clearly larger than the minimal value of about 100 wall units. For the considered Reynolds numbers, L_x/H_f was at the same time much shorter than the typical length of the large-scale structures, such that these features can be considered as infinitely long due to the missing spatial de-correlation.

To the best of the author's knowledge, the current database includes with simulation $CM850^{H3}$ ($Re_\tau \approx 828$) the highest ever obtained Reynolds number in a simulation laden with a large amount of fully-resolved particles, and with case $CL250^{H16}$ the largest number of fully-resolved particles simulated so far in a DNS-based study of sediment transport. The former simulation $CM850^{H3}$ was performed in close cooperation with M. Krayer, who organised the execution of the simulations on the supercomputers Hazel Hen and Hawk at the supercomputer centre HLRS in Stuttgart. Needless to say, the computational demands of such particle-laden simulations in large domains and/or at considerable Reynolds numbers are immense: The investigated simulation interval of around 60 bulk time units in case $CM850^{H3}$ alone consumed a total amount of approximately nine million CPU hours, including a number of around 16.8 billion grid nodes.

6.2 Fluid-bed interface and bedform dimensions

6.2.1 Extraction of the fluid-bed interface

In the following, we shall define a continuous two-dimensional fluid-bed interface that separates the sediment bed from the fluid-dominated region above it. Different concepts have been developed in the past to approximate the surface of the discretely-spaced particle set forming the sediment bed in a smooth and continuous way, for instance, by identifying the interface with the wall-normal location at which a threshold for the solid volume fraction is attained (Kidanemariam and Uhlmann, 2014a, 2017). The attractiveness of the latter approach lies in the fact that it does not require knowledge of individual particle positions which are often impossible to record in experiments, whereas instantaneous visualisations of the solid volume fraction can be obtained, for instance, by illuminating the measurement section with a laser sheet while simultaneously filling the domain with dye, such that a binary image of particle and fluid regions can be taken (Lobkovsky et al., 2008; Kidanemariam, 2016). This works quite well if the goal is, as in Kidanemariam and Uhlmann (2014a), to extract the spanwise-averaged fluid-bed interface, but has limitations if we are instead interested in the full two-dimensional surface. The problematic point is that the continuous solid volume fraction field is by definition an average over a given set of particles in a well-defined spatial subdomain. In the context of the spanwise-averaged field, the discrete particle positions can be averaged over bins of width L_z such that a continuous solid volume fraction is directly obtained. For the extraction of the full two-dimensional surface, on the other hand, it would be necessary to subdivide the whole domain into small cubic bins with dimensions sufficiently larger than D, such that each bin contains a sufficient number of samples to guarantee a more or less smooth variation of the solid volume fraction. This might not pose any restriction in systems where $D \ll H_f$, whilst in the current simulations where the domain extensions are $L_i/D = \mathcal{O}(10\text{-}100)$ $(i = x, y, z)$, it would not allow to resolve the local shape of the fluid-bed interface in a smooth way. Here, we therefore use a different technique to extract the two-dimensional surface from instantaneous particle datasets which has been developed in collaboration with M. Krayer and was first published in Scherer et al. (2020), with M. Uhlmann and A. G. Kidanemariam as coauthors.

To this end, we first decompose the wall-normal extent of the domain L_y for each point of the xz-plane in two contributions

$$L_y = h_b(x,z,t) + h_f(x,z,t), \tag{6.6}$$

where h_b and h_f are the local bed height and thickness of the fluid-dominated region, respectively. In the adopted Cartesian coordinate system, the local wall-normal position of the fluid-bed interface then collapses with $h_b(x,z,t)$. To obtain a continuous representation of the sediment bed surface from a set of discrete particle positions, we first detect a set of 'interface particles' which form the uppermost layer of the sediment bed by an algorithm that will be outlined below. From these interface particles, a continuous two-dimensional manifold is then reconstructed by means of a triangulation between the interface particles' centres followed by an interpolation to a regular uniform grid. The sorting-algorithm applied in the first step is based on the standard morphological classification in bed particles and mobile particles, from which the latter may be transported in the bedload or suspended load layer (Bagnold, 1956; van Rijn, 1984). Potential candidates for 'bed particles' should feature both a negligible kinetic energy $(|\boldsymbol{U}_p|/u_g)^2$ and a non-vanishing wall-normal contact force F_y^C/F_W, as all particles inside the densely-packed bed are necessarily in contact with at least one neighbouring particle and thus feel a non-trivial contact force. Here, $F_W = (\rho_p - \rho_f)|g|\pi D^3/6$ denotes the submerged weight of a single

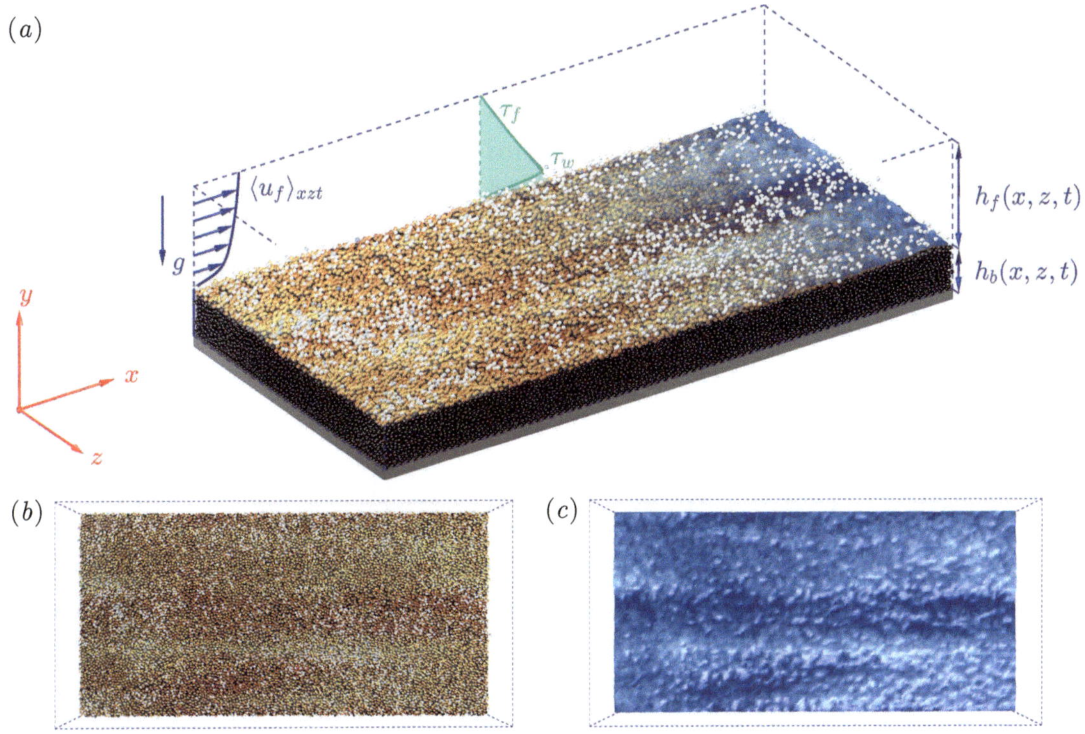

Figure 6.2: Conceptual sketch of the fluid-bed interface extraction. (*a*) Three-dimensional visualisation of the sediment particles classified into bed particles (black), interface particles (orange to yellow with increasing wall distance) and transported particles (white). At the downstream end of the domain, the sediment bed is overlain by the two-dimensional surface that represents the fluid-bed interface, the colour ranging from dark to bright blue with increasing height. (*b*) View on the sediment bed from the top of the channel. (*c*) Same view on the extracted mean fluid-bed interface.

spherical particle. Note that the normalised particle kinetic energy can be alternatively interpreted as a particle-related Shields number, therefore it is natural to choose as a threshold for $(|\boldsymbol{U}_p|/u_g)^2$ a value similar to θ_c. In practice, a velocity threshold of $(|\boldsymbol{U}_p|/u_g)^2 = 0.05$ and a force threshold $F_y^C/F_W = 10^{-5}$ are used, but the procedure has been shown to be not strongly sensitive to the choice of these two parameters.

All particles that match the formulated requirements are classified as being part of the sediment bed. From this set of bed particles, we choose in the second step only those that are part of the uppermost sediment layer, the interface particles. This is done geometrically, using the α-shape algorithm of Edelsbrunner and Mücke (1994), which is conceptually similar to a conventional convex hull around a set of points, but offers more flexibility than the classical method concerning the convexity requirements of the enclosing hyper-surface: The α-shape allows non-convexity for length scales larger than some threshold radius α_s (here taken as 1.1 times the particle diameter), while it is strictly convex for length scales smaller than this threshold. Applying the α-shape algorithm to the set of bed particles, an enclosing surface can be generated by means of triangulation. Nodes of the triangulation that represent the centres of interface particles are those which bound triangulation cells that feature an outward pointing normal with a positive y-component. In a final step, the information about the local wall-normal position of the interface is transferred to a regular equidistant Eulerian grid in the xz-plane with sampling width of $1D$ by means of linear interpolation. An exemplary visualisation of the bed, interface and mobile particles together with the generated fluid-bed interface is provided for a randomly chosen instantaneous field in figure 6.2.

6.2.2 Quantification of the bedform dimensions

In an abstract sense, we can interpret the development of sediment bedforms as growth of wave-like deformations of the just defined two-dimensional surface that represents the fluid-bed interface. In this context, we will thus quantify the wall-parallel and wall-normal bedform dimensions as wave amplitude and wavelengths, respectively. To quantify the size and evolution of streamwise and spanwise bedforms separately, we will study the amplitude and wavelength of the streamwise- and spanwise-averaged interfaces $\langle h_b \rangle_x(z,t)$ and $\langle h_b \rangle_z(x,t)$ in the following individually.

In this context, the ridge height is determined following a statistical approach based on the root mean square of the sediment bed height fluctuations (Langlois and Valance, 2007; Kidanemariam and Uhlmann, 2017), viz.

$$\sigma_{h,z}(t) = \left[\langle h_b''(z,t)\, h_b''(z,t)\rangle_z\right]^{1/2}. \tag{6.7}$$

A measure for the height of transverse bedforms is analogously defined as $\sigma_{h,x}$, based on the fluctuations of the spanwise-averaged interface. Note that the statistical ansatz chosen in the current work is only one possible way to define a characteristic height of sediment bedforms, and a number of alternative approaches is summarised in Coleman and Nikora (2011).

The lateral spacing of the sediment ridges is quantified in terms of the mean spanwise wavelength of the sediment bed, $\lambda_{h,z}$. It is defined based on the instantaneous two-point correlation coefficient

$$\rho_{hh}(\delta z,t) = \langle h_b''(z,t)\, h_b''(z+\delta z,t)\rangle_z / \sigma_{h,z}^2(t), \tag{6.8}$$

where δz denotes the spanwise separation between two positions. For a given time t, we identify the mean wavelength as twice the separation length for which ρ_{hh} attains its first and global minimum,

$$\left.\begin{array}{rcl} \rho_{hh}(\delta z_{min},t) & \leq & \rho_{hh}(\delta z,t) \quad \forall\, \delta z \in [0, L_z/2] \\ \lambda_{h,z}(t) & = & 2\delta z_{min} \end{array}\right\}. \tag{6.9}$$

6.2.3 Quantification of the particle transport

In contrast to the fluid phase, physical information of Lagrangian properties in the dispersed phase exist only at discrete locations in space. To obtain particle statistics in the current Eulerian observation framework, we apply a binning technique similar to that of Kidanemariam (2016), but generalised to the two-dimensional case. The wall-parallel periodic directions are discretised in bins of width $\Delta x_{bin} = \Delta z_{bin} \approx 1.5D$ spanning over the entire wall-normal box length L_y, resulting in a number of $N_{x,bin}$ and $N_{z,bin}$ bins in the streamwise and spanwise direction, respectively. Then, the local particle flux in the (i,k)th bin $(1 \leq i \leq N_{x,bin},\ 1 \leq k \leq N_{z,bin})$ is defined as the volumetric particle flow rate averaged over that bin, i.e. the sum of the particle velocity of all particles centred in the bin at time t times the particle volume divided by the bin base area $\Delta x_{bin}\Delta z_{bin}$, viz.

$$q_p(x_i,z_k,t) = \frac{\mathcal{V}_p}{\Delta x_{bin}\Delta z_{bin}} \sum_{l=1}^{N_p} \boldsymbol{u}_p^{(l)}(t)\; I_{(i,k)}^{(l)}(t). \tag{6.10}$$

Here, $\boldsymbol{q}_p = (q_{p,x}, q_{p,y}, q_{p,z})^T$ is the particle flux vector and $\boldsymbol{U}_p^{(l)}$ is the Lagrangian velocity vector associated with particle l, respectively. The streamwise and spanwise coordinates of the (i,k)th bin centre are $x_i = ((i-1)+i)\Delta x_{bin}/2$ and $z_k = ((k-1)+k)\Delta z_{bin}/2$, respectively. $I_{(i,k)}^{(l)}(t)$ is an indicator function that is unity if the lth particle is centred in the (i,k)th bin, i.e.

$$I_{(i,k)}^{(l)}(t) = \begin{cases} 1 & \text{if } (i-1)\Delta x_{bin} < X_p^{(l)}(t) \leq i\Delta x_{bin} \\ & \wedge\ (k-1)\Delta z_{bin} < Z_p^{(l)}(t) \leq k\Delta z_{bin} \\ 0 & \text{else} \end{cases} . \tag{6.11}$$

It should be recalled that the particle flux as an integral measure does not reflect the variation of the particle transport intensity over the channel height. For this purpose, we additionally define the corresponding streamwise particle flux density as

$$\langle \phi u_p \rangle_{xz}(y_j, t) = \frac{\mathcal{V}_p}{L_x L_z \Delta y_{bin}} \sum_{l=1}^{N_p} U_p^{(l)}(t)\, I_{(j)}^{(l)}(t), \tag{6.12}$$

where $y_j = ((j-1)+j)\Delta y_{bin}/2$ for all $1 \leq j \leq N_{y,bin}$ represents the wall-normal location of the jth bin centre and the corresponding indicator function $I_{(j)}^{(l)}(t)$ is defined in analogy to $I_{(i,k)}^{(l)}(t)$ as

$$I_{(j)}^{(l)}(t) = \begin{cases} 1 & \text{if } (j-1)\Delta y_{bin} < Y_p^{(l)}(t) \leq j\Delta y_{bin} \\ 0 & \text{else} \end{cases} . \tag{6.13}$$

Similar to the previous definitions, we have subdivided the wall-normal box length L_y into $N_{y,bin}$ bins with however smaller bin height $\Delta y_{bin} \approx 0.5D$, spanning over the entire box length and width L_x and L_z, respectively. By definition, the following relation holds (Lobkovsky et al., 2008; Chiodi et al., 2014):

$$\langle q_{p,x} \rangle_{xzt} = \int_{y=0}^{L_y} \langle \phi u_p \rangle_{xzt} \mathrm{d}y. \tag{6.14}$$

6.3 Interaction of turbulent large-scale streaks and sediment ridges

In the classical literature on morphodynamics, sediment ridges are usually described as "parallel to each other, of little relief, and of a uniform transverse spacing" (Allen, 1968). The sediment ridges in the current simulations share all these features. As can be seen from the instantaneous sediment bed snapshots in figure 6.3, the sediment bed surface is covered by laterally alternating ridges and troughs with bedform heights between one and two particle diameters, thereby confirming for the first time numerically earlier theoretical predictions that a sidewall-induced mean secondary flow is not necessary for the evolution of sediment ridges (Ikeda, 1981). The observed bedforms are essentially parallel to the streamwise direction and span over the entire streamwise domain length L_x, which is in good agreement with experimental observations according to which sediment ridges can reach streamwise extensions of up to $\mathcal{O}(10H_f - 100H_f)$ (Wolman and Brush, 1961). The number of sediment ridges that form per simulation naturally depends on the size of the computational domain. The narrow computational boxes with a lateral domain period $L_z/H_f \approx 3$ are capable of accommodating between one and

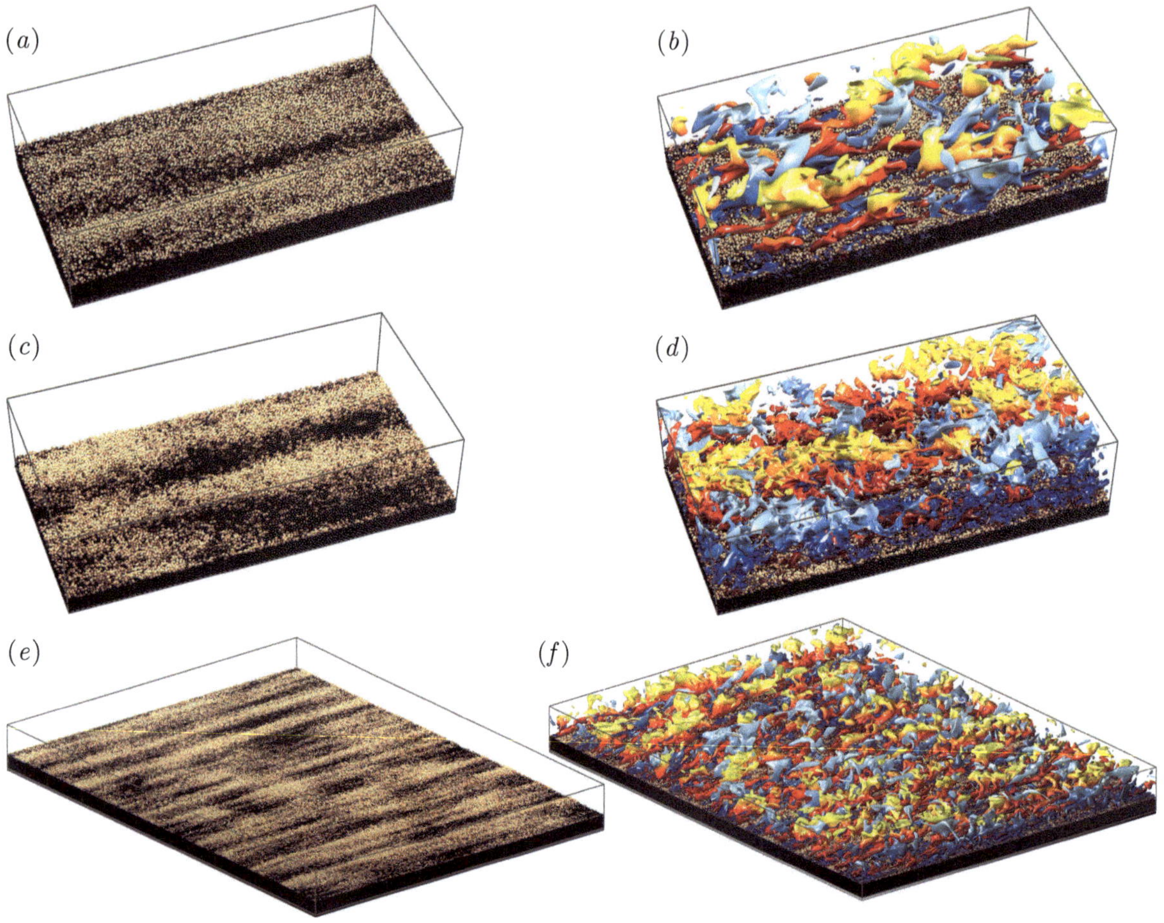

Figure 6.3: Instantaneous visualisation of (*a,c,e*) the evolved sediment ridges compared with (*b,d,f*) the instantaneous distribution of three-dimensional Reynolds stress-carrying Q^- structures. The latter are characterised as connected regions fulfilling $|-u'_f(\boldsymbol{x},t)v'_f(\boldsymbol{x},t)| > H\,u_{f,rms}(y)v_{f,rms}(y)$ with $H = 1.75$ (Lozano-Durán et al., 2012). Ejection structures are indicated by reddish colours, while sweeps are coloured in blue, with brighter colours indicating a larger distance to the bottom wall. Particles are coloured depending on their wall-normal location, ranging form dark to light brown with increasing coordinate y. For the sake of clarity, only bed and interface particles are shown, following the classification established in section 6.2.1. In each panel, flow is from bottom left to top right. (*a,b*) $CM250^{H3}$ ($t/T_b = 40$), (*c,d*) $CM850^{H3}$ ($t/T_b = 59$), (*e,f*) $CL250^{H16}$ ($t/T_b = 85$).

two ridges, whereas the large domain of $CL250^{H16}$ exhibits between nine and ten individual bedforms. We can therefore consider the small to medium domains as close to minimal in the context of the number of available ridges, which shall be favourable for the subsequent analysis in that individual ridges and their relation to turbulent coherent structures can be investigated individually, without possible merging or splitting effects between individual bedforms. The large domain simulation $CL250^{H16}$, on the other hand, contains a sufficient number of individual ridges to allow statements on the collective behaviour of the bedforms.

In order to give a first qualitative impression on how large-scale turbulent structures are organised in comparison to the spanwise locations of the bedforms, we have supplemented to figure 6.3 instantaneous visualisations of the Reynolds stress-carrying ejection ($u'_f < 0$, $v'_f > 0$) and sweep structures ($u'_f > 0$, $v'_f < 0$) (collectively termed as Q^-'s) introduced by Lozano-Durán et al. (2012) as a generalisation of the classical quadrant analysis (Wallace et al., 1972) to three dimensional objects. We follow the former study and define coherent ejection and sweep structures as connected subdomains for which $|-u'_f(\boldsymbol{x},t)v'_f(\boldsymbol{x},t)| > H\,u_{f,rms}(y)v_{f,rms}(y)$ holds with $H = 1.75$ and $u_{fi,rms} = \sqrt{\langle u'_{fi}u'_{fi}\rangle_{xzt}}$. The lateral organisation of the large-scale ejection and sweep structures correlates remarkably well with that of

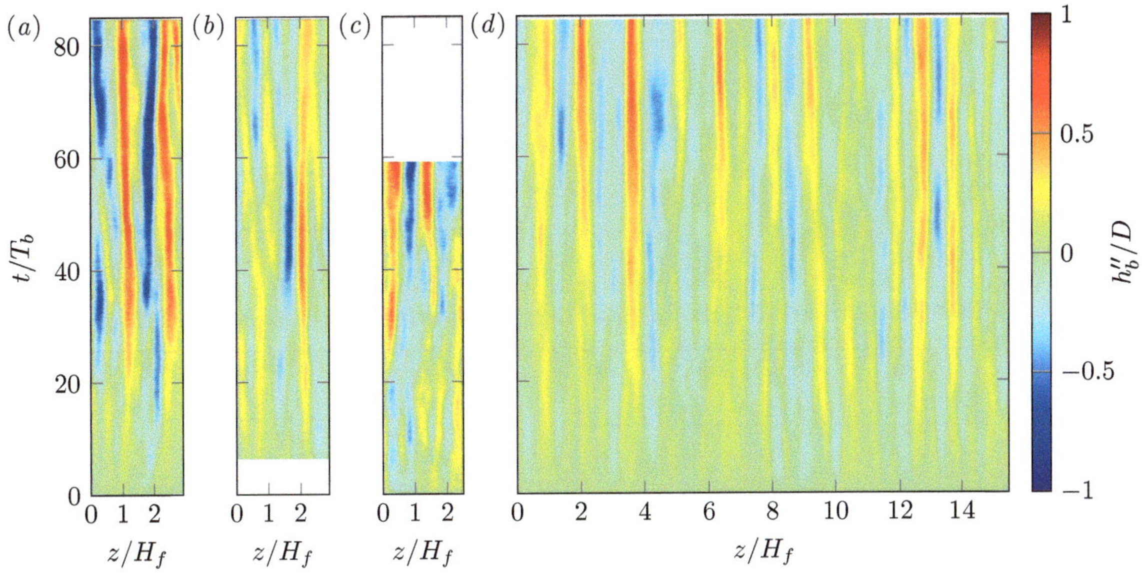

Figure 6.4: Space-time plot of the streamwise-averaged sediment bed height fluctuations $h''_b(z,t)/D$. Blue and red regions refer to troughs and crests of the streamwise-averaged fluid-bed interface profiles, respectively. Cases: (*a*) $CS250^{H3}$, (*b*) $CM250^{H3}$, (*c*) $CM850^{H3}$, (*d*) $CL250^{H16}$.

the underlying sediment ridges and troughs, with ejections predominantly located above regions of weaker erosion, i.e. ridges, and sweep structures mainly found over troughs where erosion is more intense. This is in accordance with the findings of Gyr and Schmid (1997) who concluded that particle erosion is mainly due to instantaneous sweep events that are directed towards the wall (Jiménez, 2018). Recalling that ejections and sweeps live in the longer large-scale streaks of low- and high-speed streamwise velocity, respectively, we conclude that the preferential lateral position of sediment ridges is below the large-scale high-speed streaks and *vice versa* for the troughs. Similar visualisations of these large-scale streaks that are omitted here show that this is indeed the case.

6.3.1 Sediment ridge evolution

In all simulations, the sediment ridges are seen to evolve from the initially flat sediment bed within the first few bulk time units after the onset of particle motion, solely under the action of the turbulent structures. Figure 6.4 illustrates this initial pattern growth by providing the space-time evolution of the fluctuations of the streamwise-averaged sediment bed height fluctuation, $h''_b(z,t)$. Interestingly, the initial sediment patterns evolve at different spanwise locations more or less simultaneously and thus independently from each other. During the initial phase of the first about 20 bulk time units, their mean lateral spacing is somewhat smaller than the range of values $1H_f$-$2H_f$ reported in the literature (Wolman and Brush, 1961; McLelland et al., 1999). Advancing in time, however, individual ridges are seen to merge in a kind of bedform coalescence, leading to a net reduction of the total number of bedforms and an increasing spanwise spacing which agrees now reasonably well with the experimentally determined values. Note that an exact match of the lateral separation is not necessarily expected as the ridges studied in most experiments represent quasi-asymptotic states of the sediment bed. These latter are reached after evolution phases whose length differs by several orders of magnitude from the current observation time interval, in which the initial sediment ridges are still in a transient state. Also, most experimental studies have been conducted in laboratory flumes with narrow to intermediate aspect ratio, such that the sediment ridges are additionally exposed to the influence of a sidewall-induced

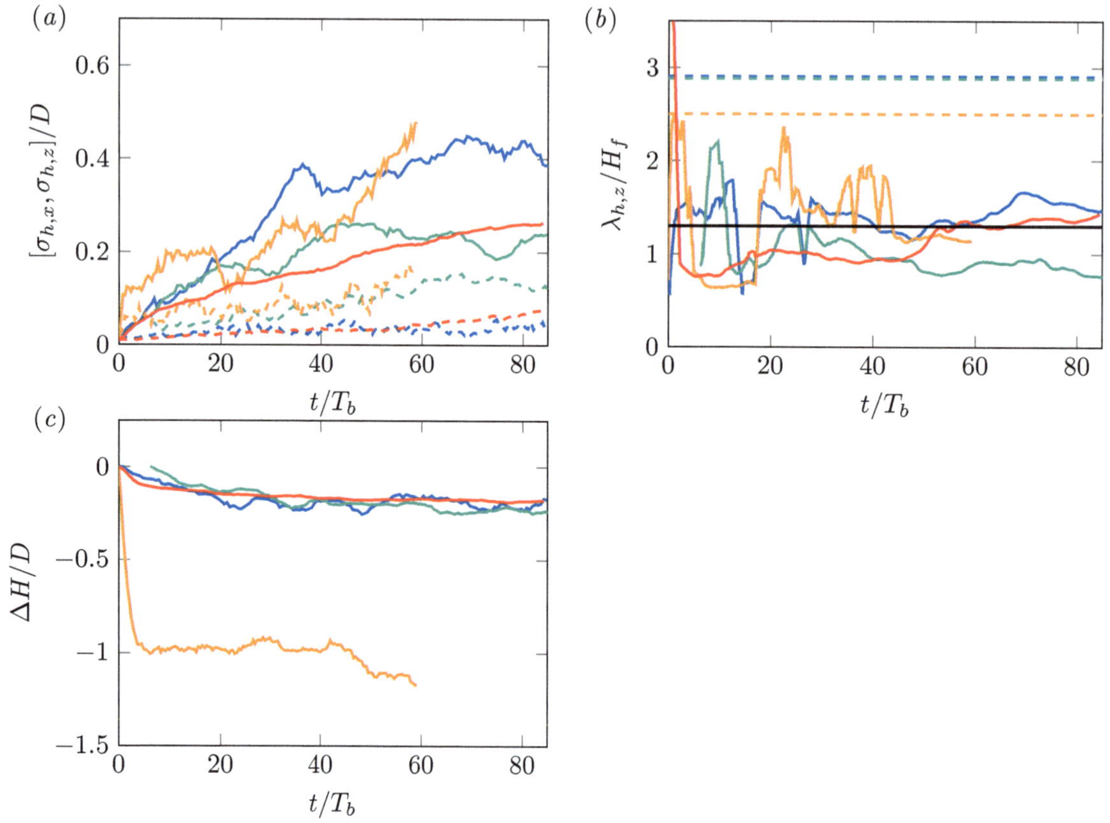

Figure 6.5: Time evolution of fluid-bed interface dynamics. (*a*) Root mean square of the fluctuation of the streamwise-averaged ($\sigma_{h,z}/D$,—) and spanwise-averaged fluid-bed interface ($\sigma_{h,x}/D$,---). (*b*) Mean sediment ridge wavelength $\lambda_{h,z}/H_f$. Horizontal dashed lines of matching colour mark the lateral domain size L_z/H_f, while the black solid line refers to the most-amplified wavelength determined in the linear stability analysis of Colombini (1993, cf. chapter 5). (*c*) Decrease of the mean sediment bed height w.r.t. the initial bed height, $\Delta H = \langle h_b \rangle_{xz}(0) - \langle h_b \rangle_{xz}(t)$, as a function of time. Cases: $CS250^{H3}$ (—), $CM250^{H3}$ (—), $CM850^{H3}$ (—), $CL250^{H16}$ (—).

secondary flow. The impact of lateral domain boundaries and the related secondary currents on the evolution of sediment ridges will be the main topic of the following chapter 7.

After approximately 40 bulk time units, the number of splitting and merging events visible in figure 6.4 clearly reduces. The remaining developed ridges feature a higher amplitude and a relatively stable mean spanwise position not only in the narrow domains, but also in the wide domain of case $CL250^{H16}$. This, in turn, highlights that the regular stable arrangement of quasi-parallel sediment ridges is a fundamental property of this class of bedforms rather than an artefact due to the limited domain size in the narrow cases.

Figure 6.5(*a*,*b*) shows the time evolution of the root mean square of the sediment bed height fluctuation $\sigma_{h,z}(t)$ and of the mean spanwise wavelength $\lambda_{h,z}(t)$, respectively, which are measures of the mean ridge height and of their lateral mean spacing (cf. the definitions in section 6.2.2). In agreement with the space-time plots in figure 6.4, it is observed that $\sigma_{h,z}$ increases with time during the first approximately 40 bulk time units. While in the first about 10 bulk time units, this growth is predominantly due to the rise of individual low-amplitude ridges, the further increase in the subsequent phase is assumed to be a consequence of pattern merging events that have been observed in the space-time visualisations. Here, the initial growth rate differs markedly between the low Reynolds number cases $CS250^{H3}$, $CM250^{H3}$ and $CL250^{H16}$ on the one hand and the high Reynolds number case $CM850^{H3}$ on the other hand. This difference most likely originates in the higher Shields number in case $CM850^{H3}$ that causes a higher erosion rate and hence a faster initial growth of the first-appearing ridges. For the sake of comparison, panel 6.5(a) provides also the time evolution of $\sigma_{h,x}$ that represents the mean amplitude of bed perturbations related to transverse ripple-like bedforms. In accordance with our earlier observations in figure 6.3, the modes related to transverse bedforms maintain a sufficiently small amplitude to have no significant influence on the discussed statistics throughout the considered time intervals.

The varying number of individual ridges in the beginning of the observation interval manifests itself in form of oscillations of the mean spanwise wavelength $\lambda_{h,z}$ for times $t/T_b < 40$ in panel 6.5(b). For $t/T_b > 40$, on the other hand, the mean wavelength settles in all four cases to a finite value without further strong oscillations, attaining final values of $1.47H_f$ ($CS250^{H3}$), $1.14H_f$ ($CM850^{H3}$) and $1.44H_f$ ($CL250^{H16}$). These values are comparable to the most amplified wavelength $\lambda_{h,z} = 1.3H_f$ determined in the 'linear response analysis' of Colombini (1993) discussed in chapter 5. Only case $CM250^{H3}$ attains a somewhat lower final value of $0.76H_f$ as it features three individual sediment ridges in the final phase of the simulation, as was already identified in the corresponding space-time plot in figure 6.4. The continuous erosion of sediment grains along the sediment bed leads to a successive reduction of the sediment bed height, as can be seen in panel 6.5(c). As expected, the higher Shields number and the consequently enhanced erosion rate in $CM850^{H3}$ leads to a stronger reduction of the sediment bed height of approximately one particle diameter within the first ten bulk time units of the simulation. In other words, the uppermost particle layer of the sediment bed is more or less entirely set in motion during this initial phase. This is contrasted by the moderate reduction of the mean sediment bed height of less than $0.25D$ observed in the remaining cases that feature a lower Shields number. After the initial phase of particle mobilisation, the sediment height settles at least temporarily to a more or less constant value, indicating a quasi-stationary state of the system in this phase.

6.3.2 Sediment transport

Simple considerations concerning the mass conservation of the dispersed phase imply that the initial reduction of the sediment bed height is necessarily correlated with an increase of the streamwise particle flux $\langle q_{p,x} \rangle_{xz}(t)$ that can be seen in figure 6.6 (for the definition of q_p see equation (6.10)). In panel 6.6(a), the particle flux is scaled in terms of the inertial scale $u_g D$ as it is common practice in the classical morphodynamic literature; the resulting non-dimensional volumetric particle flux $\langle q_{p,x} \rangle_{xz}/(u_g D)$ is also known as the Einstein number (Wong and Parker, 2006). Similar to the mean sediment bed height, the particle flux increases rapidly after the particle release before it settles at a quasi-stationary state, which is in case $CM850^{H3}$ however accompanied by marked oscillations. For simulation time intervals longer than those considered here, the particle flux will continue growing as the bottom friction further increases due to the rising transverse bedforms (Kidanemariam, 2016; Kidanemariam et al., 2022). Note that the normalisation of the particle flux with the inertial scale does not take into account the variation of the Shields number between the cases and over time, such that case $CM850^{H3}$ features a roughly ten times higher Einstein number than the lower Reynolds number cases.

In panel 6.6(b), the particle flux is therefore instead normalised by a reference flux that has been estimated by the empirical formula of Meyer-Peter and Müller (1948) in the modified form of Wong and Parker (2006), viz.

$$q_{ref}/(u_g D) = 4.93(\theta(t) - \theta_c)^{1.6}, \tag{6.15}$$

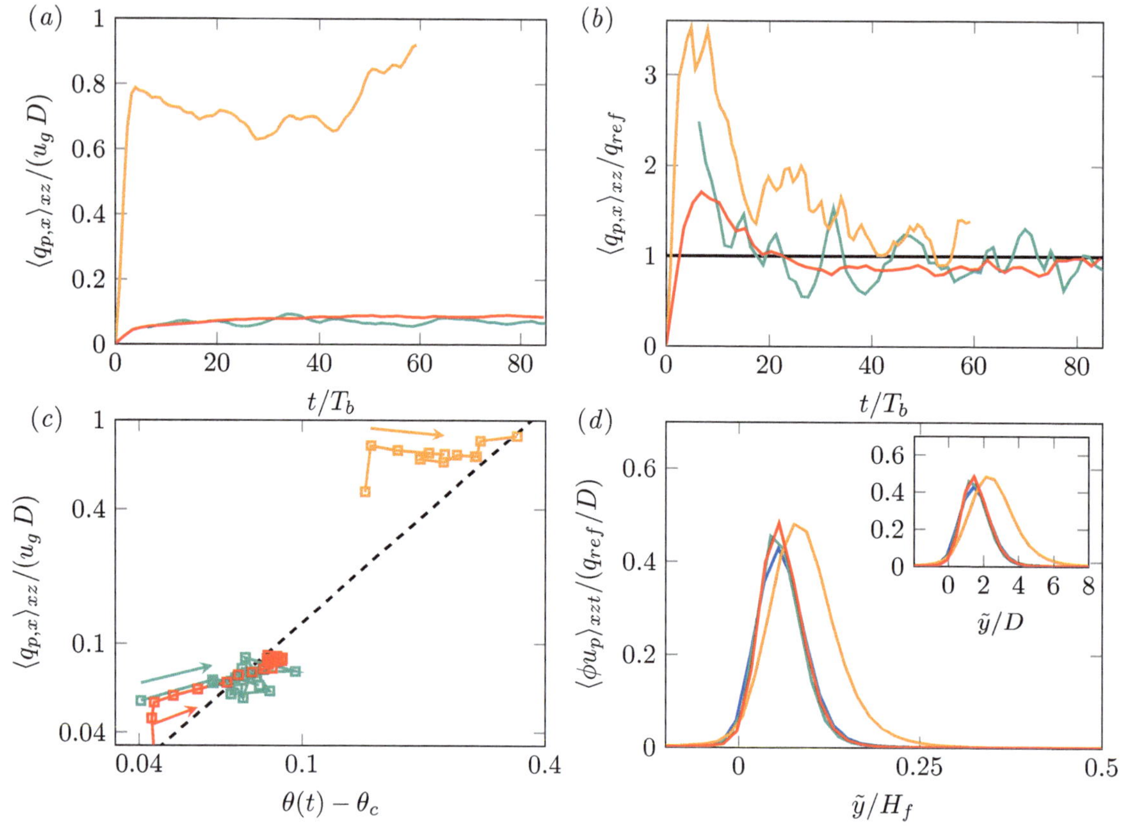

Figure 6.6: Mean streamwise particle flux $\langle q_{p,x}\rangle_{xz}$ as a function of time normalised by (*a*) the inertial scale $u_g D$ and (*b*) the reference particle flux q_{ref} computed based on the Wong and Parker (2006) version of the classical formula of Meyer-Peter and Müller (1948). That is, $q_{ref}/(u_g D) = 4.93(\theta(t)-\theta_c)^{1.6}$, with the critical Shields number $\theta_c = 0.034$ (Soulsby et al., 1997). (*c*) Streamwise particle flux $\langle q_{p,x}\rangle_{xz}/(u_g D)$ as a function of the excess Shields number $\theta(t)-\theta_c$. Each data point represents the short-time average over subintervals of $5T_b$, with time increasing along each line from left to right as indicated by the arrows. The dashed diagonal line represents the Wong and Parker (2006) reference flux $q_{ref}/(u_g D)$. (*d*) Mean particle flux density $\langle \phi u_p\rangle_{xzt}/(q_{ref}/D)$ as a function of the wall-normal distance to the mean fluid-bed interface. The inset shows the same quantity in the near-bed region, with the wall-distance scaled in terms of the particle diameter D. Cases: $CS250^{H3}$ (—), $CM250^{H3}$ (—), $CM850^{H3}$ (—), $CL250^{H16}$ (—).

where we have chosen a critical Shields number of $\theta_c = 0.034$ (Soulsby et al., 1997). While in the initial transient, the empirical relation fails to correctly predict the particle flux in the simulations by a factor up to three for case $CM850^{H3}$, it reasonably well approximates the mean particle flux in the later phase $t > 40T_b$. Kidanemariam and Uhlmann (2017) found a similarly good agreement between the particle flux estimated with relation (6.15) and their data of the mean particle flux over developed transverse patterns, even though the relation was originally developed for particle transport over macroscopically flat sediment beds.

The reason for the underestimation of the particle flux in the initial transient phase using empirical formula of Wong and Parker (2006) is seen in panel 6.6(*c*), where the particle flux $\langle q_{p,x}\rangle_{xz}/(u_g D)$ is presented as a function of the time-dependent excess Shields number $\theta(t)-\theta_c$. For the sake of clarity, particle flux and Shields number have been averaged over short time intervals of $5T_b$ length, such that each symbol represents the mean value over such a short time window. It becomes evident that the particle flux increases faster than the non-dimensional shear stress in form of the Shields number during the initial transient, with the consequence that the empirical flux that is indicated by the dashed diagonal line underpredicts the actual particle flux. Advancing in time, however, the particle flux reveals no further strong increase whereas the Shields number continues growing, such that the values

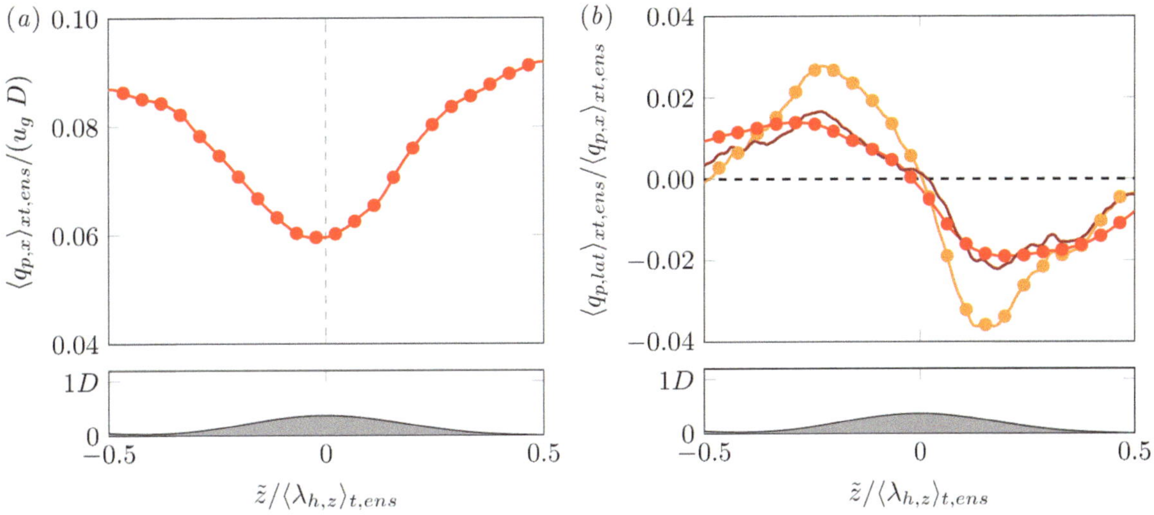

Figure 6.7: Spanwise variation of the ensemble-averaged mean streamwise and lateral particle flux, exemplary for case $CL250^{H16}$. (*a*) Streamwise particle flux $\langle q_{p,x}\rangle_{xt,ens}/(u_g D)$ (—•—). (*b*) Ratio between the particle flux component tangential to the ensemble-averaged mean fluid-bed interface and the streamwise component, $\langle q_{p,lat}\rangle_{xt,ens}/\langle q_{p,x}\rangle_{xt,ens}$ (—•—), and ratio between the tangential and the streamwise ensemble-averaged fluid velocity component $\langle w_f^{lat}\rangle_{xt,ens}/\langle u_f\rangle_{xt,ens}$ in the near bed region (—•—), extracted at a wall-normal distance of $1D$ to the ensemble-averaged mean fluid-bed interface. The data is approximated by the empirical relation (6.17) choosing a parameter $c = 0.09$ (—) which has been found to best approximate the simulation data. The presented simulation data represents the ensemble average over nine individual ridges and the time interval $t/T_b \in [20, 85]$. The corresponding mean fluid-bed interface is shown in the bottom panels. Note that for the sake of visualisation, the wall-normal extension of the ridge is therein exaggerated and thus the axes are not to scale.

observed in the simulations approach the empirically estimated ones. The fact that the Wong and Parker (2006) formula is well suited when the particle flux is in a quasi-stationary phase while it fails in highly transient situations has been recently reported by Mazzuoli et al. (2019) for oscillatory flows. In such situations, the permanently changing flow conditions lead to significant errors of the empirical relation compared to the instantaneous particle flux in the simulation.

Panel 6.6(*d*) shows the wall-normal profile of the mean particle flux density $\langle \phi u_p\rangle_{xzt}(y)$ (cf. equation (6.12)). Since the parameter points of the investigated simulations are all found in the bedload-dominated regime, sediment is predominantly transported within a layer of thickness $\mathcal{O}(D)$ above the bed in form of rolling, sliding or jumping (saltating) motion, losing the contact to the bed only for short times (van Rijn, 1984). Defining the height of the bedload layer as the global maximum of $\langle \phi u_p\rangle_{xzt}(y)$ (Chiodi et al., 2014), it is seen that the thickness of the bedload layer increases with the Shields number θ, such that in case $CM850^{H3}$ a layer of approximately $0.25H_f$ thickness above the bed reveals a non-negligible particle flux density; this is approximately $1D$ more than the value observed in the remaining lower Reynolds number cases.

The spanwise variation of the mean particle flux along the topography of the sediment bed is shown in figure 6.7. In order to increase the amount of samples for the statistical analysis, data has been additionally ensemble-averaged over nine individual ridges of case $CL250^{H16}$. Quantities that are averaged in this way are in the following indicated as $\langle\bullet\rangle_{\cdot,ens}$. A new coordinate $\tilde{z}$ is accordingly introduced, which represents the relative lateral distance to the crest (global maximum) of the ensemble-averaged mean fluid-bed interface. In that framework, we determine the contribution of the cross-stream particle flux tangential to the mean fluid-bed interface as

$$\langle q_{p,lat}\rangle_{xt,ens}(\tilde{z}) = \sin(\alpha_b)\langle q_{p,y}\rangle_{xt,ens} + \cos(\alpha_b)\,\langle q_{p,z}\rangle_{xt,ens}. \tag{6.16}$$

Let $j(\tilde{z})$ and $k(\tilde{z})$ be unit vectors normal and tangential to the ensemble-averaged mean fluid-bed interface in the cross-plane at a given spanwise location $\tilde{z}$, respectively. Then, $\alpha_b(\tilde{z})$ denotes the local inclination of the ensemble-averaged sediment bed contour at this position, that is, the angle between $k(\tilde{z})$ and the spanwise direction. The currently observed sediment ridges reveal a mild slope with typically small angles $|\alpha_b| = \mathcal{O}(1°)$ such that the deviation between $\langle q_{p,lat} \rangle_{xt,ens}$ and $\langle q_{p,z} \rangle_{xt,ens}$ is comparably small.

The lateral variation of the ensemble-averaged streamwise particle flux $\langle q_{p,x} \rangle_{xt,ens}$ is presented in panel 6.7(*a*), where it is seen that the streamwise particle transport attains its maximum over the troughs of the sediment bed profile, while it is weakest over the crest of the sediment ridge. The lateral particle flux $\langle q_{p,lat} \rangle_{xt,ens}$ shown in panel 6.7(*b*) is on average two orders of magnitude smaller than the corresponding streamwise particle flux $\langle q_{p,x} \rangle_{xt,ens}$ in all simulations. It is directed from the troughs towards the crests in consequence of the weak but finite transverse fluid motion associated with the mean secondary flow, thereby supporting the growth of local ridges and the deepening of the corresponding troughs. In theoretical studies, the spanwise particle transport is commonly determined as a fraction of the streamwise sediment flux (Blondeaux and Seminara, 1985; Colombini and Stocchino, 2012), which, in turn, is obtained using empirical relations such as the afore-discussed formulae of Meyer-Peter and Müller (1948) and Wong and Parker (2006). A popular empirical relation that estimates the ratio between spanwise and streamwise particle fluxes is, for instance, the following formula originally presented by Ikeda (1982) and further developed by Parker (1984):

$$\frac{\langle q_{p,lat} \rangle_{xt}(\tilde{z})}{\langle q_{p,x} \rangle_{xt}(\tilde{z})} = \frac{\langle w_f^{lat} \rangle_{xt}(\tilde{z})}{\langle u_f \rangle_{xt}(\tilde{z})} - \frac{c}{\sqrt{\langle \theta_l \rangle_{xt}(\tilde{z})}} \tan(\alpha_b(\tilde{z})). \tag{6.17}$$

In the previous chapter 5, the same model approach was already used in the context of the linear stability analysis of Colombini (1993). Note that several other relations of comparable form as equation (6.17) but with varying exponents and pre-factors have been proposed in the past decades, from which a few are summarised in Sekine and Parker (1992). In equation (6.17), the first term on the RHS describes the angle of the near-bed velocity w.r.t. the streamwise direction, extracted in the direct vicinity of the sediment bed. Recalling that the fluid-bed interface is determined based on the centre of gravity of the individual interface particles, the minimal distance between the fluid-bed interface and mobile particles is roughly $1D$. The near-bed velocity components are therefore extracted at a wall-normal distance of $1D$ above the ensemble-averaged mean fluid-bed interface $\langle h_b \rangle_{xt,ens}(\tilde{z})$. A local Shields number $\theta_l(x,z,t)$ is further introduced based on the total shear stress $\tau_{tot}(x,t)$ extracted along the fluid-bed interface. A rigorous definition of the total shear stress will be given in equation (6.39) in section 6.3.9 below. Relation (6.17) indicates that along a laterally sloping bed, the orientation of the mean particle flux deviates from the direction of the near-bed shear stress (represented by the first term on the RHS) as a consequence of gravitational effects which counteract the uphill particle motion driven by the secondary flow (Parker, 1984; Seminara, 2010). The amplitude of the counteracting gravity term is therein controlled by an empirical coefficient c, whose exact value remains a matter of discussion in the respective literature: While originally estimated in a range $c = 0.5$-0.6 (Engelund, 1981), most studies on linear bedform stability adopt a somewhat lower value of $c = 0.3$ (Colombini et al., 1987; Colombini, 1993), as proposed by Olesen (1983). More recently, Colombini and Stocchino (2012) used a value of $c = 0.1$ in the framework of a linear stability analysis of three-dimensional bedforms due do a better match with experimental data. As stressed by the latter authors, the exact value of c might additionally depend on the wall-normal distance to the mean fluid-bed interface at which the shear stress is extracted.

In this context, it should be noted that models of the form of relation (6.17) were also derived from theoretical considerations, evaluating the balance of all forces that act on a single spherical particle on a sloping ground, comprising contributions from drag, lift, gravitational and bottom resistive forces. For small lateral inclination angles as in the current case, linearised formulations assuming $\sin(\alpha_b) \approx \tan(\alpha_b)$ have been derived for instance by Kikkawa et al. (1976), whose model was later generalised by Parker and Andrews (1985) to obtain

$$\frac{\langle q_{p,lat}\rangle_{xt}}{\langle q_{p,x}\rangle_{xt}} = \frac{\langle w_f^{lat}\rangle_{xt}}{\langle u_f\rangle_{xt}} - \frac{1+r_b\mu_b}{\mu_b}\left(\frac{\langle u_f\rangle_{xt} - \langle u_p\rangle_{xt}}{\langle u_f\rangle_{xt}}\right)\tan(\alpha_b) \tag{6.18a}$$

$$\approx \frac{\langle w_f^{lat}\rangle_{xt}}{\langle u_f\rangle_{xt}} - \frac{1+r_b\mu_b}{\mu_b}\left(\frac{\theta_c}{\langle\theta_l\rangle_{xt}}\right)^{1/2}\tan(\alpha_b), \tag{6.18b}$$

with a coefficient r_b that describes the ratio between particle lift and drag force and a dynamic Coulomb friction coefficient μ_b associated with the motion of a spherical particle on the rough bed. For the case of arbitrarily sloping beds with non-negligible inclination angles, a more elaborate model has been derived by Parker et al. (2003) that reduces to relations (6.18a) and (6.18b) for streamwise horizontal beds with small transverse slope. Comparing equations (6.17), (6.18a) and (6.18b), we obtain a relation connecting the three parameters c, r_b and μ_b as

$$c = \sqrt{\langle\theta_l\rangle_{xt}}\,\frac{1+r_b\mu_b}{\mu_b}\,\frac{\langle u_f\rangle_{xt} - \langle u_p\rangle_{xt}}{\langle u_f\rangle_{xt}} \approx \frac{1+r_b\mu_b}{\mu_b}\sqrt{\theta_c}. \tag{6.19}$$

The deviation angle between the particle flux and the near-bed velocity is also observed for the ensemble-averaged simulation results in panel 6.7(b): The normalised transverse variation of the lateral near-bed fluid velocity $\langle w_f^{lat}\rangle_{xt,ens}/\langle u_f\rangle_{xt,ens}$ is in phase with that of the corresponding particle flux, but it attains somewhat larger amplitude values than the ratio between the particle flux components. Here, $\langle w_f^{lat}\rangle_{xt,ens}$ is the lateral fluid velocity component, defined in analogy to equation (6.16). The deviation between both ratios is seen to be strongest roughly half-way between trough and crest where the transverse slope $\tan(\alpha_b)$ is highest and, thus, the downward directed force component due to gravity attains the largest values.

In order to compare the simulation results with these empirical models, we consider equations (6.17) to (6.19) and set $\theta_c = 0.034$. Choosing the lift to drag force ratio as $r_b = 0.2$ based on the results of Chan-Braun et al. (2011) for fixed, regularly arranged particles at a comparable parameter point as in the current simulations, a parameter value $c = 0.09$ very similar to the one adopted by Colombini and Stocchino (2012) is found to best fit the simulation data (cf. figure 6.7(b)). Recomputing the corresponding friction coefficient using equation (6.19), however, leads to a much larger value $\mu_b \approx 3.2$ than the commonly applied values $\mu_b \approx 0.7$ (Parker, 1984). Choosing a value of $\mu_b = 0.7$ in equation (6.18a), on the other hand, clearly overpredicts the amplitude of the counteracting gravitational term in the simulations.

In order to clarify the reasons for the deviation between the current simulation results and the model of Parker and Andrews (1985), further simulations and a detailed investigation of the force components acting on individual mobile particles on sloping beds would be required. In particular, it would be necessary to perform a careful sensitivity analysis that ascertains the influence of the individual parameters on the predicted results; this includes in particular the chosen values of μ_b and θ_c as well as the distance to the mean fluid-bed interface at which the relevant quantities are extracted. While unfortunately out of the scope of the current work, such a detailed analysis of lateral particle transport represents an interesting question for a future study. Also, note that – strictly speaking – the above discussed models are derived for particles that are in permanent contact with the bed in form of rolling or sliding motion, while they do not capture sediment saltation. Some attempts have been made to develop empirical relations based on particle saltation models (Sekine and Kikkawa, 1992; Sekine and Parker, 1992; Niño and García, 1994; Niño et al., 1994), but the validity of some of the underlying assumptions, namely the so-called 'Bagnold hypothesis', is questionable (Seminara et al., 2002).

6.3.3 Eulerian particle force fields

In order to erode sediment from the surface of the bed, a sufficient net force or torque has to act on the individual particles. The total force exerted on a single particle comprises, aside from buoyancy effects, a hydrodynamic as well as a contact force component, reflecting the two main mechanisms that are causing mobilisation of sediment grains: hydrodynamic drag and lift forces induced by the turbulent flow (Chan-Braun et al., 2011; Cameron et al., 2020) as well as collisions with moving neighbouring particles. In the current computational approach, the collision force $\boldsymbol{F}^{C^{(l)}}(t)$ exerted on particle l originates from the applied soft-sphere collision model outlined in chapter 4. The hydrodynamic Lagrangian force $\boldsymbol{F}^{H^{(l)}}(t)$ acting on particle l, on the other hand, is determined by integration of the Eulerian force density field $\boldsymbol{f}(\boldsymbol{x},t)$ that emerges in the context of the immersed boundary formulation over the particle's surface $\mathcal{S}_p^{(l)}$ (Uhlmann, 2008), viz.

$$\boldsymbol{F}^{H^{(l)}}(t) = \int_{\boldsymbol{x}\in\mathcal{S}_p^{(l)}} \boldsymbol{f}(\boldsymbol{x},t)\mathrm{d}\boldsymbol{x}. \tag{6.20}$$

In this context, we associate $\boldsymbol{F}^{H^{(l)}}$ with the particle's centre of gravity $\boldsymbol{X}_p^{(l)}$. Note that, in the discrete setting of the currently used immersed boundary formulation, the integral in relation (6.20) transforms into a sum over the local force contributions at the discrete locations of the finite set of Lagrangian force points. Eventually, the buoyancy effect is represented by the submerged particle weight $\boldsymbol{F}_W = (0, -F_W, 0)^T$.

In the following, we map the discrete Lagrangian particle force information $\boldsymbol{F}^{tot^{(l)}}(t) = \boldsymbol{F}^{H^{(l)}}(t) + \boldsymbol{F}^{C^{(l)}}(t)$ of all particles $1 \leq l \leq N_p$ onto a single surrogate discontinuous Eulerian field $\boldsymbol{f}^{tot}(\boldsymbol{x},t)$ similar to the procedure described in Kidanemariam et al. (2022), viz.

$$\boldsymbol{f}^H(\boldsymbol{x},t) + \boldsymbol{f}^C(\boldsymbol{x},t) = \boldsymbol{f}^{tot}(\boldsymbol{x},t) = \sum_{l=1}^{N_p} \boldsymbol{F}^{tot^{(l)}}(t) I_p^{(l)}(\boldsymbol{x},t). \tag{6.21}$$

Here, $I_p^{(l)}(\boldsymbol{x},t)$ is a particle indicator function that is equal to unity if the Eulerian field location $\boldsymbol{x}$ at time t lies within the volume of particle l and zero otherwise. Further, $\boldsymbol{f}^H$ and $\boldsymbol{f}^C$ are the hydraulic and contact force components of $\boldsymbol{f}^{tot}$, respectively. Note that the two Eulerian fields $\boldsymbol{f}$ and $\boldsymbol{f}^H$ describe, despite their similar notation, two completely different quantities: The former represents a smooth force density field that contributes to the global momentum balance, whereas the latter is a surrogate discontinuous

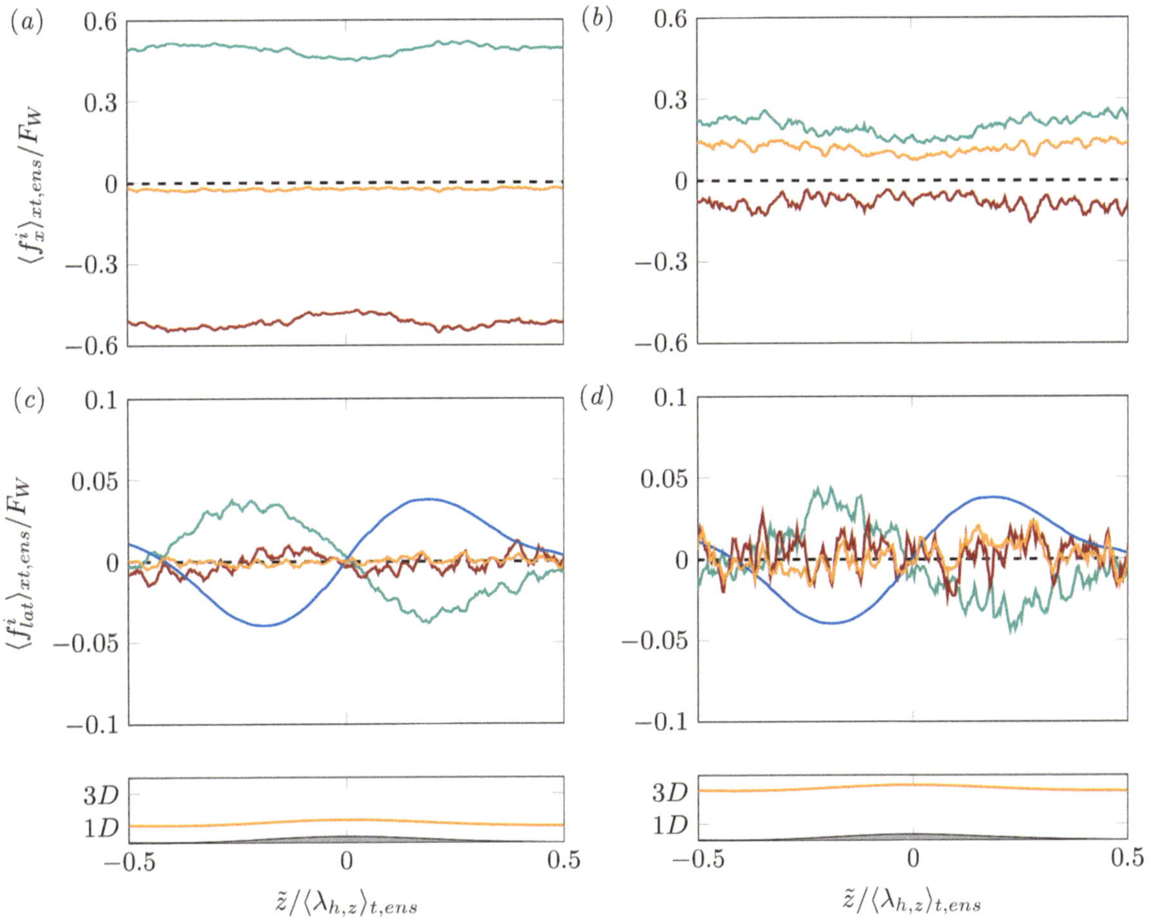

Figure 6.8: Ensemble-averaged hydrodynamic and collision forces along the fluid-bed interface over evolving ridges in case $CL250^{H16}$. *(a,b)* Streamwise force balance comprising hydrodynamic force f_x^H (—), collision force f_x^C (—) and total force $f_x^{tot} = f_x^H + f_x^C$ (—) components, respectively. *(c,d)* Lateral bed-parallel force balance comprising hydrodynamic force f_{lat}^H (—), collision force f_{lat}^C (—), and total force f_{lat}^{tot} (—) components as well as the bed-parallel contribution of the relative buoyancy force $-\boldsymbol{F}_W$ (—), respectively. The force balance has been evaluated at a wall-normal offset of *(a,c)* $1D$ and *(b,d)* $3D$ from the ensemble-averaged mean fluid-bed interface $\langle h_b \rangle_{xt,ens}(\tilde{z})$, respectively.

Eulerian force field, computed exclusively for post-processing purposes. Whilst introduced here for the particle forces, the mapping (6.21) can be generalised to arbitrary Lagrangian particle-related quantities and the corresponding Eulerian fields. For instance, the mean solid volume fraction can be obtained as

$$\langle \phi_s \rangle_x(y,z,t) = \langle \sum_{l=1}^{N_p} I_p^{(l)}(\boldsymbol{x},t) \rangle_x. \tag{6.22}$$

Note that without applying streamwise averaging, the above field would be a binary field attaining zero for $\boldsymbol{x} \in \Omega_f(t)$ and unity if $\boldsymbol{x} \in \Omega_p(t)$.

As all here considered parameter points lie in the bedload-dominated regime, the particle transport is concentrated in a layer of several particle diameters thickness above the sediment bed. Consequently, non-trivial values of the Eulerian particle force field $\boldsymbol{f}^{tot}$ predominantly occur in this region of intense particle transport, as we shall see in the following. In figure 6.8, we show for case $CL250^{H16}$ the lateral variation of the ensemble-averaged Eulerian force field $\langle \boldsymbol{f}^{tot} \rangle_{xt,ens}$ and its individual components at wall-normal distances of $1D$ and $3D$ to the ensemble-averaged mean fluid-bed interface $\langle h_b \rangle_{xt,ens}(\tilde{z})$. Comparing these locations with the wall-normal profile of $\langle \phi u_p \rangle_{xzt}$ in figure 6.6(*d*), we observe that the distance $\tilde{y} = 1D$ lies in the region of highest sediment transport intensity close to the maximum of $\langle \phi u_p \rangle_{xzt}$. The distance $\tilde{y} = 3D$, on the other hand, is found at the upper end of the particle transport layer where significantly lower values of $\langle \phi u_p \rangle_{xzt}$ are attained. Accordingly, the streamwise force balance qualitatively differs between both wall-normal locations, as can be seen by comparing figures 6.8(a) and 6.8(b). At a distance of $1D$ from the sediment bed, hydrodynamic and contact forces essentially balance each other, both attaining values around $F_W/2$ but with opposite signs. Both contributions reveal a slight variation along the bed contour, with slightly lower and higher amplitudes over the crest and troughs of the ensemble-averaged bed profile, respectively. Further away from the bed (at $\tilde{y} = 3D$) where particle collisions are much less frequent, on the other hand, the amplitude of both hydrodynamic and collision forces is reduced compared to the values in the direct vicinity of the bed, and they are no longer in equilibrium. Instead, on average, a positive net streamwise force acts on the particles in this distance to the wall, implying that particles are predominantly accelerated in the mean flow direction in this layer. The spanwise profiles of the lateral forces, on the other hand, appear to be quite similar at both distances from the sediment bed concerning both amplitude and general shape (cf. figures 6.8(c,d)). In the centre and at the top of the particle transport layer, the hydrodynamic force reveals a sinusoidal variation along the sediment bed profile, oriented in such sense that particles are accelerated towards the ridge crest. It thus acts against gravity, which causes a downward force of comparable amplitude. In both cases, therefore, hydrodynamic and gravitational contributions essentially balance each other, while lateral contact forces are apparently of lower relevance for the total force balance.

The reason for the qualitative difference in the streamwise force balance at the two different heights is not yet fully clarified, but it is suspected that it originates in the different transport mechanisms inside and outside the bedload layer. In other words, particles located inside the bedload layer feel, on average, a positive hydrodynamic force exerted on them by the turbulent flow that aims to accelerate the individual particles in the streamwise direction. On the other hand, the solid volume fraction within the bedload layer is quite high such that particles once in motion collide with other particles in all probability, whereby their momentum is significantly reduced. The collision force therefore leads, on average, to a deceleration of the particles such that the forces acting on the dispersed phase are more or less in equilibrium within the bedload layer. Whilst the majority of the particles remain in the bedload layer throughout the entire simulation time, there is a small fraction of sediment grains which are transported away from the bed due to, for instance, strong intermittently-occurring turbulent events. Those that reach regions outside the bedload layer initially possess a velocity deficit compared to the surrounding mean flow velocity that is compensated with time as the particles are accelerated due to the action of positive streamwise hydrodynamic forces (Kiger and Pan, 2002). Note that the here considered heavy particles seldom remain in these regions for longer time intervals, but they soon fall back down to the sediment bed under the action of the gravitational field as soon as the turbulent fluctuations cannot hold them in layers away from the wall anymore. The important difference to the upward motion is that particles are in this phase not slowly decelerated but their streamwise velocity reduces abruptly as they fall down to the bed.

Figure 6.9(c) shows a typical sample trajectory of such a particle in one of the later discussed long-time evolution cases $CS250^{H6}$ that is first eroded and subsequently moved out of the bedload layer into a local high-speed zone in between two low-speed regions, as shown in the instantaneous flow field cross-section in figure 6.9(a). Figure 6.9(b) shows a close-up of the particle whose trajectory is shown in figure 6.9(c) and the surrounding flow field for the same instant. We clearly see that the streamwise

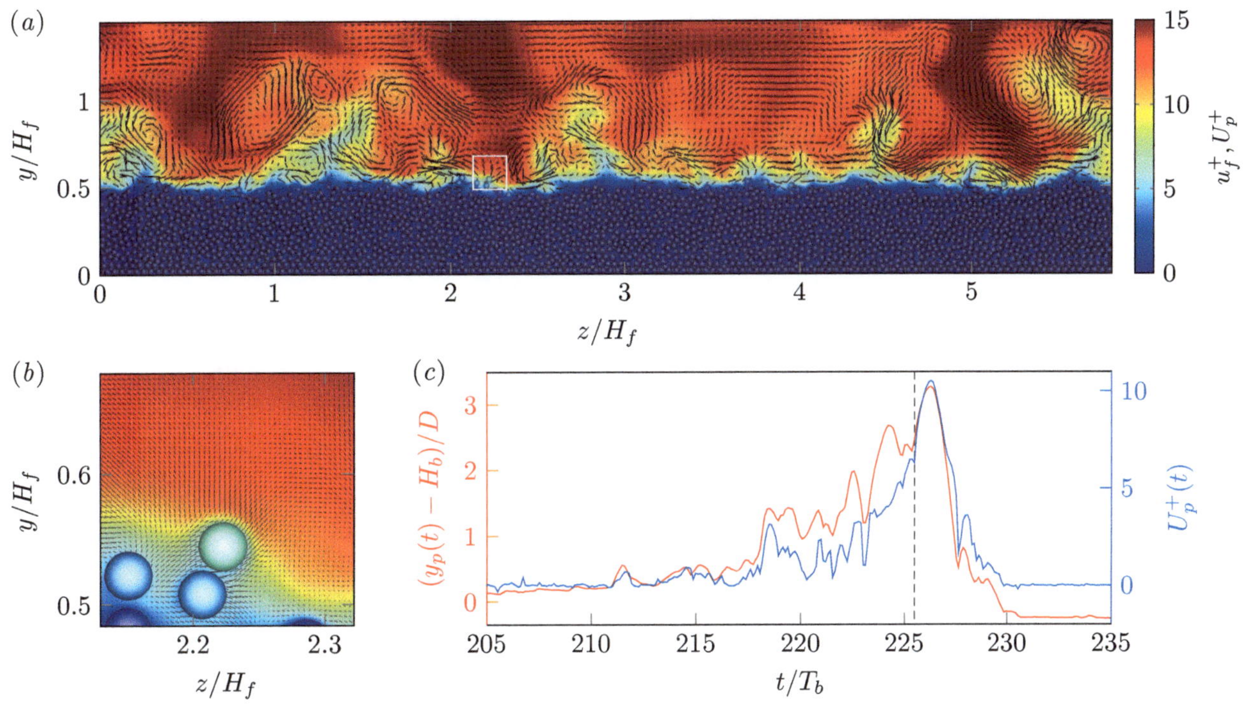

Figure 6.9: Sample particle trajectory, together with an instantaneous snapshot of the streamwise fluid and particle velocity in the cross-plane of case $CS250^{H6}$. (*a*) Instantaneous snapshot of the streamwise fluid u_f^+ and particle velocity U_p^+ in the cross-plane at a fixed streamwise position $x/H_f = 1.45$ of case $CS250^{H6}$. The fluid velocity is presented as colour plot, while the cross-stream flow field $(v_f^+, w_f^+)^T$ is represented in form of a vector plot. The particle velocity is indicated by the sphere's colour, i.e. the surface has been coloured depending on the streamwise particle velocity U_p^+ following the same colour scheme as for the fluid. Note that only those particles are shown whose volumes intersect the chosen (y, z)-plane. (*b*) Close-up of the data shown in (*a*) in the region marked by the white rectangle. (*c*) Temporal evolution of the wall-normal location $(y_p(t) - H_b)/D$ (—) and streamwise velocity $U_p^+(t)$ (—) for the uppermost particle in (*b*). The vertical grey line indicates the instant at which the instantaneous flow field in (*a,b*) has been extracted.

particle velocity is roughly the same as that of the fluid in its direct vicinity due to the no-slip boundary condition at the particle surface, but it is clearly lower than the streamwise fluid velocity further away from its centre, indicating an instantaneous velocity deficit between the particle and the fluid phase.

In figure 6.9(*c*), it is seen that the investigated particle is initially more or less in rest except for two short phases during which it is slightly lifted up, indicating that the net force acting on the particle was to weak or its duration too short to effectively mobilise the particle (Chan-Braun, 2012). Then, about ten time bulk time units before the shown snapshot of the flow field, the sediment grain is mobilised and starting from $t \approx 218T_b$, the particle is seen to be completely eroded from the sediment bed. During the following approximately ten bulk time units, it successively gains in height and its streamwise velocity $U_p(t)$ increases at an almost identical rate, even though partly at a small time delay, which agrees fairly well with the above conceived mechanism. The particle reaches its maximum streamwise velocity at the same time as its trajectory is at the maximum distance to the bed. However, the flow cannot hold the heavy particle aloft for a longer time interval, and so the particle falls back down to the sediment bed in a time clearly shorter than the one it took to transport it to that height. Simultaneously, the particle velocity $U_p(t)$ decreases rapidly to zero as the particle reaches the sediment bed again.

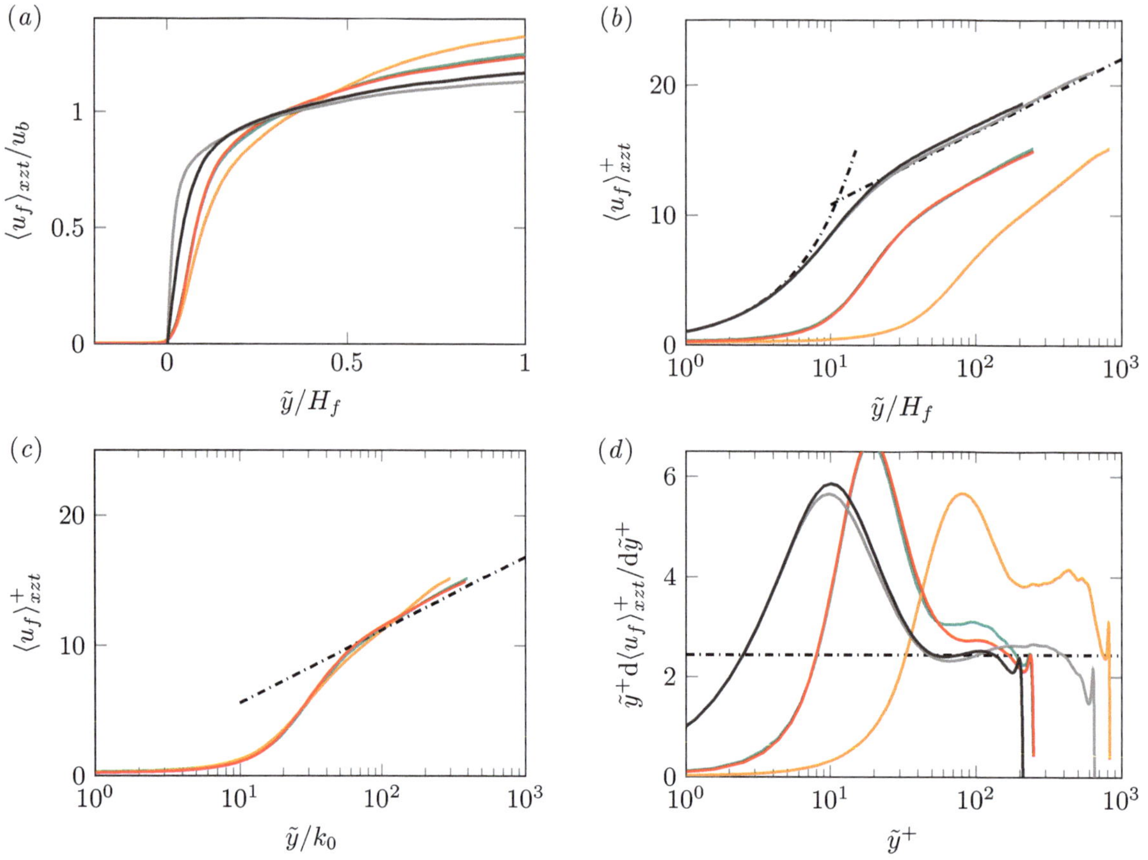

Figure 6.10: Wall-normal profiles of the double-averaged mean velocity *(a)* $\langle u_f \rangle_{xzt}/u_b$ as a function of $\tilde{y}/H_f$ (linear scaling) and *(b)* $\langle u_f \rangle^+_{xzt}$ as a function of $\tilde{y}^+$ (semi-logarithmic scaling). The dashed-dotted line shows the theoretically predicted profiles for the buffer and log-layer as $\langle u_f \rangle^+_{xzt} = \tilde{y}^+$ and $\langle u_f \rangle^+_{xzt} = 1/\kappa \ln(\tilde{y}^+) + B_u$ ($\kappa = 0.41$, $B_u = 5.2$), respectively. *(c)* Same as *(b)*, but with the wall-distance scaled in terms of the hydraulic roughness k_0. The dashed line indicates the relation $\langle u_f \rangle^+_{xzt} = 1/\kappa \ln(\tilde{y}/k_0)$. The single-phase cases are accordingly not included. *(d)* Log-layer indicator function $\tilde{y}^+ \mathrm{d}\langle u_f \rangle^+_{xzt}/\mathrm{d}\tilde{y}^+$. The horizontal dashed line marks the inverse of the usually considered value of v. Kármán's constant, $1/\kappa = 1/0.41$. $CM250^{H3}$ (—), $CM850^{H3}$ (—), $CL250^{H16}$ (—), $CM650^{H3}_{smooth}$ (—), $CL250^{H16}_{smooth}$ (—).

6.3.4 Turbulent mean flow

The presence of a rough lower domain boundary together with the mobile sediment severely modify the mean flow profile $\langle \boldsymbol{u}_f \rangle_{xzt}(y)$, as can be seen in figure 6.10(*a*,*b*) where we compare the mean velocity profile for smooth-wall and sediment-laden simulations in double-linear outer and semi-logarithmic inner scaling, respectively. Especially the region close to the virtual wall is markedly altered compared to the single-phase flows: In the particle-laden cases, the mean shear $S = \mathrm{d}\langle \boldsymbol{u}_f \rangle_{xzt}/\mathrm{d}y$ is of negligible amplitude near the bed whereas it attains its maximum directly at the smooth wall in single-phase simulations. Consequently, it is responsible for the entire wall shear stress in the single-phase configurations, while in the multiphase cases, the bottom shear stress along the fluid-bed interface is clearly dominated by the stress resulting from particle-fluid interactions (Kidanemariam and Uhlmann, 2017). Note that the mean velocity profile in the sediment-laden simulations additionally reveals an inflection point in the vicinity of the bed due to the porosity of the underlying sediment bed, which is susceptible to a Kelvin-Helmholtz instability (Jiménez et al., 2001) and which is naturally absent in the single-phase cases. Also, the increase of $\langle \boldsymbol{u}_f \rangle_{xzt}$ with increasing distance to the bottom wall is significantly slower in the particle-laden simulations, causing a velocity deficit between single- and multiphase simulations that is a common feature of turbulent flows over rough boundaries as a consequence of an enhanced friction along the wall (Flores and Jiménez, 2006; Chan-Braun et al., 2011). To maintain the constant bulk flow rate q_f, the locally reduced flow rate along the bed is compensated by an increased flow rate in the upper half of the clear-fluid region, with the consequence that the mean velocity is higher in the particle-laden simulations in this region. The redistribution of the mean flow velocity results in a different slope of the mean velocity profile across the logarithmic and outer layer in the sediment-laden cases. In the low Reynolds number cases, this deviation is still comparably weak allowing to compensate most of it by the conventional shift of the log-layer velocity profile by an offset ΔU^+ defined in rough wall turbulence as (Jiménez, 2004a)

$$\langle u_f \rangle_{xzt}^+ = \frac{1}{\kappa} \ln(\tilde{y}^+) + B_u - \Delta U^+. \tag{6.23}$$

Using a standard choice of von Kármán's constant $\kappa = 0.41$ and of the coefficient $B_u = 5.2$ (Pope, 2000, p.274), we determine the values of the roughness function as $\Delta U^+ = 3.59$ ($CM250^{H3}$) and $\Delta U^+ = 3.73$ ($CL250^{H16}$), respectively. These values are larger than the value $\Delta U^+ = 1.03$ determined by Chan-Braun et al. (2011) for open channel flow over a layer of fixed spheres in square arrangement at comparable values of $Re_b = 2900$ and $D^+ = 10$, while they are closer to the value $\Delta U^+ = 4.85$ found for $Re_b = 2900$ and $D^+ = 50$ in the same study. Relation (6.23) can be reworked into the following expression for the velocity profile over rough walls

$$\langle u_f \rangle_{xzt}^+ = \frac{1}{\kappa} \ln(\tilde{y}/k_0). \tag{6.24}$$

Therein, k_0 is the hydrodynamic roughness length that is often used interchangeably with ΔU^+ to quantify the modulation of the mean flow profile due to wall roughness (Raupach et al., 1991). In figure 6.10(c), the inner-scaled velocity profiles of the particle-laden cases are repeated, with the distance to the mean fluid-bed interface now scaled in terms of the hydrodynamic roughness length k_0 that was determined as $k_0/D = 0.07$ ($CM250^{H3}$), $k_0/D = 0.10$ ($CM850^{H3}$) and $k_0/D = 0.07$ ($CL250^{H16}$), respectively. Whilst the low Reynolds number cases reasonably well collapse with the idealised relation (6.24), the slope of the mean velocity profile in the high Reynolds number case $CM850^{H3}$ is seen to deviate clearly from the theoretically predicted line.

This is further highlighted by the log-layer indicator function $\tilde{y}^+ \mathrm{d}\langle u_f \rangle_{xzt}^+ / \mathrm{d}\tilde{y}^+$ shown in figure 6.10(d), which indicates the wall-normal range in which the logarithmic velocity profile in equation (6.23) is applicable. It is clearly discernible that all cases reveal, within the bounds of the relatively low Reynolds numbers, a more or less developed logarithmic layer. Compared to the smooth-wall cases, however, the region in which the velocity profile in the particle-laden simulations can be considered to vary logarithmically is further away from the virtual wall, especially in case $CM850^{H3}$ where the mentioned region is reached at 200 wall units distance to the mean sediment bed height. In agreement with the previous observation that the slope of the velocity profile varies between smooth-wall and particle-laden simulations, the classical von Kármán's constant $\kappa = 0.41$ is seen to correctly describe the profile outside the buffer layer only in the smooth-wall cases. While for the low Reynolds number cases, κ is only slightly lower than in the smooth-wall cases, a value of $\kappa \approx 0.25$ has to be chosen to correctly match the profile of case $CM850^{H3}$. As outlined above, the reason for this discrepancy is that the high particle concentration within the bedload layer clearly dampens the fluid flow rate near the bed, which is compensated by an accordingly higher shear in the outer flow.

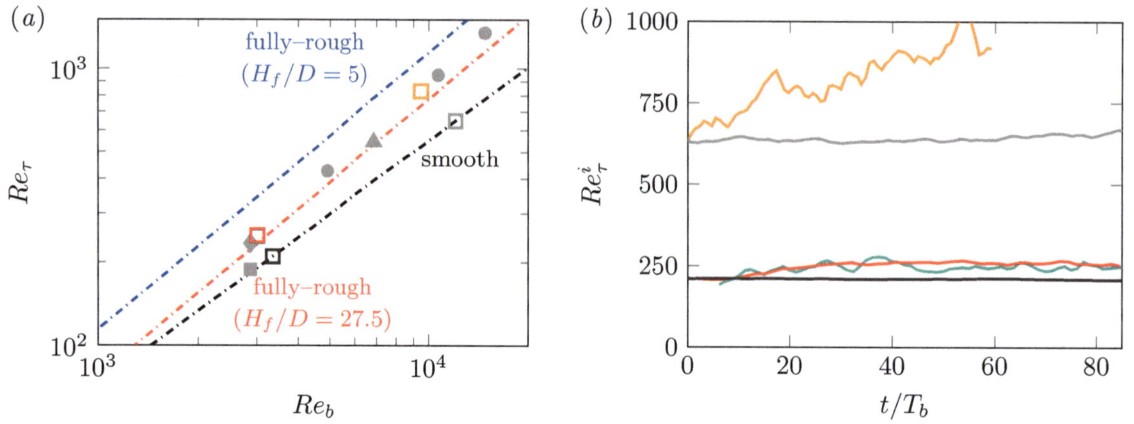

Figure 6.11: (*a*) Variation of the mean friction Reynolds number Re_τ as a function of the bulk Reynolds number Re_b. The black dashed-dotted line refers to the relation $Re_\tau = 0.166 Re_b^{0.88}$ obtained for smooth-wall channels (Pope, 2000). The red and blue dashed-dotted curves represent the relation between Re_b and Re_τ under fully-rough conditions for $H_f/D = 27.5$ (-··-) and $H_f/D = 5.0$ (-··-), respectively, based on the following relation for the friction coefficient $c_f = 2u_\tau^2/u_b^2 \approx 2(\ln(H_f/D)/\kappa + 8.5 - 3/(2\kappa))^{-2}$, with $\kappa = 0.41$ (Pope, 2000). Additional data points are from the following studies on open channel flow over a layer of fixed spheres ($H_f/D = 2.5$-18.3): DNS studies of Chan-Braun et al. (2011) (■: $D^+ = 10$, $H_f/D = 18.3$; ◆: $D^+ = 50$, $H_f/D = 5.6$) and Mazzuoli and Uhlmann (2017) (▲: $D^+ = 120$, $H_f/D = 5.4$), both in square arrangement; experimental study of Amir et al. (2014) (●: $D^+ = 170$-538, $H_f/D = 2.5$), hexagonal arrangement. (*b*) Time evolution of the instantaneous friction Reynolds number Re_τ^i. $CM250^{H3}$ (—), $CM850^{H3}$ (—), $CL250^{H16}$ (—), $CM650^{H3}_{smooth}$ (—), $CL250^{H16}_{smooth}$ (—).

The intense particle transport in the vicinity of the sediment bed causes an increase of the mean friction Reynolds number Re_τ which is shown in figure 6.11(*a*) as a function of the bulk Reynolds number Re_b. While for the single-phase cases, the relation between the bulk and friction Reynolds numbers is well described by the analytically obtained function $Re_\tau = 0.166 Re_b^{0.88}$ (Pope, 2000), the particle-laden cases attain clearly higher values of Re_τ for comparable values of Re_b. In particular, a comparison with the red dashed-dotted line reveals that the here-considered transitionally-rough mobile bed cases attain friction Reynolds numbers roughly equivalent to those expected on a fixed but fully-rough sediment bed. The overview is supplemented with datasets from experiments (Amir et al., 2014) and DNS-based studies (Chan-Braun et al., 2011; Mazzuoli and Uhlmann, 2017) of open channel flow at low to medium relative submergence ($H_f/D = 2.5$-18.3) over a wall-mounted layer of regularly arranged spherical particles (hexagonal and square arrangements) in the transitionally and fully-rough regime. For the sake of completeness, it should be stressed that the exact location of the virtual wall in these references slightly differs from ours, which is however not expected to have a relevant influence on the observations: While our fluid-bed interface is reconstructed from the spheres' centre locations, Chan-Braun et al. (2011) and Mazzuoli and Uhlmann (2017) locate the virtual wall roughly $0.15D$-$0.2D$ below the particles' highest points. Amir et al. (2014), in turn, place it at the particles' tops.

A comparison of the current simulation results with these reference datasets underlines that the mobility of the sediment bed has a similar effect on the friction Reynolds number as larger relative roughness heights D^+ and D/H_f for the same bulk Reynolds number. As pointed out by Kidanemariam et al. (2022), the here observed increase in friction over a mobile sediment bed is of relevance, for instance, in the context of theoretical stability analysis (Colombini, 1993, 2004; Fourriere et al., 2010), in which the influence of roughness on the flow is often estimated by empirical formulae for flow over fixed roughness even in studies on sediment bedform evolution.

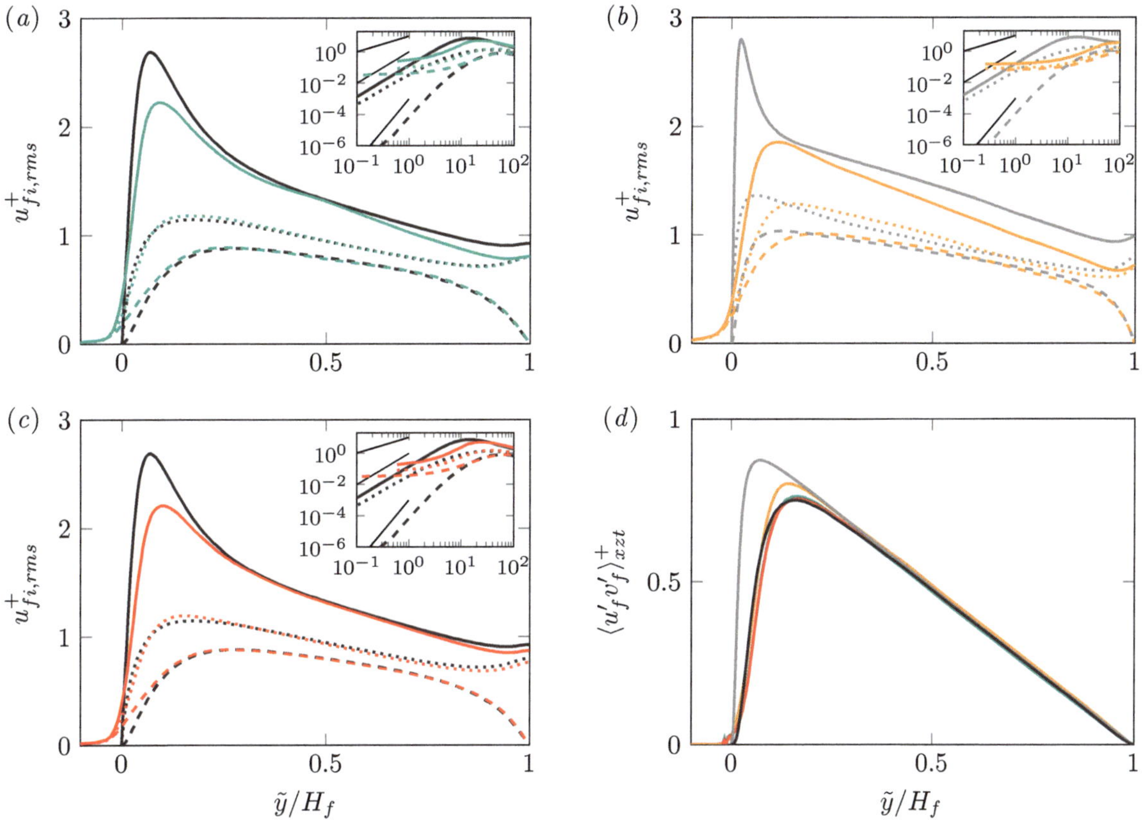

Figure 6.12: Wall-normal profiles of the mean Reynolds stresses in cases (*a*) $CM250^{H3}$ (—), (*b*) $CM850^{H3}$ (—) and (*c*) $CL250^{H16}$ (—). The small insets show close-ups of the same data close to the (virtual) origin as a function of the wall-normal distance $\tilde{y}^+$ in double-logarithmic scaling. Solid grey lines therein indicate decay rates as $\sim \tilde{y}$, $\sim \tilde{y}^2$ and $\sim \tilde{y}^4$, respectively. $u_{f,rms}$ (—), $v_{f,rms}$ (- - -), $w_{f,rms}$ ($\cdots$), where $u^+_{fi,rms} = (\langle u'_{fi} u'_{fi} \rangle_{xzt})^{1/2}/u_\tau$ ($i = x, y, z$). (*d*) Wall-normal profiles of the Reynolds shear stresses $\langle u'_f v'_f \rangle_{xzt}/u_\tau^2$. Reference data from companion single-phase flow simulations are (*a,c*) $CL250^{H16}_{smooth}$ (—) and (*b*) $CM650^{H3}_{smooth}$ (—), respectively.

Figure 6.11(*b*) illustrates the temporal evolution of an equivalent instantaneous friction Reynolds number $Re^i_\tau(t) = \sqrt{(\tau^i_w(t)/\rho_f)}\langle h_f \rangle_{xz}(t)/\nu_f$, revealing that the bottom friction is continuously growing as the particle transport is increasing. In case of the single-phase simulations, on the other hand, $Re^i_\tau(t)$ is essentially constant throughout the entire observation interval, revealing only low-amplitude oscillations around the mean Reynolds number Re_τ. Note that the instantaneous wall shear stress $\tau^i_w(t)$ in above definition has been determined by evaluating equation (6.3) for the instantaneous plane-averaged fluid velocity and velocity fluctuation covariances of each instantaneous flow field separately.

The modified distribution of the mean shear directly influences the intensity and distribution of the Reynolds stresses, as can be seen in figure 6.12. Note that the statistics in the particle-laden cases are not assumed to be fully converged, since they are gathered over a relatively short time interval during which the system is still in a transient state (cf. table 6.2). While it is assumed that the dynamics of the short-living small-scale structures are well captured, the observed differences between the single- and multiphase simulations in the vicinity of the free surface are claimed to be the result of the relatively short observation time interval T_{obs}, compared to the mean lifetime of the large-scale structures. In all particle-laden simulations visualised in figure 6.12, the peak of the streamwise normal stress $\langle u'_f u'_f \rangle_{xzt}$ representative of the buffer layer is markedly reduced compared to the corresponding single-phase smooth-wall cases due to the lower mean shear (Chan-Braun et al., 2011), which represents the main source of energy for the fluctuating field. In the low Reynolds number cases $CM250^{H3}$ and $CL250^{H16}$, the peak in $\langle u'_f u'_f \rangle_{xzt}$ is still detectable at roughly the same wall distance as in the smooth-wall cases, even though it is less pronounced. In the high Reynolds number case $CM850^{H3}$, on the other hand, the roughness of the sediment bed and the intense bedload transport are responsible for the complete disappearance of the typical buffer-layer peak of $\langle u'_f u'_f \rangle_{xzt}$. The same phenomenon is commonly reported for flows over fully-rough walls where the mean height of the roughness elements is much larger than the viscous length scale such that the self-sustained near-wall cycle is destroyed (Jiménez, 2004a; Flores and Jiménez, 2006; Mazzuoli and Uhlmann, 2017). In the studied low Reynolds number cases, on the other hand, the particles are sufficiently small compared to the viscous length scale ($D \approx 10\delta_\nu$) to be transported by the buffer-layer structures (Kidanemariam et al., 2013) rather then to destroy the near-wall cycle, explaining the persistence of the near-wall peak in $\langle u'_f u'_f \rangle_{xzt}$.

The remaining normal stresses $\langle v'_f v'_f \rangle_{xzt}$ and $\langle w'_f w'_f \rangle_{xzt}$ show weaker deviations between particle-laden and smooth-wall statistics, which predominantly appear in a layer of thickness $\tilde{y} \approx 0.11 H_f$ ($\tilde{y} \approx 3D$) above the virtual wall for the low Reynolds number cases and $\tilde{y} \approx 0.17 H_f$ ($\tilde{y} \approx 4.5D$) in the high Reynolds number case, respectively. The decaying behaviour of the Reynolds stresses in the vicinity of the (virtual) wall is shown in the insets of figures 6.12(*a*-*c*). The absence of the no-slip boundary condition at the mean fluid-bed interface allows the Reynolds stresses in the particle-laden cases to settle at small but finite levels when entering the bed, whereas the stresses in the smooth-wall simulations are seen to decay at the theoretically predicted decay rates when approaching the physical wall (cf. Pope, 2000, p.284). The velocity covariances $\langle u'_f v'_f \rangle_{xzt}$ shown in figure 6.12(*d*) reveal in all cases the expected linear profile away from the wall and for the low Reynolds number simulations, the curves for single- and multiphase simulations essentially collapse. For the higher Reynolds number case, though, the profile reaches its peak and the subsequent decay further away from the virtual wall than in the corresponding smooth-wall case. This further strengthens the earlier observation that the buffer layer effectively disappears in this case and inside the bedload layer, the pressure gradient is predominantly balanced by the fluid-particle interaction term, while viscous and turbulent stresses are of secondary importance for the total shear stress budget (cf. the detailed discussion on the shear stress balance in section 6.3.9).

The absence of the buffer-layer regeneration cycle in case $CM850^{H3}$ and in natural flows over fully-rough beds indicates that a general mechanism of sediment ridge formation cannot be due to the action of the small-scale buffer layer structures. We therefore concentrate in the following on the dynamics of the large-scale flow structures as main origin for the onset of sediment ridge formation. To this end, we first study how the distribution of the kinetic energy among the scales in the outer flow is affected by the presence of the mobile sediment and of sediment ridges in particular. Figure 6.13 shows the premultiplied streamwise energy spectra for smooth-wall and particle-laden simulations in the centre of the clear-fluid region ($\tilde{y}/H_f = 0.5$, left column) as well as at the free surface ($\tilde{y}/H_f = 1.0$, right column). Here, we have defined the instantaneous streamwise energy spectra at a given wall-normal location y as

$$\phi_{uu}(k_x, y, k_z, t) = \hat{u}_f \hat{u}_f^*, \tag{6.25}$$

where $\hat{u}_f(k_x, y, k_z, t) = \mathcal{F}(u_f - \langle u_f \rangle_{xz})$ is the Fourier transform of the instantaneous fluctuating field in the two periodic directions. While the asterisk indicates complex conjugation, $k_x = 2\pi/\lambda_x$ and $k_z = 2\pi/\lambda_z$ are the streamwise and spanwise wavenumber-wavelength pairs, respectively. Velocity spectra for the remaining velocity components are defined accordingly. In figure 6.13, it is seen that the spectral patterns in the multiphase simulations agree well with those in the single-phase cases, apart from some

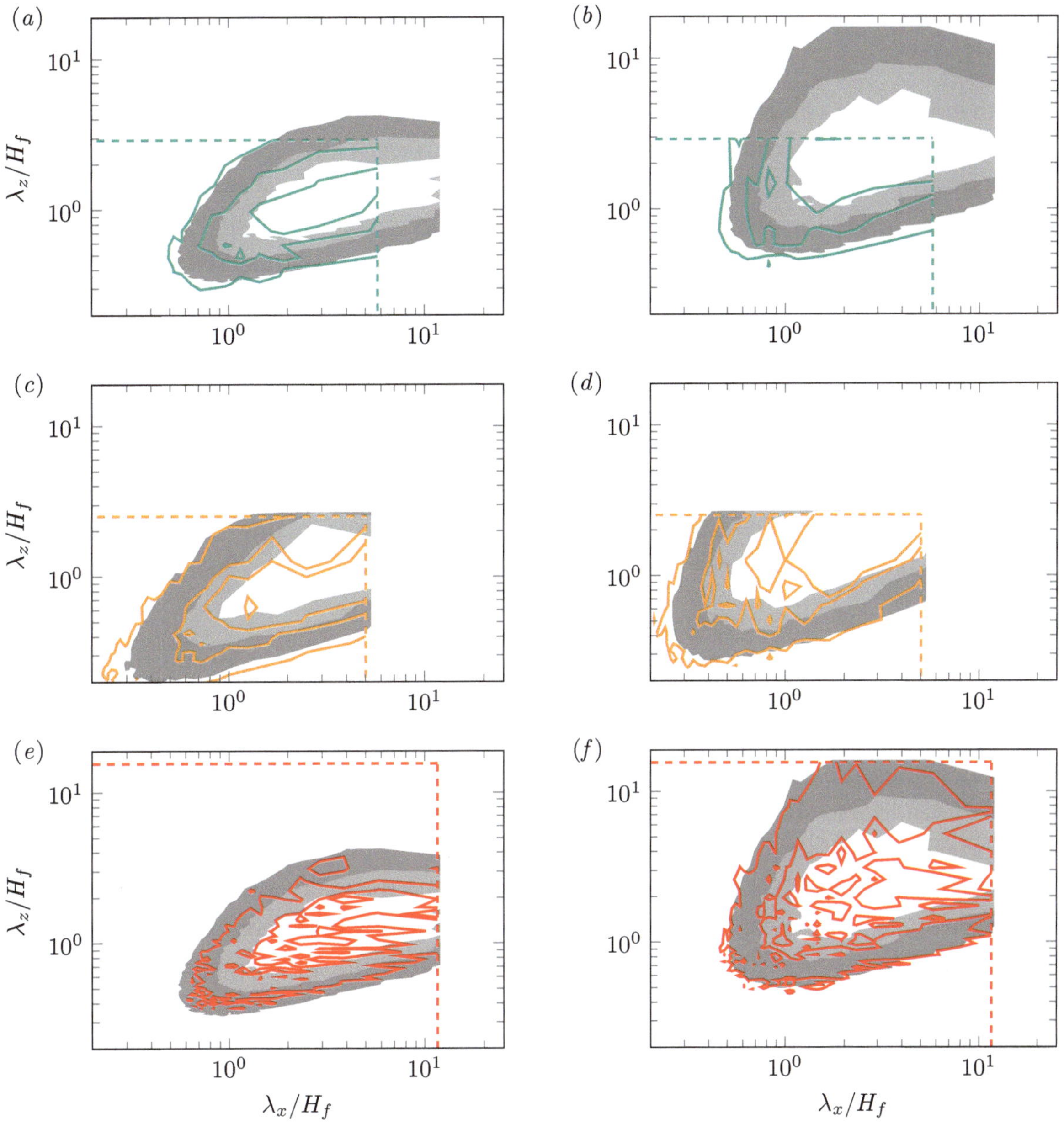

Figure 6.13: Time-averaged premultiplied streamwise energy spectra $k_x k_z \langle \phi_{uu} \rangle_t(\lambda_x, y, \lambda_z)\, H_f^2/u_\tau^2$ at the wall-normal reference locations *(a,c,e)* $\tilde{y}/H_f = 0.5$ and *(b,d,f)* $\tilde{y}/H_f = 1$. *(a,b)* $CM250^{H3}$, *(c,d)* $CM850^{H3}$, *(e,f)* $CL250^{H16}$. Coloured isolines are 0.2(0.2)0.6 times the maximum value of the respective energy spectra. Grey-shaded areas represent the same quantity evaluated for the smooth-wall reference simulations *(a,b,e,f)* $CL250^{H16}_{smooth}$ and *(c,d)* $CM650^{H3}_{smooth}$, respectively. The relative streamwise and spanwise domain periods of the particle-laden simulations L_x/H_f and L_z/H_f are highlighted by dashed lines in the respective colours.

expected fluctuations due to the limited time period during which statistics have been gathered. The limited domain size in the medium boxes allows to accommodate only part of the full spectra, but this part reveals a fairly good agreement with the smooth-wall reference simulations. We conclude that the presence of a mobile sediment bed does not significantly modify the spectral energy distribution in the channel core and that, consequently, there is no fundamental difference between large-scale streaks over developed ridges and those over smooth walls. For flows over fixed roughness elements, there is a general consensus that the influence of the roughness is limited to a layer of several multiples of their height above the bottom wall and that, thus, the organisation of the large-scale structures is more or less unaffected from the exact roughness configuration (Townsend, 1976; Jiménez, 2004a; Mazzuoli and Uhlmann, 2017). Here, we observe that the same conclusion holds also in the case of a mobile sediment bed.

Also, in both smooth-wall and particle-laden cases, a significant increase in the intensity of the largest scales is noticed between the channel centre and the free surface. This becomes evident when comparing the two spectra for case $CL250^{H16}$: While at $\tilde{y}/H_f = 0.5$, the width of the domain L_z/H_f is more than sufficient to capture the essential part of the kinetic energy, this is not the case at the free surface, where the spanwise scales are roughly twice as wide as those in the channel centre at $\tilde{y}/H_f = 0.5$ and even the large box slightly cuts the spectra. That kinetic energy resides in significantly wider scales near the free surface than it is the case in the channel centre is nowadays well-known for single-phase flows (Handler et al., 1993). More recently, however, Bauer et al. (2022) also showed in a comparison with spectra of closed smooth-wall channel flow simulations from Del Álamo and Jiménez (2003) that such an increase in spanwise scale does not occur for closed channel flows and that the effect is thus exclusive to free-surface flows. In the hydraulic context, these observations might be of relevance since earlier studies have tried to reconstruct the lateral spacing of subaqueous sediment ridges and secondary flow cells based on the width of free-surface structures (Kinoshita, 1967; Nezu and Nakagawa, 1993).

Figures 6.14 and 6.15 provide the premultiplied streamwise-integrated energy spectra associated with the streamwise (left column) and wall-normal velocity component (right column) as functions of the lateral wavelength λ_z and the wall-normal distance $\tilde{y}$ in inner and outer scales, respectively. Each spectrum is normalised by its maximum over the entire fluid height rather than by the local mean at the given wall distance to show how the energy is distributed across the different wall-parallel layers in single- and multiphase simulations, respectively. First, let us point out that the lateral width of the medium domains has been chosen sufficiently wide to capture the main part of the wall-normal energy spectra and are as such at the limit of being minimal in the vicinity of the free surface (Jiménez, 2013b), supporting our expectations from section 6.1. In agreement with our previous considerations, it is seen that the spectrum of the high Reynolds number case $CM850^{H3}$ lacks the conventional spectral peak at $\lambda_z^+ \approx 100$ associated with the buffer-layer structures, whereas for the low Reynolds number particle-laden cases the near-wall peak is maintained. In case $CM850^{H3}$, kinetic energy focuses instead further away from the wall somewhat above the bedload layer, where particle transport is significantly less intense. Comparing the spectra of $CM850^{H3}$ with those of the smooth-wall reference simulation, it turns out that the outer high-energetic spectral pattern is connected to the outer energy peak that is also visible in the smooth-wall simulation at $\lambda_z/H_f \approx$ 1-1.5 ($\lambda_z^+ \approx 1000$) and which reflects the fraction of kinetic energy that resides in the large-scale structures.

As expected, the kinetic energy associated with the wall-normal velocity concentrates in the bulk of the flow away from both the lower wall and the free surface, where the applied boundary conditions force the wall-normal velocity to vanish. As for the streamwise energy spectra, the wall-normal spectra of the low Reynolds sediment-laden cases almost match those of the corresponding smooth-wall case apart from slight variations as a consequence of the somewhat higher value of Re_τ, suggesting that the distribution of the wall-normal kinetic energy is not significantly altered by the presence of a mobile sediment bed. In the high Reynolds number case $CM850^{H3}$, there is a significant decrease of the wall-normal kinetic energy within the bedload layer compared to the single-phase simulations, while the most energetic scales are exclusively found above it. The general ellipsoidal shape and the inclination of the spectral patterns, however, are similar to those in the smooth-wall and low Reynolds number cases.

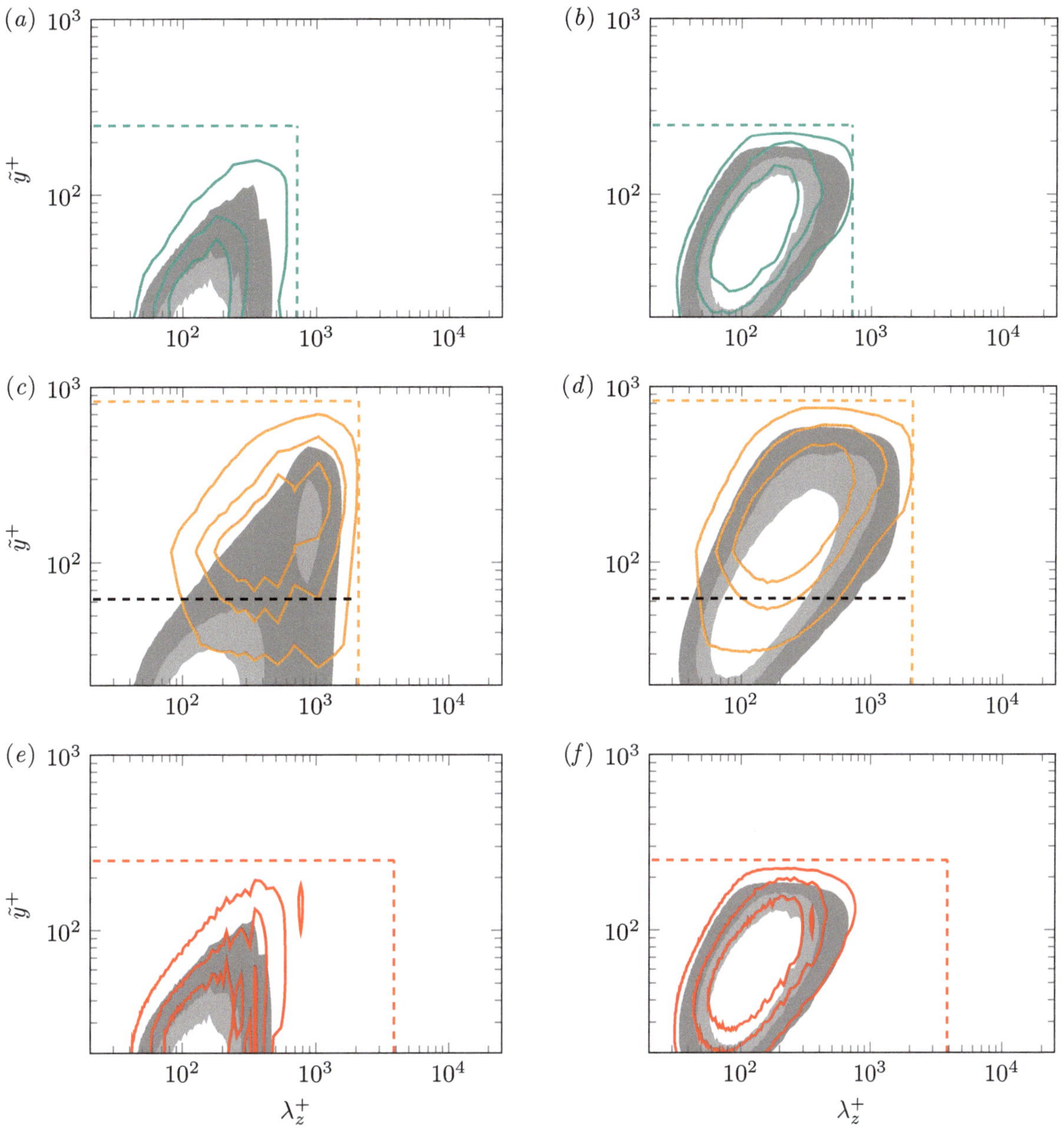

Figure 6.14: Time-averaged streamwise-integrated premultiplied energy spectra as a function of the inner-scaled spanwise wavelength λ_z^+ and wall distance $\tilde{y}^+$: $k_z \int \langle \phi_{\alpha\alpha} \rangle_t (\lambda_x, y, \lambda_z) \mathrm{d}k_x \, H_f^2/u_\tau^2$. (*a,c,e*) Streamwise $\langle \phi_{uu} \rangle_t$ and (*b,d,f*) wall-normal $\langle \phi_{vv} \rangle_t$ energy spectra for cases (*a,b*) $CM250^{H3}$, (*c,d*) $CM850^{H3}$ and (*e,f*) $CL250^{H16}$. Coloured isolines are 0.2(0.2)0.6 times the maximum value of the respective energy spectra, while grey-shaded areas indicate the same quantities determined for the smooth-wall reference simulations (*a,b,e,f*) $CL250^{H16}_{smooth}$ and (*c,d*) $CM650^{H3}_{smooth}$, respectively. The mean fluid height $H_f^+ = Re_\tau$ and the spanwise domain period L_z^+ of the particle simulations are marked by coloured dashed lines. The dashed black line refers to the wall-normal distance at which the mean particle flux density $\langle \phi u_p \rangle_{xzt}$ attains its maximum (cf. figure 6.6(*d*)).

Considering the same spectra in terms of the outer-scaled length scales as presented in figure 6.15(*b*), the spectrum is seen to almost collapse with that of the smooth-wall case in the largest scales in the channel bulk, providing further evidence that the structures of this size are essentially unaffected by the sediment activity in the bedload layer. Similar conclusions can be made for the outer-scaled streamwise energy spectra in figure 6.15(*a*).

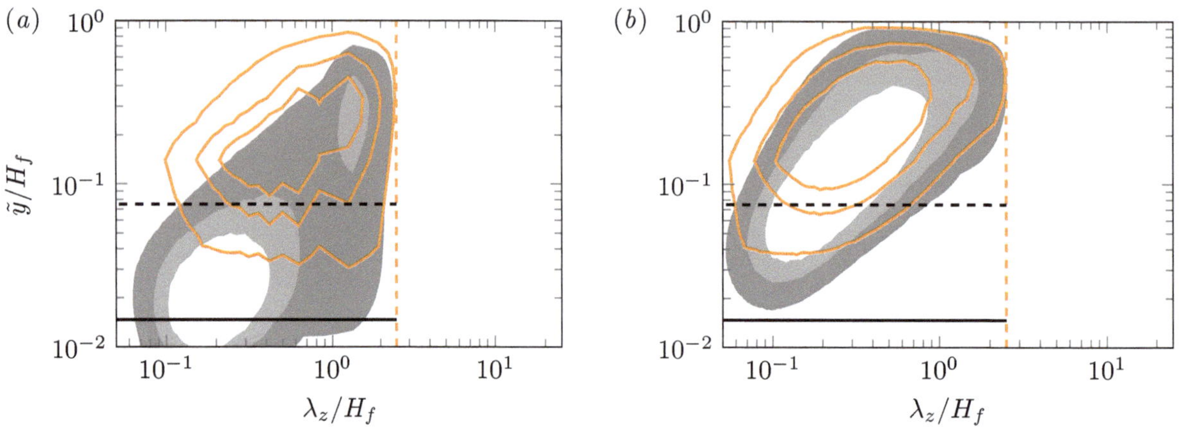

Figure 6.15: Time-averaged and streamwise-integrated premultiplied energy spectra for cases $CM850^{H3}$ and $CM650^{H3}_{smooth}$. Data is identical to that in figure 6.14(*c*,*d*), but the lateral wavelength λ_z and the wall-normal distance $\tilde{y}$ are scaled in outer units. As an additional reference height, the crest height of the mean fluid-bed interface averaged over $t/T_b \in [40, 85]$ is shown as black solid line. (*a*) streamwise spectra $\langle\phi_{uu}\rangle_t$, (*b*) wall-normal spectra $\langle\phi_{vv}\rangle_t$.

6.3.5 Large-scale flow organisation

The investigation of the energy spectra has revealed that there is little difference between the energy distribution among the scales in single- and multiphase cases. In the following, we will analyse the individual large-scale velocity streaks and their dynamics in physical space. In figure 6.16, the space-time evolution of the streamwise-averaged streamwise velocity fluctuations $u''_f(z,t)$ is shown for the channel centres, $\tilde{y}/H_f = 0.5$. Note that in this context, streamwise averaging serves as an indirect filter for laterally meandering structures, while it retains the signature of large streamwise-elongated structures.

For the here considered cases, the streamwise-averaged low- and high-speed regions evolve comparably straight in the space-time plane, which means that these zones exhibit only weak lateral propagation in the investigated time-interval. In this regard, no qualitative difference is noted between single- and multiphase cases. This indicates that even laterally homogeneous canonical flows feature a substantial spanwise modulation of the mean flow at least over intermediate time scales due to the long lifetime of the large-scale streaks (Jiménez, 2013b), while statistical homogeneity is obtained if the flow field is averaged over sufficiently long time intervals. A comparison between the simulations in the medium boxes with that in the wide domain of case $CL250^{H16}$ shown in figure 6.16(*d*) underlines that the weak lateral meandering and the regular lateral organisation of the streaks are not due to the narrow box width $L_z/H_f \approx 3$ in the medium domains, but that the flow in the wide domain $L_z/H_f \approx 16$ shows essentially the same large-scale dynamics. The pronounced signatures of low- and high-speed zones are furthermore indicative of large-scale streaks that are of significant length compared to the streamwise domain length $L_x/H_f \approx 12$, otherwise they would not appear in the streamwise average. Even though the lateral position of the high- and low-speed regions is relatively time-persistent, their amplitude is seen to intermittently reduce and the clear signature is lost, for instance at approximately $t = 40T_b$ in case $CM650^{H3}_{smooth}$ or in a period $10 \lesssim t/T_b \lesssim 20$ in case $CM850^{H3}$ (cf. figure 6.16(*b*,*c*)). It is expected that at least some of these 'events' reflect intermittently occurring bursting events similar to those in the buffer layer during which the large streaks bend and eventually break – a mechanism that has been observed for log-layer streaks likewise (Flores and Jiménez, 2010). After such 'events', the laterally organised high- and low-speed zones are seen to recover, but only in some situations the streaks form at essentially the same lateral positions as before. In others, the flow reorganises and partly even

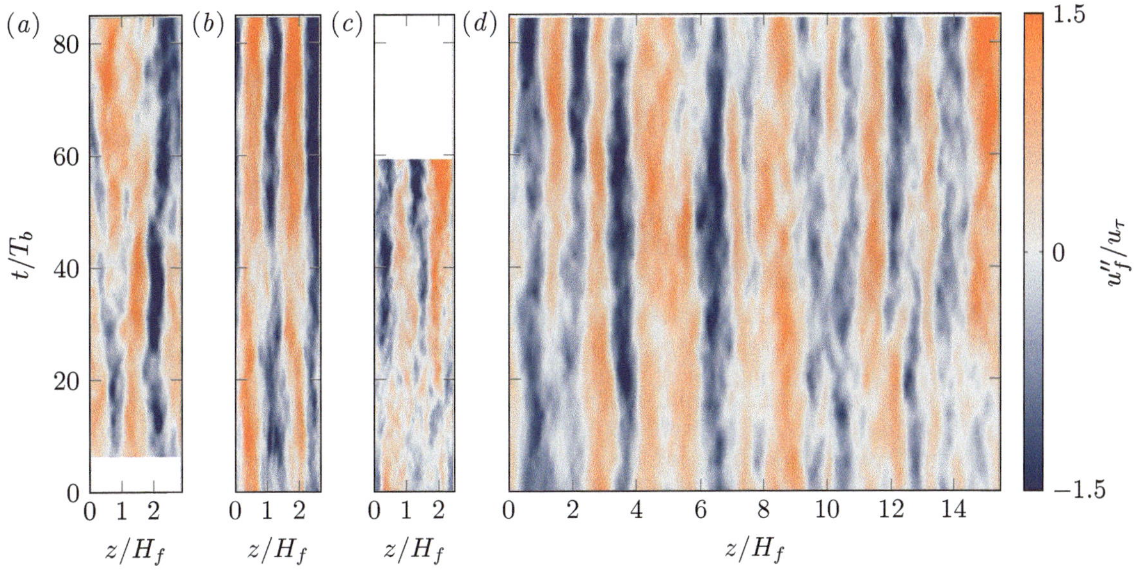

Figure 6.16: Space-time plot of the streamwise-averaged fluctuations of the streamwise velocity component u''_f/u_τ extracted at $\tilde{y}/H_f = 0.5$. Blue and red regions refer to streamwise-averaged low- and high-speed regions, respectively. Cases: (*a*) $CM250^{H3}$, (*b*) $CM650^{H3}_{smooth}$, (*c*) $CM850^{H3}$, (*d*) $CL250^{H16}$.

the number of streaks changes as in case $CM250^{H3}$ after $t = 45T_b$ (cf. figure 6.16(*a*)). The influence of these phases of flow reorganisation on the dynamics of the sediment ridges shall be scrutinised in the discussion of figure 6.21 below.

The striking similarity between the organisation of the outer flow in single- and multiphase flows supports our earlier conjecture that the large-scale streaks over sediment ridges are effectively the same structures as those in smooth-wall channels. Flores et al. (2007) similarly concluded for flow over laterally homogeneous roughness that the self-similar vortex clusters and the associated velocity streaks show no major differences to their counterparts over smooth walls outside the roughness layer. What remains to be discussed is the question of causality: Considering the results that have been presented up to this point, it is tempting to assume that the regular spanwise organisation of low- and high-speed regions causes a laterally varying erosion of the sediment bed once particles are released and, that way, triggers the rise of the initial ridges and not *vice versa*. Of course, this does not rule out that in later stages of their lifetime, developed sediment ridges might also affect the dynamics and meandering tendency of large-scale streaks. In the following, we will verify this assumption by investigating the interaction of large-scale streaks and the wall shear stress that is directly linked to the sediment erosion rate.

The large-scale structures in the logarithmic and outer layer are – in contrast to the smooth buffer layer streaks – fully-turbulent objects (Jiménez, 2013b), and the associated large-scale instantaneous quasi-streamwise rollers are, in general, almost impossible to detect by the classical vortex detection techniques such as the λ_2-criterion of Jeong and Hussain (1995), since the velocity gradients are much weaker then those of the near-wall quasi-streamwise vortices. In the following, we will therefore use a filtering technique to detect large-scale coherent structures in varying distances to the wall. Filtering can be done either indirectly as in the context of conditional averaging (see, for instance, Del Álamo et al., 2006; Lozano-Durán et al., 2012) and in some sense in our own streamwise-averaging process, or in a direct way by applying, for instance, a Gaussian filter to the instantaneous flow field as done in Motoori and Goto (2019, 2021). The use of a direct filtering approach has the obvious asset that

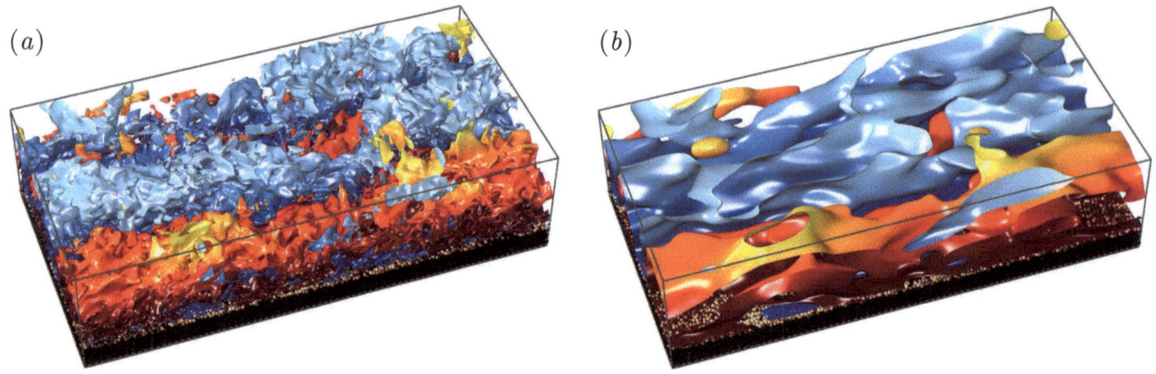

Figure 6.17: Instantaneous visualisation of large-scale streamwise velocity streaks before and after applying a Gaussian filter with filter widths $[\Delta_x, \Delta_y, \Delta_z]/H_f = [0.6, 0.2, 0.3]$ exemplary for case $CM850^{H3}$. The shown instant is the same as that presented in figure 6.3(*d*). High- and low-speed streaks are coloured in red/yellow and blue, respectively, characterised by connected regions fulfilling $|u_f'^{+}(\boldsymbol{x},t)| > H$ with (*a*) $H = 1.5$ and (*b*) $H = 0.75$.

information loss is avoided for all scales larger than the chosen filter widths. In the following, we study the low-pass filtered field related to the *i*th velocity component, $\widetilde{u}_{f,i}(\boldsymbol{x},t)$, that is obtained by the following convolution of the field with an anisotropic Gaussian kernel (Lozano-Durán et al., 2016):

$$\widetilde{u}_{f,i}(\boldsymbol{x},t) = \iiint_V G \cdot u_{f,i}(\boldsymbol{x}-\boldsymbol{x}',t) \exp\left(-\left(\frac{\pi x'}{\Delta_x}\right)^2 - \left(\frac{\pi y'}{\Delta_y}\right)^2 - \left(\frac{\pi z'}{\Delta_z}\right)^2\right) \mathrm{d}x'\mathrm{d}y'\mathrm{d}z'. \qquad (6.26)$$

Therein, Δ_i ($i = x, y, z$) is the filter cut-off width in the three Cartesian directions and $\boldsymbol{x}'$ is the inner-convolutional coordinate. The volume over which the field is filtered is denoted by V, and G is a constant coefficient normalised such that the integral of the kernel over V is unity. We treat the boundary conditions as proposed in Lozano-Durán et al. (2016), that is, the flow field is periodically extended in the two wall-parallel directions and mirrored vertically at the bottom wall, while the sign of the wall-normal velocity component is reversed to ensure the incompressibility of the resulting field. The free surface is treated analogously to fulfil the free-slip and impermeability boundary conditions along the plane. By applying the filtering technique to the multiphase simulations, a layer of thickness Δ_y/H_f of the filtered flow field above the bed will be affected by the velocity field inside the latter. In the subsequent investigations, we therefore focus either on filtering results that are obtained in regions at a sufficient distance to the sediment bed, or use the technique in the context of smooth-wall simulations where there is no such effect. We conventionally choose, if not otherwise declared, filter widths $[\Delta_x, \Delta_y, \Delta_z]/H_f = [0.6, 0.2, 0.3]$ that obey the typical aspect ratio of the vortex clusters found by Del Álamo et al. (2006). Two-dimensional filtering in analogy to equation (6.26) in wall-parallel xz-planes is frequently used whenever the focus is on the structure's wall-parallel extensions. In these cases, filter widths $[\Delta_x, \Delta_z]/H_f = [0.6, 0.3]$ will be conventionally used. The effect of the Gaussian filter on a turbulent velocity field is visualised in figure 6.17, where the large-scale streamwise velocity streaks in an instantaneous field of case $CM850^{H3}$ are compared for the filtered and unfiltered state. While the essential macroscopic features of the large-scale structures including their shape, size and orientation remain essentially unchanged, the filtered streaks are much smoother as the small-scale variations have been removed by the filtering.

In the following, we make use of the introduced filtering technique to study the relation between large-scale coherent structures and the bottom-wall shear stress. As recently discussed by Bagherimiyab and Lemmin (2018), ejections and sweeps cause a local reduction or increase of the wall shear stress along the sediment bed, respectively, indicating that the regular organisation of both large-scale streaks and

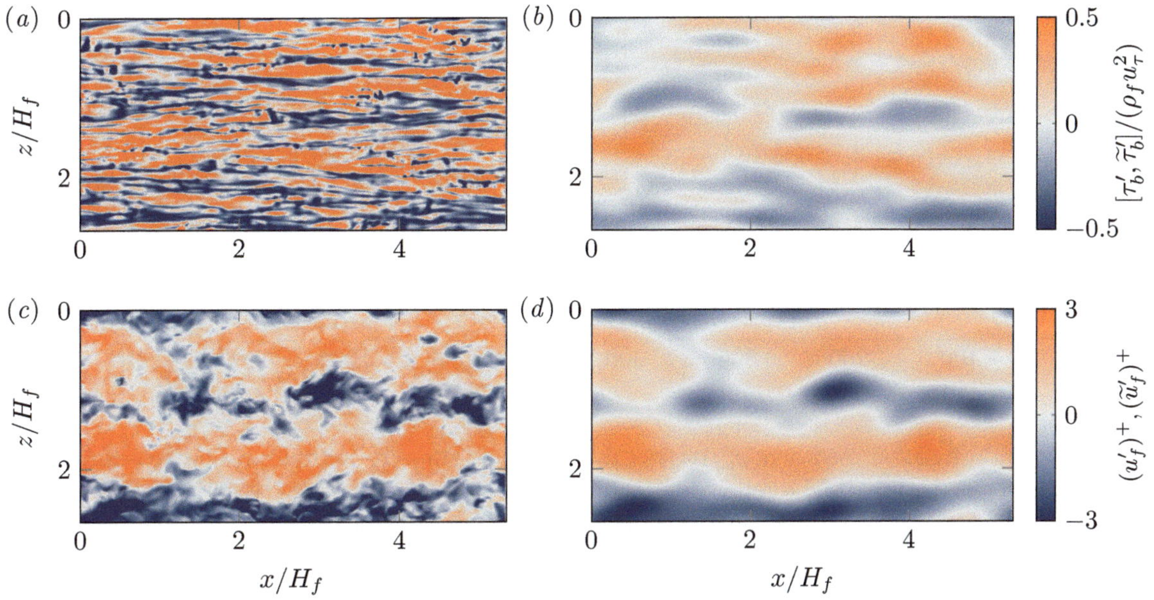

Figure 6.18: *(a,b)* Instantaneous wall shear stress fluctuations $\tau_b'(x,z,t)/(\rho_f u_\tau^2)$ for the smooth-wall simulation $CM650^{H3}_{smooth}$ *(a)* without applied filtering and *(b)* after a 2D Gaussian cut-off filter has been applied in the two homogeneous directions with filter widths $\Delta_x = 3y_{ref}$ and $\Delta_z = 1.5y_{ref}$, respectively. The reference value has been set to the height $y_{ref} = 0.3H_f$. *(c,d)* Corresponding streamwise velocity fluctuation field $(u_f')^+$ extracted at $\tilde{y}/H_f = 0.3$ *(c)* unfiltered and *(d)* filtered with the same filter size as in *(b)*. The time at which all snapshots have been extracted corresponds to the second marker point in figure 6.23*(b)*, $t/T_b = 83$.

Reynolds stress-carrying structures might clearly affect the organisation of the wall shear stress and, that way, the erosion of sediment. These considerations are in accordance with today's understanding that the buffer layer contains, apart from the typical small-scale structures of size $\mathcal{O}(100\delta_\nu)$, also large-scale patterns of dimensions $\mathcal{O}(H_f)$ that are correlated to the structures in the logarithmic and outer layer (Del Álamo and Jiménez, 2003; Hutchins and Marusic, 2007b). Although it is well known that the self-sustaining regeneration cycle in the buffer layer itself is capable of functioning autonomously (Jiménez and Pinelli, 1999), it has nonetheless been observed that large-scale effects also alter the flow organisation in the near-wall region (Jiménez et al., 2004). A continuous mutual interaction of small near-wall structures and large outer-scaled structures was proposed by Toh and Itano (2005), where continuity requires that the small-scale structures agglomerate below the large-scale ejections (Jiménez, 2018). Zhou et al. (2022), in turn, found only evidence for an outer-layer (top-down) influence on the near-wall pattern, but not for the bottom-up effect of the small-scale structures on the large scales.

The organisation of the inner-scaled structures of the buffer layer with dimensions $\mathcal{O}(100\delta_\nu)$ into larger streaks of scale $\mathcal{O}(H_f)$ is clearly discernible in figure 6.18. Here, an instantaneous snapshot of the wall shear stress along the smooth bottom wall of case $CM650^{H3}_{smooth}$ is shown before and after applying the Gaussian filter, supplemented with the unfiltered and corresponding spatially-filtered streamwise velocity field at a reference height $\tilde{y}/H_f = 0.3$. For the sake of clarity, we have introduced the instantaneous bottom shear stress

$$\tau_b(x,z,t) = \rho_f \nu_f \left.\frac{\mathrm{d}u_f}{\mathrm{d}y}\right|_{y=0} \tag{6.27}$$

to avoid confusion with the mean wall shear stress τ_w. In figure 6.18, the high-speed buffer-layer streaks, whose position can be inferred by their induced zones of locally increased shear, are seen to cluster in two streamwise-elongated streak-like structures: One of those structures spans the entire box length in a slightly meandering way, while the other one is roughly half as long and laterally shifted

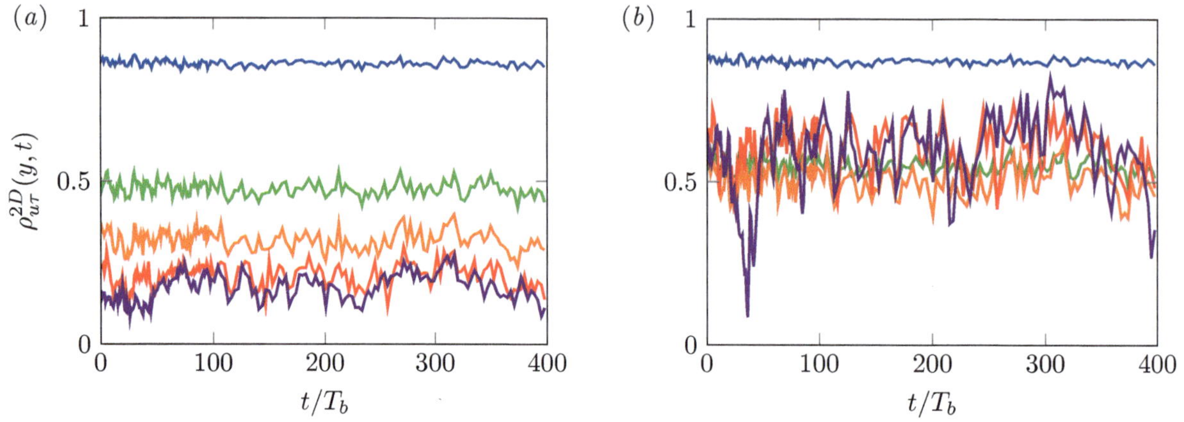

Figure 6.19: Two-point correlation coefficient $\rho_{u\tau}^{2D}(y,t)$ of the wall-shear stress τ_b' and the streamwise velocity fluctuations u_f' at different distances to the bottom wall in the smooth-wall reference case $CM650_{smooth}^{H3}$: (*a*) Without applied filtering and (*b*) after a 2D Gaussian cut-off filter has been applied in the two homogeneous directions with filter widths $\Delta_x = 3y_{ref}$ and $\Delta_z = 1.5y_{ref}$, respectively. Here, the reference height y_{ref} is set equal to the respective wall distance: $y^+ = 10$ (—), $y^+ = 30$ (—), $y/H_f = 0.1$ (—), $y/H_f = 0.3$ (—) and $y/H_f = 0.5$ (—).

by an offset somewhat larger than H_f. A comparison with the large-scale velocity structures reveals that these two regions are located below the large-scale high-speed streaks at $\tilde{y}/H_f = 0.3$ and feature a comparable lateral extent, while the streamwise length is seen to deviate in particular for the patchy structure whose centre is found at approximately $z/H_f = 0.5$. A qualitatively similar correlation between large-scale flow structures and the shear stress in the vicinity of the sediment bed is also found for the sediment-laden simulations in the period just before particles are released.

In order to further quantify the mutual correlation of the wall shear stress and the velocity structures, the temporal evolution of the instantaneous two-point correlation coefficient

$$\rho_{u\tau}^{2D}(y,t) = \max_{\delta x \in [-L_x/2, L_x/2)} \frac{\langle \tau_b'(x,z,t)\, u_f'(\boldsymbol{x}+\boldsymbol{\delta},t)\rangle_{xz}}{\left[\langle (\tau_b'(x,z,t))^2\rangle_{xz}\, \langle (u_f'(\boldsymbol{x}+\boldsymbol{\delta},t))^2\rangle_{xz}\right]^{1/2}} \tag{6.28}$$

is shown in figure 6.19(*a*) for selected wall-normal distances y, where $\boldsymbol{\delta} = (\delta x, 0, 0)^T$ is the spatial separation vector with the streamwise component δx. In definition (6.28), we determine the correlation coefficient as the maximum over all $\delta x \in [-L_x/2, L_x/2)$ to take care of a possible phase shift between the wall shear stress structures and the large-scale streaks arising from the different propagation speeds of structures at different distances to the wall (Del Álamo and Jiménez, 2009).

The results presented in figure 6.19(*a*) underline the high correlation between the wall shear stress and the buffer layer structures, while the correlation decreases with increasing distance to the wall due to the growing scale difference between the wall shear stress structures and the streamwise velocity streaks, the latter scaling self-similarly with y (Townsend, 1976; Sillero, 2014). In order to study in the following only the correlation between the large-scale structures of both fields, $\rho_{u\tau}^{2D}(y,t)$ is recomputed replacing τ_b' and u_f' in equation (6.28) by their spatially-filtered analogues $\tilde{\tau}_b'$ and $\tilde{u}_f'$, respectively. Having removed the small scales at each height, a constantly high mean correlation of $\langle \rho_{u\tau}^{2D}\rangle_t \approx 0.5$ is observed in figure 6.19(*b*) for all wall-distances $30\delta_\nu \lesssim \tilde{y} \lesssim 0.5H_f$, highlighting the significant dependency between the large-scale organisation of the wall shear stress and the large-scale velocity structures in the channel bulk.

6.3.6 Streak-ridge interaction

In the context of sediment ridge formation, the observed correlation between the large-scale flow organisation and that of the wall shear stress provides a causal connection between the local erosion rate at the fluid-bed interface and the large-scale velocity structures. While it appears inevitable that particle motion in case $CM850^{H3}$ is connected to the large-scale structures due to the lack of the buffer-layer cycle, particles in low Reynolds number cases are known to be transported also by the buffer-layer streaks and vortices (Kidanemariam et al., 2013). The observed link between the large-scale streaks and the local wall shear stress might thus explain why the same ridge spacing is observed for both cases.

In the following, let us first discuss in more detail the relation between the flow organisation in large-scale streaks and the local erosion rate along the sediment bed. To this end, we define the sets of all particles that are eroded or deposited between two instants $t_m = m\Delta t$ and $t_{m+1} = (m+1)\Delta t$ as

$$\mathcal{P}_{erod}(t_{m+1/2}) = \{p \mid p \in \mathcal{P}_b(t_m) \wedge p \in \mathcal{P}_m(t_{m+1})\} \tag{6.29a}$$

$$\mathcal{P}_{depos}(t_{m+1/2}) = \{p \mid p \in \mathcal{P}_m(t_m) \wedge p \in \mathcal{P}_b(t_{m+1})\}\,, \tag{6.29b}$$

respectively, where $t_{m+1/2} = (t_m + t_{m+1})/2$ and $m \in [1, N_t - 1]$ with N_t being the number of simulated time steps. Here, $\mathcal{P}_b(t_m)$ and $\mathcal{P}_m(t_m)$ are the sets of bed and mobile particles at time t_m, respectively. To locate regions of strong erosion or deposition, we discretise the wall-parallel directions making use of the same binning technique as in the context of the particle flux (cf. section 6.2.3), choosing bin widths $\Delta x_{bin} = \Delta z_{bin} = 1.5D$. The number of eroded and deposited particles in the (i,k)th bin at a given intermediate time $t_{m+1/2}$ $\forall\ m \in [1, N_t - 1]$ is then determined as

$$N_p^{erod}(x_i, z_k, t_{m+1/2}) = \sum_{l=1}^{N_p} I^{(l)}_{(i,k)}(t_m)\; I^{(l)}_{erod}(t_{m+1/2}) \tag{6.30a}$$

$$N_p^{depos}(x_i, z_k, t_{m+1/2}) = \sum_{l=1}^{N_p} I^{(l)}_{(i,k)}(t_{m+1})\; I^{(l)}_{depos}(t_{m+1/2})\,. \tag{6.30b}$$

As the eroded and deposited sediment grains in general do not reside in the same bin at times t_m and t_{m+1}, it was decided in the above definition to associate eroded sediment grains with the bin they are eroded from and deposited particles with the one they are deposited in. As a consequence, the indicator functions on the RHS of equations (6.30a) and (6.30b) that will be defined in the following are evaluated at two different time steps. For the lth particle $p^{(l)}$, these latter are defined as

$$I^{(l)}_{erod}(t) = \begin{cases} 1 & \text{if } p^{(l)} \in \mathcal{P}_{erod}(t) \\ 0 & \text{else} \end{cases}, \quad I^{(l)}_{depos}(t) = \begin{cases} 1 & \text{if } p^{(l)} \in \mathcal{P}_{depos}(t) \\ 0 & \text{else} \end{cases}. \tag{6.31}$$

The streamwise-averaged erosion-deposition balance or equivalently the particle loss of the sediment bed per unit length and time in the kth bins over the period $t \in [t_a, t_b]$ $(1 \le a < b \le N_t)$ is then defined as

$$\Phi(z_k, t_a, t_b) = \frac{1}{L_x(t_b - t_a)} \sum_{m=a}^{b-1} \sum_{i=1}^{N_{x,bin}} \left[N_p^{erod}(x_i, z_k, t_{m+1/2}) - N_p^{depos}(x_i, z_k, t_{m+1/2})\right]. \tag{6.32}$$

In figure 6.20, relation (6.32) has been evaluated for the initial time interval $[0T_b, 5T_b]$ starting with

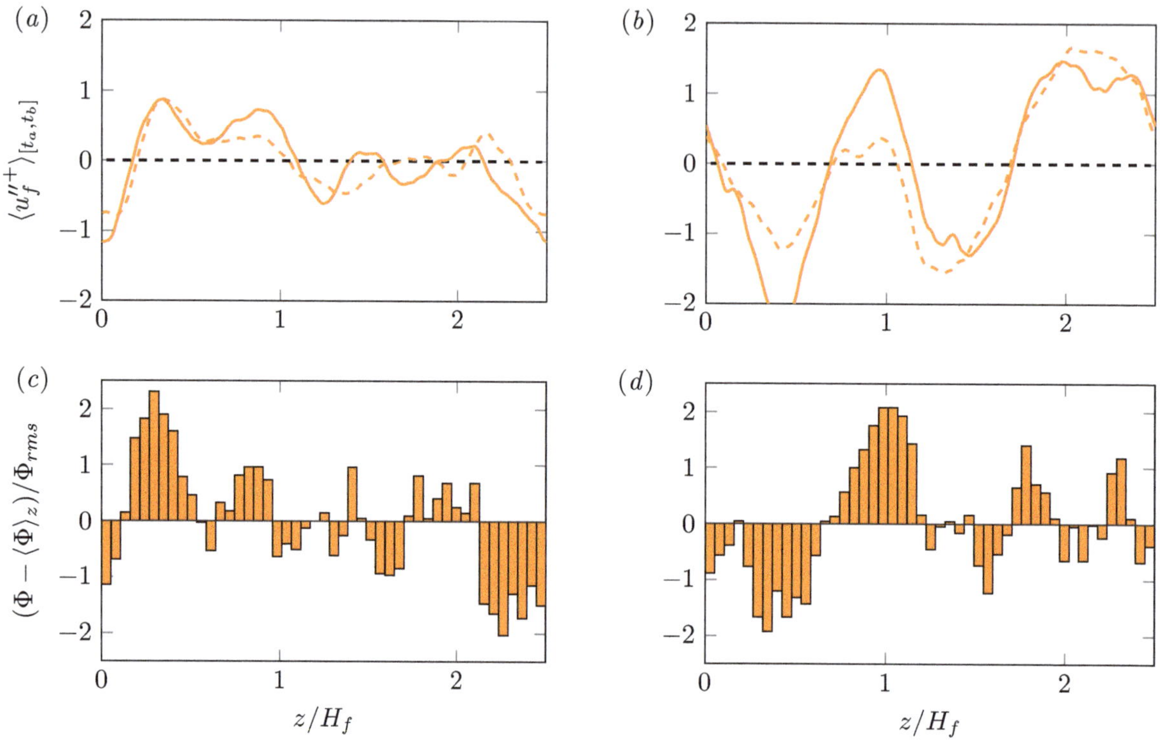

Figure 6.20: Erosion-deposition balance during the initial and final phase of bed evolution in comparison with the velocity field, exemplary for case $CM850^{H3}$. *(a,b)* Streamwise-averaged velocity fluctuations u''_f at $\tilde{y}/H_f = 0.3$ (solid) and $\tilde{y}/H_f = 0.5$ (dashed), averaged over the time interval $[t_a, t_b]$. *(c,d)* Fluctuation of the balance between eroded and deposited particles per unit length and time during the interval $[t_a, t_b]$, $(\Phi - \langle\Phi\rangle_z)$, normalised by the corresponding r.m.s. value $\Phi_{rms} = [\langle(\Phi - \langle\Phi\rangle_z)^2\rangle_z]^{1/2}$. *(a,c)* Initial time interval $[t_a, t_b] = [0T_b, 5T_b]$, *(b,d)* final time interval $[t_a, t_b] = [55T_b, 60T_b]$.

the particle release as well as for the final time interval $[55T_b, 60T_b]$ of simulation $CM850^{H3}$. The data is compared with the lateral profiles of the streamwise velocity fluctuation field at $\tilde{y}/H_f = 0.3$ and $\tilde{y}/H_f = 0.5$, respectively. In order to highlight the correlation between velocity and erosion activity, we show in figure 6.20 the lateral fluctuation of the erosion-deposition balance $\Phi - \langle\Phi\rangle_z$, rather than the absolute values. The absolute values are quite different between the initial phase $[0T_b, 5T_b]$ and the final phase of bedform evolution $[55T_b, 60T_b]$: In the former, a marked erosion excess is seen over the entire width of the channel, indicating that the sediment bed height clearly decreases at all spanwise locations. This is in agreement with our observations in figure 6.5(*c*) that the mean bed height reduces by approximately one particle diameter in this phase. Even though erosion dominates deposition at all spanwise locations, there is a significant variation of the erosion excess over the span of the channel such that the decrease of the sediment bed height is strongly heterogeneous, as can be seen in figure 6.20(*c*). Consequently, initial troughs form in regions of stronger erosion by a 'carving' process, while initial sediment ridges represent regions of weaker erosion activity. The regions of stronger and weaker erosion excess match reasonably well the regions of higher and lower streamwise velocity in the bulk of the channel (cf. figure 6.20(*a*)), indicating that the local erosion rate is indeed directly correlated with the large-scale flow structures. A similarly good correlation between the erosion activity and regions of high and low streamwise velocity is also observed during the final time interval shown in figure 6.20(*b,d*). In contrast to the initial phase, however, erosion and deposition are globally almost in equilibrium and alternating zones dominated either by erosion or deposition are observed. Comparing these regions with the bed contour at the same time, it turns out that regions of dominant erosion

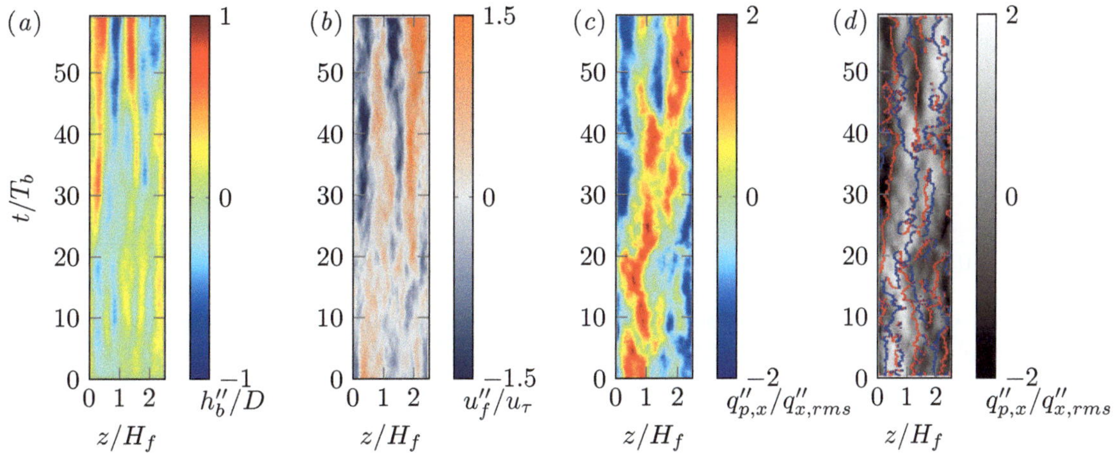

Figure 6.21: Space-time evolution of (*a*) the sediment-bed height fluctuation of the streamwise-averaged bed for case $CM850^{H3}$ (identical to figure 6.4(*c*)). (*b*) Same for the streamwise-averaged streamwise velocity fluctuation u_f''/u_τ at $\tilde{y}/H_f = 0.5$ (identical to figure 6.16(*c*)) and (*c*) for the streamwise-averaged particle flux fluctuation $q_{p,x}''/q_{x,rms}''$, with $q_{x,rms}''(t) = (\langle q_{p,x}'' q_{p,x}'' \rangle_z)^{1/2}$. (*d*) Same data as in (*c*) is shown in the background as grey map, supplemented with red (blue) dots marking regions of lateral bed growth (decrease), that are, regions with vanishing lateral particle flux (i.e. $q_{p,z}'' = 0$) and negative (positive) lateral gradient $\partial_z q_{p,z}''(z,t)$.

are those over a developed trough while deposition dominates near sediment ridges, supporting our earlier statements.

These observations provide further evidence that particle erosion and hence bed topography are indeed linked with the large-scale structures via their near-bed effects and the local bed shear stress. In figure 6.21, the close correlation between high- and low speed regions on the one hand and the organisation of the sediment ridges and of the mean particle flux on the other hand is clearly identifiable. Therein, we compare the space-time evolution of the sediment bed fluctuation $h_b''(z,t)$ and the respective fluctuation of the streamwise velocity component at $\tilde{y}/H_f = 0.5$ (which are repeated here for convenience) with that of the streamwise and spanwise particle flux components $q_{p,x}''$ and $q_{p,z}''$, respectively. The relations between the different quantities can be summarised as follows: Regions of enhanced streamwise momentum cause a locally increased erosion activity that leads to a stronger decrease of the sediment bed in these regions due to turbulent sweep events that are located in the high-speed streaks. The enhanced erosion results in a positive streamwise particle flux fluctuation $q_{p,x}''$ in these regions. The lateral particle flux $q_{p,z}''$ is seen to predominantly transport sediment from regions of high streamwise particle flow rate ($q_{p,x}'' > 0$) to those with lower streamwise particle flux ($q_{p,x}'' < 0$), as is indicated by the red and blue lines in figure 6.21(*d*). In other words, sediment is on average transported from the troughs to the sediment ridge crests. Analogous relations hold for the opposite case of low-speed regions, local sediment ridges and regions of low streamwise particle transport ($q_{p,x}'' < 0$). Recalling that the spanwise particle flux is two orders of magnitude smaller than its streamwise counterpart, however, it appears likely that the initial formation of sediment ridges and troughs is predominantly driven by a laterally varying sediment erosion in the streamwise direction, related to the presence of elongated streamwise high- and low-speed streaks. The lateral particle transport is, in turn, assumed to be of less importance in this initial phase, while it might play a more important role in the further growth of the initial sediment ridges in the phase when erosion and deposition are more or less in equilibrium (cf. Figures 6.5(*c*) and 6.20(*d*)).

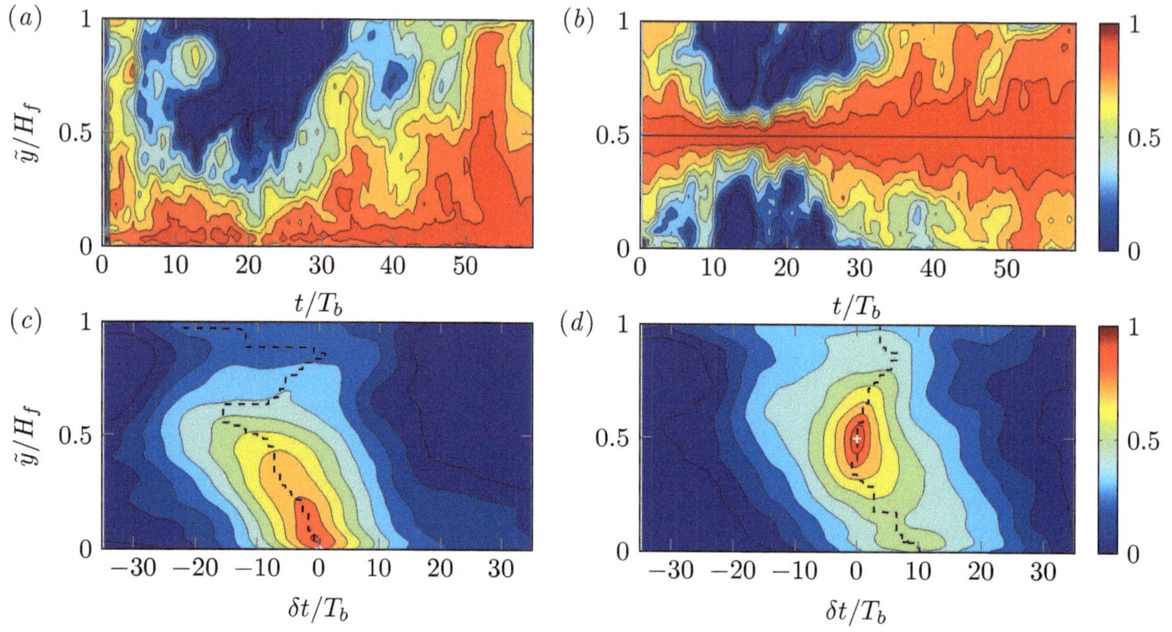

Figure 6.22: (*a,b*) Two-point cross-correlations and cross-time correlations between the streamwise-averaged flow field and the sediment bed contour in case $CM850^{H3}$. Time-averaged two-point cross-correlation between $u''_f(y,z,t)$ and: (*a*) the sediment bed height fluctuations, $-\rho_{uh_b}(y,t)$; (*b*) the velocity fluctuations at the reference height $u''_f(\tilde{y}/H_f = 0.5, z, t)$, $\rho_{uu}(y,t)$. (*c,d*) Two-point cross-time correlations between $u''_f(y,z,t+\delta t)$ and: (*c*) the sediment bed height fluctuations, $-\rho^t_{uh_b}(y,\delta t)$; (*d*) the velocity fluctuations at the reference height $u''_f(\tilde{y}/H_f = 0.5, z, t)$, $\rho^t_{uu}(y,\delta t)$. In (*c,d*), the reference time and reference wall-normal position are indicated by a white cross. Red (blue) regions represent a strong (weak) correlation of the compared quantities. Black dashed lines connect the maximum correlation values at each wall-normal distance $\tilde{y}/H_f$. Contours separating the coloured areas are 0(0.1)1 in all panels.

The visually identified correlation between large-scale velocity structures and the bed contour in the multiphase simulations is in the following quantified based on two-point cross-correlations of the streamwise-averaged fields

$$\rho_{uh_b}(y,t) = \langle h''_b(z,t)\, u''_f(y,z,t)\rangle_z / \left[\langle (h''_b(z,t))^2\rangle_z \langle (u''_f(y,z,t))^2\rangle_z\right]^{1/2} \tag{6.33a}$$

$$\rho_{uu}(y,t) = \langle u''_f(H_f/2,z,t)\, u''_f(y,z,t)\rangle_z / \left[\langle (u''_f(H_f/2,z,t))^2\rangle_z \langle (u''_f(y,z,t))^2\rangle_z\right]^{1/2}, \tag{6.33b}$$

where ρ_{uh_b} is the correlation between the bed contour and the flow field and ρ_{uu} quantifies the correlation between the flow organisation at a reference height $\tilde{y}/H_f = 0.5$ and that at other distances to the wall.

The time evolution of both quantities is shown in figures 6.22(*a,b*), the correlation between bed and flow field being presented premultiplied with a global factor -1 to take into account that a locally higher velocity is related to a local decrease of the sediment bed height and *vice versa*. In accordance with the previous findings, it is seen in figure 6.22(*a*) that the bed contour is highly correlated with the flow field over most of the simulation time, at least up to the bulk flow centreline $\tilde{y}/H_f = 0.5$ reaching values of more than $\rho_{uh_b} \approx 0.6$. This is, however, not the case in the time interval $t \in [10T_b, 30T_b]$, during which the sediment bed contour and the organisation of the flow in sufficient distance to the bed are essentially uncorrelated. Similarly, we conclude based on the results provided in figure 6.22(*b*) that the flow in the channel bulk centreline at $\tilde{y}/H_f = 0.5$ is most of the time clearly correlated with the structure of the flow field both near the bed and close to the free surface. Interestingly, the high correlation of the flow organisation at $\tilde{y}/H_f = 0.5$ with that in the rest of the channel is lost in the same period $t \in [10T_b, 30T_b]$, during which only a thin layer in the core of the clear-fluid region is still highly correlated with the flow at the centreline.

The weak correlation between the large-scale streaks and the location of sediment ridges in this intermediate time interval is also discernible in figures 6.21(*a,b*). Therein, it is seen that between $t = 10T_b$ and $t = 20T_b$, the amplitude of the streamwise-averaged velocity fluctuations decreases and the before clearly separated high- and low-speed regions become blurred. It has been proposed earlier in this study that such an 'event' might be related to a break up of the velocity streaks similar to the bursting of streaks in the logarithmic layer observed by Flores and Jiménez (2010). Two large-scale streaks redevelop shortly afterwards between $t = 20T_b$ and $t = 25T_b$ at slightly different transverse positions. During the reorganisation of the large-scale streaks in this time period, the lateral positions of ridges and troughs first remain essentially unchanged, but once the high- and low-speed regions have recovered, the bedforms are seen to adapt to the altered organisation of the flow field with a noticeable time lag, laterally 'propagating' towards the regions below the low-speed zones.

The revealed time delay between large-scale streak organisation and bed evolution is an important indication of how physical information propagates across the channel. In the following, we aim to quantify the observed time delay by means of a two-time cross-correlation function

$$\rho^t_{ua}(y, \delta t) = \langle u''_f(y, z, t + \delta t)\, a''(z, t)\rangle_{zt} / \left[\langle (u''_f(y, z, t + \delta t))^2\rangle_{zt}\, \langle (a''(z, t))^2\rangle_{zt}\right]^{1/2}, \qquad (6.34)$$

which is a measure for the correlation of an arbitrary physical quantity a to the organisation of the streamwise-averaged flow field at an earlier or later time. In figure 6.22(*c*), the cross-time correlation between the sediment bed contour and the flow field is presented ($a'' = h''_b(z, t)$), while in figure 6.22(*d*), the cross-time correlation between the flow field at varying wall distances and that at a reference height $\tilde{y}/H_f = 0.5$ is shown ($a'' = u''_f(\tilde{y}/H_f = 0.5, z, t)$). The correlation presented in figure 6.22(*c*) corroborates the observation that sediment ridges do not instantaneously adapt to a change in the arrangement of the large-scale structures in the channel bulk, but they do so with a time lag of around $10T_b$ (equivalent to approximately $0.9H_f/u_\tau$ eddy turnover times). Similarly, it is seen in figure 6.22(*d*) that a time lag of comparable size exists between the flow dynamics in the bulk and those in the vicinity of the sediment bed. The identified time delay signifies that the evolution of the initial sediment ridges and troughs is indeed controlled by the dynamics of the large-scale streaks in the core of the clear-fluid region, which interact in a 'top-down mechanism' with the fluid layers closer to the bed.

Of special interest are therefore the previously discussed intermittently occurring situations during which the clear signatures of high- and low-speed regions in the channel bulk disappear and thus the 'information input' from the channel bulk to the near-bed region is lost. In the discussion of figure 6.16, we have conjectured that the large-scale streaks may break up in these phases, such that their statistical footprints in the streamwise average are not visible anymore. In the following, we try to clarify how the structural organisation of the flow changes during these 'events', foregoing streamwise averaging in order to highlight that the modulation of u''_f in these situations is indeed related to a break-up of the large-scale streaks.

To this end, the most-energetic Fourier modes of the streamwise energy spectra $\phi_{uu}(k_x, y, k_z, t)$ and their time evolution are studied individually in figure 6.23, exemplary for wall-parallel planes at a distance $\tilde{y}/H_f = 0.5$ to the sediment bed. Modes with wavelength pairs $\lambda_x = L_x/i$ and $\lambda_z = L_z/k$ ($i, k \in \mathbb{N}_0$)

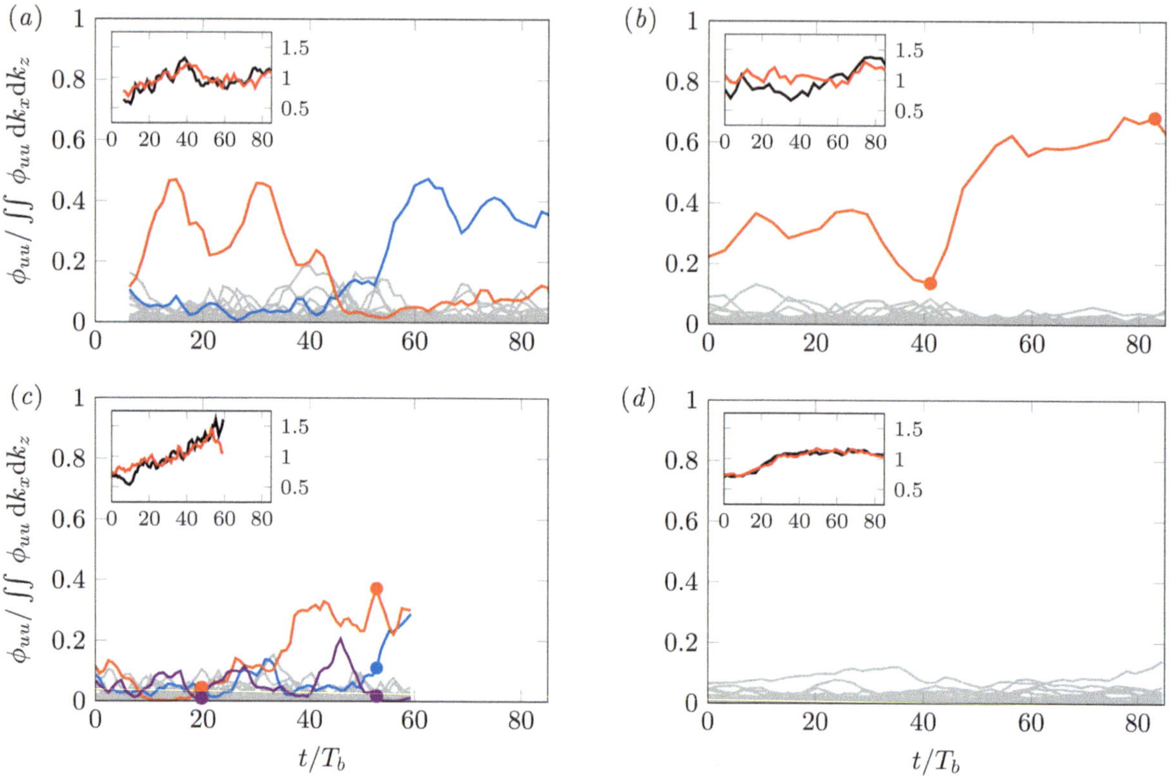

Figure 6.23: Time evolution of the energy distribution between individual modes of the streamwise velocity spectrum $\phi_{uu}(k_x, \tilde{y}, k_z, t)$, evaluated at a wall-normal position $\tilde{y}/H_f \approx 0.5$. The spectrum at each time has been normalised such that the contributions of all individual modes sum up to unity. Note that the most dominant modes (those modes that carry more than 20% of the total energy at least once during the observation time) are highlighted using the following colour scheme: $(0,1)$-mode (—), $(0,2)$-mode (—) and $(1,2)$-mode (—). The smaller insets show the time evolution of the total streamwise and wall-normal fluctuation energy at the same wall-normal position $\langle (u_f - \langle u_f \rangle_{xz})^2 \rangle_{xz}$ (—) and $\langle (v_f - \langle v_f \rangle_{xz})^2 \rangle_{xz}$ (—), respectively, normalised by their time-averaged mean. (*a*) $CM250^{H3}$, (*b*) $CM650^{H3}_{smooth}$, (*c*) $CM850^{H3}$, (*d*) $CL250^{H16}$.

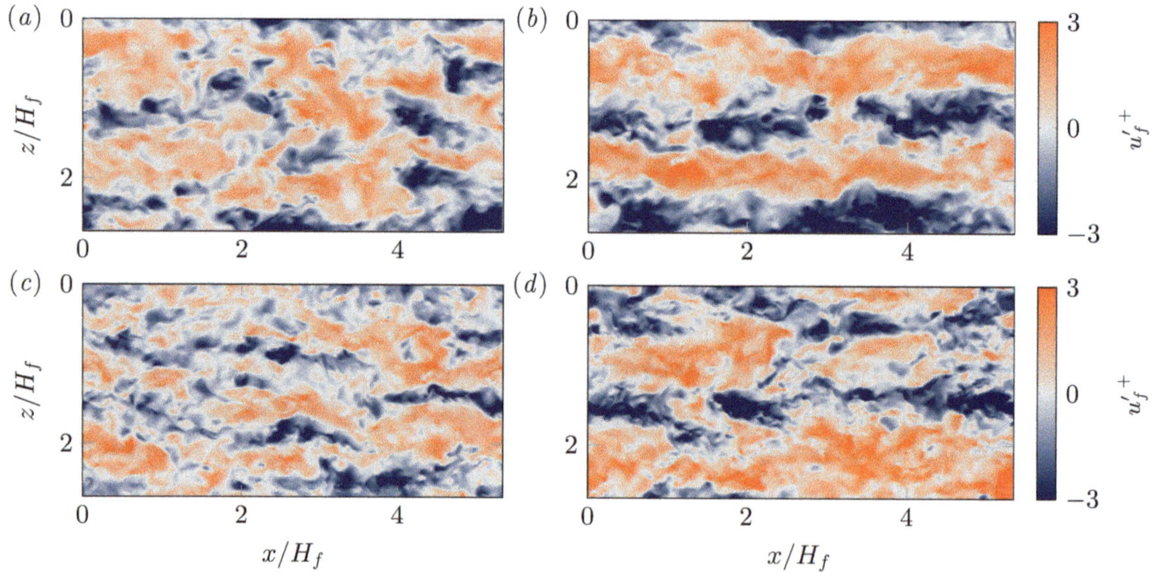

Figure 6.24: Instantaneous visualisations of the wall-parallel planes of the streamwise velocity fluctuation $u_f'^+$, extracted at $\tilde{y}/H_f \approx 0.5$ for cases (*a*,*b*) $CM650^{H3}_{smooth}$ and (*c*,*d*) $CM850^{H3}$ at two different times. The flow fields refer to the instances marked by filled circles in figure 6.23(*b*) and (*c*), respectively.

will be denoted as (i,k)-mode in the remainder to keep notation short. The curves presented in figure 6.23 are normalised by the total streamwise kinetic energy contained in the fluctuating field at the given time and wall-parallel plane, respectively, such that the sum of all contributions is unity. Such normalisation appears suitable recalling the transient nature of the flow under consideration. As a consequence of the increasing friction along the sediment bed, the absolute values of the streamwise $\langle(u_f - \langle u_f\rangle_{xz})^2\rangle_{xz}$ (black) and wall-normal kinetic energy $\langle(v_f - \langle v_f\rangle_{xz})^2\rangle_{xz}$ (red) contained in the specific wall-parallel plane initially increase in all particle-laden cases. The limited box size in the medium domains allows to accommodate only a few large-scale modes, such that a large portion of the total kinetic energy resides in a few dominant modes which are indicated by different colours in figure 6.23. Note that we have classified modes as dominant if they are seen to carry more than 20% of the total perturbation energy at least once during the simulation. In the medium domains, we find three different highly energetic dominant modes, from which two feature the streamwise zero mode. These modes are of infinite streamwise extent and thus do not appear in the premultiplied energy spectra. In physical space, on the other hand, these modes refer to the large alternating high- and low-speed streaks that span over the entire streamwise domain and consequently appear as high- and low-speed regions in the streamwise average. As opposed to the medium box simulations, the large domain of $CL250^{H16}$ accommodates a much wider range of harmonic wavelengths both in the streamwise and spanwise direction. The kinetic energy is consequently more or less evenly distributed among the different wavelengths, without individual modes dominating the spectra.

A comparison of the evolution of individual modes in the medium domain simulations with the corresponding space-time visualisations in figure 6.16 discloses that the intermittent disappearances of the streamwise-averaged high- and low-speed regions correspond to times during which the spectra exhibits no clearly dominant mode. In particular, the energy contained in the infinitely long modes is significantly reduced during these phases, attaining values that are comparable to those in the remaining non-dominant modes. In case $CM250^{H3}$, for instance, figure 6.23(a) shows that starting from approximately $t = 50T_b$, the number of large-scale streak pairs reduces from two to one as has been also observed in figure 6.16(a). In this phase, not only the fraction of energy contained in the $(0,2)$-mode decreases but also the total streamwise and wall-normal fluctuating kinetic energy decay as seen in the inset to figure 6.23(a), implying that the two streak pairs break up at this time. In a similar way, the dominant $(0,2)$-mode in the smooth-wall case $CM650^{H3}_{smooth}$ shows a sudden decrease at approximately $t = 40T_b$, but in contrast to case $CM250^{H3}$, the same mode recovers maintaining the two streak pairs rather than to change to a one-pair state. Corresponding flow fields in figure 6.24(a,b) extracted at the times indicated by the two symbols in figure 6.23(b) highlight the different organisation of the flow in physical space in phases that do and do not feature a clear dominant energetic mode: If pronounced large-scale streaks exist, they reveal no relevant lateral meandering but are roughly aligned with the streamwise direction and thus correspond to the streamwise zero mode. In the corresponding sediment-laden case $CM850^{H3}$ for which the time evolution of the individual modes is shown in figure 6.23(c), the streamwise zero mode contains only a slightly larger fraction of the total kinetic energy than the remaining modes at $t = 0T_b$. Readily after the onset of sediment erosion, the energy further reduces until there is no visually detectable difference in the energy carried by the individual modes, agreeing with the space-time plots in figure 6.16 and the corresponding flow field visualisation in figure 6.24(c) that highlights the absence of dominant large-scale streaks in this phase. After the break-up of the 'old' streaks, two new streak pairs are seen to develop both in physical space (cf. figure 6.16(c))

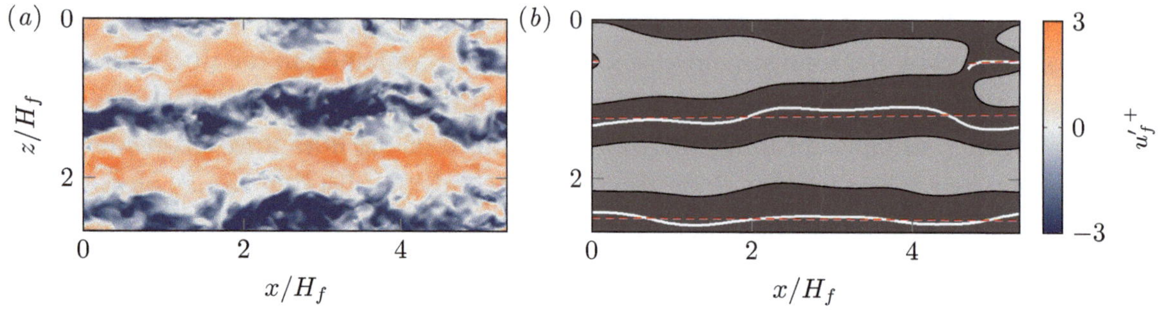

Figure 6.25: Detection of large-scale low-speed streaks (cf. Schoppa and Hussain, 2002; Kevin et al., 2019b). (*a*) Instantaneous wall-parallel plane at $\tilde{y}/H_f \approx 0.5$ of a randomly-chosen streamwise velocity fluctuation field $u_f'^+$ in case $CM650^{H3}_{smooth}$. (*b*) Same data, after applying a 2D Gaussian cut-off filter with filter widths $\Delta_x = 3y_{ref}$ and $\Delta_z = 1.5y_{ref}$, $y_{ref} = 0.5H_f$. Regions shaded in dark grey indicate filtered low-speed streaks, bright grey areas refer to the corresponding high-speed streaks. White solid lines represent the detected centrelines of the low-speed streaks, connecting the local minima of the streamwise velocity fluctuations w.r.t. the spanwise direction. Linear fits to the individual centrelines are shown as red dashed lines. The inclination $\hat{\varphi}$ is the angle between the linear fit and the streamwise direction.

and in the spectra in form of the $(0,2)$-mode that gains in energy first moderately, later more strongly until it becomes the dominant mode for $t \gtrsim 40T_b$ (cf. also figure 6.24(*d*)).

6.3.7 Lateral streak meandering

At first glance, the large-scale streaks in figure 6.24(*b*,*d*) exhibit a rather weak lateral meandering over smooth bottom wall and developed ridges likewise. In the following, we will scrutinise the meandering tendency of velocity streaks of varying size and its dependency on the domain size and the presence or absence of sediment ridges. A suitable methodology for such an investigation has been developed by Schoppa and Hussain (2002) to study the dynamics of individual buffer-layer streaks. Later, Kevin et al. (2019b) used a variant thereof to study the meandering behaviour of coherent structures in the log-layer of canonical boundary layers at different distances from the wall and recently, Bae and Lee (2021) adopted a similar strategy to investigate the dynamics and shape of three-dimensional streak structures. A sketch of the basic concept of this methodology is provided in figure 6.25: For a given wall-parallel slice of the flow field, the local minima z_{min} of the spanwise velocity profile $u_f'(x,z,t)|_{x,t\,fix.}$ are determined for each position x. A set of such points that is connected (in the streamwise direction) then represents the centreline of a single streak. The smoothness of the low-speed streaks in the buffer layer on which the original methodology of Schoppa and Hussain (2002) was focused allows to forego a smoothing or filtering, while the rough appearance of log-layer streaks does not. Kevin et al. (2019b) used a standard Gaussian filter, while we make once more use of the two-dimensional version of the anisotropic Gaussian cut-off filter as defined in equation (6.26). In the following, we measure the streamwise length of a streak $\hat{x}$ as the extent of the projection of the linear fit upon the streamwise x-axis. In this context, let us stress that streaks can overlap the domain boundaries due to the periodic boundary conditions and thus might reach lengths longer than the streamwise domain size L_x. In order to quantify the meandering amplitude and the current orientation of a streak, we fit a straight line to each streak centreline based on a least-square fitting and measure the streak inclination w.r.t. the x-direction as the angle $\hat{\varphi}$ between the linear fit and the streamwise axis. The local meandering amplitude $\hat{A}_s(x,t)$ of a given streak is introduced as the local distance between the streak centreline and the linear fit in the z-direction at each position x. The root mean square over $\hat{A}_s$ at all streamwise positions x of a single streak is in the following denoted as $\hat{\sigma}_s$.

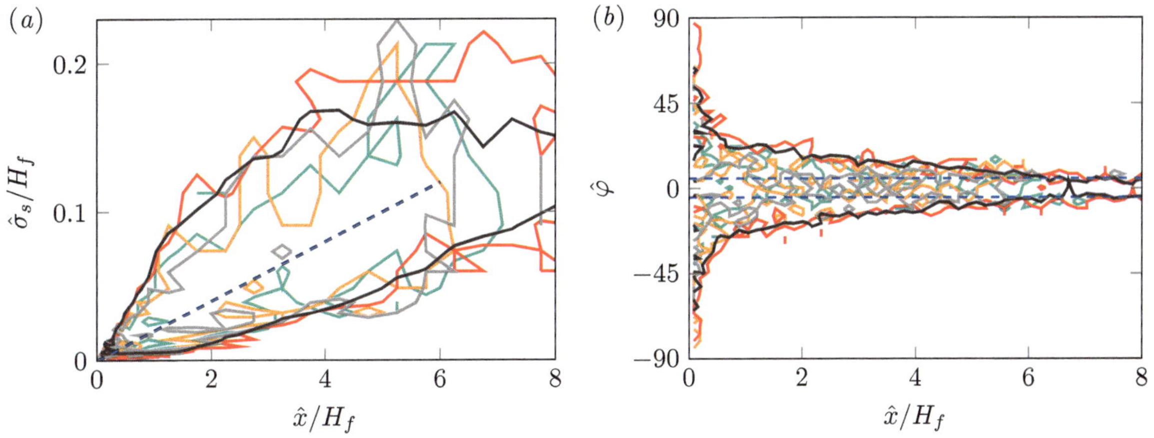

Figure 6.26: Joint p.d.f. of the projected streamwise streak length $\hat{x}/H_f$ with (*a*) the root mean square of the spatial meandering amplitude $\hat{\sigma}_s/H_f$ and (*b*) the orientation angle $\hat{\varphi}$ of the linear fit w.r.t. the *x*-direction (in degree). In (*a*), the blue dashed line refers to the linear relation $\hat{\sigma}_s/H_f = 0.02\hat{x}/H_f$, while the horizontal dashed lines in (*b*) denote angles of $\pm 5°$. Contours include 95% of the total mass of the joint p.d.f.s. Note that due to the heterogeneous distribution of the underlying dataset, the cells for the computation of the joint p.d.f. are accordingly non-homogeneously distributed. Cases: $CM250^{H3}$ (—), $CM650^{H3}_{smooth}$ (—), $CM850^{H3}$ (—), $CL250^{H16}_{smooth}$ (—), $CL250^{H16}$ (—).

In figure 6.26(*a*), the joint probability density function (joint p.d.f.) of the root mean square of the streak meandering amplitude $\hat{\sigma}_s$ and the streak length $\hat{x}/H_f$ is shown, whereas figure 6.26(*b*) provides the joint p.d.f. of the inclination angle $\hat{\varphi}$ and $\hat{x}/H_f$. Note that due to the short observation time interval combined with moderate domain sizes, only a comparably small number of individual streaks was available as samples for the construction of the joint p.d.f.s such that they still reveal some scatter. Nonetheless, the presented data clearly shows that the orientation and meandering tendency of the low-speed streaks are functions of their size: Longer structures show – in a statical sense – stronger lateral oscillations, but their mean axes are quasi-parallel to the mean flow direction. Smaller structures are, as expected, more isotropically arranged as indicated by the wide range of inclination angles $\hat{\varphi}$ that are attained by structures of short streamwise lengths. For all structures shorter than $\hat{x}/H_f \lesssim 6$, the meandering amplitude seems to decrease linearly with their streamwise extension, while for longer structures, the increase of the mean amplitude with $\hat{x}$ reduces noticeably. This is in line with the findings of Kevin et al. (2019b) who observed that the mean streak meandering amplitude for a given wall-parallel plane increased when they excluded small-scale streaks from the analysis. On the other hand, the authors reported that the meandering of individual structures depended on their distance to the bottom wall, with stronger lateral motion for the larger streaks further away from the wall and reduced spanwise oscillations for the smaller ones closer to the wall.

The results shown in figure 6.26 highlight that the amplitude of the lateral meandering is essentially the same for streaks of the same size in all simulations irrespective of domain size and flow configuration, leading to the conclusion that at least at the considered parameter point, the observed spatial meandering is a robust feature of turbulent streaks and is neither significantly amplified nor damped by the presence of sediment ridges. Also, it underlines that lateral meandering is not artificially suppressed when studied in the relatively narrow domains of cases $CM250^{H3}$, $CM650^{H3}_{smooth}$ and $CM850^{H3}$.

6.3.8 Large-scale streaks and mean secondary flow

As reviewed in section 3.6, sediment ridges are often suspected to 'lock' the spanwise location of large-scale coherent structures such as the velocity streaks, such that the latter leave their footprint in the streamwise and time-averaged fields in form of up- and downwelling of the mean velocity profile, accompanied by depth-spanning mean secondary currents (Nezu, 2005). The results presented in the previous section, on the other hand, have underlined that within the here considered intermediate time intervals $\mathcal{O}(10T_b)$, large-scale streaks are even in the single-phase smooth-wall simulations more or less well aligned in the streamwise direction without marked lateral meandering. A substantial fraction of the turbulent kinetic energy was moreover seen to reside in some distinct streamwise-elongated modes. Let us recall at this point that the organisation of the large-scale streamwise velocity streaks is closely coupled with that of other coherent structures in the flow, such as Reynolds stress-carrying ejection and sweep structures as well as individual vortex clusters, as was documented by Del Álamo et al. (2006) and Lozano-Durán et al. (2012). By conditionally averaging individual structures at all scales of the logarithmic layer, these authors were able to determine the preferential mutual organisation of these different types of coherent structures. The resulting idealised self-similar configuration consists of a pair of elongated high- and low-speed streaks, with a shorter ejection inducing an upflow inside the low-speed streak and a shorter sweep that marks a downflow inside the high-speed streak (cf. also figure 3.3). The ejection object is additionally associated with a vortex cluster (Del Álamo et al., 2006; Jiménez, 2018), with the upflow occurring in the centre of the cluster. That way organised, the up- and downward fluid motion inside the two streaks cause a conditional quasi-streamwise secondary roller to emerge in between the two structures.

Even though the described mutual organisation of the individual structures is an idealised configuration which is as such unlikely to appear in exactly this way in instantaneous flow fields, the here studied instantaneous structures share an astonishing similarity with this idealised configuration. Figure 6.27 shows instantaneous snapshots of the three-dimensional organisation of the flow field in case $CM850^{H3}$ and the therein embedded large-scale velocity streaks (cf. figure 6.27(*a*)), ejection and sweep structures (cf. figure 6.27(*b*)) and vortex clusters (cf. figure 6.27(*c*)). Following Del Álamo et al. (2006), instantaneous vortex clusters are therein defined as connected subdomains for which the root mean square of the discriminant of the velocity gradient tensor, $\mathcal{D}$, exceeds a given threshold of 0.02. Strictly speaking, $\mathcal{D}$ is not the discriminant of the tensor field $\nabla \boldsymbol{u}_f$, but of the characteristic polynomial used to determine the eigenvalues of the former. Subject to the incompressibility constraint, this latter reduces to a depressed cubic polynomial such that the discriminant is obtained as (Chong et al., 1990, 1998)

$$\mathcal{D}(\boldsymbol{x},t) = \frac{27}{4}\mathcal{Q}^2 + \mathcal{R}^3, \tag{6.35}$$

where $\mathcal{Q}$ and $\mathcal{R}$ are the second and third invariant of $\nabla \boldsymbol{u}_f$. Depending on the sign of $\mathcal{D}$, $\nabla \boldsymbol{u}_f$ possesses either three real or one real and two complex conjugated eigenvalues. In the latter case which corresponds to the condition $\mathcal{D} > 0$, the rotational part of $\nabla \boldsymbol{u}_f$ is dominant and the respective location is classified as lying inside a vortex (Blackburn et al., 1996).

In figure 6.27, we observe that ejection and sweep structures are organised in the same streamwise-elongated way as the longer large-scale velocity structures, with ejections being stringed together inside the low-speeds streaks and *vice versa* for the sweeps. In complete agreement with the results of Del Álamo et al. (2006) and the visualisations of instantaneous closed channel flow simulations by

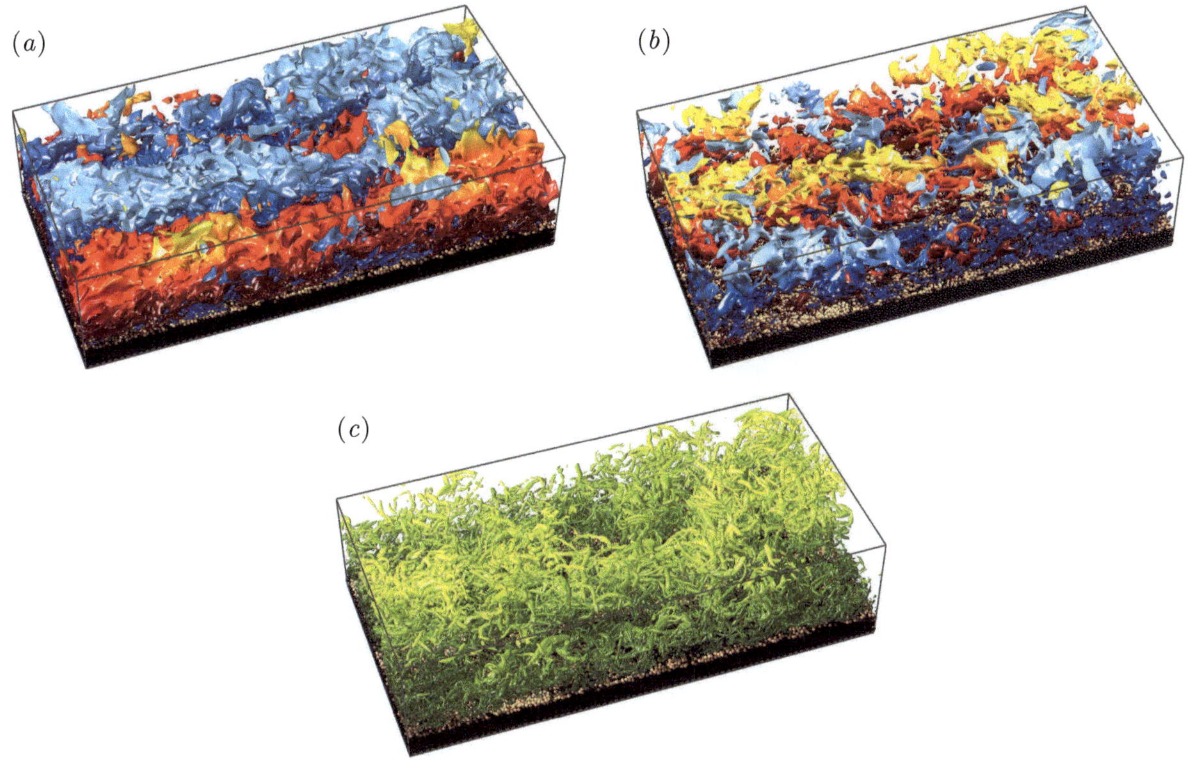

Figure 6.27: Instantaneous visualisation of the large-scale organisation of (*a*) streamwise velocity streaks, (*b*) Reynolds stress-carrying structures and (*c*) vortex clusters. Panel (*a*) and (*b*) are identical to figures 6.17(*a*) and 6.3(*d*), respectively, and are repeated here for convenience. (*a*) High-speed streaks: red, low-speed streaks: blue; (*b*) ejections: red, sweeps: blue. Vortex clusters shown in panel (*c*) are detected as connected regions for which the root mean square of the discriminant of the velocity gradient tensor $\mathcal{D}$ (Chong et al., 1998) exceeds a value of $0.02\mathcal{D}_{rms}(y)$ (Del Álamo et al., 2006). In all panels, bright colours indicate a larger distance to the bottom wall than dark-coloured regions of the structures.

Jiménez (2013a), larger vortex clusters away from the bed are seen to concentrate almost exclusively in the same regions as the large-scale ejections, reaching therein up to the free surface of the channel. This is inasmuch remarkable as individual vortices are usually much smaller and locally more or less isotropically distributed, whereas the vortex clusters reveal a clearly anisotropic organisation. Jiménez (2018) argued that the "residual anisotropy of the vorticity organisation in the presence of the shear of the mean velocity profile" eventually leads to the formation of the conditional quasi-streamwise secondary rollers that connect the conditioned velocity streaks in the cross-plane and are thus closely related to the here observed secondary currents, as we shall see below.

Owing to their relatively regular streamwise and spanwise organisation, the statistical footprint of the large-scale coherent structures is clearly identifiable in the streamwise and time-averaged fields. Figure 6.28 shows the p.d.f. of the occurrence of different quadrant events (Wallace et al., 1972) in the cross-section of case $CM850^{H3}$, which represents a measure for the preferential location of the Reynolds stress-carrying structures in the cross-section. In agreement with our earlier observations, ejections dominate in the regions above sediment ridges (figure 6.28(*b*)), whereas the domain above the troughs is predominantly populated by sweep structures (figure 6.28(*d*)). Inward and outward directed contributions that counteract the mean Reynolds stress occur, as expected, with less probability (Jiménez, 2018) and are of secondary importance in most regions of the cross-section, the only exception being the flow in the direct vicinity of the trough, where locally outward events dominate the p.d.f.. In this context, it should be cautioned that due to the curved fluid-bed interface, wall-plane averaging 'cuts'

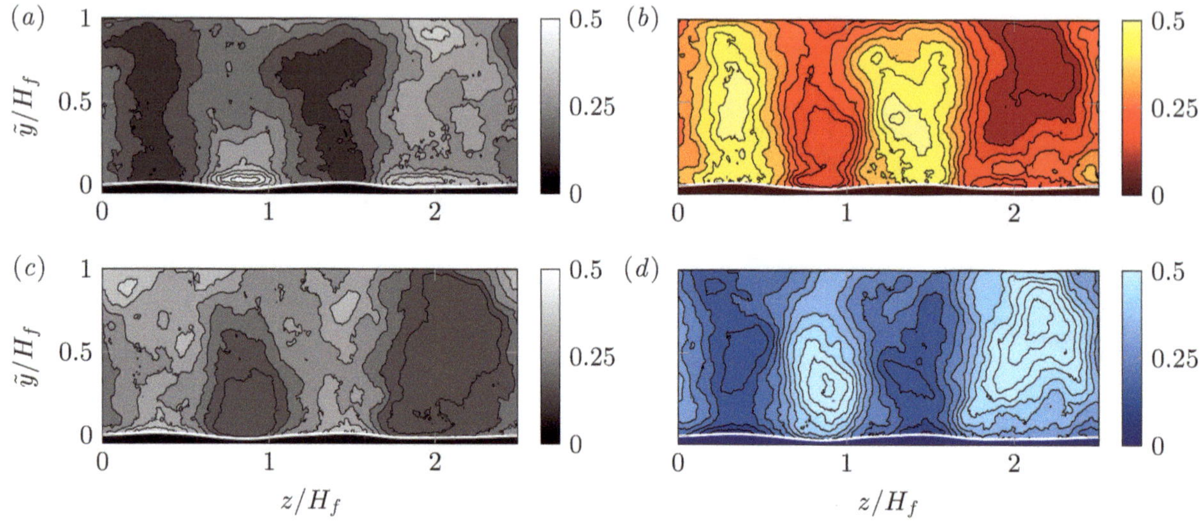

Figure 6.28: Streamwise-averaged p.d.f. of the occurrence of quadrant events in the cross-plane of the open channel flow case $CM850^{H3}$ in the time interval $t/T_b \in [40, 60]$. (*a*) Q1-events (outward: $u'_f > 0$, $v'_f > 0$), (*b*) Q2-events (ejection: $u'_f < 0$, $v'_f > 0$), (*c*) Q3-events (inward: $u'_f < 0$, $v'_f < 0$), (*d*) Q4-events (sweep: $u'_f > 0$, $v'_f < 0$). The individual p.d.f.'s are normalised such that all four contributions sum up to unity pointwise. The time-averaged fluid-bed interface profile is indicated by the white solid curve.

through the ridges where both velocity components are trivially close to zero while they attain non-zero values over the troughs at the same height. Nonetheless, the observations imply that below the large sweep-dominated region, there is a thin layer in which outward motions directed away from the bed are of relevance and might be able to transport particles away from the latter.

The clear dominance of sweeps and ejections over troughs and ridges, respectively, is closely linked to the appearance of mean secondary currents of Prandtl's second kind, as these structures induce at their preferential lateral positions a mean upflow in the low-speed regions over the sediment ridges and accordingly a mean downflow over the troughs. The resulting mean secondary flow fields computed over short time intervals (between 20 and 80 bulk time units) are provided in figure 6.29. The secondary mean flow is indicated in terms of the mean secondary flow streamfunction $\langle\psi\rangle_{xt}(y,z)$ that can be determined by integration of one of the two cross-stream velocity components, since

$$\nabla_\perp \langle\psi\rangle_{xt}(y,z) = (\partial_y\langle\psi\rangle_{xt}, \partial_z\langle\psi\rangle_{xt})^T = (-\langle w_f\rangle_{xt}, \langle v_f\rangle_{xt})^T, \tag{6.36}$$

where $\nabla_\perp = (\partial_y, \partial_z)^T$ is the two-dimensional nabla operator that acts only in the two cross-stream directions. In all cases, the cross-plane is filled with pairs of depth-spanning counter-rotating secondary mean flow vortices. In the ridge-featuring simulations (cf. figure 6.29(*a*,*c*,*d*)), the streamwise mean flow exhibits an alternating up- and downwelling over ridge crests and troughs, respectively, with accordingly oriented secondary currents transporting high-momentum fluid towards the free surface over the crests and *vice versa* over troughs (Nezu and Nakagawa, 1993). Remarkably, also the statistically spanwise homogeneous smooth-wall simulation $CM650^{H3}_{smooth}$ in figure 6.29(*b*) features over the here considered averaging time interval mean secondary currents that are of striking similarity to their counterparts over developed sediment ridges. In particular, all simulations show roughly the same lateral spacing of the secondary currents, which is equivalent to the preferential spanwise wavelength of the instantaneous large-scale streaks, i.e. $\lambda_z/H_f \approx$ 1-1.5. Case $CM250^{H3}$ represents in this context an exception since in the period over which the statistics for figure 6.29 have been gathered, this latter

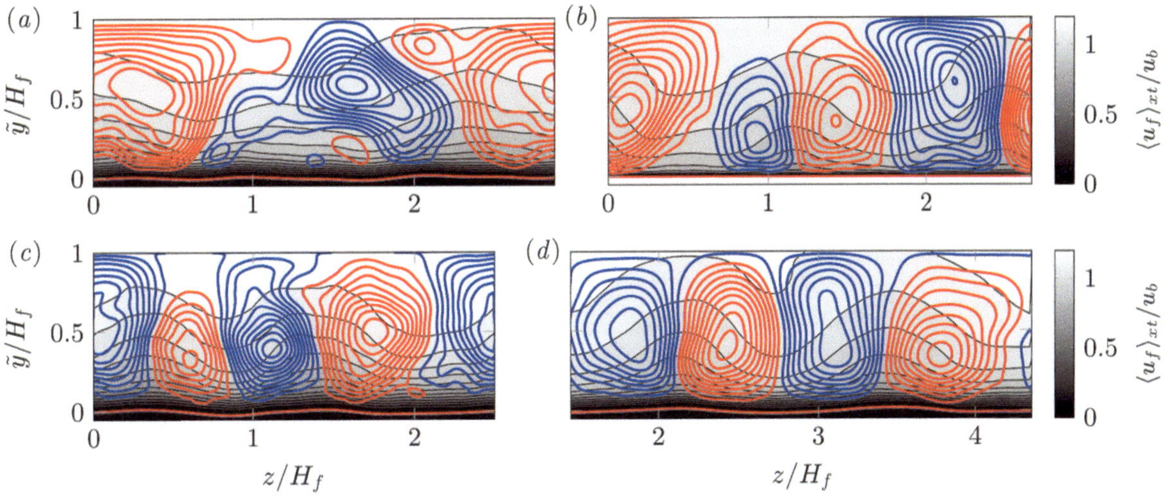

Figure 6.29: Mean primary $\langle u_f \rangle_{xt}/u_b$ and secondary flow field $(\langle v_f \rangle_{xt}, \langle w_f \rangle_{xt})^T/u_b$ for cases (*a*) $CM250^{H3}$, (*b*) $CM650^{H3}_{smooth}$, (*c*) $CM850^{H3}$ and (*d*) $CL250^{H16}$. Note that for the sake of comparison, in panel (*d*) we have arbitrarily chosen a subdomain of the entire cross-section of case $CL250^{H16}$ with a lateral width similar to L_z/H_f in the medium domain cases. The time-averaged fluid-bed interface profile is indicated by a red curve. The time window over which data has been accumulated can be seen in figure 6.30. Isolines of the primary flow $\langle u_f \rangle_{xt}/u_b$ are shown in intervals 0.1(0.1)1.2, while the secondary flow pattern $(\langle v_f \rangle_{xt}, \langle w_f \rangle_{xt})^T$ is shown in terms of the secondary mean flow streamfunction $\langle \psi \rangle_{xt}$. Clockwise (counterclockwise) secondary flow rotation is indicated by red (blue) contours. The shown contours represent 20 equally spaced subintervals of the range $[\min_{y,z} \langle \psi \rangle_{xt}, \max_{y,z} \langle \psi \rangle_{xt}]$.

intermittently features only a single pair of low- and high-speed streaks, as has been already discussed in the context of figure 6.16(*a*).

In addition to the size and shape of the secondary currents, also the secondary flow amplitude turns out to be of comparable size for all simulations: Figure 6.30 shows the time-dependent secondary flow intensity $u_\perp(t)$ defined from the cross-plane averaged kinetic energy contained in the streamwise-averaged cross-flow field $(\langle v_f \rangle_x, \langle w_f \rangle_x)^T$, viz.

$$u_\perp(t) = \left[\frac{1}{L_z \, \langle h_f \rangle_{xz}(t)} \int_0^{L_z} \int_{\langle h_b \rangle_{xz}(t)}^{L_y} \left(\langle v_f \rangle_x^2 + \langle w_f \rangle_x^2 \right) \mathrm{d}y\mathrm{d}z \right]^{1/2} . \tag{6.37}$$

Relation (6.37) is a variation of a formulation for the secondary flow intensity proposed by Sakai (2016) in the context of sidewall-induced mean secondary flow in open and closed ducts, adapted to the current flow configuration. The intensity of the mean secondary flow $u_\perp(t)$ is observed to oscillate between 1.5% and 2.5% of the bulk velocity in figure 6.30 for all cases. The lowest value is attained in case $CL250^{H16}$, which features a larger averaging ensemble in that a larger number of ridges and associated secondary flow cells is accommodated in the wide domain (by a factor of three to four). As expected, the intensity of the time-averaged cross-flow $(\langle v_f \rangle_{xt}, \langle w_f \rangle_{xt})^T$

$$U_\perp = \left[\frac{1}{L_z \, H_f} \int_0^{L_z} \int_{H_b}^{L_y} \left(\langle v_f \rangle_{xt}^2 + \langle w_f \rangle_{xt}^2 \right) \mathrm{d}y\mathrm{d}z \right]^{1/2} , \tag{6.38}$$

indicated by horizontal dashed lines in figure 6.30, is systematically weaker than the instantaneous value $u_\perp(t)$, but turns out to be comparable to the amplitude of sidewall-induced mean secondary flows usually reported for turbulent open duct flows. Sakai (2016), for instance, determined the intensity of the secondary flow within a distance $z = 1H_f$ from the sidewalls of a smooth-wall open duct

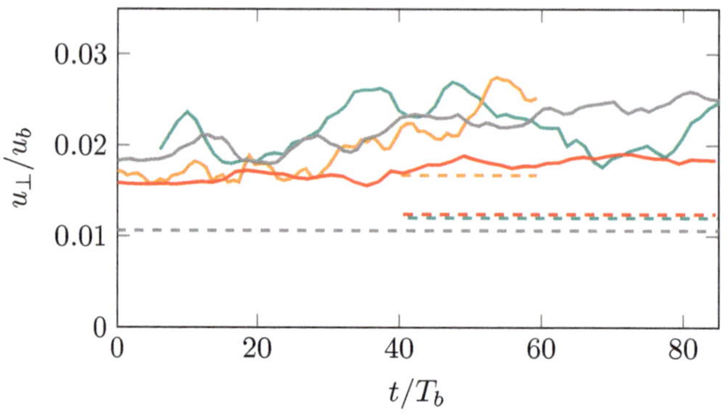

Figure 6.30: Instantaneous secondary flow intensity $u_\perp/u_b$ as a function of time (solid lines) and the corresponding value determined for the time-averaged fields in figure 6.29 (dashed lines) following the definition in equation (6.37). The length of the dashed lines indicates the time window over which the fields shown in the respective panels of figure 6.29 have been averaged: $CM250^{H3}$ (—), $CM650^{H3}_{smooth}$ (—), $CM850^{H3}$ (—), $CL250^{H16}$ (—).

in a range between 1.1% and 1.3% of the bulk velocity. In the following chapter, we will further investigate the impact of developed ridges on the sidewall-induced secondary flow pattern and the mutual relation between the mean secondary flow streamfunction $\langle\psi\rangle_{xt}$ and the mean streamwise vorticity $\langle\omega_{f,x}\rangle_{xt}$ (cf. section 7.2.4), which are directly related by a Poisson equation (Pinelli et al., 2010).

In figure 6.31, structures that carry a significant portion of streamwise vorticity are seen to preferentially cluster inside the mean secondary currents of similar sense of rotation for both the flow over the smooth wall and that over developed sediment ridges. In other words, these structures are predominantly located at the border between low- and high-speed streaks or, equivalently, ejection and sweep structures. Along the bottom wall of case $CM650^{H3}_{smooth}$, mean mirror vortex cells similar to those in the open duct case (cf. section 7.2.4) appear as a consequence of the no-slip boundary condition at the wall. Whilst similar mean vortex patterns are also seen along the mean fluid-bed interface of case $CM850^{H3}$, vorticity tends to zero when approaching the bed as a consequence of the small velocity gradients in direct vicinity of the sediment bed. The mean distribution of the streamwise vorticity generally resembles that observed in the conditionally-averaged datasets of Lozano-Durán et al. (2012), for which $\omega_{f,x}$ was investigated by Jiménez (2013a). The main conditional roller observed in the latter work was similarly sitting between the conditional ejection and sweep object, and its intensity was with $\mathcal{O}(u_\tau/y_{cg})$ of the same order as the mean shear S at its centre of gravity y_{cg}, which is $S \approx u_\tau/\kappa y_{cg}$ inside the logarithmic layer. For the here considered outer mean vorticity cells, the mean shear at their centre of gravity is approximately $u_\tau/(0.5H_f\,\kappa) \approx 5u_\tau/H_f$ assuming the logarithmic velocity profile extends up to this height, which is indeed comparable to the maximum values of $|\langle\omega_{f,x}\rangle_{xt}|H_f/u_\tau \approx$ 3-5 that are observed in the centre of the secondary flow cells in figure 6.31.

6.3.9 Total shear stress

The appearance of a mean secondary motion in the cross-plane of the channel goes hand in hand with a modulation of the mean total shear stress $\langle\tau_{tot}\rangle_{xt}$, which is determined by integration of the streamwise momentum balance over the streamwise direction and time, viz. (Nikora et al., 2007)

$$\langle\tau_{tot}\rangle_{xt}(y,z) = \underbrace{\rho_f \nu_f \frac{\partial\langle u_f\rangle_{xt}}{\partial y}}_{\langle\tau_{visc}\rangle_{xt}} - \underbrace{\rho_f\langle u_f''' v_f'''\rangle_{xt}}_{\langle\tau_{turb}\rangle_{xt}} - \underbrace{\rho_f\langle u_f'''' v_f''''\rangle_{xt}}_{\langle\tau_{disp}\rangle_{xt}} + \underbrace{\int_y^{L_y}\langle f_x\rangle_{xt}(s,z)\,\mathrm{d}s}_{\langle\tau_{part}\rangle_{xt}}, \tag{6.39}$$

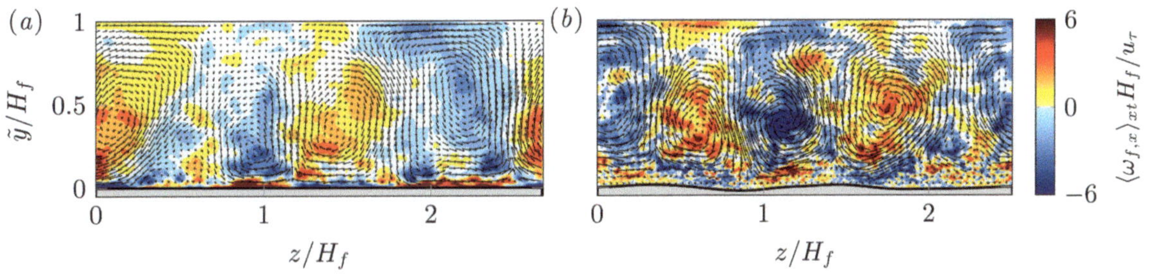

Figure 6.31: Short-time averaged mean streamwise vorticity distribution $\langle\omega_{f,x}\rangle_{xt}H_f/u_\tau$ for cases (*a*) $CM650^{H3}_{smooth}$ and (*b*) $CM850^{H3}$. For the sake of comparison, the intensity and orientation of the mean secondary flow field $(\langle v_f\rangle_{xt}, \langle w_f\rangle_{xt})^T/u_b$ is shown as vector plot. The coloured areas refer to values of $\langle\omega_{f,x}\rangle_{xt}H_f/u_\tau$ in the interval $[-5.4, 5.4]$ with an increment of 1.2. The time-averaged fluid-bed interface profile is indicated by the black solid curve. The time window over which data has been accumulated is of length $\mathcal{O}(10T_b)$ and can be seen in figure 6.30.

where f_x is the streamwise component of the Eulerian force field $\boldsymbol{f}$ arising in the framework of the immersed boundary formulation due to fluid-particle interactions. Unfortunately, $\langle f_x\rangle_{xt}$ is not directly accessible in the here considered statistics as it is not stored alongside the simulation run. In practice, the integrand of $\langle\tau_{part}\rangle_{xt}$ is therefore approximated as

$$\langle f_x\rangle_{xt} \approx -\langle f_x^H\rangle_{xt}\langle\phi_s\rangle_{xt}/\mathcal{V}_p. \tag{6.40}$$

As was shown by Uhlmann (2008), relation (6.40) approximates the fluid-particle force term reasonably well in absence of strong gradients of the solid volume fraction $\langle\phi_s\rangle_{xt}$, which occur in the current system solely in the direct vicinity of the mean fluid-bed interface. It might be further noted that, strictly speaking, the integration of the streamwise momentum balance that leads to equation (6.39) should be performed comprising both fluid and particle phase and thus contain the composite velocity rather than the fluid-phase velocity (Uhlmann, 2008; Kidanemariam, 2016). However, Kidanemariam et al. (2022) recently showed that the difference between both formulations is negligible due to the small difference between fluid and composite velocities, such that we stick with relation (6.39) in the remainder of this work.

The total mean shear stress τ_{tot} defined therein comprises four different contributions, that are, the viscous (τ_{visc}), turbulent (τ_{turb}) and dispersive stresses (τ_{disp}) as well as the shear stress resulting from particle-fluid interactions (τ_{part}). Note that τ_{turb} and τ_{disp} are off-diagonal entries of the two velocity fluctuation covariance tensors $\langle\boldsymbol{u}_f''' \otimes \boldsymbol{u}_f'''\rangle_{xt}$ and $\langle\boldsymbol{u}_f'''' \otimes \boldsymbol{u}_f''''\rangle_{xt}$, which base upon the following flow field decompositions

$$\boldsymbol{u}_f'''(\boldsymbol{x},t) = \boldsymbol{u}_f(\boldsymbol{x},t) - \langle\boldsymbol{u}_f\rangle_t(\boldsymbol{x}) \tag{6.41a}$$

$$\boldsymbol{u}_f''''(\boldsymbol{x}) = \langle\boldsymbol{u}_f\rangle_t(\boldsymbol{x}) - \langle\boldsymbol{u}_f\rangle_{xzt}(y) = \langle\boldsymbol{u}_f'\rangle_t(\boldsymbol{x}). \tag{6.41b}$$

In equation (6.41a), $\boldsymbol{u}_f'''(\boldsymbol{x},t)$ represents the fluctuation of the velocity w.r.t. its local time-averaged value $\langle\boldsymbol{u}_f\rangle_t(\boldsymbol{x})$ at a point $\boldsymbol{x}$ and $\boldsymbol{u}_f''''(\boldsymbol{x})$ in equation (6.41b) is the fluctuation of the time-averaged velocity field w.r.t. its plane-average.

Thus, the turbulent stresses refer to standard Reynolds-averaging and originate from pure temporal fluctuations at a given spatial position, whereas dispersive stresses emanate from spanwise variations in the time-averaged flow field (Nikora et al., 2001). In this regard, dispersive stresses are related to a deflection of the streamlines from the mean flow direction in the time-averaged flow field. A possible

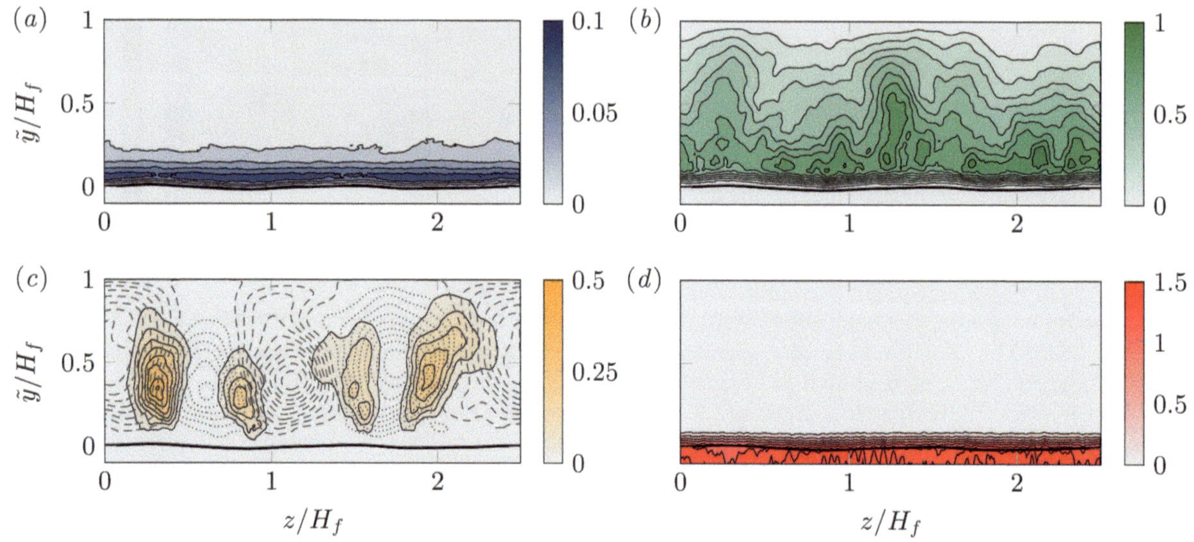

Figure 6.32: Cross-plane distribution of the different contributions to the mean total shear stress $\langle \tau_{tot} \rangle_{xt}$: (*a*) Viscous stresses $\langle \tau_{visc} \rangle_{xt}^+ = (\nu_f \rho_f \partial \langle u_f \rangle_{xt} / \partial y)^+$, (*b*) turbulent stresses $-\langle \tau_{turb} \rangle_{xt}^+ = -(\rho_f \langle u_f''' v_f''' \rangle_{xt})^+$, (*c*) dispersive stresses $-\langle \tau_{disp} \rangle_{xt}^+ = -(\rho_f \langle u_f'''' v_f'''' \rangle_{xt})^+$ and (*d*) approximation of the stresses due to fluid-particle interaction $\langle \tau_{part} \rangle_{xt}^+$ using relation (6.40). The time-averaged fluid-bed interface profile is indicated by the black solid curve. For the sake of comparison, (*c*) contains in addition isocontours of the mean secondary flow streamfunction $\langle \psi \rangle_{xt}$ analogous to figure 6.29(*c*); clockwise (counterclockwise) rotating secondary currents are indicated by dotted (dashed) lines.

reason for such a deviation of the streamlines' orientation from the x-direction are deformed domain boundaries, as in flows over developed transverse bedforms (Kidanemariam et al., 2022) or over irregularly arranged bottom topographies (Jelly and Busse, 2018; Busse and Jelly, 2020), in which context dispersive stresses are sometimes termed 'form-induced stresses'. In the current situation, however, dispersive stresses arise due to the spanwise heterogeneity of the mean turbulent flow field and the secondary currents which, in turn, are the consequence of the afore-discussed regular organisation of the large-scale coherent structures. In other words, dispersive stresses quantify in the present context the contribution of well-organised large-scale coherent structures to the mean shear stress budget, whereas the contributions of smaller-scale short-living or disordered structures are collected in the turbulent stresses. It is worth noting that this implies dispersive stresses to attain non-trivial values even in the absence of mobile sediment or a curved lower domain boundary. Indeed, we will see below that τ_{disp} significantly contributes to the total shear stress budget in the smooth-wall case $CM850^{H3}$ as well during the here considered intermediate time intervals of $\mathcal{O}(10T_b)$.

In figure 6.32, we show exemplary for case $CM850^{H3}$ how the four different contributions to $\langle \tau_{tot} \rangle_{xt}$ are distributed over the channel cross-section. As expected, the contribution of the viscous stresses $\langle \tau_{visc} \rangle_{xt}$ to the total shear stress budget (cf. figure 6.32(*a*)) is of relevance only in a thin layer slightly above the sediment bed. In contrast to a solid impermeable wall at which the wall shear stress is purely viscous, the viscous stresses in the direct vicinity of the mobile permeable sediment bed do not contribute more than around 10% to the total mean shear stress at the given parameter point. The turbulent stresses $\langle \tau_{turb} \rangle_{xt}$ (cf. figure 6.32(*b*)), on the other hand, represent the dominant contribution in the clear-fluid region. Interestingly, $\langle \tau_{turb} \rangle_{xt}$ reveals a clear spanwise variation that is in phase with the mean up- and downflow associated with the secondary currents: Above the sediment ridges where the mean secondary flow is directed from the bed towards the free surface, regions of high turbulent stress extend further into the outer flow than over the troughs, where such regions appear to be more restricted to the near-bed region. Recalling that $\langle \tau_{turb} \rangle_{xt}$ represents the contribution to the total shear stress from all Reynolds stress-carrying structures but the largest which leave their footprint predominantly in the dispersive stresses, this suggests that smaller-scale structures inside the large-scale low-speed streaks can reach closer to the free surface than their counterparts in the corresponding high-speed streaks due to the preferential downward motion (Motoori and Goto, 2021). The dispersive stresses $\langle \tau_{disp} \rangle_{xt}$ shown in figure 6.32(*c*), by contrast, contribute to the total stress merely in the up- and downflow regions of the mean secondary flow. As such, they represent the statistical footprint of the there predominantly residing large-scale ejection and sweep structures, respectively. While of negligible size both near the sediment bed and the free surface, the dispersive stresses are of relevant size roughly at the height at which the centre of gravity of the secondary currents is located. As expected, the total mean shear stress is thus dominated by pure fluid stresses above the bedload layer, whereas inside the region of intense particle transport, the mean shear and the velocity fluctuations decrease markedly. The decay of the pure fluid stresses is compensated by the increasing amplitude of the stresses due to fluid-sediment interactions, $\langle \tau_{part} \rangle_{xt}$, which eventually become the only non-trivial component inside the sediment bed (cf. figure 6.32(*d*)).

The spatial variability of the turbulent and dispersive stresses agrees well with the measurements of Zampiron et al. (2020a) for the flow over artificial triangular ridges with a similar lateral spacing as the here considered self-formed sediment ridges. Also, a similar organisation of the dispersive stresses in regions of mean up- and downward motion has been reported by Vowinckel et al. (2017b) who observed secondary currents to arise in their numerical simulations after particles have agglomerated into streamwise-aligned sediment clusters. Recently, Zampiron et al. (2021) reanalysed the data of Zampiron et al. (2020a), evaluating the individual terms of the mean kinetic energy budget as well as of the transport equations related to the double-averaged, turbulent and dispersive stresses, respectively. The authors concluded that the dispersive kinetic energy contained in the secondary mean flow is fed from the cross-stream components of the purely turbulent field, rather than directly from the double-averaged mean flow, and that the energy supply resides predominantly in structures near the lower domain boundary. This observation is inasmuch conclusive as the fixed artificial ridges were used in the aforementioned study to generate secondary currents and in their absence no such mean secondary flow was observed, such that there has to be a causal connection between the contour of the lower domain boundary and the secondary mean flow. In the current case, on the other hand, sediment ridges have been seen to follow the dynamics of the large-scale flow features rather than the other way round, such that the kinetic energy transport between the different fields might be markedly different. In particular, the earlier in this chapter observed 'top-down mechanism' in which large-scale streaks and sediment ridges interact and the observation of secondary currents in the smooth-wall simulations clearly contradict a control of the secondary currents from the sediment bed.

In order to recover the standard linear variation of the plane- and time-averaged mean shear stress profile $\langle \tau_{tot} \rangle_{xzt}(y)$, we average equation (6.39) over the spanwise direction and obtain

$$\langle \tau_{tot} \rangle_{xzt}(y) = \langle \tau_{visc} \rangle_{xzt} - \langle \tau_{turb} \rangle_{xzt} - \langle \tau_{disp} \rangle_{xzt} + \underbrace{\int_y^{L_y} \langle f_x \rangle_{xzt}(s) \mathrm{d}s}_{\langle \tau_{part} \rangle_{xzt}} . \tag{6.42}$$

In case of a fully-developed statistically stationary flow, the force exerted on the two-phase system by the mean driving pressure gradient $\langle \Pi \rangle_t$ is entirely balanced by the counteracting total shear stress,

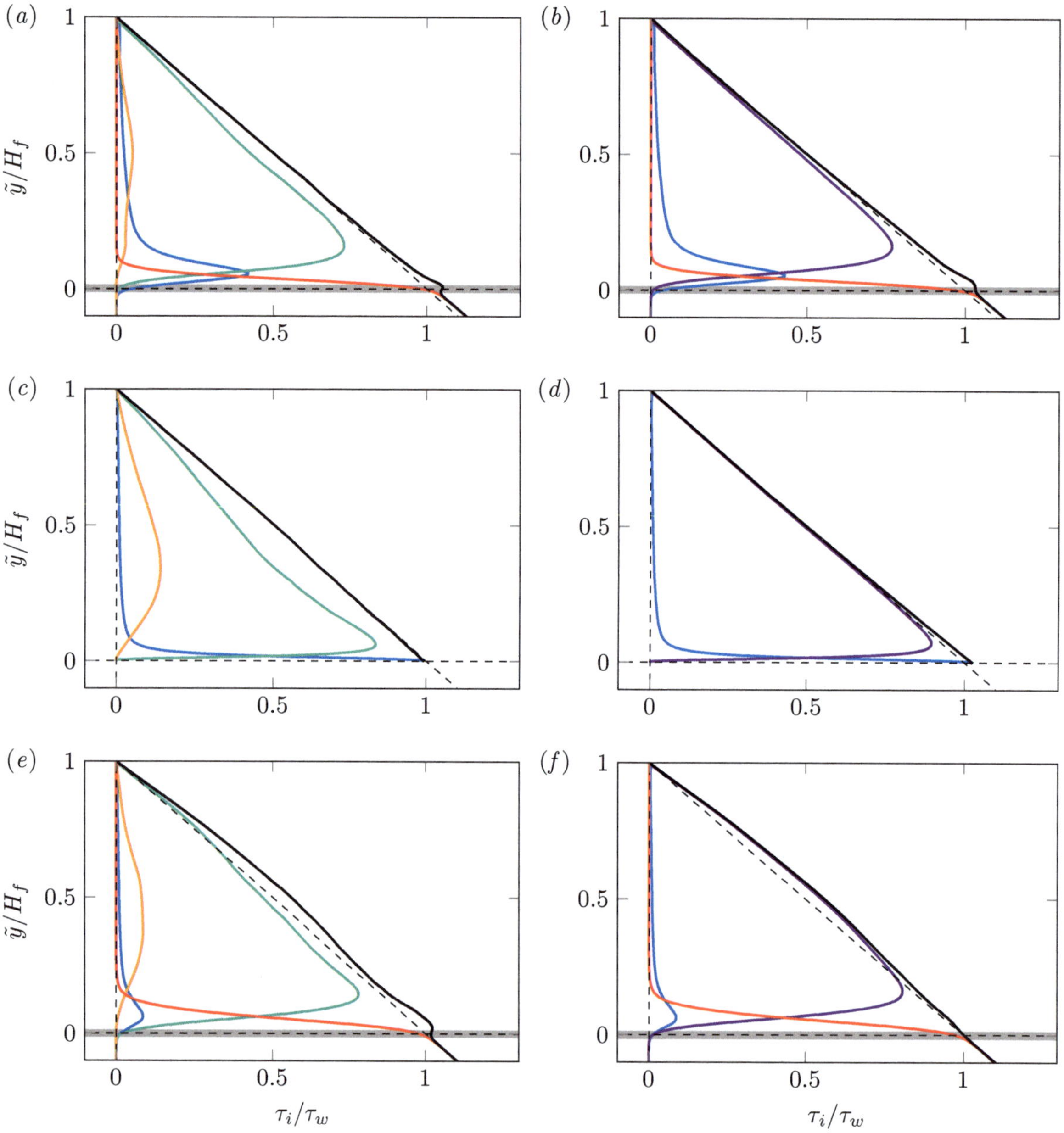

Figure 6.33: Wall-normal profiles of the time- and plane-averaged mean total shear stress $\langle \tau_{tot} \rangle_{xzt}$ and its individual contributions. *(a,c,e)* Time- and plane-averaged total shear stress $\langle \tau_{tot} \rangle_{xzt}$ determined based on the pure fluid velocity obtained from individual flow fields, evaluating equation (6.42). *(b,d,f)* Same as in *(a,c,e)*, but with $\langle \tau_{tot} \rangle_{xzt}$ determined based on the composite velocities collected as runtime statistics, evaluating equation (6.44). The individual stress contributions are $\langle \tau_{visc} \rangle^+_{xzt}$ (—), $-\langle \tau_{turb} \rangle^+_{xzt}$ (—), $-\langle \tau_{disp} \rangle^+_{xzt}$ (—), $-\langle u'_f v'_f \rangle^+_{xzt}$ (—), $\langle \tau_{part} \rangle^+_{xzt}$ (—), $\langle \tau_{tot} \rangle^+_{xzt}$ (—). Cases: *(a,b)* $CM250^{H3}$, averaging interval $t/T_b \in [20, T_{obs}]$; *(c,d)* $CM650^{H3}_{smooth}$, $t/T_b \in [0, T_{obs}]$; *(e,f)* $CM850^{H3}$, $t/T_b \in [20, T_{obs}]$. For cases $CM250^{H3}$ and $CM850^{H3}$, the grey-shaded area indicates the interval between the minimum and maximum of the streamwise- and time-averaged fluid-bed interface, viz. $[\min_z(\langle h_b \rangle_{xt}), \max_z(\langle h_b \rangle_{xt})]$.

i.e. $-\langle\Pi\rangle_t(L_y - y) = \langle\tau_{tot}\rangle_{xzt}(y)$. The number of terms on the RHS of equation (6.42) can be further reduced taking into account that (Raupach and Shaw, 1982)

$$\langle u'_f v'_f\rangle_{xzt} = \langle u'''_f v'''_f\rangle_{xzt} + \langle u''''_f v''''_f\rangle_{xzt} = 1/\rho_f\langle\tau_{turb}\rangle_{xzt} + 1/\rho_f\langle\tau_{disp}\rangle_{xzt}. \tag{6.43}$$

Making use of equation (6.43), relation (6.42) can be rewritten as

$$\langle\tau_{tot}\rangle_{xzt}(y) = \langle\tau_{visc}\rangle_{xzt} - \rho\langle u'_f v'_f\rangle_{xzt} - \langle\tau_{part}\rangle_{xzt}. \tag{6.44}$$

While in the continuous case mathematically equivalent, the two formulations are slightly differently evaluated in the following, as a consequence of the way statistics are accumulated in the context of the numerical code we are using. When evaluating relation (6.42), the individual terms are computed based on individual snapshots of the flow field which were stored in intervals of 1-3T_b length, as the dispersive and turbulent stresses are not stored in form of runtime statistics alongside the simulation. The double-averaged Reynolds stresses $\langle u'_f v'_f\rangle_{xzt}$ which represent the sum of the former two contributions, on the other hand, are available as runtime statistics meaning that the data is accumulated over the entire simulation period with much shorter intermediate time intervals. Also, $\langle f_x\rangle_{xzt}$ is directly accessible in the runtime statistics such that in the following, we can forego using approximation (6.40).

The results obtained when evaluating both relations are visualised in figure 6.33, exemplary for cases $CM250^{H3}$, $CM650^{H3}_{smooth}$ and $CM850^{H3}$. Taking into account the relatively short averaging intervals considered here together with the fact that some of the particle-laden cases are still in a transient phase, the variation of the total shear stress $\langle\tau_{tot}\rangle_{xzt}$ shown in figure 6.33 reasonably well recovers the expected linear variation across the channel height, which can be seen as a measure for the state of statistical convergence in the respective case. While the deviation due to not fully-converged statistics is seen all over the channel height, we also observe a rather localised deviation from the linear profile in the vicinity of the mean fluid-bed interface, especially for case $CM250^{H3}$ (cf. figure 6.33(*a,b*)). We expect this deviation to be the consequence of the fact that in the context of the current finite difference immersed boundary method, the analytically derived momentum equation is fulfilled only up to a truncation error. The latter is assumed to be of negligible size in the absence of particles as in the channel bulk as well as in regions where the velocity of both phases is vanishingly small as inside the sediment bed. Similarly, the error is assumed to be negligible in case of dilute particle suspensions in which particles travel at a similar velocity as the surrounding fluid phase (Uhlmann, 2008; Kidanemariam et al., 2013). On the other hand, the deviation will be more pronounced in regions of stronger differences between particle and fluid velocity as it is the case along the fluid-bed interface, where the fluid mobilises particles that are essentially in rest. With increasing height of the studied bedforms, the observed deviation is smeared over a larger wall-normal interval and is thus almost not to detect in case of the higher-amplitude transverse bedforms studied, for instance, in Kidanemariam and Uhlmann (2017). For the here considered low-amplitude sediment ridges, by contrast, the interval between the minimum and maximum of the bed profile is relatively narrow, such that the deviations concentrate in the vicinity of the mean fluid-bed interface.

In agreement with our earlier observations, the viscous stresses $\langle\tau_{visc}\rangle_{xzt}$ in figure 6.33 are of significance mainly in the region close to the bed, whereas most of the clear-fluid region is dominated by the turbulent stresses $\langle\tau_{turb}\rangle_{xzt}$. In this context, let us point out that the contribution of the viscous stresses to the total shear stress $\langle\tau_{tot}\rangle_{xzt}$ is fairly different for the three shown cases: While in the smooth-wall case $CM650^{H3}_{smooth}$, the viscous shear stress naturally is the only contribution directly a the solid impermeable bottom wall due to the there applied boundary conditions, its contribution to the total mean shear stress is drastically reduced in the particle-laden cases (attaining only up to 40% of $\langle \tau_{tot} \rangle_{xzt}$ in case $CM250^{H3}$ and less than 10% in case $CM850^{H3}$). In both cases, instead, the stresses due to fluid-particle interactions clearly dominate the mean shear stress in the near-bed region, where the particle transport is most intense. In all three cases, dispersive stresses are seen to markedly contribute to the total shear stress budget in the bulk of the channel, attaining values between 15% and 25% of $\langle \tau_{tot} \rangle_{xzt}$. Remarkably, it is the smooth-wall case $CM650^{H3}_{smooth}$ that reveals the highest contribution of dispersive stresses of all three cases, underlining that the dispersive stresses are indeed a consequence of the mean secondary flow, rather than being caused by the sinusoidal deformation of the sediment bed. In the context of developed ripple-like bedforms where $\langle \tau_{disp} \rangle_{xzt}$ is indeed a form-induced stress, on the other hand, its contribution is of relevance only in the vicinity of the sediment bed and of negligible size in the outer flow (Kidanemariam et al., 2022).

The earlier discussed experimental measurements of flow over fixed artificial ridges at a lateral spacing of $\lambda_{h,z}/H_f \approx 1.6$ performed by Zampiron et al. (2020a), in turn, reveal similar wall-normal profiles of the individual shear stress components as in the here considered simulations. The contribution of the dispersive stresses $\langle \tau_{disp} \rangle_{xzt}$ to the total stress is, however, somewhat stronger in their case, partly attaining values of comparable size as the turbulent stresses $\langle \tau_{turb} \rangle_{xzt}$. The reason for the stronger contribution of form-induced stresses to the total shear stress budget in these experiments is believed to be twofold: On the one hand, the relative height of their artificial ridges is more than twice the amplitude of the sediment ridges in the current study. On the other hand, their artificial ridges are spatially fixed, as opposed to the self-formed mobile ridges considered in the current study. Both effects are assumed to cause an enhance of the secondary flow intensity and could thus lead to the stronger contribution of $\langle \tau_{disp} \rangle_{xzt}$ to the total shear stress.

6.3.10 Longtime-evolution of sediment ridges

6.3.10.1 Streamwise-minimal simulations

In the previous sections, we have discussed the role of large-scale streaks in generating initial sediment ridges on a macroscopically flat sediment bed, and we have pointed that the secondary flow cells of Prandtl's second kind are the statistical footprints of the large-scale coherent structures. In the following section, we are interested in the long-time evolution of sediment ridges in the phase subsequent to the initial formation interval. It shall be shown that the mean height of the ridges does not further grow in this phase, but ridges nevertheless undergo regular transformations as well as merging and splitting events.

As has been set out in section 6.1, long-time observations of ridges free from the influence of transverse bedforms are (at the current parameter points) only possible in streamwise-minimal boxes such as that of $CS250^{H3}$, in which the limited streamwise box length $L_x/D = 51.2$ effectively hinders the growth of transverse patterns (Scherer et al., 2020). In the context of the current study, we have carried out further streamwise-minimal simulations with varying lateral extent to allow for the evolution of different numbers of ridges. For the sake of clarity, the supplementary simulations are classified in three groups depending on their respective characteristic parameters (cf. tables 6.3 and 6.4): Cases that belong to the first group (—) have been carried out at the same parameter point as case $CS250^{H3}$

	Case	Re_b	Re_τ	ρ_p/ρ_f	Ga	D^+	H_f/D	H_b/D	H_f/H_b	θ
—	$CS250^{H1}$	3011	256.13	2.5	28.37	9.69	26.44	11.96	2.21	0.12
—	$CS250^{H1.5}$	3011	262.14	2.5	28.37	9.91	26.46	11.94	2.22	0.12
—	$CS250^{H2}$	3011	261.37	2.5	28.37	9.88	26.46	11.94	2.22	0.12
—	$CS250^{H3}$	3011	259.43	2.5	28.37	9.81	26.43	11.97	2.21	0.12
—	$CS250^{H4}$	3011	260.69	2.5	28.37	9.85	26.46	11.94	2.22	0.12
—	$CS250^{H6}$	3011	260.78	2.5	28.37	9.86	26.46	11.94	2.22	0.12
—	$CS250^{H8}$	3011	262.16	2.5	28.37	9.90	26.49	11.91	2.22	0.12
—	$CS850$	9483	836.35	2.5	56.68	29.24	28.60	6.96	4.11	0.27
—	$CS200^{H1.5}$	3008	207.29	2.5	11.54	4.00	51.87	12.13	4.28	0.12
—	$CS200^{H2}$	3008	202.53	2.5	11.54	3.90	51.89	12.11	4.28	0.11
—	$CS200^{H4}$	3008	201.78	2.5	11.54	3.89	51.92	12.08	4.30	0.11

Table 6.3: Physical parameters of long-time streamwise-minimal open channel flow simulations. The parameters are defined in accordance with those in table 6.1.

	Case	$[L_x \times L_z]/H_f$	$[L_x \times L_z]/D$	$D/\Delta x$	$\min(\Delta y^+)$	N_p	T_{obs}/T_b
—	$CS250^{H1}$	1.94×0.97	51.2×25.6	10	0.97	14 673	677
—	$CS250^{H1.5}$	1.94×1.45	51.2×38.4	10	0.99	22 020	1014
—	$CS250^{H2}$	1.93×1.93	51.2×51.2	10	0.99	29 359	1014
—	$CS250^{H3}$	1.94×2.91	51.2×76.8	10	0.98	43 730	678
—	$CS250^{H4}$	1.94×3.87	51.2×102.4	10	0.99	58 742	676
—	$CS250^{H6}$	1.94×5.81	51.2×153.6	10	0.99	88 096	676
—	$CS250^{H8}$	1.93×7.73	51.2×204.8	10	0.99	117 368	675
—	$CS850$	2.49×2.49	71.1×71.1	36	0.81	45 612	48
—	$CS200^{H1.5}$	0.99×1.48	51.2×76.8	10	0.40	43 730	586
—	$CS200^{H2}$	0.99×1.97	51.2×102.4	10	0.39	58 742	586
—	$CS200^{H4}$	0.99×3.94	51.2×204.8	10	0.39	117 368	585

Table 6.4: Numerical parameters of long-time streamwise-minimal open channel flow simulations. The parameters are defined in accordance with those in table 6.2.

($Re_\tau \approx 260$, $H_f/D \approx 26$, $D^+ \approx 10$) while varying the spanwise dimension of the computational domain in a range $L_z/H_f \in [1, 8]$. To differentiate between the individual cases of the simulation series, the value of L_z/H_f is added as a superscript to the simulation name. The single simulation $CS850$ (—) reveals essentially the same parameter set as the high-Reynolds number case $CM850^{H3}$ that has been investigated in the previous section ($Re_\tau \approx 836$, $H_f/D \approx 28$, $D^+ \approx 29$), but it features a shorter streamwise box length $L_x/D = 71.1$ in order to exclude transverse pattern evolution. In contrast to the remaining simulations in this section, it was not possible to maintain case $CS850$ for a comparably long observation time interval of several hundred bulk time units due to the enormous computational costs of this simulation. Eventually, a third set of simulations (—) has been performed in which the mean fluid height has been doubled compared to the first series while maintaining the bulk Reynolds number and the particle size. Consequently, the relative submergence has increased to $H_f/D \approx 52$, while the particle Reynolds number is reduced to $D^+ \approx 4$. The Galileo number of the individual simulations

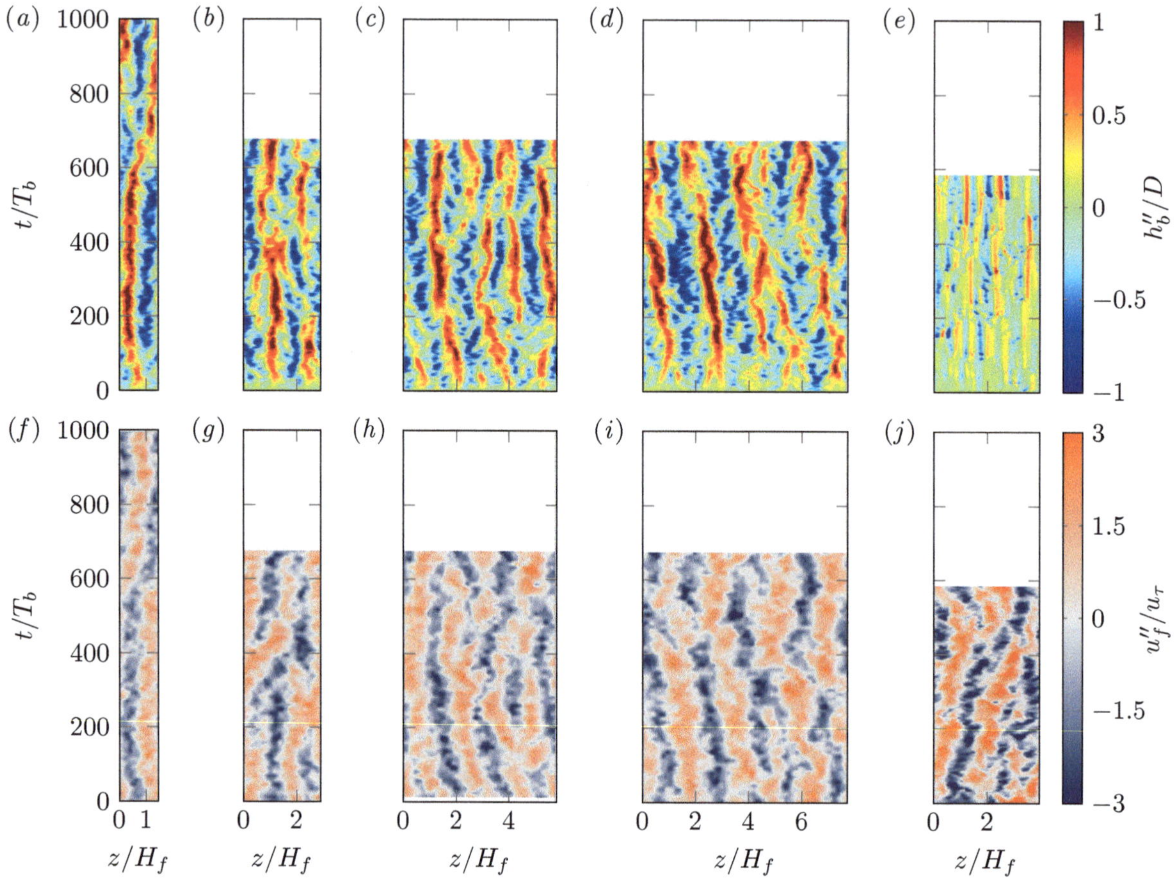

Figure 6.34: Space-time plot of the streamwise-averaged (*a-e*) sediment bed height fluctuations $h_b''(z,t)/D$ and (*f-j*) fluctuations of the streamwise velocity component $u_f''(y,z,t)/u_\tau$ at $\tilde{y}/H_f = 0.5$ in streamwise-minimal simulations. (*a,f*) $CS250^{H1.5}$, (*b,g*) $CS250^{H3}$, (*c,h*) $CS250^{H6}$, (*d,i*) $CS250^{H8}$ (all $H_f/D \approx 26$, $D^+ \approx 10$); (*e,j*) $CS200^{H4}$ ($H_f/D \approx 52$, $D^+ \approx 4$).

has been adapted accordingly to ensure a Shields number that is sufficiently larger than the critical shear stress required for sediment erosion. Note that all simulations except for case $CM850^{H3}$ feature a streamwise box length that is just long enough to maintain turbulence in the channel (Jiménez and Moin, 1991), which has been additionally verified by continuously monitoring the turbulent kinetic energy during the simulation run.

6.3.10.2 Evolution of developed sediment ridges

For some selected cases, the long-time evolution of the sediment ridges is presented in figure 6.34 together with the streamwise-averaged position of large-scale high- and low-speed regions in form of space-time visualisations. It is seen that the domain of case $CS250^{H1.5}$ shown in figure 6.34(*a*) is just wide enough to accommodate a single sediment ridge and a single associated large-scale high- and low-speed streak pair, whereas the channel width in cases $CS250^{H3}$, $CS250^{H6}$ and $CS250^{H8}$ shown in figures 6.34(*b-d*) allows to host a group of several individual ridges simultaneously. For none of the four cases, the ridge-covered sediment bed reaches a quasi-stationary state: Instead, their time evolution is rather characterised by different phases of lateral propagation, merging and splitting, alternating with time intervals during which the lateral positions of the developed sediment ridges do not significantly vary. In case $CS250^{H1.5}$, for instance, a single isolated ridge is seen to reside at essentially the same lateral position during the interval $150 \lesssim t/T_b \lesssim 600$, followed by a phase of roughly 200 bulk time units length during which the ridge migrates in positive spanwise direction until it settles again at a new preferential position. Lateral propagation of ridges is also visible in some of the remaining simulations such as case $CS250^{H8}$ (cf. panel 6.34(*d*)), where a family of ridges simultaneously propagates in negative spanwise direction at a however smaller and roughly constant propagation speed. The more pronounced lateral propagation of the bedform in case $CS250^{H1.5}$ is attributed to the fact that the isolated ridge has more freedom to laterally meander than sediment ridges do when they are laterally bounded by other neighbouring bedforms. Comparing the evolution of the sediment ridges with the dynamics of the large-scale low- and high-speed regions, it is recognised that both are closely correlated implying that even after the initial phase of pattern evolution, the spanwise arrangement of sediment ridges and troughs strongly depends on the organisation of the large-scale streaks. Also, it is found in agreement with the observations of the previous sections that sediment ridges intermittently disappear or reorganise in situations in which the arrangement of the large-scale low- and high-speed regions changes.

A qualitatively different evolution of the sediment bed is seen for the cases with higher relative submergence and lower particle Reynolds number, from which case $CS200^{H4}$ is exemplary shown in figure 6.34(*e*). The therein observed bedforms are of lower amplitude and their lateral spacing is clearly smaller than those of the sediment ridges in the remaining simulations. In contrast to all previously investigated cases, the space-time evolution of the sediment bed strikingly differs from that of the streamwise-averaged flow field (cf. figure 6.34(*j*)), which shows a qualitatively similar development as in the remaining cases. The only difference here is that the lateral positions of high- and low-speed regions seem to vary more strongly with time, which is assumed to be a consequence of the more severe limitations of the streamwise box length in case $CS200^{H4}$, for which $L_x/H_f \approx 1$. The differing lateral spacing of sediment ridges and large-scale velocity streaks in case $CS200^{H4}$ implies that the bedform evolution is in this case decoupled from the large-scale flow organisation in the bulk region. As a consequence, the mechanism that leads to the evolution of these sediment patterns is assumed to clearly differ from the above discussed 'top-down mechanism'. A key difference between case $CS200^{H4}$ and the remaining simulations is the quite small particle Reynolds number, that is with $D^+ \approx 4$ smaller than the thickness of the viscous sublayer. Gyr (1998) observed for fine sediment at such low particle Reynolds numbers ($D^+ \lesssim 5$) that sediment grains agglomerate in form of streamwise-aligned 'sediment riblets', with a characteristic spacing comparable to the characteristic wavelength of the buffer-layer structures $\lambda^+_{h,z} \approx 100$. When discussing the impact of these 'sediment riblets' on the flow field, Gyr (1998) suspected that the latter might possess a similar drag-reducing effect as artificial riblets (Goldstein and Tuan, 1998; Jiménez, 2004a). According to the conceptual mechanism claimed by Gyr (1998), sufficiently small sediment grains are eroded from the bed by the action of intense near-wall sweep events, while the deposition is controlled by the quasi-streamwise vortices of the buffer layer, explaining the similar lateral spacing of riblets and buffer-layer structures. In this context, it might be worth noting that, in contrast to the larger fully-developed sediment ridges observed in the remaining cases, the patterns visible in case $CS200^{H4}$ reveal a shorter lifetime, which could be another indication that they are formed by near-bed structures whose characteristic time scale is shorter than those of the large-scale velocity streaks.

Analogously to figure 6.5, figure 6.35 provides the long-time evolution of the mean amplitude and wavelength of sediment ridges in the streamwise-minimal simulations. The time evolution of $\sigma_{h,z}/D$ presented in figure 6.35(*a*) shows that the sediment ridges in the simulation series with higher relative submergence and lower particle Reynolds number indeed reach amplitudes that are, on average,

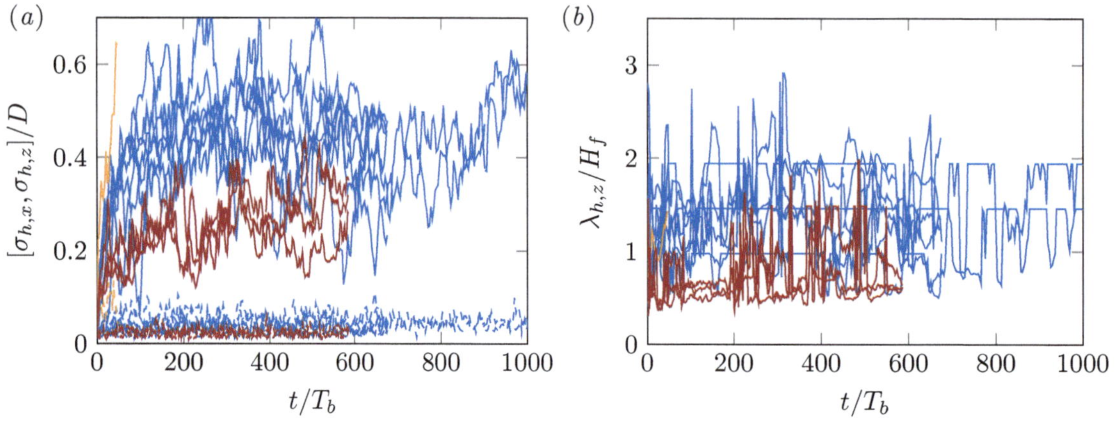

Figure 6.35: Time evolution of fluid-bed interface dynamics in streamwise-minimal simulations. (*a*) Root mean square of the fluctuation of the streamwise-averaged ($\sigma_{h,z}/D$,—) and spanwise-averaged fluid-bed interface ($\sigma_{h,x}/D$,---). (*b*) Mean wavelength of the fluctuation of the streamwise-averaged fluid-bed interface $\lambda_{h,z}/H_f$. Cases: (—) $H_f/D \approx 26$, $D^+ \approx 10$; (—) $H_f/D \approx 28$, $D^+ \approx 29$; (—) $H_f/D \approx 52$, $D^+ \approx 4$.

smaller by a factor of two compared to those of developed ridges in the remaining simulations. As has been discussed in the context of figure 6.5 before, the initial growth rate of $\sigma_{h,z}$ depends on the value of the Shields number, being strongest in the high-Reynolds number simulation *CS*850. In the phase subsequent to the initial growth period that starts approximately $150T_b$ after the onset of particle motion, the mean ridge amplitude settles in all cases without exhibiting a further global increase, featuring however pronounced oscillations around this level. The rather strong fluctuations are assumed to be a consequence of the continuous transformations, splitting and merging events that the sediment ridges undergo throughout the simulations, while the rather stable long-time mean value implies that these processes do not significantly affect the mean height of the patterns. The time evolution of $\sigma_{h,x}/D$ that is also visualised in figure 6.35(*a*) additionally verifies that the reduction of the streamwise domain length L_x/D successfully suppresses the growth of perturbations associated with transverse bedforms such that the sediment bed contour can be considered as statistically stationary in the streamwise direction.

The mean lateral spacing of the sediment ridges $\lambda_{h,z}$ presented in figure 6.35(*b*) is markedly reduced in the simulations with higher relative submergence and lower particle Reynolds number when scaled in terms of the mean fluid height H_f, in accordance with the corresponding space-time evolution plots in figure 6.34. Similar to the mean pattern height, the mean pattern spacing is subject to permanent oscillations related to the continuous changes of the mean fluid-bed interface during the simulation. In particular in simulations featuring a small domain width L_z/H_f, the pattern wavelength is restricted by the domain size and the system can choose only between a narrow range of discrete harmonics of the domain width, explaining the partly strong jumps between several harmonics occurring predominantly in these cases. In the wide domains, on the other hand, the range of available harmonics is large enough that the system is less restricted in the choice of the pattern wavelength.

To assess the scaling properties of the sediment ridges as well as the influence of the box size on their mean pattern wavelength, the mean pattern spacing $\langle \lambda_{h,z} \rangle_t$ is shown in figure 6.36 as a function of the lateral box length L_z scaled in terms of the outer, inner and characteristic particle length scale, respectively. For the sake of comparison, data points from experiments of Wolman and Brush (1961), Hirano and Ohmoto (1988) and Nezu et al. (1988) have been added to the parameter plane, from which the latter two are provided in table 1 of McLelland et al. (1999). Let us first focus on the variation

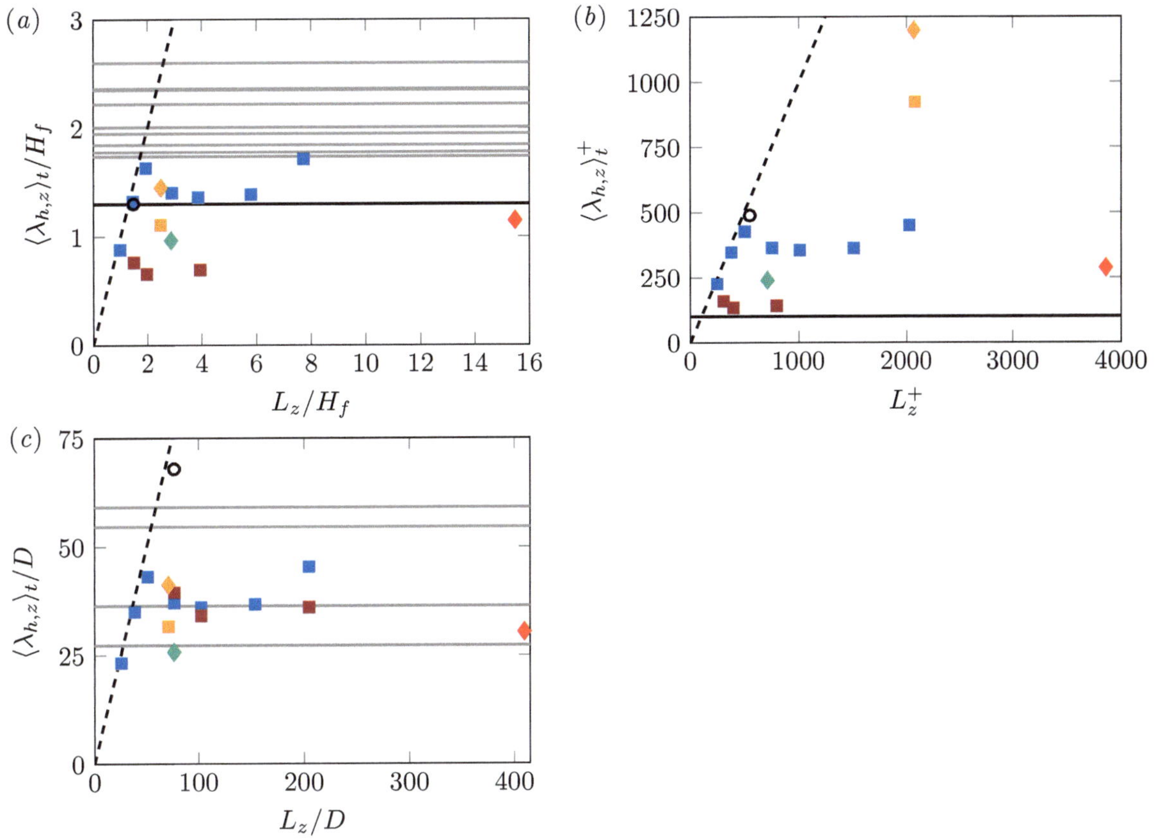

Figure 6.36: Time-averaged mean lateral ridge wavelength $\lambda_{h,z}$ as a function of the spanwise box extent L_z scaled with *(a)* the mean fluid height H_f, *(b)* the viscous length scale δ_ν and *(c)* the particle diameter D, respectively. The averaging interval comprises the entire simulation period excluding the initial phase of approximately 25 bulk time units. Horizontal grey lines represent ridge wavelengths measured in the experiments of Wolman and Brush (1961), Hirano and Ohmoto (1988) and Nezu et al. (1988), respectively. The values observed in the latter two studies are summarised in table 1 of McLelland et al. (1999). The black solid line in *(a)* marks the most-amplified wavelength determined in the linear stability analysis of Colombini (1993), while in *(b)* it indicates the characteristic spanwise wavelength of the buffer-layer coherent structures of $\lambda_z^+ = 100$. In each panel, a black dashed line marks the relation $\lambda_{h,z} = L_z$. The individual symbols indicate the simulations (series): $H_f/D \approx 26$, $D^+ \approx 10$ (■); $H_f/D \approx 28$, $D^+ \approx 29$ (■); $H_f/D \approx 52$, $D^+ \approx 4$ (■); $CM250^{H3}$ (♦); $CM850^{H3}$ (♦); $CL250^{H16}$ (♦). An additional data point represents case $CH2D102^2$ from Scherer et al. (2020), for which $Re_\tau \approx 375$, $H_f/D \approx 52$ and $D^+ \approx 7.2$ (○).

of the mean pattern wavelength in the first simulation series at lower relative submergence $H_f/D \approx 26$ (blue squares in figure 6.36): When scaled in terms of the mean fluid height, cases with a small lateral box length $L_z < 2H_f$ can accommodate only a single sediment ridge, thereby attaining the maximum possible wavelength $\lambda_{h,z} \approx L_z$. The related large-scale velocity structures are subject to the same restrictions (cf. figure 6.34(*f*)), such that these systems accordingly feature only a single pair of large-scale high and low-speed zones at a however smaller lateral wavelength as those in sufficiently large domains. These domains are thus minimal boxes in the sense of Flores and Jiménez (2010). Cases featuring a domain width $L_z > 2H_f$, on the other hand, feature at least two individual ridges and pairs of high- and low-speed regions, respectively. The pattern wavelengths for these domains attain values comparable to the most-amplified wavelength $\lambda_{h,z}/H_f \approx 1.3$ determined in the linear stability analysis of Colombini (1993) (cf. chapter 5). They also show a good agreement with the experimental datasets, even though the exact values in most simulations are slightly lower than in the experiments. As has been pointed out earlier, such slight deviations between simulation and experiment are in this context not unexpected, since even for the current long-time observations, the simulation time is still much shorter than the observation time in the experiment. Apart from that, the presented datasets have been investigated in laboratory flumes with a medium but finite aspect ratio, such that an influence of the lateral sidewalls may additionally influence the measured wavelengths.

Turning our attention to the second series of simulation with the higher relative submergence $H_f/D \approx 52$ (red squares in figure 6.36), it is seen that the attained values of the mean pattern wavelength are clearly smaller than the theoretically predicted wavelength of $\lambda_{h,z}/H_f \approx 1.3$ and the experimentally determined values, differing by a factor two to the former and a factor three to four to the latter, respectively. If the wavelength is alternatively scaled in inner units (cf. figure 6.36(*b*)), on the other hand, the simulations of the latter series reach values that are with $\lambda^+_{h,z} \approx 150$ very close to the conventional spacing of the buffer-layer quasi-streamwise vortices, as postulated for sedimentary riblets by Gyr (1998). These cases thus qualitatively differ from the remaining simulations which feature wavelengths of at least several hundred wall units, ranging up to those in the high Reynolds number cases which attain values that are one order of magnitude larger than the buffer-layer streak spacing. This, once more, highlights that the fully-developed sediment ridges studied in the previous sections scale in outer units.

When scaling the mean wavelength in terms of the particle diameter D (cf. figure 6.36(*c*)), eventually, all currently observed cases are found to develop a mean wavelength in the range $\lambda_{h,z}/D \in [25, 50]$ comparable to the values measured in the experimental studies. For the sake of comparison, we have added an additional parameter point to figure 6.36 that represents case $CH2D102^2$ from Scherer et al. (2020). This simulation features the same relative submergence $H_f/D \approx 52$ as the runs in the second series, but at a higher bulk ($Re_b = 5013$), friction ($Re_\tau = 375$) and particle Reynolds number ($D^+ \approx 7.2$). For additional information on this simulation, the reader is referred to the original work. Note that the here presented parameter set associated with case $CH2D102^2$ differs slightly from the one originally reported in Scherer et al. (2020), because we have, for the sake of consistency with the remaining simulations, recomputed most of them according to the here adopted definition of the mean fluid-bed interface. The evolution of ripple-like features was in this case not suppressed, as the streamwise domain length attains supercritical values $L_x/D \approx 102 > 76.8$. Therefore, in contrast to the simulations with $H_f/D \approx 52$ carried out in the current work, case $CH2D102^2$ has been observed to feature a clearly developed ridge with a wavelength comparable to L_z that is mutually interacting or concurring with transverse bedforms (cf. figure 7 in Scherer et al. (2020)). Ignoring for the moment the intermittently occurring transverse patterns, it turns out that the lateral spacing in case $CH2D102^2$ is with $\lambda_{h,z}/D \approx 68$ clearly larger than in the remaining simulations. When scaled in outer units (cf. figure 6.36(*a*)), on the other hand, it attains a value of $\lambda_{h,z}/H_f \approx 1.3$ that almost perfectly collapses with the wavelength predicted by means of linear stability analysis, highlighting that the mean pattern wavelength does not scale with the particle diameter.

The discussed different scaling properties give a first hint that streamwise-elongated bedforms may be developing due to the action of coherent structures of different scale, possibly depending on the relative size of the sediment with respect to the characteristic fluid length scales, i.e. D^+ and D/H_f. For a verification of this hypothesis, however, it requires further efforts to explore the parameter space in form of simulations with sediment grains at sufficiently small particle Reynolds number $D^+ \lesssim 5$ and simultaneously high friction Reynolds numbers such that a proper scale separation is given, which is unfortunately not the case in the here investigated simulations. Apart form that, it would be similarly interesting to perform a series of numerical experiments in the spirit of Jiménez and Pinelli (1999), artificially damping all turbulent fluctuations outside the buffer layer such that the influence of the remaining near-wall structures on the sediment bed evolution can be assessed. If the above hypothesis holds, it should be possible to observe patterns with wavelengths $\lambda_{h,z}^{+} \approx 100$ for fine sediment with small particles Reynolds numbers $D^{+} \lesssim 5$, whereas the bed should remain macroscopically flat for particles much larger than the viscous sublayer thickness, $D^{+} \gg 5$, due to the absence of large-scale structures in the outer flow.

6.4 Discussion of the observed formation mechanism

In the previous sections, it has been shown that the organisation of the turbulent flow field in large-scale streamwise velocity streaks is the main cause for the emergence of the characteristic regular pattern of streamwise-aligned quasi-parallel sediment ridges on an initially flat sediment bed. These large-scale coherent flow structures correspond to the large-scale streaks or large-scale motions well known from canonical channel and boundary layer flows and with dimensions $\mathcal{O}(H_f)$, they are the largest members of a family of self-similar streaks that populate the logarithmic layer. Along the sediment bed, the large-scale streaks induce a laterally varying shear stress distribution, which, in turn, leads to non-homogeneous sediment erosion across the channel width from the moment on when particles are released. Below large-scale high-speed streaks, troughs form rapidly after the onset of particle motion by virtue of the locally higher bottom shear stress and sediment erosion rate, whereas sediment ridges evolve in regions of weaker erosion below the corresponding low-speed streaks. The spanwise varying erosion rate naturally results in a sinusoidal oscillation of the particle transport that is accordingly higher in troughs and lower above the crest of the sediment ridges. Our observations reveal that the initial formation of sediment ridges and troughs on the sediment bed is controlled by the dynamics of the large-scale velocity streaks in the bulk of the channel which do interact in a 'top-down mechanism' with the sediment patterns. This causal connection could be verified in figure 6.22 by evaluation of the two-time two-point cross-correlations between the organisation of large-scale high- and low-speed regions in the channel centre and that of the sediment ridges and troughs on the sediment bed. For the therein investigated case $CM850^{H3}$, a time lag between the evolution of the flow structures and that of the sediment bed could be estimated as approximately $10T_b$, or $0.9H_f/u_\tau$ in multiples of the eddy turnover time.

6.4.1 Comparison with the linear instability process

The outlined 'top-down mechanism' between instantaneous large-scale streaks and sediment ridges fundamentally differs from the linear instability process proposed by Colombini (1993) that has been revisited in the previous chapter 5. Effectively, the roles of cause and effect in the model of Colombini (1993) are reversed compared to the formation process observed in our simulations: In the linearised model, a strictly one-dimensional turbulent velocity base profile is perturbed with a sinusoidal modulation of infinitesimal amplitude due to an infinitesimal lateral variation of the otherwise flat sediment bed. The lateral oscillations of the wall-normal and spanwise velocity components then directly result in secondary flow cells, whose amplitude is thus of the same order as the perturbation of the velocity field. Let us recall that the linearised equations of fluid motion in the linear model are quasi-stationary and decoupled from the sediment bed continuity equation due to the assumption that the flow adapts quasi-instantaneously to changes of the lower domain boundary. In chapter 5, it was accordingly shown that the most-amplified wavelength in the system is determined by the linearised Navier-Stokes operator, subject to the steady forcing in form of the bottom wall undulation. In other words, sediment bed undulations can appear at a variety of different wavelengths, but the flow field's response in form of depth-spanning secondary currents is 'optimal' (meaning a maximisation of the secondary flow intensity) for a wavelength around $\lambda_{h,z} \approx 1.3H_f$. In contrast to the 'top-down process' observed in our simulations, this linear mechanism can thus be interpreted as a 'bottom-up process' in which the development of the secondary currents is triggered by the topography of the lower domain boundary (Nezu and Nakagawa, 1993).

As we have discussed in section 5.4, the perturbation of the sediment bed contour is existential for the functioning of the linearised model which depends on some kind of external forcing in order to establish the persisting secondary currents, as the turbulent base profile itself is linearly stable (Reynolds and Tiederman, 1967) and the missing time-dependence of the governing equations hinders a transient growth of individual modes. In the context of the full time-dependent Navier-Stokes system considered in our simulations, on the other hand, finite amplitude variations of the turbulent velocity field are naturally present in form of large-scale streamwise velocity streaks and Reynolds stress-carrying structures, respectively. The secondary currents are in this context the statistical footprint of the latter structures. Most importantly, the dynamics of these features are more or less independent of the sediment bed development as they appear in flows over smooth walls, macroscopically flat sediment beds and developing ridges in essentially the same way. Let us emphasise that our findings concerning the 'top-down interaction' do not contradict the existence of the instability mechanism discussed by Colombini (1993), but they nevertheless underline that a linear instability starting from an initially infinitesimal amplitude is less likely to be of relevance in a fully-turbulent flow, in which intense intermittently occurring large-scale velocity streaks can cause sediment ridge formation too.

In spite of the fact that causality is reversed in the linearised model compared to the 'top-down process' observed in the current chapter, the resulting secondary currents share several fundamental similarities. As we have discussed earlier in this chapter, the mean lateral spacing of the sediment ridges in the linear model and in the multiphase simulations are in a very similar range $\lambda_{h,z}/H_f \approx$ 1-1.5, that is moreover comparable to values $\lambda_{h,z}/H_f \approx$ 1-3 measured in experiments (Wolman and Brush, 1961; Ikeda, 1981; McLelland et al., 1999). The secondary currents in both situations reveal a qualitatively similar structure concerning both the location of their centre of rotation and their shape, as can be seen when comparing figures 5.4 and 6.29, respectively. This is in the end not unexpected, as it has been stated earlier that in both situations the dominant lateral wavelength of the secondary currents and of the sediment ridges is chosen solely by the Navier-Stokes operator: in the theoretical model in a linearised streamwise-independent and stationary form and in the DNS in its full non-linear time-dependent form. Recalling the analogy of the linear model with those in transient growth analysis (Del Álamo and Jiménez, 2006; Pujals et al., 2009) that was highlighted in chapter 5, the match of the wavelength in the model of Colombini (1993) with the current DNS data is as good as the match between the strongest lateral wavelengths observed in transient growth analysis and that in canonical closed channel flows (Jiménez, 2018). This eventually explains the good match of the predicted wavelength with the DNS data.

6.4.2 Comparison with conceptual models in canonical turbulent flows

In section 6.3.8, it was observed that the large-scale velocity streaks and Reynolds stress-carrying structures in the considered open channel flow simulations share many similarities with the conditionally-averaged coherent structures in the logarithmic layer of canonical closed channel flows, studied by

Del Álamo et al. (2006) and Lozano-Durán et al. (2012) and further analysed in Jiménez (2013a). Based on the therein presented results, Jiménez (2018, § 5.6 and references therein) proposed a conceptual model of causality in canonical wall-bounded flows, according to which the main purpose of the solid wall is to provide the mean shear from which turbulence is fed with the necessary kinetic energy. In contrast to models whose basic element are individual hairpin vortices that are born near the wall and advected outwards where they can cluster in larger groups to form large-scale structures (Adrian et al., 2000; Adrian, 2007), the model of Jiménez bases on the concept that coherent structures can form at all distances from the wall independently. So, while the former class of models naturally features a preferred 'direction of causality' given in form of the motion of hairpin vortices from the wall towards the bulk of the domain, the freedom of structures to develop at arbitrary wall distance in Jiménez' model reposes the question of cause and effect and whether structures of different scales interact with each other in a preferential direction. Revisiting data from the minimal log-layer simulations of Flores and Jiménez (2010), Jiménez (2018) indeed found such a preferential direction of information propagation across the logarithmic layer that is directed from the bulk of the domain towards the wall; indicating a preferred 'top-down interaction' of outer large-scale structures with those near the wall also in canonical flows. Recently, Zhou et al. (2022) have shown by means of DNS in minimal and larger domains that, indeed, 'top-down interactions' are the dominant processes in canonical channel flows, whereas no evidence for the existence of counter-oriented 'bottom-up interactions' was found.

In addition to these findings, important arguments contradicting the control of large-scale structures from the vicinity of the wall come, amongst others, from numerical experiments in which the self-sustained process in the buffer layer is effectively destroyed, for instance, by replacing the smooth channel walls by a fully-rough boundary (Flores et al., 2007), by artificially suppressing the entire buffer layer (Mizuno and Jiménez, 2013; Kwon and Jiménez, 2021) or by alternatively considering over-damped large eddy simulations in which the smallest scales of turbulence are filtered out (Hwang and Cossu, 2010a, 2011). In none of these cases, the logarithmic and large-scale outer layer structures were significantly affected by the severely perturbed flow in the near-wall region, which further implies that these structures are indeed independent of the buffer-layer processes. The essential independence of structures of different scales and wall-distances from each other that represents an integral element of Jiménez' model is further supported by a number of studies that agree in the claim that a self-sustaining process similar to that observed in the buffer layer (Jiménez and Moin, 1991; Hamilton et al., 1995; Schoppa and Hussain, 2002) between streaks and some kind of quasi-streamwise rollers might exist for streaks of all scales likewise, from the buffer to the logarithmic and outer layer (Flores and Jiménez, 2010; Cossu and Hwang, 2017; Kevin et al., 2019b; Lozano-Durán et al., 2020; Motoori and Goto, 2021).

The results of our particle-laden simulations are in line with these findings, showing that even the intense particle motion in the bedload layer and the related destruction of the buffer layer structures in the vicinity of the sediment bed seem to have only a minor influence on the dynamics of the large-scale velocity streaks. The discussed 'top-down interaction' between these large-scale structures and the organisation of the sediment patterns supports the observations of a preferential direction of information propagation towards the bottom wall by Jiménez (2018) and Zhou et al. (2022). Also, the large-scale structures over both smooth walls and mobile sediment beds undergo phases of reorganisation during which they are seen to break up intermittently (see, for instance, the interval $t/T_b \in [10, 20]$ in case $CM850^{H3}$), resembling the log-layer streaks in the minimal domain simulations of Flores and Jiménez (2010) that disappear during regularly occurring bursting phases. In complete agreement with the idea that coherent structures can form at arbitrary wall-normal positions autonomously due to the local shear only, large-scale streaks have been observed to first develop in the channel centre after such break-up events, essentially uncorrelated with the flow organisation in the near-bed/near-wall region. Only after the discussed time lag (cf. figure 6.22), the flow in the vicinity of the sediment bed adapts to the new organisation of the large-scale streaks owing to the 'top-down process'.

6.5 Summary and conclusion

The main purpose of the current chapter was to clarify the role of turbulent large-scale velocity streaks in the formation process of subaqueous sediment ridges and to elucidate how these coherent flow structures correlate with the appearance of mean secondary currents of Prandtl's second kind. To this end, a series of direct numerical simulations of turbulent open channel flow over a thick bed of fully-resolved mobile sediment particles has been performed, supplemented with two reference simulations in smooth-wall single-phase open channel flow at comparable parameter values. In this context, the friction Reynolds number was varied in a range $200 \lesssim Re_\tau \lesssim 830$ to scrutinise the impact of the Reynolds number and the scaling properties of the sediment patterns in domains of different size: While narrow domains with lateral periods $L_z/H_f \approx$ 2-3 allowed to study the evolution and mutual interaction of one or two isolated sediment ridges and large-scale structures, computational boxes with spanwise dimensions up to $L_z/H_f \approx 16$ provided the possibility to investigate collective effects between up to ten individual sediment ridges as well as the dynamics of large-scale streaks in an essentially laterally unconstrained environment.

Based on the so created database, it was possible to verify theoretical predictions according to which sediment ridges can evolve in the complete absence of lateral sidewalls solely due to an interaction of the mobile sediment bed with turbulent structures (Ikeda, 1981). The observed sediment ridges feature with $\lambda_{h,z}/H_f \approx$ 1-1.5 a mean lateral spacing that is comparable to corresponding values $\lambda_{h,z}/H_f \approx$ 1-3 that were determined by measurements in laboratory flumes (Wolman and Brush, 1961; McLelland et al., 1999). A similarly good match was observed with the most unstable wavelength $1.3H_f$ of the linear stability analysis of Colombini (1993) which was revisited in the previous chapter, even though the physical mechanism that underlies the latter model was shown to conceptually differ from the 'top-down formation process' observed in the current simulations.

The characteristic regular lateral spacing of the sediment ridges turns out to be the footprint of the turbulence organisation in the bulk of the channel: As in canonical wall-bounded flows, large-scale high- and low-speed streaks at lateral wavelengths comparable to those of the bedforms populate the centre of the channel. A comparison of the premultiplied energy spectra for single-phase smooth-wall and particle-laden flows reveals that the general properties of these large-scale streaks are essentially unaffected by the mobile particles, that are almost exclusively transported in the bedload layer in the vicinity of the sediment bed. In this regard, the results are in accordance with similar observations in flows over fully-rough walls, for which the statistics of the coherent structures in sufficient distance to the roughness layer were found to be essentially the same as over smooth walls (Flores and Jiménez, 2006; Flores et al., 2007).

Even though none of the here considered sediment beds themselves would be classified as fully-rough due to the relatively low particle Reynolds numbers $10 \lesssim D^+ \lesssim 30$ and bedform amplitudes, the mobility of the particles inside the bedload layer leads to a significant increase in bottom friction, comparable to that observed in single-phase flows over fully-rough bottom walls. For particle grains whose relative diameter is with $D^+ \approx 30$ (case $CM850^{H3}$) not significantly smaller than the mean size of the buffer-layer structures, we observe the complete destruction of the self-sustained process near the bottom: The buffer layer cannot be maintained in the presence of an intense bedload transport layer with particles of these dimensions, quite similar to the case of a fully-rough wall (Jiménez, 2004a; Cameron et al., 2008).

Where sediment grains are predominantly eroded depends merely on the distribution of the bottom shear stress along the sediment bed. From canonical wall-bounded flows, it is well known that large-scale coherent structures penetrate deep into the near-wall region, where their footprint can be seen in form of outer-scaling patches of locally enhanced or reduced wall shear stress. These latter are correlated with the large-scale streaks in the outer layers (Del Álamo and Jiménez, 2003; Jiménez et al., 2004). In the current system, the consequence of this process is a laterally varying erosion of sediment grains from the initially macroscopically flat sediment bed, with troughs developing in regions of stronger erosion below large-scale high-speed streaks and sediment ridges evolving below the corresponding low-speed streaks. Intense instantaneous sweep events that are naturally associated with streaks of relatively higher velocity contribute to the locally enhanced particle erosion (Gyr and Schmid, 1997). The laterally alternating erosion rate goes hand in hand with a spanwise oscillating particle flux that is accordingly higher in the trough region and reduced in the low-speed regions over the sediment ridge crests. In this context, different in the hydraulic community established empirical relations were tested for their ability to correctly predict the streamwise and lateral sediment transport based on the local bed shear stress. While the model of Wong and Parker (2006) for the streamwise and that of Ikeda (1982) and Parker (1984) for the lateral particle transport performed relatively well outside the initial transient, the model of Parker and Andrews (1985) required the choice of unrealistically high values of the Coulomb friction in order to match our simulation data.

It was thereafter shown that, as strongly suggested by the previous observations, the formation and spanwise organisation of sediment ridges is indeed controlled from the outer layer trough the action of the large-scale streaks and Reynolds stress-carrying structures. Based on two-time cross-correlations, it could be verified that the evolution of the sediment bed indeed lags behind that of the large-scale structures in the channel bulk by about 10 bulk time units in case $CM850^{H3}$, indicating that the interaction of these large-scale structures with the sediment bedforms is a kind of 'top-down process'. A similar preferred direction of information propagation from the channel centre towards the wall was recently proposed in the conceptual model of Jiménez (2018, § 5.6 and references therein), who observed that the Reynolds stress in the vicinity of the wall was correlated to that further away from the wall at earlier times in the minimal log-layer simulations of Flores and Jiménez (2010). Recently, the dominant 'top-down mechanism' in canonical channel flows was further supported by the findings of Zhou et al. (2022). The here observed dynamics of the large-scale coherent structures agree fairly well with the observations in Jiménez (2018) and the model described therein. In particular, the concept of the aforementioned model that coherent structures can form at arbitrary wall-normal layers autonomously only due to the local shear is supported by our results: The large-scale streaks in the current open channel flows break up intermittently, consistent with the bursting of log-layer streaks reported by Flores and Jiménez (2010), whereupon new structures first evolve in the channel bulk, later expanding down to the sediment bed.

In a last step, we closed the loop to the mean secondary currents by analysing the flow field when averaged over the streamwise direction and intermediate time-intervals of $\mathcal{O}(10T_b)$ length. The here discussed large-scale flow structures are well aligned in the streamwise direction and reveal only a weak tendency of lateral meandering over both smooth walls and mobile particle beds, such that they leave a clear statistical footprint in the streamwise and time-averaged fields: While regions of up- and downwelling of the mean streamwise velocity refer to the preferential locations of the large-scale streamwise streaks, the mean up- and downflow regions of the cellular mean secondary flow cells represents the collective effect of large-scale ejections and sweeps, respectively. The idea that secondary currents arise due to the regular organisation of the large-scale coherent structures is further strengthened by the observation of very similar secondary flow patterns with comparable shape, size and intensity in the smooth-wall cases when averaged in the same way. In both flow configurations, instantaneous coherent structures and averaged flow fields were found to share many similarities with the conditionally-averaged coherent structures scrutinised by Del Álamo et al. (2006) and Lozano-Durán et al. (2012). In this context, the mean secondary flow cells in the current open channel flow simulations represent the analogue to their conditional rollers which, in turn, are assumed to take the place of the buffer layer quasi-streamwise vortices in the regeneration process of larger velocity streaks in the logarithmic and outer layer (Jiménez, 2018; Kevin et al., 2019b).

Outstanding remains the question of the long-time behaviour of ridges and secondary flow cells for observation times much longer than considered here, as they are typical for laboratory experiments. In smooth-wall single-phase flows, a mean secondary flow pattern can be maintained for such long time intervals only if the system features a marked spanwise inhomogeneity as in the case of a square duct (Pinelli et al., 2010) or for the flow over spanwise varying roughness stripes at the bottom wall. While in the former case, lateral sidewalls hinder the mobility of large-scale structures as will be shown in the following chapter, the heterogeneous bottom roughness pattern in the latter case is usually assumed to 'lock' the spanwise position of instantaneous large-scale structures such that their statistical footprint is visible in form of secondary currents even in the long-time average (Kevin et al., 2019a). For self-forming sediment ridges, the situation is more complex as the bedforms themselves are mobile: In the initial formation period, the bedforms are seen to follow the organisation of the outer flow most of the time. In most of the domains considered to investigate the initial formation of sediment ridges, the evolution of dominant transverse ripple-like bedforms hinder us in studying the development of sediment ridges for time intervals much longer than 100 bulk time units. To this end, a second series of streamwise-minimal simulations was performed which are known to artificially suppress the growth of transverse bedforms (Kidanemariam and Uhlmann, 2017; Scherer et al., 2020), such that the evolution of sediment ridges could be integrated over time intervals of up to 1000 bulk time units. Based on these simulations, it was possible to shed some light on a later stage of sediment ridge development, in which the bedforms do, on average, not further grow but still undergo regular phases of reorganisation. As in the initial phase, sediment ridges are controlled from the large-scale flow structures in the channel bulk and generally follow the organisation of the flow field in this layer.

An exception from this general behaviour was found for simulations, in which the relative size of the particles was reduced to $D^+ \approx 4$ or equivalently $D/H_f \approx 0.02$. In these cases, the evolving sediment bedforms organised themselves in a way that is markedly different from the arrangement of the large-scale streaks in the outer flow, attaining a smaller size than the sediment ridges in the remaining simulations concerning both amplitude and wavelength. The latter was with $\lambda^+_{h,z} \approx 150$ only slightly larger than the typical wavelength of the buffer-layer coherent structures, but clearly smaller than that of the large-scale outer-scaling streaks. The results however agree with the findings of Gyr (1998), according to which particles with diameter smaller than the height of the viscous sublayer, i.e. $D^+ \lesssim 5$, agglomerate in 'sediment riblets' of roughly the same spacing as the buffer layer streaks since they are eroded, transported and deposited by these structures.

Whether or not the smaller bedforms in the latter case scale indeed in inner units and develop due to the formation process proposed by Gyr (1998) could not be finally clarified, as the Reynolds number considered in this simulation series was not large enough to allow for a clear scale separation. Therefore, it would be of interest to extend the current database in the future to higher flow and particle Reynolds numbers to study whether different types of bedforms do emerge, depending on the size of individual sediment grains and that of the turbulent structures they should be transported by. In the context of the outer-scaling sediment ridges, a first step towards higher Reynolds numbers was gone by carrying out two simulations at friction Reynolds numbers up to $Re_\tau \approx 830$ which allowed valuable insights into the dynamics of structures of different scale. Due to the enormous costs of these simulations and the massive amount of data generated therein, it is today not yet realisable to perform a full parameter study for these higher Reynolds numbers, but with increasing computing power such campaigns should become feasible in the next decade. Once this is the case, it would be desirable to detect, possibly in comparison with accompanying experiments at comparable Reynolds numbers, parameter points similar to classical experiments such as those of Wolman and Brush (1961), at which the sediment bed can maintain sediment ridges over very long time intervals without being disturbed by the development of larger transverse ripple-like bedforms. To achieve this, however, a larger number of individual simulations will have to be carried out at different parameter points and over quite long physical time intervals, both of which made such an undertaking impossible to conduct in the current work.

On the other hand, numerical simulations at moderate Reynolds numbers $Re_b = \mathcal{O}(10^3)$ have become affordable even in the context of multiphase simulations. This offers the possibility to now conduct numerical experiments comparable to those successfully used in canonical flows in order to get a better understanding of the interaction between individual turbulent structures and sediment beds. A good example for such an experiment is the simulation campaign of Kidanemariam and Uhlmann (2017), in which the authors sought the minimal domain size necessary to accommodate transverse bedforms in a similar way as the minimal flow unit of Jiménez and Moin (1991). In a similar vein, in order to investigate the impact of coherent structures of different scales on the sediment transport, simulations with appropriate filtering techniques could be performed to artificially suppress different parts and structures of the flow in analogy to, for instance, the studies by Jiménez and Pinelli (1999) and Kwon and Jiménez (2021).

Chapter 7

Coherent structures and sediment ridges in open duct flow

The previous chapters have proven from a theoretical and a numerical point of view that formation mechanisms exist, in virtue of which sediment ridges and secondary currents can evolve in the absence of lateral sidewalls. It is thus clear that a sidewall-induced mean secondary flow is not a necessary condition for the evolution of sediment ridges, but the question remains how the presence of lateral sidewalls and the associated mean secondary flow interact with a mobile sediment bed.

In this chapter, we endeavour to clarify how the presence of sidewall-induced secondary currents influences the formation of sediment ridges in narrow and wide open duct flows and, *vice versa*, which impact the mobile sediment has on the organisation of the mean secondary flow as well as on instantaneous small- and large-scale coherent structures. For this purpose, a number of fully-resolved open duct simulations over both smooth bottom walls and mobile sediment beds composed of fully-resolved particles at different aspect ratios and Reynolds numbers have been performed in the course of this chapter. Based on the data obtained in these simulations, it will be shown that sediment ridges in the core of wide open duct flows in sufficient distance to the sidewalls do form owing to the essentially same 'top-down mechanism' as their counterparts in laterally periodic open channel flows. In these regions, large-scale turbulent streaks can freely develop without being constrained by the lateral sidewalls and thus induce, as in the open channel case, a laterally varying bed shear stress and erosion rate, respectively. We will present evidence that the situation in narrow domains, where the sidewalls strongly affect all regions of the cross-section, differs markedly from the afore described pattern evolution in sufficient distance to sidewalls. In the former case, only a single high-speed region can be accommodated by the narrow cross-section and it will turn out that smaller but outer-scaling flow structures closer to the bottom wall control the initiation of sediment bed structures in a generally comparable 'top-down interaction'. The mean flow field as well as instantaneous coherent structures over such mobile sediment beds will be shown to be qualitatively similar to those in smooth open ducts, even though predominantly near the bed the organisation of the mean secondary flow and mean vorticity patterns will be seen to deviate. The latter effect can be attributed to the different organisation of instantaneous quasi-streamwise vortices along the sediment bed compared to that over a smooth impermeable bottom wall. Finally, a possible connection between the preferential organisation of instantaneous vortex clusters (Del Álamo et al., 2006) and the mean secondary flow field is discussed in comparison with the results of Kawahara et al. (2012a) for instantaneous streamwise rolls.

The chapter is organised in analogy to the previous one and starts with a presentation of the physical multiphase system that will be analysed in the following. The relevant physical measures required for the subsequent investigation of the sediment bed and the flow field will be defined. The section is followed by the analysis of the simulation data, which again first focuses on the temporal evolution of the sediment ridges and the associated development of the sediment transport. The mean primary and secondary flow field will be scrutinised in the following and the impact of the mobile sediment bed on the secondary flow intensity is discussed in detail. After having established the causal connection between the above mentioned smaller flow structures and the sediment bedforms in narrow domains which is indicative of the 'top-down interaction', we will turn our attention to the investigation of two-point correlation functions and statistics of the preferential organisation of individual coherent structures. Specifically, we are going to analyse the geometry and preferential organisation of streamwise velocity streaks, quasi-streamwise vortices and small- and large-scale vortex clusters. Where possible, statistics of both sediment-laden and single-phase cases are compared to highlight the influence of the mobile sediment bed on the organisation of these structures.

7.1 Flow configuration and computational setup

7.1.1 Flow configuration

The database analysed in this chapter comprises 14 individual simulations, 13 of which are simulations of turbulent open duct flow, supplemented with a single open channel reference case $CL250^{H3}$ in a computational domain of similar size. The latter open channel simulation shares the parameter point of case $H12^1$ presented in Kidanemariam and Uhlmann (2017), but is a new simulation conducted in the course of this study from a comparable initial condition. The open duct simulations include three different flow and bedform configurations: Seven cases feature a mobile sediment bed on which sediment ridges readily evolve after the onset of sediment motion, two simulations were performed over a stationary sediment bed and the remaining four simulations cover three different data points of single-phase open duct flow enclosed by smooth walls. The particle-laden simulations were performed with the finite-difference immersed boundary code presented in section 4.2. The single-phase smooth-wall simulations, on the other hand, were discretised and integrated in time using the pseudo-spectral method, developed and validated in Uhlmann et al. (2007) and Pinelli et al. (2010) for closed duct flow and in Sakai (2016) for open duct flow (cf. section 4.3.2). A validation of the finite-difference method against the pseudo-spectral solver in single-phase smooth-wall open duct flow is provided in appendix C.

In the pseudo-spectral framework, flow variables are expanded as truncated Fourier series on a set of uniformly distributed grid points in the homogeneous streamwise direction, whereas a Chebyshev polynomial expansion on a set of Gauss-Lobatto points is employed in the wall-normal and lateral directions. Note that the long-time statistics in these simulations were provided by Y. Sakai (cf. Sakai (2016) for a complete overview of the database), to whom we are very grateful for sharing his data with us. For the purpose of the following analysis, we have furthermore conducted two new simulations in this setting by ourselves, starting from two initial conditions included in the dataset of Sakai (2016) using the same numerical method.

A sketch of the physical system under consideration in the multiphase simulations can be seen in figure 7.1. As in the corresponding open channel simulations, physical fields are expressed in components with respect to a Cartesian basis, with the origin of the corresponding coordinate system $\mathbf{x} = (x, y, z)^T$ being placed in the lower left corner of the duct cross-section, such that the unit vectors are pointing

Figure 7.1: Sketch of the physical system analysed in the multiphase open duct simulations. Mean flow and gravity are pointing in positive x- and negative y-direction, respectively. A vertical and horizontal cut of the two-dimensional streamwise mean flow field $\langle u_f\rangle_{xt}(y,z)$ are shown in blue at the upstream boundary of the domain. Bed, interface and transported sediment particles are coloured in black, orange and white, respectively, following the definitions in section 6.2.1. At the downstream end of the domain, the mean fluid bed interface is indicated as blueish surface. Note that for the sake of visualisation, the sidewall in the foreground has been partially removed.

in the streamwise (x), bottom wall-normal (y) and lateral (z) directions, respectively. The fluid velocity field is accordingly described by the three components of $\boldsymbol{u}_f = (u_f, v_f, w_f)^T$. When switching to index notation, the equivalent expressions x_i and $u_{f,i}$ will be used, where $i = 1,2,3$ refers to the x-, y- and z-component of the respective vector fields, respectively.

While the boundary conditions in the streamwise and wall-normal directions are identical to those in the doubly-periodic open channel, the statistical spanwise homogeneity is broken in open duct turbulence due to the presence of solid sidewalls at $z = 0$ and $z = L_z$, respectively. The standard decomposition of flow variables in a mean and fluctuating field is therefore understood w.r.t. the single homogeneous x-direction and time, such that the fluctuation of the streamwise velocity is defined as

$$\boldsymbol{u}'_f(\boldsymbol{x},t) = \boldsymbol{u}_f(\boldsymbol{x},t) - \langle \boldsymbol{u}_f\rangle_{xt}(y,z). \tag{7.1}$$

Note that there is some degree of arbitrariness in choosing a sophisticated decomposition of the flow field, but the above averaging scheme is the natural choice in that it ensures that both the mean and the fluctuating field fulfil the physical boundary conditions individually and it is conventionally applied to turbulent wall-bounded flows (Pope, 2000). Occasionally, we will consider in addition to definition (7.1) a decomposition of the instantaneous streamwise-averaged flow field into its lateral mean and the fluctuation w.r.t. the former, consistent with equations (6.2) in the open channel context, viz.

$$\boldsymbol{u}''_f(y,z,t) = \langle \boldsymbol{u}_f\rangle_x(y,z,t) - \langle \boldsymbol{u}_f\rangle_{xz}(y,t). \tag{7.2}$$

In a similar way as for the corresponding open channel simulations, single- and multiphase open duct simulations alike are driven by a time-dependent streamwise pressure gradient $\Pi(t)$ that is adjusted at each time step, ensuring a constant mass flow rate q_f throughout the simulation interval. It might be noted that q_f is in the current cases of unit volume per time or velocity times area, whereas in the channel flow q_f has been defined as the mass flow per unit length in the spanwise direction, and is thus of unit velocity times length. Bulk velocity and bulk Reynolds number are consequently defined as

$$u_b = \frac{q_f}{H_f L_z} \tag{7.3a}$$

$$Re_b = \frac{q_f}{\nu_f L_z} = \frac{u_b H_f}{\nu_f}, \tag{7.3b}$$

respectively. The organisation of the mean streamwise and secondary flow is in general a function of the aspect ratio of the duct cross-section (Nezu and Rodi, 1985; Vinuesa et al., 2014a), which we introduce here for consistency with the notation adopted in Sakai (2016) as $AR = L_z/(2H_f)$. The latter definition slightly differs from the definition L_z/H_f conventionally used in most hydraulic studies, as it is chosen in such a way that the hydraulic diameter

$$D_{hy} = 4\frac{H_f\, AR}{1+AR} = 4\frac{H_f L_z}{2H_f + L_z} \tag{7.4}$$

is identical for open and closed ducts with the same aspect ratio (in the latter case $AR = L_z/L_y$). In this context, a corresponding open channel flow might be interpreted as the asymptotic case in which $AR \to \infty$.

Note that the hydraulic diameter as introduced above represents the ratio between the cross-sectional area and the wetted perimeter in the cross-section and therefore, strictly speaking, the hydraulic radius would depend on the curvature of the evolving sediment bed in the particle-laden cases. The difference between L_z and the curve length of the fluid-bed interface is, however, of negligible size as a consequence of the small amplitude of the developed ridges, such that replacing the curve length of the fluid-bed interface by L_z appears justified.

In an open channel flow, the force exerted by the driving pressure gradient on the system is balanced by the friction force along the bottom wall as well as by the net streamwise force on the particles. Shortly after the onset of sediment erosion during which a considerable amount of energy is spent to mobilise particles, the system reaches an equilibrium state in which the driving force is almost entirely balanced by the resistive force of the few stationary particles closest to the bottom wall, which are included in the simulations to ensure a minimum of bottom roughness (Kidanemariam et al., 2022). Considering the flow in an open duct, on the other hand, additional force contributions appear in the considered balance in form of frictional resistive forces along the two lateral sidewalls. The instantaneous balance between the driving pressure gradient Π and the counteracting resistive force contributions then reads

$$\begin{aligned}
\Pi(t) \quad = \quad & -\frac{1}{L_x L_y L_z}\sum_{l=1}^{N_p}\left(F_x^{H(l)}(t) + F_x^{C(l)}(t)\right) \\
& -\frac{1}{L_z}\left(\frac{1}{L_y}\int_0^{L_y}\rho_f\nu_f\left[\frac{\partial\langle u_f\rangle_x}{\partial z}\right]_{z=0}\mathrm{d}y + \frac{1}{L_y}\int_0^{L_y}\rho_f\nu_f\left[\frac{\partial\langle u_f\rangle_x}{\partial z}\right]_{z=L_z}\mathrm{d}y\right) \\
& -\frac{1}{L_y}\left(\frac{1}{L_z}\int_0^{L_z}\rho_f\nu_f\left[\frac{\partial\langle u_f\rangle_x}{\partial y}\right]_{y=0}\mathrm{d}z\right),
\end{aligned} \tag{7.5}$$

where $F_x^{H(l)} + F_x^{C(l)}$ is the total force acting on the lth particle comprising contributions from hydrodynamic and collision forces. The mean driving pressure gradient $\langle\Pi\rangle_t$ is obtained as an average of

$\Pi(t)$ over most of the simulation time interval excluding merely the first few bulk time units of each run during which there are still strong changes of the pressure gradient due to the onset of sediment motion.

Taking into account that the wall-normal velocity gradients $\partial\langle u_f\rangle_x/\partial \boldsymbol{n}$ (where $\boldsymbol{n}$ denotes the inward pointing normal vector at the respective wall) are of negligible amplitude in the domain below the mean sediment bed height H_b, equation (7.5) can be approximated by considering only the velocity gradients along the two sidewalls above the sediment bed, viz.

$$\begin{aligned}\Pi(t) \approx & -\frac{1}{L_x L_y L_z}\sum_{l=1}^{N_p}\left(F_x^{H(l)}(t)+F_x^{C(l)}(t)\right)\\ & -\frac{1}{L_z}\left(\frac{\rho_f \nu_f}{H_f}\int_{H_b}^{L_y}\left[\frac{\partial\langle u_f\rangle_x}{\partial z}\right]_{z=0}\mathrm{d}y+\frac{\rho_f \nu_f}{H_f}\int_{H_b}^{L_y}\left[\frac{\partial\langle u_f\rangle_x}{\partial z}\right]_{z=L_z}\mathrm{d}y\right).\end{aligned} \tag{7.6}$$

Note that the relative deviation of expression (7.6) compared to the full equation (7.5) has been verified to be of negligible size ($\mathcal{O}(10^{-3})$ for case $DS250^{H2}$ exemplary), such that in the following, the mean pressure gradient shall be approximated by the latter formulation. Assuming that the flow is fully-developed, the mean wall shear stress τ_w can be obtained by evaluation of the global momentum balance over the domain as

$$\tau_w = \rho_f u_\tau^2 \approx -\langle\Pi\rangle_t \frac{H_f L_z}{2H_f + L_z} = -\langle\Pi\rangle_t \frac{D_{hy}}{4}. \tag{7.7}$$

That way defined, τ_w and accordingly u_τ represent averages over the shear stress contributions along the sediment bed and the smooth sidewalls and are thus used to define the friction Reynolds number $Re_\tau = u_\tau H_f/\nu_f$. However, the actual mean shear stress that is exerted on the mobile sediment bed by the turbulent flow is naturally higher than τ_w due to the relatively lower friction along the smooth sidewalls compared to the mobile sediment bed. For the use of the mean shear stress in the context of particle transport models, we therefore additionally introduce a mean bed shear stress τ_{bed} that is defined in a similar way as τ_w, but without considering the frictional force contributions that arise along the lateral sidewalls:

$$\tau_{bed} = \frac{H_f}{L_x L_y L_z}\sum_{l=1}^{N_p}\left\langle\left(F_x^{H(l)}(t)+F_x^{C(l)}(t)\right)\right\rangle_t. \tag{7.8}$$

Note that due to the absence of sidewall related contributions in equation (7.8), this latter expression is identical to the corresponding formulation in open channel flows (Kidanemariam et al., 2022).

The definition of the mean fluid height H_f and the mean sediment bed height H_b are identical to those in section 6.1, and the detection procedure following Scherer et al. (2020) that has been outlined in section 6.2.1 is applied in a comparable way here. The only exception is that the interface is not periodically repeated across the lateral domain boundaries during the interpolation and smoothing phase, but mirrored at the solid sidewalls such that $\partial h_b(x,z,t)/\partial z = 0$ holds at $z \in \{0, L_z\}$. As in the previous chapter, we shall regularly refer to the shifted wall-normal coordinate $\tilde{y} = y - H_b$ that measures the wall-normal distance from the mean location of the fluid-bed interface.

	Case	Re_b	Re_τ	ρ_p/ρ_f	Ga	D^+	H_f/D	H_f/H_b	θ_{bed}	AR
—	$DS250^{H1}$	2999	272.11	2.5	28.37	10.11	26.93	2.35	0.20	0.48
—	$DS250^{H2}$	3000	261.95	2.5	28.37	9.80	26.73	2.29	0.16	0.96
—	$DS250^{H4}$	2999	253.69	2.5	28.37	9.53	26.61	2.26	0.14	1.92
—	$DL240^{H2}$	2992	243.74	2.5	40.12	8.90	27.39	2.49	0.06	0.93
---	$DL220^{H2}_{fix}$	2992	225.76	2.5	40.12	8.29	27.23	2.44	0.05	0.94
—	$DL250^{H2}$	2992	259.26	2.5	28.37	9.38	27.65	2.57	0.15	0.93
—	$DL500^{H2}$	5986	514.69	2.5	47.49	18.39	27.99	2.69	0.22	0.91
—	$DL250^{H16}$	2998	249.56	2.5	28.37	9.42	26.50	2.23	0.12	7.73
---	$DL210^{H16}_{fix}$	2999	207.41	2.5	28.37	7.87	26.34	2.18	0.08	7.77
—	$DL200^{H2}_{smooth}$	2969	197.46	-	-	-	-	-	-	1.00
—	$DL400^{H2}_{smooth}$	6964	415.77	-	-	-	-	-	-	1.00
—	$DL150^{H16}_{smooth}$	2188	143.59	-	-	-	-	-	-	8.00
—	$CL250^{H3}$	3011	252.26	2.5	28.37	9.58	26.32	2.18	0.11	-

Table 7.1: Physical parameters of open duct flow simulations, where Re_b, Re_τ, D^+, Ga and ρ_p/ρ_f are the bulk, friction and particle Reynolds numbers as well as the Galileo number and the density ratio, respectively. Further non-dimensional numbers include the relative submergence H_f/D, the relative sediment bed height H_b/D, the fluid to bed height ratio H_f/H_b, the Shields number θ_{bed} as well as the duct aspect ratio $AR = L_z/(2H_f)$. Note that only Re_b, Ga and ρ_p/ρ_f are imposed a priori, whereas the remaining quantities are determined a posteriori (cf. table 7.2).

7.1.2 Preparation of the simulations

The initial states of the sediment bed for the current multiphase simulations were generated from macroscopically flat sediment beds in open channel flow cases, which, in turn, have undergone the preparation process described in section 6.1.2. In order to adapt these particle fields to the current configuration with a non-periodic lateral direction, particles overlapping the lateral target domain boundaries were removed and the resulting reduced particle set was kept settling under the action of gravity until a new quasi-stationary state had developed. In cases in which this caused a drop of the sediment bed height in the direct vicinity of the sidewalls, the resulting gap was filled by letting settle a small amount of particles in these regions. Despite the careful preparation, there remained in some cases a slight but unavoidable drop of the mean fluid-bed interface of less than one diameter depth in the direct vicinity of the sidewalls. This gap is readily filled up in simulations with intense particle transport, while it remains detectable for longer time intervals in simulations that are either characterised by a weak particle transport or feature a stationary sediment bed. The initial state of the flow field was then established by disturbing a laminar flow pattern over a short time interval of $\mathcal{O}(1T_b\text{-}10T_b)$ with a finite amplitude perturbation to trigger the transition towards a fully-developed turbulent flow state (Kawahara et al., 2012b). It was then waited until the flow had reached the desired fully-developed state, in the meanwhile holding the sediment particles fixed in their current spatial positions. Once this flow state was reached, particles were released and the onset of sediment mobility was, as before, defined as $t = 0$.

	Case	$[L_x \times L_z]/H_f$	$[L_x \times L_z]/D$	$D/\Delta x$	$\min(\Delta x_i^+)$	N_p	T_{obs}/T_b
—	$DS250^{H1}$	1.90×0.95	51.2×25.6	10	1.01	13 988	976
—	$DS250^{H2}$	1.92×1.92	51.2×51.2	10	0.98	28 670	990
—	$DS250^{H4}$	1.92×3.85	51.2×102.4	10	0.95	58 057	999
—	$DL240^{H2}$	11.22×1.87	307.2×51.2	10	0.89	164 540	159
---	$DL220^{H2}_{fix}$	11.28×1.88	307.2×51.2	10	0.83	164 540	237
—	$DL250^{H2}$	11.11×1.85	307.2×51.2	10	0.94	157 955	72
—	$DL500^{H2}$	10.98×1.83	307.2×51.2	20	0.92	158 124	49
—	$DL250^{H16}$	11.59×15.45	307.2×409.6	10	0.94	1 402 849	75
---	$DL210^{H16}_{fix}$	11.66×15.55	307.2×409.6	10	0.79	1 402 849	102
—	$DL200^{H2}_{smooth}$	$8\pi \times 2.00$	-	-	0.03	0	9082
		$4\pi \times 2.00$	-	-	0.03	0	9082[1]
—	$DL400^{H2}_{smooth}$	$8\pi \times 2.00$	-	-	0.01	0	4765[2]
—	$DL150^{H16}_{smooth}$	$8\pi \times 16.00$	-	-	0.04	0	8140
—	$CL250^{H3}$	11.67×2.92	307.2×76.8	10	0.96	263 412	73

Table 7.2: Numerical parameters of open duct flow simulations. The streamwise and spanwise extents of the computational domain $L_x \times L_z$ are presented in terms of the mean fluid height and the particle diameter, respectively. While the domain in the particle-laden simulations is discretised using a uniform finite difference grid with mesh width Δx, the pseudo-spectral Fourier-Chebyshev-Chebyshev method used to simulate the smooth-wall configurations features a non-uniform distribution of the grid/collocation points in the wall-normal and transverse directions. N_p is the total number of particles per simulation and T_{obs} is the total observation time of each simulation, starting from the release of the mobile particles at $t = 0$. Time dependent physical and numerical parameters in tables 7.1 and 7.2 are computed as an average over the entire simulation period. [1,2] Statistics for the reference smooth-wall simulations in case $DL400^{H2}_{smooth}$ and case $DL200^{H2}_{smooth}$ are from Sakai (2016). Investigations on instantaneous flow fields, on the other hand, are performed based on new simulations performed in the course of this study at identical parameter points. In case $DL400^{H2}_{smooth}$, the current simulation features a however shorter domain compared to the original case. The newly simulated time intervals comprise $286T_b$ for case $DL400^{H2}_{smooth}$ and $272T_b$ for case $DL200^{H2}_{smooth}$.

7.1.3 Simulation parameter values

The simulations performed in the course of this study cover a range of different parameter points concerning domain length L_x/H_f, aspect ratio AR, Reynolds number Re_τ and Shields number θ_{bed}. An overview of all performed simulations and their relevant physical and numerical parameters is provided in tables 7.1 and 7.2, respectively. Note that, referring to the discussion above, we define the Shields number in the current study based on the sediment bed-related shear stress rather than τ_w, such that $\theta_{bed} = (\tau_{bed}/\rho_f u_g^2)$.

Based on the experience gained in previous studies on transverse pattern evolution (Kidanemariam and Uhlmann, 2017; Scherer et al., 2020), we have performed three short domain simulations (S) with a streamwise domain size $L_x/D = 51.2$ ($L_x/H_f \approx 2$) and aspect ratios $AR \approx \{0.5, 1, 2\}$ which are too short to accommodate any unstable mode that would give rise to transverse ripple-like patterns. These simulations thus allow us to study the evolution of sediment ridges exclusively over time intervals of lengths $\mathcal{O}(10^3 T_b)$. In all three simulations, the inner-scaled streamwise and spanwise domain lengths are with $L_x^+ \approx 500$ and $L_z^+ \approx \{250, 500, 1000\}$ sufficiently larger than the minimal streamwise and spanwise extent required to accommodate a single buffer-layer streak with its associated quasi-streamwise vortices (Jiménez and Moin, 1991; Pinelli et al., 2010; Sakai, 2016).

The remaining simulations (L) feature a clearly longer streamwise extent of $L_x/H_f \approx$ 11-12 in the particle-laden cases and $L_x/H_f = 8\pi$ in the smooth-wall cases in order to capture also structures of larger streamwise dimension. For the sake of completeness, it should be mentioned that the streamwise domain sizes in the particle-laden simulations are not large enough to allow for a complete decorrelation of the velocity statistics in the streamwise direction, and thus Sakai (2016) considered domains that are roughly twice as long. Taking into account the significantly higher computational costs for the simulation of one physical time unit with the IBM-finite difference solver compared to the pseudo-spectral code, we decided to perform the multiphase simulations in a somewhat shorter domain as a compromise between computational expenses and the goal to reduce the effect of a finite domain size as much as possible. In order to scrutinise the influence of the domain length on the statistics of interest, we have conducted an additional smooth-wall simulation at the same parameter point as case $DL200^{H2}_{smooth}$ of Sakai (2016), featuring a reduced length of $L_x/H_f = 4\pi$.

As discussed earlier, enlarging the streamwise domain size comes at a price: After an initial phase during which ridges are essentially the only developed sediment patterns and thus can be studied exclusively, ripple-like features arise and eventually dominate the surface of the sediment bed. The growth rate and thus the exact length of the period required by transverse patterns to grow to a relevant amplitude might depend on different factors, from which the value of the Shields number is one. In order to stall the evolution of transverse bedforms and thus to increase the time window of exclusive ridge formation, we have assessed the effect of a decrease of the Shields number by roughly factor two between cases $DL250^{H2}$ and $DL240^{H2}$. It turns out that the amplitude of the observed ridges is, as expected, smaller in the case with lower Shields number, while at the same time the period of ridges being the dominant bedforms was more than twice as long as in the case with higher Shields number.

The parameter range of the performed simulations was chosen in such a way that the simulations cover both the narrow and the wide duct regime. Open ducts are conventionally defined as 'narrow' if they feature an aspect ratio of $AR \lesssim 2.5$ (Nezu and Rodi, 1985; Nezu and Nakagawa, 1993), such that the presence of the lateral sidewalls can be assumed to influence the full domain of the flow. On the other hand, open ducts are typically considered as 'wide' if $AR \gtrsim 5$ (Rodríguez and García, 2008), for which the influence of the sidewall boundary layers on the flow in the duct centre is assumed to be weak. The here conducted simulations lie with aspect ratios of $AR \approx$ 0.5-2 and $AR \approx 8$ well below and well above these limits, respectively. Eventually, smooth-wall and particle-laden simulations were analysed for different Reynolds numbers in a range $140 \lesssim Re_\tau \lesssim 515$, from which the highest Reynolds number flow thus features a scale separation of roughly $1.5H_f/(100\delta_\nu) \approx 7.5$ between the characteristic size of large-scale outer-scaling structures and that of the small-scale buffer layer structures.

7.2 Sediment ridge formation in the presence of lateral sidewalls

The evolution of streamwise-elongated sediment ridges has been observed in many laboratory flume experiments since the seminal studies of Casey (1935) and Vanoni (1946), including, amongst others, the works of Wolman and Brush (1961), Nezu and Nakagawa (1989) and McLelland et al. (1999). In all these studies, groups of streamwise-elongated sediment ridges were observed to cover the entire cross-section of the flumes from the near-sidewall region to the duct core. The here investigated simulations show a qualitatively similar behaviour in that they all allow for the evolution of at least two parallel

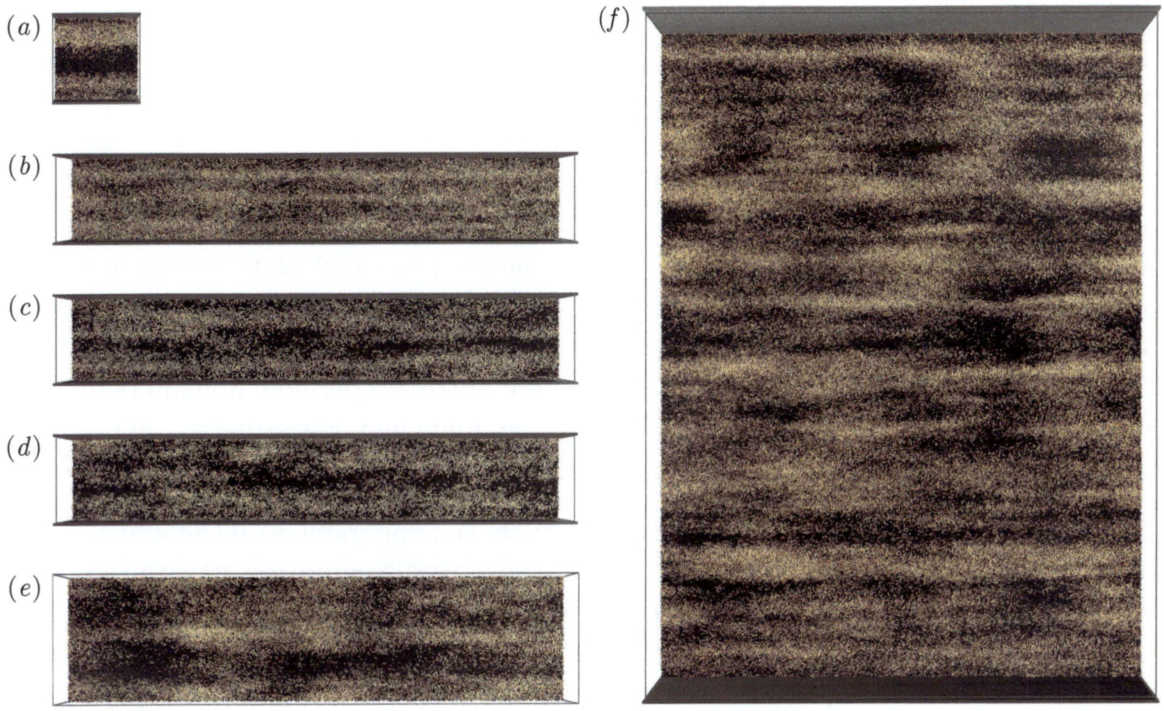

Figure 7.2: Instantaneous visualisation of evolved sediment ridges as seen from the free surface looking downward. Colouring of the particles indicates their wall-normal location, ranging form dark to light brown with increasing coordinate y. For the sake of clarity, only bed and interface particles are shown while mobile bedload is not. Flow is from left to right in each panel. (*a*) $DS250^{H2}$ ($t/T_b = 66$), (*b*) $DL240^{H2}$ ($t/T_b = 74$), (*c*) $DL250^{H2}$ ($t/T_b = 39$), (*d*) $DL500^{H2}$ ($t/T_b = 23$), (*d*) $CL250^{H3}$ ($t/T_b = 73$) and (*f*) $DL250^{H16}$ ($t/T_b = 75$).

streamwise-aligned ridges. However, the shapes and amplitudes of these bedforms differ considerably between the different cases, as the instantaneous snapshots of the evolving sediment beds in figure 7.2 indicate. The three simulations at higher Shields number θ_{bed} and low aspect ratio $AR \approx 1$ ($DS250^{H2}$, $DL250^{H2}$ and $DL500^{H2}$) shown in figures 7.2(*a*,*c*,*d*) develop a qualitatively similar pattern that consists of a global trough, preferentially located in the duct centre, and two evolved ridges flanking the former on either side. In the streamwise-minimal simulation $DS250^{H2}$, the two sediment ridges and the trough in between are well separated and perfectly aligned with the mean flow direction. In addition, they feature the largest amplitude among the three cases, which is assumed to be a consequence of the severe spatial constraints: The limited domain extension together with the laterally bounding sidewalls entirely suppress the lateral mobility of the flow and a few quasi-infinitely long flow structures dominate the flow, such that the region of stronger erosion in the duct centre is relatively persistent in space. In the simulations with a longer domain $L_x/H_f \approx 11$ ($DL250^{H2}$, $DL500^{H2}$), the two sediment ridges show a weak but visible lateral meandering tendency across the domain, but they still span the entire box length. This supports our findings in the previous chapters and observations in experimental studies, according to which sediment ridges can easily reach streamwise extents of $\mathcal{O}(10H_f\text{-}100H_f)$ (Vanoni, 1946; Wolman and Brush, 1961). Interestingly, the arrangement of the sediment patterns does not significantly differ between the case with lower ($DL250^{H2}$) and the one with higher Reynolds number ($DL500^{H2}$), indicating that again outer-scaling flow structures are responsible for the generation of the sediment ridges.

In hydraulic experiments, sediment ridges are merely observed when the sediment transport rate is too low for the formation of developed transverse bedforms such as ripples or dunes (Wolman and Brush, 1961). Indeed, Kleinhans et al. (2002) concluded based on flume experiments and field measurements that sediment ridges can form even under strong limitations of the sediment supply, as it is the case over armoured gravel beds (Venditti et al., 2019) or in flume experiments at low sediment transport rate. As soon as a sufficient amount of mobile sediment is available in the flow, however, they are assumed to transform into larger bedforms such as barchans, ripples or dunes (Kleinhans et al., 2002). In order to study the impact of the sediment transport intensity on the formation of sediment bedforms, the Shields number in case $DL240^{H2}$ was significantly decreased to a value that is much closer to the critical value for the onset of sediment motion than in the remaining cases. As can be seen in figure 7.2(*b*), the cross-section accommodates in this case three individual sediment ridges, which are of however lower amplitude than those in the remaining simulations as a consequence of the limited availability of mobile sediment. In some regions such as close to the upstream periodic domain boundary, these bed features are only hardly distinguishable from the remaining bed. Nevertheless, all three sediment ridges span over the entire length of the domain, just as their counterparts at higher Shields numbers.

The width of the sediment ridges found in narrow ducts is comparable to those which form in the region close to the sidewalls of the high-aspect ratio case $DL250^{H16}$ in figure 7.2(f): The box accommodates between nine and eleven more or less regularly spaced individual sediment ridges. In addition to these features, the sediment bed reveals here already first 'nuclei' for the development of larger transverse bedforms at different locations of the domain. The further evolution of these localised disturbances and their transition into larger transverse ripple-like bedforms will be the main topic of the subsequent chapter 8. The sediment ridges that evolve in the centre of the duct in relatively large distance to the sidewalls reveal a somewhat larger spanwise width than their counterparts near the sidewalls, more resembling the ridges in the doubly-periodic open channel observed in figure 7.2(*e*). The latter case shows a very similar sediment bed organisation as in the open channel simulations discussed in figure 6.3 of the previous chapter: In particular, it features two individual sediment ridges with streamwise dimensions identical to the domain length L_x and a mean lateral wavelength that is comparable to the values H_f-$2H_f$ which have been observed for the remaining channel simulations in this work.

7.2.1 Sediment ridge evolution

The discussed properties of the instantaneous sediment ridges give a first hint that the formation mechanism of sediment ridges in the core of a high aspect ratio duct could be very similar to the one discussed in chapter 6 for infinitely wide open channels. The evolution of sediment ridges in the direct vicinity of a lateral sidewall, on the other hand, is subject to the particular flow organisation in this region and, thus, their development might be controlled by different flow structures. Indeed, it turns out that many of the characteristics observed in the instantaneous sediment bed visualisations in figure 7.2 are relatively robust and can be detected in a similar way in the space-time evolution of the streamwise-averaged sediment bed profiles visualised in figure 7.3 as well. The narrow duct cases with higher Shields number ($DS250^{H2}$, $DL250^{H2}$, $DL500^{H2}$) whose bed evolution is depicted in figures 7.3(*a,c,d*) accommodate two individual ridges whose crests are located roughly half-way between the duct bisector and the respective sidewall, separated by a pronounced trough in their middle. A very similar mean arrangement of two sediment ridges with a comparable amplitude of approximately $1D$ enclosing a single trough has been observed by McLelland et al. (1999) in an open flume experiment at only slightly larger aspect ratio of $AR = 1.5$, using a weakly bimodal sediment mixture. Remarkably, it

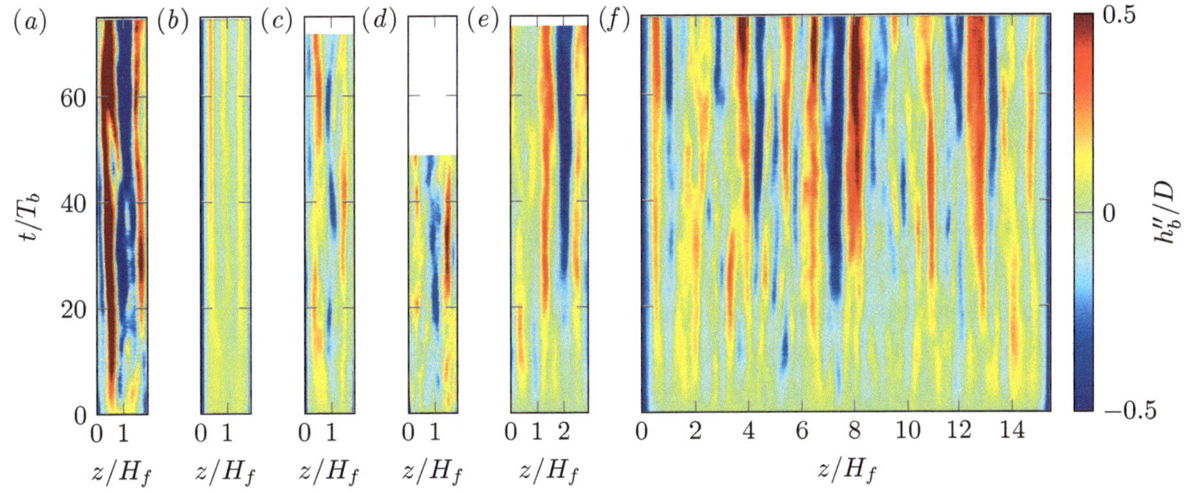

Figure 7.3: Space-time plot of the streamwise-averaged sediment bed height fluctuations $h_b''(z,t)/D$ in open duct simulations. Blue and red regions refer to troughs and crests of the streamwise-averaged fluid-bed interface profiles, respectively. Cases: *(a)* $DS250^{H2}$, *(b)* $DL240^{H2}$, *(c)* $DL250^{H2}$, *(d)* $DL500^{H2}$, *(e)* $CL250^{H3}$, *(f)* $DL250^{H16}$.

turns out that the instantaneous sediment ridges seen in our numerical experiments are not necessarily symmetrically arranged: The most probable configuration rather seems to consist of two sediment ridges from which one is dominant in that it features a higher amplitude than its weaker neighbour. In case $DL250^{H2}$, it is furthermore seen that the two sediment ridges change their roles intermittently, i.e. their amplitudes increase and decrease in time such that either the left or the right one is dominant in the above sense. In all three cases, the sediment ridges and the enclosed trough also exhibit a certain degree of lateral meandering in time with a maximum amplitude $\mathcal{O}(0.1H_f)$. A possible explanation for both effects is that the turbulent velocity structures above the sediment bed undergo a similar intermittently changing flow organisation, such that zones of stronger and weaker erosion along the bed alternate accordingly. This hypothesis shall be further scrutinised in figure 7.25 below.

The sediment ridges that form at lower Shields number $\theta_{bed} = 0.06$ in case $DL240^{H2}$ (cf. figure 7.3(*b*)) reveal a markedly different evolution compared with those in the previously discussed simulations, characterised by the formation of three essentially parallel low-amplitude sediment ridges, more or less equally spaced across the duct span. The three sediment ridges in case $DL240^{H2}$ exhibit essentially no lateral propagation throughout the entire simulation interval. This reduced bedform mobility is assumed to be a consequence of the reduced ability of the flow to mobilise sediment grains, leading to a stabilisation of the sediment patterns: Once formed, the bedforms are less affected by the intermittent dynamics of the turbulent flow. On the other hand, the reduced availability of mobile sediment retards the evolution not only of transverse patterns, but also the growth rate of the here considered sediment ridges is reduced compared to the remaining simulations. It is thus likely that the variation of the Shields number generally affects the time scale of the sediment bed evolution. The reason for the evolution of an additional sediment ridge in the duct centre from approximately $t/T_b = 20$ on cannot be fully clarified based on the sediment bed data alone, such that we will postpone the discussion of and possible explanations for this phenomenon to section 7.2.5 below.

If the aspect ratio of the open duct is sufficiently large as in case $DL250^{H16}$ (cf. figure 7.3(*f*)), a large number of initial sediment ridges forms quasi-simultaneously throughout the entire cross-section shortly after the onset of particle erosion. The formation process and the resulting ridge pattern is qualitatively similar to those observed for the corresponding open channel flow configuration $CL250^{H16}$ in section 6.3.1. In a similar way as in the latter case, the sediment patterns undergo a phase of coalescence during which individual bedforms merge, eventually leading to an arrangement of between eight and twelve evolved ridges whose mean amplitude and lateral width are larger than that of the early-stage initial ridges. Note that a comparably fast evolution of sediment ridges at different locations of the cross-section has been also observed in the experiments of Nezu and Nakagawa (1989), even though the comment of the authors that the first ridges next to the sidewalls evolved slightly earlier than their counterparts in the duct centre cannot be confirmed for the current cases, where all features appear quasi-simultaneously. As has been already seen in the instantaneous sediment bed visualisation in figure 7.2(*f*), the lateral width of the sediment ridges seems to vary with their distance to the lateral sidewalls. Sediment bedforms located in the duct centre such as the sediment ridges in the region $z/H_f \in [6,8]$ are wider and typically feature a higher amplitude than their counterparts in the vicinity of the sidewalls, whose development is more constrained by the lateral domain boundary. In fact, the higher and wider sediment ridges in the duct centre resemble in their shape those in the reference channel simulation $CL250^{H3}$ provided in panel 7.3(*e*). In good agreement with the cases investigated in section 6.3.1 of the previous chapter, the two sediment ridges that can be accommodated by the doubly-periodic domain are seen to develop without strong lateral meandering at a transverse spacing of approximately $\lambda_{h,z} \approx 1.5H_f$.

Eventually, let us remark that the initially slightly lower sediment bed height in the closest proximity of the sidewalls that was unavoidable in the bed preparation process evolves quite differently in time for the different cases: In the low aspect ratio cases with higher Shields number ($DS250^{H2}$, $DL250^{H2}$, $DL500^{H2}$), there is a sufficient lateral sediment input into the sidewall regions such that the initially existing gap readily fills up. For case $DL240^{H2}$, on the other hand, the lateral particle flux can be assumed to be too weak to do so, with the consequence that the signature of the initially slightly lower sediment bed remains visible. Interestingly, also in the high aspect ratio case $DL250^{H16}$ the lateral particle transport into the bed-sidewall corners is apparently not strong strong enough to fill the initial gap in the here considered time interval. As we shall see in the following chapter 8 in which we investigate the subsequent ripple-dominated bed evolution phase, however, the near-sidewall region will eventually fill up with sediment grains.

It should be kept in mind that the initial offset of the bed height in these regions is with $\mathcal{O}(0.1D)$ relatively small, but due to the also small amplitude of the sediment ridges especially in the narrow ducts, the effect is visible in the statistics of the bed evolution. For the mean pattern amplitude evolution shown in figure 7.4(*a*) in terms of the r.m.s. of the streamwise-averaged sediment bed height fluctuations, its influence appears in form of a finite initial value $\sigma_{h,z}(t=0)$ in the duct cases, whereas the respective value in the channel case is initially of negligible size. In order to assess the contribution of the initial bed variation near the wall on the statistics, we have evaluated $\sigma_{h,z}$ again, excluding however the region close the sidewalls (a lateral distance of $2D$ from either sidewall). It is seen that in all cases except for $DL240^{H2}$ which features the significantly lower Shields number, the influence of the initial bed variation is relevant only in the beginning of the simulations, while the evolution of $\sigma_{h,z}$ is in later stages indeed predominantly related to the evolving sediment ridges.

After the release of the particles, the strongest increase and largest pattern amplitude is seen to occur for the streamwise-minimal simulation $DS250^{H2}$, which is again assumed to be caused by the strong spatial constraints due to the small domain size: As explained above, the short and narrow domain

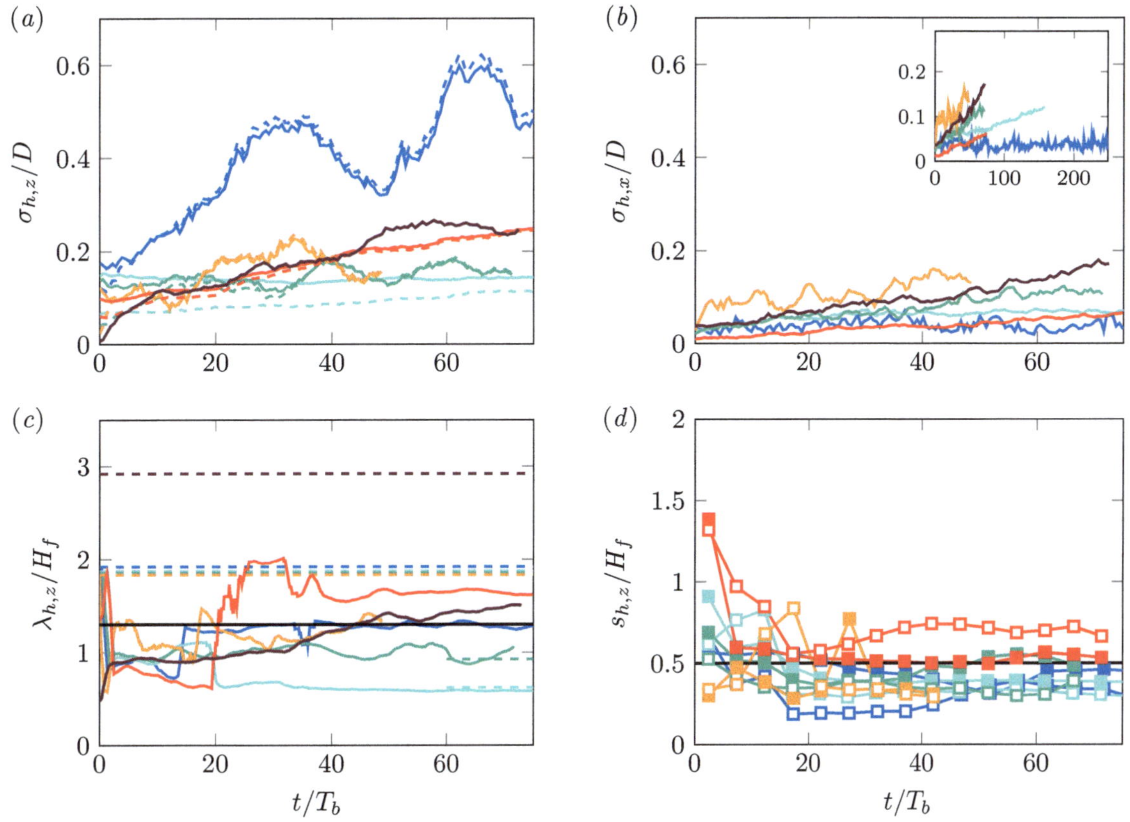

Figure 7.4: Time evolution of fluid-bed interface dynamics. Root mean square of the fluctuation of the (*a*) streamwise-averaged ($\sigma_{h,z}/D$) and (*b*) spanwise-averaged fluid-bed interface ($\sigma_{h,x}/D$). In (*a*), the solid lines show $\sigma_{h,z}/D$ evaluated for the entire spanwise width of the duct $z \in [0, L_z]$, while the dashed lines represent the same quantity determined excluding the closest vicinity of the sidewalls, i.e. over the range $z \in [2D, L_z - 2D]$. In (*b*), the inset shows the same data as in the main plot but for a longer time interval. (*c*) Mean spanwise pattern spacing $\lambda_{h,z}$. The black solid and the long coloured horizontal dashed lines mark the most-amplified wavelength in the linear stability analysis of Colombini (1993) and the relative lateral domain sizes L_z/H_f, respectively. Short horizontal dashed lines represent fractions of the entire domain width L_z in the respective case: (---) $L_z/3$; (---) $L_z/2$. (*d*) Minimum spanwise distance between a sidewall and the closest sediment ridge crest, $s_{h,z} = \min_k(|z_k - z_{wall}|)$, where z_k is the lateral position of the kth of n_{ridges} local maxima of $h_b''(z,t)$. (■): left sidewall at $z_{wall} = 0$; (□): right sidewall at $z_{wall} = L_z$. Each data point represents an average over a time interval of approximately $5T_b$. Cases: $DS250^{H2}$ (—), $DL240^{H2}$ (—), $DL250^{H2}$ (—), $DL500^{H2}$ (—), $DL250^{H16}$ (—), $CL250^{H3}$ (—).

leads to the formation of a single quasi-infinitely long high-speed region in the centre of the duct associated with a strong erosion, which results in the formation of almost perfectly streamwise-aligned trough and crest regions, whose signature in the streamwise average is consequently quite pronounced. In the long domains at the same low aspect ratio ($DL250^{H2}$ and $DL500^{H2}$), the initial growth rate and the eventually attained amplitude are smaller than in case $DS250^{H2}$. This is most likely due to a larger freedom to meander in the lateral direction of both sediment ridges and flow structures in these cases, leading to a reduced pattern amplitude in the streamwise average. Interestingly, the increase of $\sigma_{h,z}$ in these cases is not monotonous, but it exhibits a low amplitude oscillation whose local extrema are directly correlated to the different phases of bedform evolution seen in the space-time evolution plots in figure 7.3. Considering, for instance, case $DL250^{H2}$, it is observed that local maxima of $\sigma_{h,z}$ at approximately $t/T_b \in \{20, 40, 60\}$ correspond to situations in which the 'dominant' of the two sediment ridges reaches its maximum amplitude, while local minima at approximately $t/T_b \in \{30, 50\}$ are characterised by the 'transition' state in which both ridges are of comparable amplitude. In case $DL240^{H2}$, which features as opposed to the remaining cases a relatively low Shields number, we also observe a general increase of the sediment ridge amplitude at a however clearly reduced growth rate as a consequence of the weak particle transport intensity. Also, as discussed above, there is no sufficient sediment supply to the corner regions to fill up the initial gap which therefore clearly contributes to $\sigma_{h,z}$ during the entire simulation interval. The high aspect ratio duct case $DL250^{H16}$ and the laterally periodic channel flow $CL250^{H3}$, eventually, reveal a qualitatively and quantitatively very similar evolution of the sediment ridge height, featuring an almost monotonic growth of $\sigma_{h,z}$ during the considered time window. The very similar evolution in these two cases further strengthens our previous claim that the majority of the sediment ridges in the wide duct form in a similar process as that observed in open channel flows.

For the sake of comparison, figure 7.4(*b*) provides the time evolution of the corresponding spanwise-averaged sediment bed amplitude, $\sigma_{h,x}$, which is a measure for the mean height of transverse ripple-like patterns (Coleman and Nikora, 2009; Kidanemariam and Uhlmann, 2017). It confirms that in the considered time intervals, the mean sediment ridge amplitude is always higher than that of the ripple-like patterns, indicating that sediment ridges are in all situations the dominant bedform type. In agreement with the open channel simulations of Kidanemariam and Uhlmann (2017) and Scherer et al. (2020), the streamwise-minimal domain of case $DS250^{H2}$ effectively suppresses the growth of the modes related to transverse bedforms also in the presence of lateral sidewalls, whereas all remaining cases eventually develop ripple-like bedforms, even though at clearly different growth rates. While the low growth rate of the 'ripple-mode' in case $DL240^{H2}$ is again a consequence of the low Shields number, this explanation does not apply to the situation in case $DL250^{H16}$ that features a comparable non-dimensional shear stress as the remaining cases. Indeed, time-resolved visualisations of the bed evolution in this latter case verify that the spanwise-averaged framework fails short to resolve spatially-localised initial transverse disturbances in this very large domain (cf. figure 7.2(*f*)) that represent the onset of ripple-like bedform evolution. These initial transverse bed undulations occur at quasi-random positions across the domain and are thus partly hidden in the spanwise-average over long time intervals, until they merge with other localised sediment patches to form larger, essentially two-dimensional transverse patterns. A more detailed investigation of this phase of the sediment bed evolution is presented in the following chapter 8. In contrast to this more complex formation process, initial disturbances in narrow open channel and duct flows are artificially constrained to remain essentially two-dimensional without a clear dependence on the transverse direction due to the limited lateral extension of the domain. In such cases, spanwise averaging can capture all essential features of the emerging transverse bedforms from the very beginning.

The development of the mean lateral width of the sediment ridges is presented in figure 7.4(*c*). Note that the general definition of $\lambda_{h,z}$ in equation (6.9) can still be applied in the case of a non-periodic lateral direction, but in contrast to the open channel case it has to be determined in physical rather than in Fourier space. Accordingly, the previously used term of a lateral wavelength is here rather misleading, in particular for narrow duct cases which accommodate not more than two sediment ridges. For this reason, we will henceforth refer to $\lambda_{h,z}$ as the averaged lateral distance or spacing between two neighbouring sediment ridges. After an initial transient of approximately 20 bulk time units during which the value of $\lambda_{h,z}$ partly oscillates strongly, most cases choose a preferred lateral ridge spacing, which they maintain until the end of the simulation intervals. The difference between the evolution of $\lambda_{h,z}$ in the initial 20 bulk time units and that in the subsequent phase is most pronounced in the wide domain of $DL250^{H16}$, where a strong increase of $\lambda_{h,z}$ is detected within a relatively short time-interval between 20 and 25 bulk time units after the particle release. This abrupt change of the preferred lateral spacing is related to the merging or dying of the initial smaller sediment ridges in this phase that has been observed in the previous figure 7.3(*f*). Note that the corresponding wide open channel flow in case $CL250^{H16}$ showed a similar increase of the mean ridge spacing during a comparable bedform merging phase, but in this latter case the increase of $\lambda_{h,z}$ appeared more gradually than the sudden jump seen in case $DL250^{H16}$.

Except for that in case $DL240^{H2}$, all finally attained ridge spacings $\lambda_{h,z}$ are in the range of values typically reported in flume experiments (e.g. McLelland et al., 1999). The eventually attained values of $\lambda_{h,z}$ in the low aspect duct cases differ somewhat between the different simulations, but they are systematically smaller than those in cases $CL250^{H3}$ and $DL250^{H16}$. We have seen earlier that the sediment ridges can spread over a wider lateral distance if they are in sufficient distance to a lateral sidewall, which is in agreement with the generally higher values of $\lambda_{h,z}$ in the latter two cases. Case $DL240^{H2}$ accommodates three individual low-amplitude ridges and accordingly reveals a mean lateral spacing of roughly $L_z/3$, whereas cases $DS250^{H2}$ and $DL500^{H2}$ show larger values of comparable size as the lateral wavelength $1.3H_f$ that has been theoretically predicted in the limit of an infinitely wide open channel (Colombini, 1993). For case $DL250^{H2}$, in turn, the mean spacing is found to attain a value of $\lambda_{h,z} \approx L_z/2$ that is somewhat lower than those found in cases $DS250^{H2}$ and $DL500^{H2}$, implying that the lateral distance $s_{h,z}$ between the two sidewalls and their neighbouring ridges is higher in this case. The time evolution of $s_{h,z}$ is provided in figure 7.4(*d*): As already observed in the space-time plots in figure 7.3, the sediment ridges in the narrow duct simulations are seen to slightly meander in time, but their distance to the respective neighbouring sidewall always remains in a range $(0.25\text{-}0.55)H_f$. Interestingly, even in the high aspect ratio case $DL250^{H16}$, the first and last ridges typically do not depart from the sidewall region, but reside at an only slightly larger distance $0.5 < s_{h,z}/H_f < 0.8$ to the domain boundaries.

7.2.2 Sediment transport

In the previous section, we have seen how the reduced Shields number in case $DL240^{H2}$ has caused a reduction of the growth rate of the sediment ridge amplitude under, apart from θ_{bed}, comparable conditions as in case $DL250^{H2}$. The link between the non-dimensional shear stress and the bedform evolution is the particle flux which is, amongst others, a function of the Shields number. Aim of a number of engineering-type models is to provide an empirical estimate of this functional relationship between the particle transport rate and the Shields number, from which the most popular one is surely the relation of Meyer-Peter and Müller (1948). As in section 6.3.2, the latter is in the following discussed in the revised version of Wong and Parker (2006).

Figures 7.5(*a,b*) show the temporal development of the streamwise particle flux $\langle q_{p,x}\rangle_{xz}(t)$ in the current open duct simulations, scaled in inertial scales (using u_g and D as reference scale) and in terms of a reference flux q_{ref}. The latter is the particle flux computed based on the empirical relation of Wong and Parker (2006) in the same way as in section 6.3.2, except for the fact that we restrict ourselves here to the investigation of the particle flux in the core region of the duct, that is, $z \in [0.5H_f, L_z - 0.5H_f]$. The concentration on regions away from the sidewalls is justified by the fact that both the original and the modified version of the Meyer-Peter and Müller (1948) formula were not conceived for the near-wall region. In fact, applying these models over the entire duct width including the near-sidewall regions leads to a poorer prediction of the particle flux observed in our simulations (plots omitted).

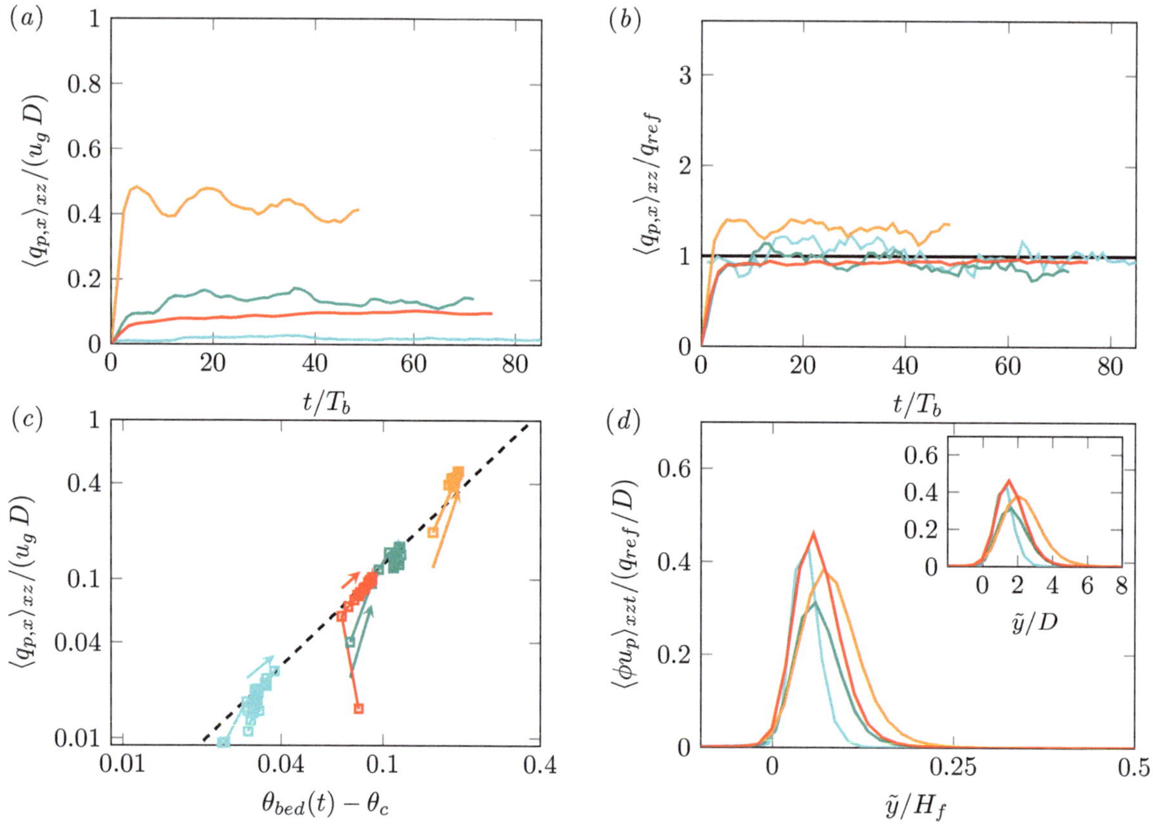

Figure 7.5: Mean streamwise particle flux $\langle q_{p,x}\rangle_{xz}$ as a function of time normalised by (*a*) the inertial scale $u_g D$ and (*b*) the reference particle flux q_{ref} computed based on the Wong and Parker (2006) version of the empirical Meyer-Peter and Müller (1948) formula: $q_{ref}/(u_g D) = 4.93(\theta_{bed}(t) - \theta_c)^{1.6}$, with the critical Shields number $\theta_c = 0.034$ (Soulsby et al., 1997) and θ_{bed} being computed based on the shear stress contribution due to the sediment bed only. (*c*) Streamwise particle flux $\langle q_{p,x}\rangle_{xz}/(u_g D)$ as a function of the excess Shields number $\theta_{bed}(t) - \theta_c$. Each data point represents the short-time average over subintervals of $5T_b$ length, with time increasing along each line from left to right, as indicated by the arrows. The dashed diagonal line represents the Wong and Parker (2006) reference flux $q_{ref}/(u_g D)$. Note that in (*a-c*), $\langle q_{p,x}\rangle_{xz}$ is averaged over the central part of the duct only, that is, the spanwise averaging is performed $\forall z \in [0.5H_f, L_z - 0.5H_f]$. (*d*) Mean particle flux density $\langle \phi u_p\rangle_{xzt}/(q_{ref}/D)$ as a function of the wall-normal distance to the mean fluid-bed interface. The inset shows the same quantity in the near-bed region, with the wall-distance scaled in terms of the particle diameter D. Cases: $DL240^{H2}$ (—), $DL250^{H2}$ (—), $DL500^{H2}$ (—), $DL250^{H16}$ (—).

Additional empirical sidewall correction steps as they are often included in experiments (cf. Wong and Parker, 2006, and references therein) were also considered, without succeeding in a significantly better prediction of the particle flux. Excluding, however, a region of width $0.5H_f$ next to both sidewalls in the analysis results in a good approximation of the particle flux after an initial transient of roughly 10 bulk time units. A slight but systematic under-prediction is seen only for case $DL500^{H2}$.

The good match of the predicted and the observed particle flux in narrow and wide duct configurations likewise is also seen in figure 7.5(*c*), where $\langle q_{p,x}\rangle_{xz}(t)$ is shown as a function of the current excess Shields number, averaged over short time intervals of 5 bulk time units lengths. For all cases, the Shields number and the particle flux increase simultaneously due to the increasing bed friction after the release of the sediment particles at $t = 0$, soon reaching a mutual ratio that agrees well with the Wong and Parker (2006) relation.

As for the open channel simulations, higher Shields numbers cause an increase of the bedload layer thickness, as can be seen from the wall-normal profiles of the particle flux density $\langle \phi u_p\rangle_{xzt}$ shown in figure 7.5(*d*): In case $DL240^{H2}$, sediment transport occurs almost exclusively in a relatively thin layer

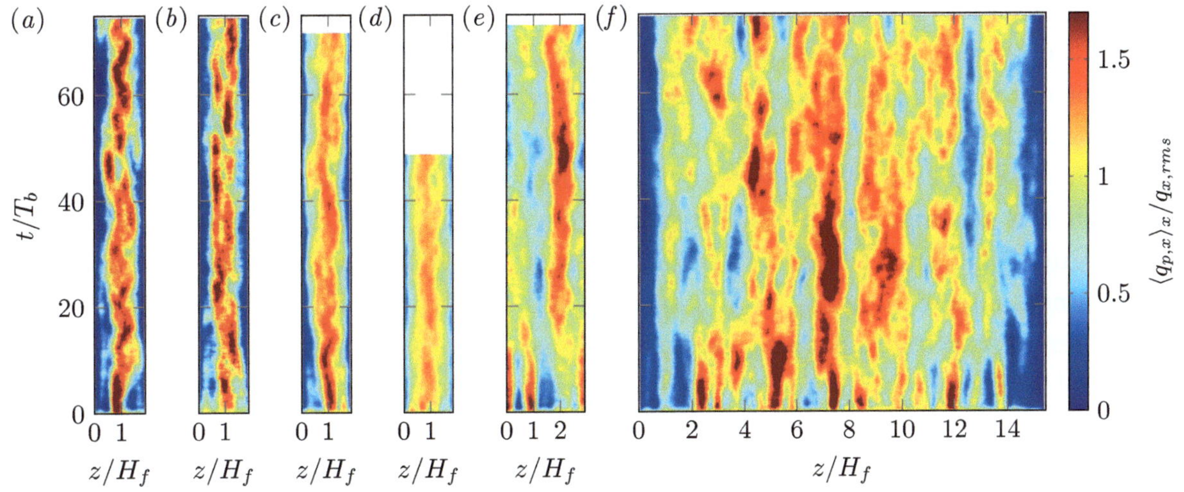

Figure 7.6: Space-time plot of the streamwise-averaged streamwise particle flux $\langle q_{p,x}\rangle_x(z,t)/q_{x,rms}$ in open duct simulations, with $q_{x,rms}(t) = (\langle\langle q_{p,x}\rangle_x^2\rangle_z)^{1/2}$. Red and blue regions refer to zones of strong and weak streamwise particle transport, respectively. Cases: (*a*) $DS250^{H2}$, (*b*) $DL240^{H2}$, (*c*) $DL250^{H2}$, (*d*) $DL500^{H2}$, (*e*) $CL250^{H3}$, (*f*) $DL250^{H16}$.

of approximately $2D$ thickness above the sediment bed. For case $DL500^{H2}$ that features the highest studied Shields number, on the other hand, a non-negligible fraction of the transported sediment mass reaches distances of up to $4D$ above the mean sediment. The thickening of the bedload transport layer with increasing Shields number is a direct consequence of the fact that the sediment saltation height increases with the Shields number, such that individual particles can reach into layers further away from the sediment bed (Niño et al., 1994).

From the very beginning of all simulations, the streamwise particle flux $\langle q_{p,x}\rangle_x$ is seen to vary considerably in the spanwise direction of the duct on account of the heterogeneous distribution of the streamwise velocity, which shall be examined in the following in more detail. For the three narrow duct simulations at lower Reynolds number, the time evolution of the streamwise-averaged lateral particle flux profile shown in figure 7.6(*a-c*) indicates that the vast majority of the sediment grains is transported in a comparably narrow region in the centre of the domain. On the other hand, $\langle q_{p,x}\rangle_x$ decreases strongly when approaching the sidewall region, leading to an almost negligible particle load in the flow near the sidewalls. Interestingly, the narrow region of most intense particle transport is seen to laterally meander during the simulations, approximately in phase with the lateral oscillations of the sediment ridges and troughs in the corresponding panels in figure 7.3(*a-c*). It can be suspected that this intermittent meandering tendency of the sediment ridges and troughs on the one hand and the regions of intense particle transport on the other hand originates in similar lateral oscillations of the turbulent flow field, as proposed earlier. The validity of this assumption will be investigated in the context of figure 7.25 below. In case $DL240^{H2}$ (cf. figure 7.6(*b*)), the lateral profile of $\langle q_{p,x}\rangle_x$ partly features two distinct local maxima for times $t/T_b > 30$, which could explain why in this case three ridges and two troughs form after the initial phase, instead of the ridge-trough-ridge arrangement that is observed in the remaining simulations with $AR \approx 1$.

The lateral distribution of the streamwise particle transport in the higher Reynolds number case $DL500^{H2}$ (cf. panel 7.6(*d*)) qualitatively differs from the previously discussed cases in that $\langle q_{p,x}\rangle_x$ is more homogeneously distributed across the duct span, attaining non-negligible values even close to both sidewalls. Nonetheless, the maximum particle transport still occurs in the duct core around the bisector, revealing a comparable lateral meandering in time as in the remaining cases. In the channel reference case $CL250^{H3}$ (cf. panel 7.6(*e*)), initially two regions of intense and two of weak particle transport form after the release of the particles, correlating with the rise of two initial small-amplitude ridges and troughs at corresponding positions in figure 7.3(*e*). In close correlation with the fluid-bed interface evolution depicted therein, the initial regions of strong and weak particle transport disappear after a while. They are displaced by a new, more pronounced region of intense particle flux evolving at $z \approx 2H_f$, which is accompanied by the formation of a deeper trough at essentially the same lateral position in figure 7.3(*e*), surrounded by two pronounced ridges. In contrast to the situation in the laterally constrained narrow ducts, $\langle q_{p,x} \rangle_x$ organises in sufficiently wide ducts such as case $DL250^{H16}$ in large-scale laterally varying regions of relatively stronger and weaker particle transport with a mean width of $(1\text{-}2)H_f$, again quite similar to the corresponding channel case $CL250^{H16}$. In this context, the distance to the sidewalls at which the first local maxima of $\langle q_{p,x} \rangle_x$ arise correspond to the first troughs next to the sidewalls. In regions less than H_f away from the lateral domain boundaries, however, the particle transport significantly reduces due to the lower streamwise velocity in these regions (cf. figure 7.23 below).

The different organisation of the particle transport in the cross-section of the low aspect ratio duct cases depending on the Shields and Reynolds number is clearly visible in figure 7.7. Therein, the lateral variation of the mean streamwise and transverse particle flux is shown together with the mean sediment bed profile for cases $DL240^{H2}$, $DL250^{H2}$ and $DL500^{H2}$, respectively. In agreement with the time-resolved visualisations in figure 7.6, it is seen that the profile of $\langle q_{p,x} \rangle_{xt}$ in case $DL500^{H2}$ (cf. figure 7.7(*e*)) is wider than in the remaining cases and even in the bins next to the sidewalls, it still attains values of approximately 30% of the overall mean flux $\langle q_{p,x} \rangle_{xzt}$. In cases $DL240^{H2}$ and $DL250^{H2}$ (cf. figure 7.7(*a*,*c*)), on the other hand, the near-sidewall regions are characterised by an almost vanishing streamwise particle transport. The mean transverse particle flux shown in figures 7.7(*b*,*d*,*f*) is in all cases oriented in such a way that sediment is, on average, transported from the duct centre towards the sidewalls, that way supporting the growth of the two ridges that form in all cases between sidewalls and duct bisector. In case $DL240^{H2}$, $\langle q_{p,x} \rangle_{xzt}$ attains a local minimum in the duct centre and $\langle q_{p,z} \rangle_{xzt}$ is almost zero at the same location (cf. figures 7.7(*a*,*b*)), which indicates that less sediment is eroded in this region combined with a vanishing lateral transport that could move eroded particles out of the duct centre. The consequence is the evolution of the earlier discussed 'extra ridge' in the centre of this case.

Close to the sidewalls, all cases reveal a net lateral particle flux of a few percent of the mean streamwise particle transport, directed from the near-wall side of the respective ridge towards the bed-sidewall corner. For cases with a relatively high Shields number and thus a pronounced streamwise particle transport, this sediment input into the corners suffices to readily fill the small initial gap between the bed and the sidewalls, while it does not for the relatively weak streamwise sediment transport in case $DL240^{H2}$. As stated earlier, this observed net particle transport into the near-wall region eventually causes, for sufficiently long time intervals, larger particle agglomerations in these regions, as those seen in the long-time experiments of McLelland et al. (1999).

In accordance with the observation of a mean sediment flux from the duct centre towards the left and right half of the domain, figure 7.8(*a*,*c*,*e*) shows in terms of the time-averaged erosion-deposition balance that the adjacent sediment troughs and ridges indeed correspond to regions of preferential particle erosion and deposition, respectively. Note that we restrict the analysis here on the time interval $t \in [5T_b, T_{obs}]$ as in the initial phase of each simulation, erosion naturally is the dominant process

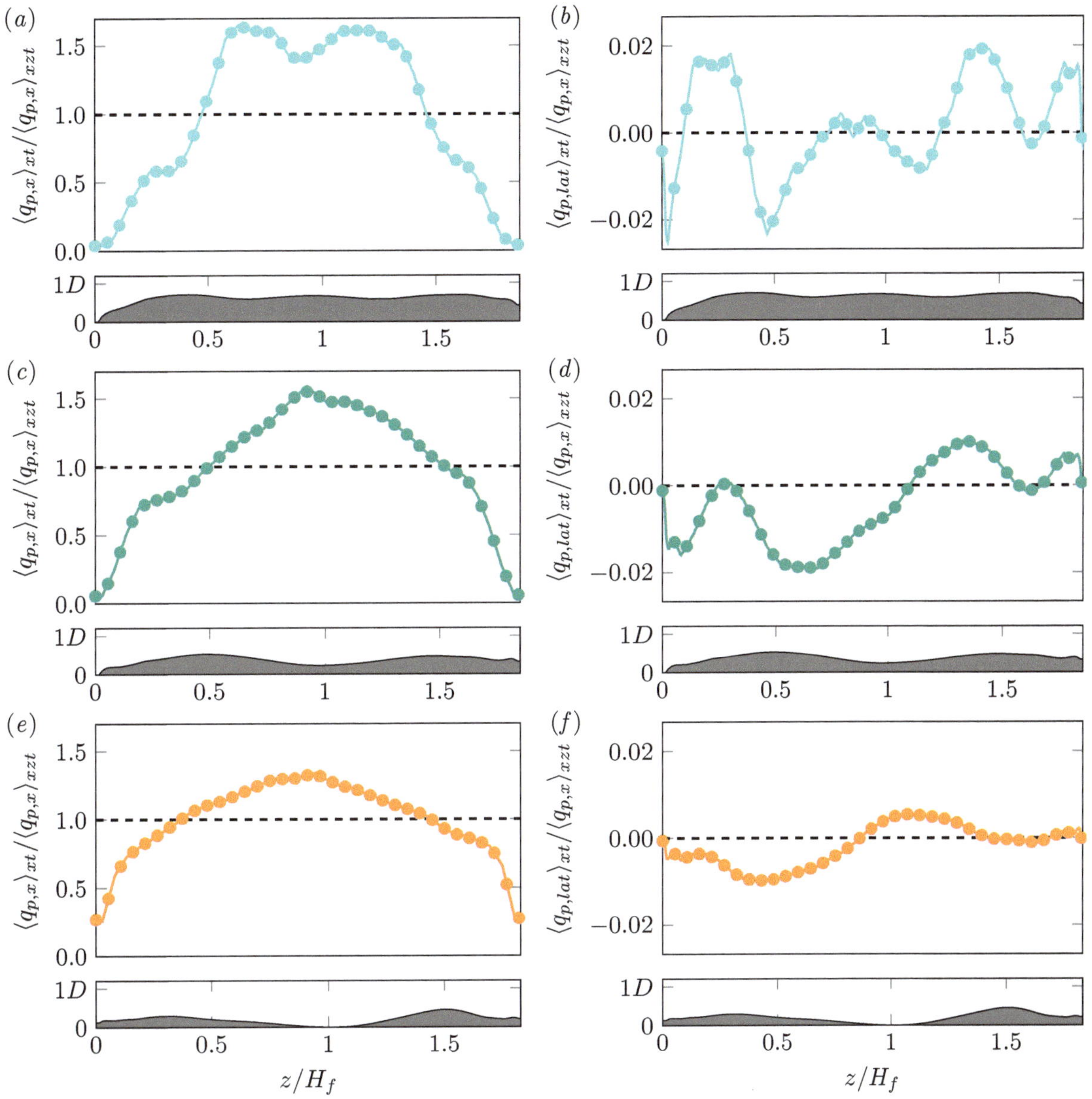

Figure 7.7: Spanwise variation of the mean streamwise and lateral particle flux. (*a,c,e*) Streamwise particle flux $\langle q_{p,x}\rangle_{xt}/\langle q_{p,x}\rangle_{xzt}$, (*b,d,f*) lateral particle flux tangential to the mean fluid-bed interface, normalised by the streamwise component $\langle q_{p,lat}\rangle_{xt}/\langle q_{p,x}\rangle_{xzt}$. Positive (negative) values of $\langle q_{p,lat}\rangle_{xt}$ indicate particle transport in positive (negative) z-direction. Note the two different ordinate scales for the streamwise and spanwise particle flux. The mean fluid-bed interface is visualised below each panel, the wall-normal extension of the ridge being exaggerated for the sake of visualisation. (*a,b*) $DL240^{H2}$, (*c,d*) $DL250^{H2}$, (*e,f*) $DL500^{H2}$.

everywhere in the cross-section, even though its amplitude strongly varies along the lateral direction. Apart from the regions where sediment ridges evolve, zones of preferential sediment deposition are also found in the regions near both sidewalls for cases $DL250^{H2}$ and $DL500^{H2}$, in agreement with the net lateral particle flux into the corner region that has been discussed above. For case $DL240^{H2}$, on the other hand, erosion and deposition are in the corner regions essentially in equilibrium, explaining why the initial gap between sediment bed and sidewall is not filled up in the course of this simulation. Also, the presented dataset underlines that case $DL240^{H2}$ features an additional zone of preferential deposition in the duct centre that causes the evolution of the 'additional ridge'.

The causal connection between the here discussed regions of preferential sediment erosion and deposition and the turbulent flow field will be the subject of the following sections. In order to give

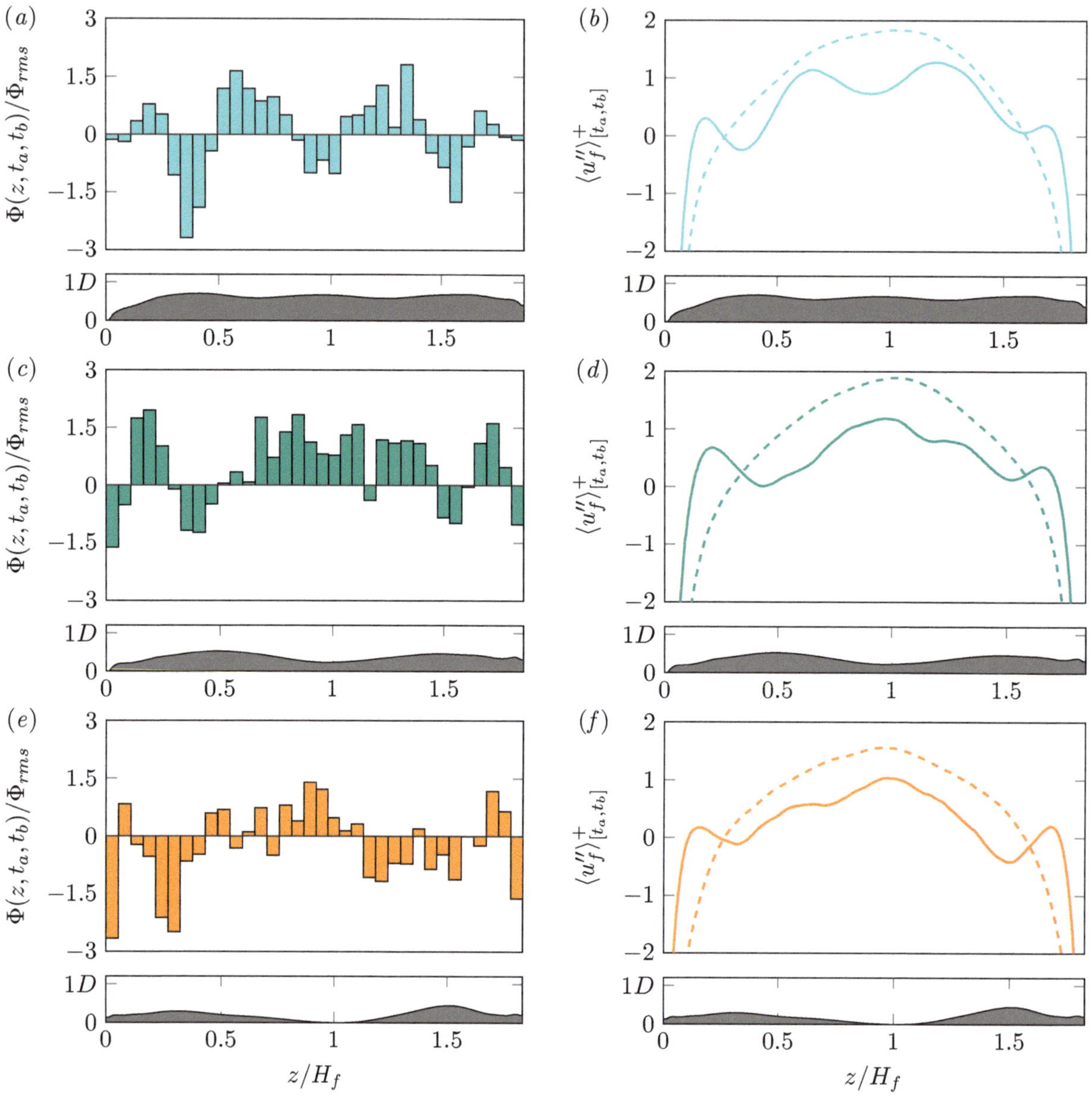

Figure 7.8: Mean erosion-deposition balance in comparison with the velocity field in narrow open duct simulations, averaged over the time interval $[t_a, t_b] = [5T_b, T_{obs}]$ of cases *(a,b)* $DL240^{H2}$, *(c,d)* $DL250^{H2}$ and *(e,f)* $DL500^{H2}$. *(a,c,e)* Balance between eroded and deposited particles per unit length and time, Φ, as defined in section 6.3.6, normalised by the corresponding r.m.s. value $\Phi_{rms} = [\langle(\Phi - \langle\Phi\rangle_z)^2\rangle_z]^{1/2}$. *(b,d,f)* Streamwise-averaged velocity fluctuations u_f'' at $\tilde{y}/H_f = 0.2$ (solid) and $\tilde{y}/H_f = 0.5$ (dashed), averaged over the same time interval $[t_a, t_b]$.

the reader already here a first impression on the mutual organisation of the mean streamwise velocity and the sediment ridges, figures 7.8*(b,d,f)* show profiles of the streamwise-averaged mean flow at wall-normal distances $\tilde{y}/H_f = 0.2$ and $\tilde{y}/H_f = 0.5$ above the fluid-bed interface. The spanwise velocity profile extracted at $\tilde{y}/H_f = 0.5$ increases for all cases monotonically towards its global maximum which is attained at approximately $z \approx L_z/2$. In contrast to that, the profile at $\tilde{y}/H_f = 0.2$ reveals a more complex shape, featuring local minima and maxima at different spanwise locations which appear to correlate quite well with regions of stronger deposition and erosion, respectively. This close correlation suggests that the flow organisation in the latter region might be of higher relevance for the formation of sediment ridges in these narrow duct cases.

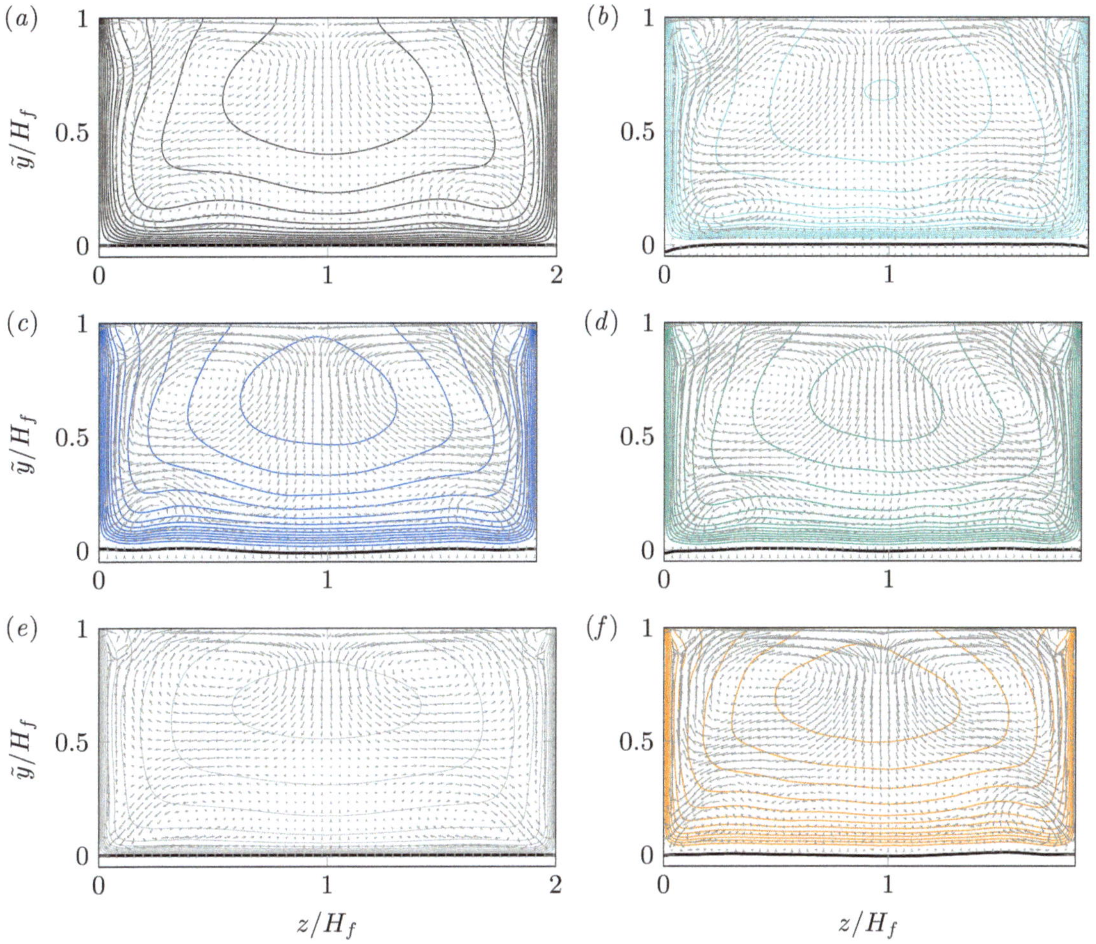

Figure 7.9: Mean streamwise velocity $\langle u_f \rangle_{xt}/u_b$ in open duct simulations with aspect ratio $AR \approx 1$. Intensity and orientation of the mean cross-stream secondary flow field $(\langle v_f \rangle_{xt}, \langle w_f \rangle_{xt})^T/u_b$ are indicated as vector plots in the background. Isocontours of $\langle u_f \rangle_{xt}/u_b$ are shown for values $0.1(0.1)1.3$ as coloured lines, whereas the time-averaged fluid-bed interface profile is indicated by a black solid curve. Cases: (*a*) $DL200^{H2}_{smooth}$, (*b*) $DL220^{H2}_{fix}$, (*c*) $DS250^{H2}$, (*d*) $DL250^{H2}$, (*e*) $DL400^{H2}_{smooth}$, (*f*) $DL500^{H2}$.

7.2.3 Turbulent mean flow

Before discussing in more detail the correlation of the flow field organisation and the evolution of sediment ridges, we first turn our attention to the cross-sectional variation of the primary mean flow $\langle u_f \rangle_{xt}(y,z)$ in the smooth-wall and multiphase duct cases, respectively, which is shown for selected cases in figure 7.9. Keeping in mind that the time interval over which statistics could be gathered differs by two orders of magnitude between single-phase and multiphase cases, the mean flow fields over smooth walls (figures 7.9(*a,e*)), stationary (figure 7.9(*b*)) and mobile sediment beds featuring developed sediment ridges (figures 7.9(*c,d,f*)) are very similar in amplitude and structure. This is in particular remarkable for the streamwise-minimal simulation $DS250^{H2}$ as it indicates that, provided the averaging time interval is sufficiently long, the characteristic mean streamwise and secondary flow field is recovered – even if the flow field accommodates only modes of either short ($\lambda_x \leq 2H_f$) or infinite streamwise wavelength. We observe that the characteristic features of the mean flow are fairly well developed for both single-phase and particle-laden cases, including the inner vortex in the mixed corner between sidewalls and free surface (Grega et al., 2002) and the velocity dip phenomenon (Stearns, 1883; Nezu and Nakagawa, 1993), which will be discussed in the context of figure 7.11 below. The local maxima of the mean fluid-bed interface that represent the preferential location of the sediment ridges in cases $DS250^{H2}$, $DL250^{H2}$ and $DL500^{H2}$ (cf. figures 7.9(*c,d,f*)) more or less collapse with the local low-speed region of the mean flow field, induced by the mean bottom vortex that transports low momentum fluid away from the wall. The bottom vortex and the associated modulation of the mean streamwise velocity field $\langle u_f \rangle_{xt}$ are a characteristic feature of the near-wall region in open duct flows (Tominaga et al., 1989) and can be detected in the corresponding single-phase duct flows as well (cf. figures 7.9(*a,e*)), albeit its intensity and size depend on the Reynolds number as well as the bed configuration. The peculiar structure of the mean secondary flow of Prandtl's second kind as well as the influence of a mobile sediment bed on the former deserves a more detailed investigation, which will be provided in the following section 7.2.4. Lastly, let us mention that in order to obtain fully-converged statistics that also fulfil the expected statistical mirror symmetry w.r.t. the duct bisector, significantly longer simulation time intervals with a length of at least $\mathcal{O}(10^3 T_b)$ would be necessary in the particle-laden simulations, which clearly exceeds the time window during which sediment ridges can be studied exclusively in the current simulations.

The lateral inhomogeneity of the mean flow field in open and closed ducts manifests itself also in a variation of the mean wall shear stress (cf. Pinelli et al., 2010; Sakai, 2016, and the discussion of figure 7.14 below). Part of this lateral inhomogeneity can be compensated using local inner units for the scaling of the wall-normal velocity profile in order to recover the law of the wall also locally (Gavrilakis, 1992; Pirozzoli et al., 2018). Here, we consequently normalise the velocity profile with the local friction velocity obtained as $\sqrt{\langle \tau_{tot} \rangle_{xt}(\langle h_b \rangle_{xt}(z), z)/\rho_f}$, where the mean total shear stress $\langle \tau_{tot} \rangle_{xt}(y, z)$ is evaluated along the sediment bed $y = \langle h_b \rangle_{xt}(z)$. Note that $\langle \tau_{tot} \rangle_{xt}$ will be properly defined in equation (7.11) below, where we will also discuss its lateral variation in the context of figure 7.14. Quantities normalised in local inner units will be henceforth marked as $(\bullet)^{+_l}$, representing the local equivalent to inner scaling in global inner units by u_τ and ν_f.

Figure 7.10 shows wall-normal profiles of the mean streamwise velocity, scaled in local inner units at various spanwise positions throughout the duct cross-section up the corner bisectors $\tilde{y} = z$ and $\tilde{y} = L_z - z$, respectively. This means in particular that only part of the domain is considered for which the bottom wall represents the closest of the three solid sidewalls. For the smooth wall cases, the presented profiles collapse throughout the buffer layer and only small deviations are observed in the outer layer when normalised in terms of local inner units. Along the duct bisector, a quasi perfect match with the theoretically predicted velocity profile is obtained, deviating from the idealised profile only in the vicinity of the free surface in the context of the velocity dip phenomenon.

Possibly due to the relatively small averaging ensemble in the particle-laden cases, normalisation in local inner coordinates does not yield to a complete collapse of the profiles in these cases. However, the general shape and slope of the individual curves are again comparable. As for the open channel simulations in figure 6.10, the velocity profiles over the evolving sediment bed deviate from those in the smooth-wall simulations. In case $DL240^{H2}$, the deviation is weakest as a consequence of the relatively low particle transport intensity, which in turn leads only to a mild increase in bed friction. Case $DL500^{H2}$, on the other hand, features a clearly higher Shields number and thus a markedly higher bedload transport intensity as well as a larger particle Reynolds number of $D^+ \approx 18$. As a consequence, the near-bed flow is strongly disturbed by the particles moving inside the dense bedload layer which causes a significant enhancement of the bed friction, resulting in a much stronger deviation between the velocity profiles in case $DL500^{H2}$ and those in the smooth-wall simulations. In agreement with the situation in the open channel flow simulations, a simple wall-normal shift of the velocity profiles'

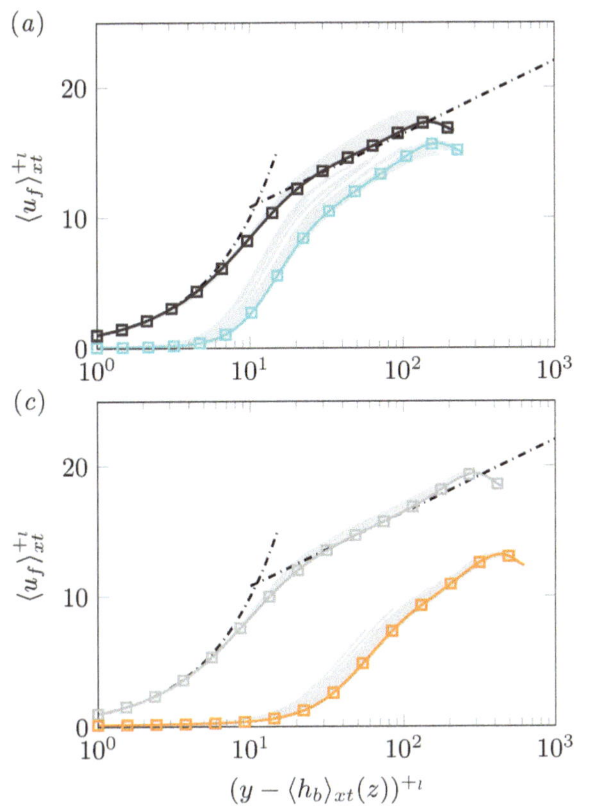

Figure 7.10: Wall-normal profiles of the mean streamwise velocity $\langle u_f \rangle_{xt}^{+\iota}(y,z)$ at different lateral positions scaled in local inner units as a function of the local wall/bed distance $(y - \langle h_b \rangle_{xt}(z))^{+\iota}$, based on the local mean wall shear stress at each lateral position. Grey lines represent individual wall-normal profiles at various spanwise positions z/H_f, while only the profile along the duct bisector at $z = L_z/2$ is highlighted by a coloured line. The dashed-dotted line shows the classically predicted velocity profiles for the buffer and log-layer, respectively, as $\langle u_f \rangle_{xzt}^{+\iota} = \tilde{y}^+$ and $\langle u_f \rangle_{xzt}^{+\iota} = 1/\kappa \ln(\tilde{y}^+) + B_u$ ($\kappa = 0.41$, $B_u = 5.2$). Profiles are shown only below the corner bisectors $\tilde{y} = z$ and $\tilde{y} = L_z - z$, respectively. Cases: *(a)* $DL240^{H2}$ (—), *(b)* $DL250^{H2}$ (—), *(c)* $DL500^{H2}$ (—), *(d)* $DL250^{H16}$ (—). Each panel additionally features the same quantity in the respective smooth-wall reference case $DL200^{H2}_{smooth}$ (—), $DL400^{H2}_{smooth}$ (—) or $DL150^{H16}_{smooth}$ (—), respectively.

origins (Jiménez, 2004a) does not lead to a satisfactory collapse of the velocity profiles over mobile sediment beds and smooth bottom walls, since the slopes of the profiles outside the buffer layer generally differ between the two configurations. As in the open channel case, it is argued that the larger slope of the local velocity profile in the outer flow of particle-laden simulations is a necessary consequence of the reduced fluid mass flow rate in the near-bed region, which the flow compensates by an increased mean shear in the outer layer. Eventually, a pronounced velocity dip at the duct bisector is observed for all duct cases at low aspect ratio $AR \approx 1$, whereas the profiles in the wide duct cases $DL150^{H16}_{smooth}$ and $DL250^{H16}$ lack such a feature, which is in agreement with earlier studies on this topic (Nezu and Rodi, 1985; Sakai, 2016). However, considering also the profiles away from the duct bisector, it is seen that there are indeed other spanwise locations at which the maximum of the local velocity profile is attained below the free surface in both cases, indicating the existence of a set of 'local', 'off-centre' velocity dips at varying spanwise locations.

For the low aspect ratio cases, figure 7.11 shows the velocity dip d in outer and inner units as a function of the bulk and friction Reynolds number, respectively. Here, d represents the distance between the free-slip plane and the wall-normal location at which $\langle u_f \rangle_{xt}$ attains its maximum along the bisector, viz. $d = L_y - \underset{y}{\operatorname{argmax}}(\langle u_f \rangle_{xt}(y, L_z/2))$. For both outer and inner scaling, the evolution of the velocity dip

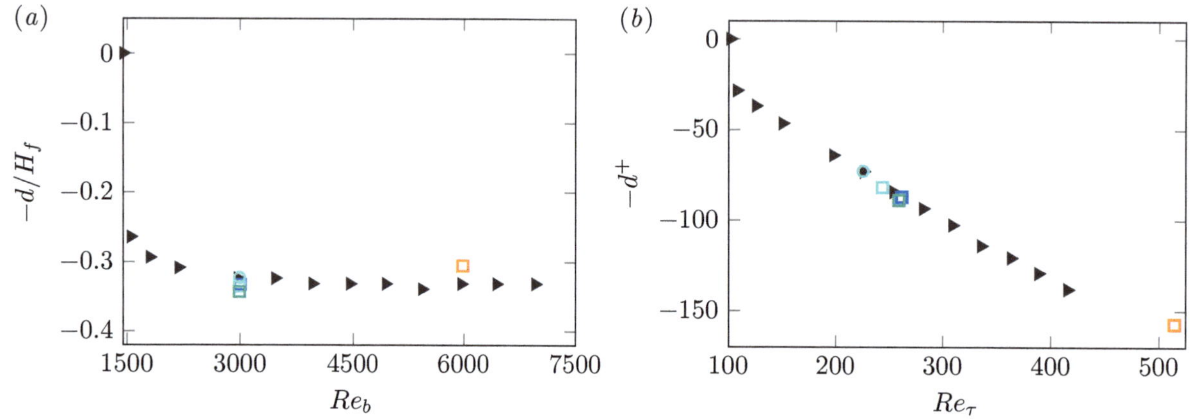

Figure 7.11: Velocity dip $-d$ of the mean streamwise velocity component of open duct cases with $AR \approx 1$ scaled in (*a*) outer units as a function of Re_b and (*b*) inner units as a function of Re_τ. Symbols represent data from current open duct simulations over fixed (○) and mobile sediment beds (□) as well as smooth wall reference data from Sakai (2016) (▶). The velocity dip d is determined as the wall-normal distance between the location at which the mean streamwise velocity attains its maximum along the duct bisector $z = L_z/2$ and the free surface, i.e. $d = L_y - \underset{y}{\operatorname{argmax}}(\langle u_f \rangle_{xt}(y, L_z/2))$.

as a function of the bulk and friction Reynolds number follows the general trend revealed by the series of smooth-wall duct simulations investigated by Sakai (2016), exhibiting only small deviations from the values found therein: For non-marginal Reynolds numbers $Re_b > 3000$, the velocity dip essentially scales in outer units, attaining a more or less constant value of $d/H_f \approx 0.3$. When scaled in inner units, on the other hand, d^+ seems to increase almost linearly with Re_τ. The slight deviations between d in the particle-laden cases and that in the corresponding smooth-wall simulations are small enough to represent the statistical error that arises as a consequence of the limited averaging time interval in the sediment bed cases. The results therefore generally imply that the mobile sediment bed and the evolving sediment ridges have no particular influence on the size of the velocity dip phenomenon.

The increase in bottom friction caused by the presence of the mobile sediment bed manifests itself in a higher friction Reynolds number in the particle-laden cases. For the case of a smooth-wall closed duct, Jones (1976) proposed an empirical relation to determine the friction coefficient $c_f = 2u_\tau^2/u_b^2$, which has been proven to reasonably well predict the effectively attained values in direct numerical simulations of marginal to moderate Reynolds number closed and open duct flows (Uhlmann et al., 2007; Sekimoto, 2011; Sakai, 2016). In Sakai (2016), the original equation presented in Jones (1976) was generalised to the rectangular open duct case, viz.

$$\frac{1}{\sqrt{4c_f}} = 2\log_{10}\left(\phi^* \frac{4AR}{(1+AR)} Re_b \sqrt{4c_f}\right) - 0.8, \tag{7.9}$$

where a shape factor $\phi^*(AR)$ is to be determined, for which Jones (1976) proposed the approximation

$$\phi^* \approx \frac{2}{3} + \frac{11}{24}\frac{H_f}{L_z}\left(2 - \frac{H_f}{L_z}\right) = \frac{2}{3} + \frac{11}{24}\frac{1}{2AR}\left(2 - \frac{1}{2AR}\right). \tag{7.10}$$

As can be seen in figure 7.12(*a*), the relation between the friction Reynolds number Re_τ and the bulk Reynolds number Re_b in the single-phase smooth-wall simulations is fairly well captured by relation (7.9) for narrow and wide duct simulations alike, as has been concluded earlier by Sakai (2016). In the presence of a stationary sediment bed in case $DL220^{H2}_{fix}$ and $DL210^{H16}_{fix}$, Re_τ attains only slightly

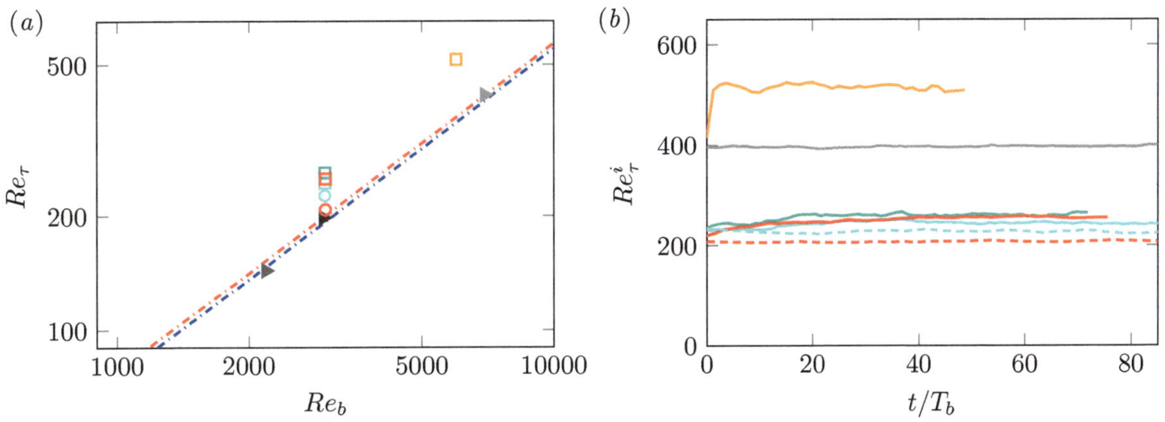

Figure 7.12: (*a*) Variation of the mean friction Reynolds number Re_τ as a function of the bulk Reynolds number Re_b. The dashed-dotted lines refer to the empirical relation (7.9) (Jones, 1976; Sakai, 2016), evaluated for aspect ratios $AR = 1$ (-··-) and $AR = 16$ (-··-), respectively. (*b*) Time evolution of the instantaneous friction Reynolds number $Re_\tau^i(t)$. Cases: $DL240^{H2}$ (—,□), $DL250^{H2}$ (—,□), $DL220^{H2}_{fix}$ (---,○), $DL500^{H2}$ (—,□), $DL250^{H16}$ (—,□), $DL210^{H16}_{fix}$ (---,○), $DL200^{H2}_{smooth}$ (▶), $DL400^{H2}_{smooth}$ (—,▶) and $DL150^{H16}_{smooth}$ (▶).

higher values due to the relatively small roughness height in both inner ($D^+ < 8$) and outer scaling ($D/H_f < 0.04$).

For the cases featuring a mobile sediment bed, on the other hand, we observe a more pronounced increase in bed friction compared to the single-phase smooth-wall cases, whose size depends on the respective Shields number θ_{bed} and the relative size of the particles D^+: At the lowest Shields number in case $DL240^{H2}$, the weak particle transport causes the friction Reynolds number to increase mildly to a value of $Re_\tau \approx 240$, which is somewhat smaller than the values $Re_\tau \approx 250$ attained in cases $DL250^{H2}$ and $DL250^{H16}$ under otherwise similar conditions (i.e. values of D^+ and D/H_f, respectively). Case $DL500^{H2}$, in turn, features with $\theta_{bed} = 0.22$ and $D^+ \approx 18$ the highest Shields and particle Reynolds number of all considered simulations, respectively. The consequence is that Re_τ and c_f increase by roughly 35% and about 80%, respectively, compared to both the values predicted by the relation of Jones (1976) and the corresponding smooth-wall reference DNS at essentially matching value of Re_b of Sakai (2016).

The impact of sediment mobility on the overall friction coefficient becomes clear in the time evolution of the instantaneous friction Reynolds number $Re_\tau^i(t)$ provided in figure 7.12(*b*), which is recomputed from the instantaneous driving pressure gradient $\Pi(t)$. In both single-phase and stationary sediment bed simulations, Re_τ^i is essentially stationary exhibiting merely small-amplitude oscillations around a long-time mean. In the mobile sediment bed cases, on the other hand, Re_τ^i increases starting from the release of the particles at $t = 0$ at more or less the same rate as the particle flux discussed in figure 7.5. In the low Reynolds number mobile sediment cases, the growth of Re_τ^i is rather gradually. On the contrary, in the higher Reynolds number case $DL500^{H2}$ its value rapidly increases within a few bulk time units to a quasi-stationary plateau of $Re_\tau^i \approx 510$, in line with the strong initial increase of the particle flux $\langle q_{p,x}\rangle_{xz}$ in the same time interval. The fact that Re_τ^i increases predominantly in the first five to ten bulk time units of simulation $DL500^{H2}$ underlines that the increase in bed friction is mainly due to the mobilised sediment grains, since the evolving sediment ridges are not yet developed in this phase and can be thus considered to have little effect on the bed friction.

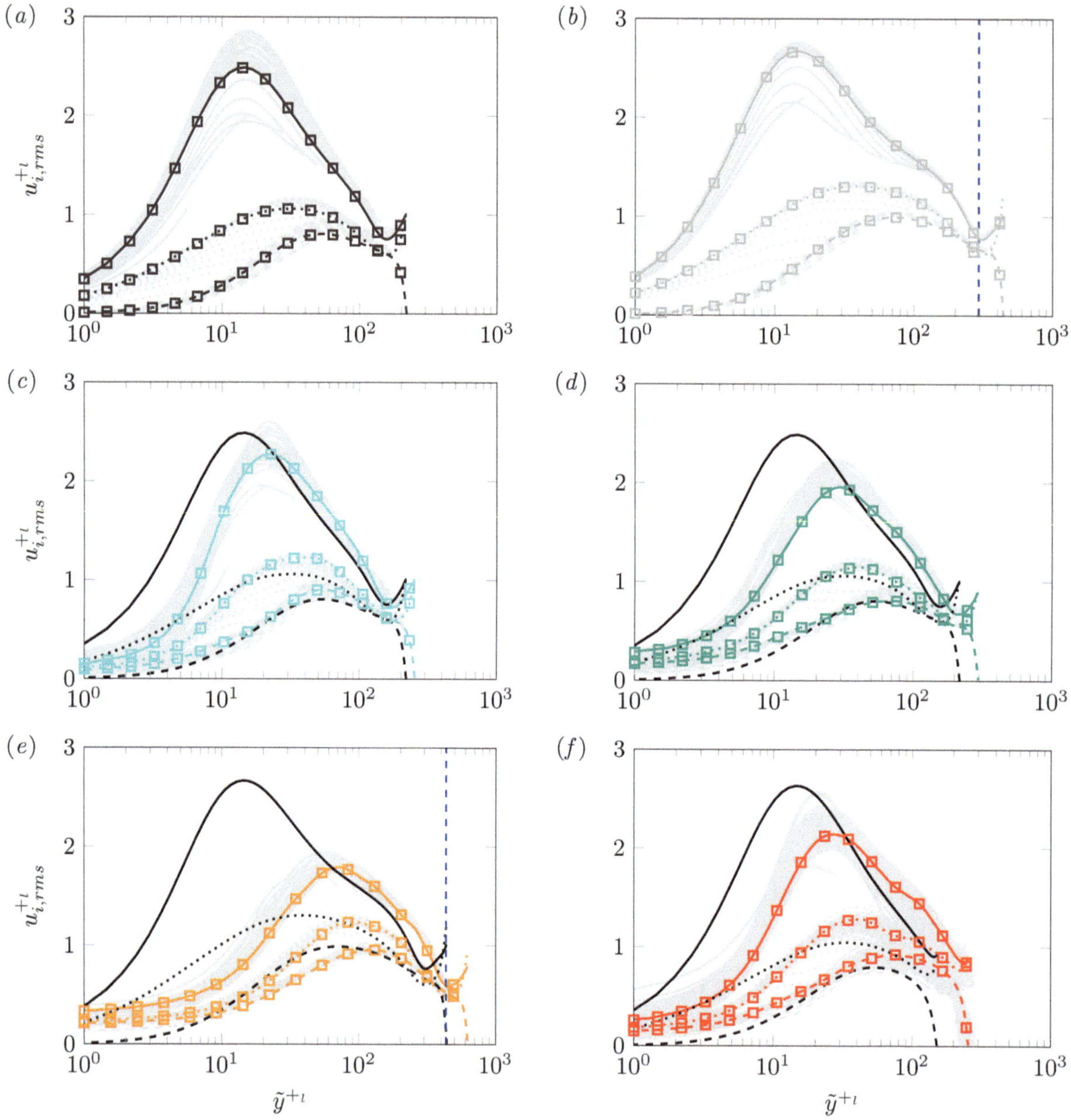

Figure 7.13: Wall-normal profiles of the mean normal Reynolds stresses $u_{fi,rms}{}^{+_l} = (\langle u'_{fi}u'_{fi}\rangle_{xt}^{1/2})^{+_l}$ ($i = 1,2,3$) at different lateral positions, scaled in local inner units: $u_{f,rms}$ (—), $v_{f,rms}$ (---) and $w_{f,rms}$ ($\cdots$). The coloured profile is that extracted along the duct bisector $z = L_z/2$, while the wall-normal profiles at the remaining spanwise positions are shown as light grey lines in behind. For the sake of clarity, profiles are shown in spanwise intervals of $0.05H_f$ ($0.15H_f$ for $DL250^{H16}$, $DL150^{H16}_{smooth}$) for each component only below the corner bisectors $\tilde{y} = z$ and $\tilde{y} = L_z - z$, respectively. In those panels that contain data from particle-laden simulations, the turbulence intensities of the corresponding smooth-wall reference simulation are repeated for the sake of comparison (solid/dashed/dotted black lines). Cases: (*a*) $DL200^{H2}_{smooth}$, (*b*) $DL400^{H2}_{smooth}$, (*c*) $DL240^{H2}$, (*d*) $DL250^{H2}$, (*e*) $DL500^{H2}$ and (*f*) $DL250^{H16}$. In panels (*b*,*e*), the vertical blue dashed lines additionally indicate the wall-normal position of the velocity dip normalised in local inner units, $(L_y - d)^{+_l}$.

In a similar way as the mean flow field, the Reynolds stress fields reveal a clear variation along the lateral direction. Wall-normal profiles of the streamwise, wall-normal and spanwise turbulence intensities are provided in figure 7.13, scaled in local inner units. The local inner scaling cannot compensate for the whole lateral heterogeneity of the turbulence intensities, but it still leads to a better match than a normalisation in global inner units (figures omitted). The presented profiles reveal a qualitatively similar shape as their counterparts in canonical open channel flows: In both narrow smooth-wall open duct cases (cf. figures 7.13(*a*,*b*)), $u_{f,rms}^{+l}$ exhibits the characteristic energetic peak at around $\tilde{y}^{+l} = 15$ that is related to the intense buffer layer streaks (Smits et al., 2011; Lee and Moser, 2015) and then decays towards the free surface. Of interest is the development of a local minimum of $u_{f,rms}^{+l}$ near the free surface that turns out to occur roughly at the same position as the maximum velocity along the duct bisector, i.e. it essentially collapses with the velocity dip (Nezu and Nakagawa, 1993). From the three components, the individual profiles of $w_{f,rms}^{+l}$ appear, as one might expect, to vary most strongly along the lateral direction. From those profiles that more or less collapse with the profile along the duct bisector, most reveal a maximum at approximately $\tilde{y}^{+l} \in [30, 50]$, again comparable to the values observed for a large range of Reynolds numbers in canonical wall-bounded flows (Lee and Moser, 2015). For $v_{f,rms}^{+l}$, eventually, the profiles reveal a relatively weak variation between the different spanwise locations. In contrast to the remaining components, the wall-normal location at which the profiles of $v_{f,rms}^{+l}$ attain their maxima seems to move outwards when the Reynolds number is increased, also this having been observed for canonical wall-bounded flows in a similar way (Lee and Moser, 2015).

In the particle-laden cases (cf. figures 7.13(*c*-*f*)), the general shape of the intensity profiles discussed in the smooth-wall reference cases is maintained, even though with some remarkable differences: Similar to the open channel cases, it is seen that with increasing intensity of the sediment transport, the peak of the streamwise component $u_{f,rms}^{+l}$ is more and more damped, while its position is shifted to regions further away from the bed. Nonetheless, the peak is even in the highest Reynolds number case $DL500^{H2}$ clearly detectable, indicating that the self-sustaining buffer layer regeneration cycle is not completely destroyed, as it was the case in the open channel simulation $CM850^{H3}$. Case $DL240^{H2}$, on the other hand, shows the weakest deviation from the smooth-wall reference case concerning both the general shape of the profile and the amplitude of the near-bed peak. This indicates that also the fluctuating velocity field is rather weakly affected by the relatively low sediment transport rate. For the wall-normal and transverse fluctuations $v_{f,rms}^{+l}$ and $w_{f,rms}^{+l}$, respectively, the profiles in the particle-laden cases similarly attain peak values at larger distances to the (virtual) wall compared to the smooth-wall reference data. There is, however, no clear trend concerning the amplitude of these maxima: While the peak values of $v_{f,rms}^{+l}$ and $w_{f,rms}^{+l}$ are in case $DL500^{H2}$ comparable to those in the corresponding smooth-wall case, the maxima in the low Reynolds number simulations $DL240^{H2}$ and $DL250^{H2}$ are somewhat higher than in the corresponding smooth-wall case $DL200_{smooth}^{H2}$. Finally, figure 7.13(*f*) shows the turbulence intensity profiles in the high aspect ratio cases $DL250^{H16}$ and $DL150_{smooth}^{H16}$, respectively, which feature essentially the same general characteristics as their low aspect ratio counterparts. In sufficient distance to the sidewalls, however, the statistics are expected to closely resemble those in an open channel, which is confirmed by the presented data.

In contrast to the spanwise statistically homogeneous open channel, the bottom wall shear stress in open duct flows varies considerably along the duct span. Figure 7.14 provides the lateral wall shear stress distribution along the bottom wall and the sediment bed, respectively, for low and high aspect ratio cases separately. While the total wall shear stress along the bottom wall of the smooth-wall duct cases is given by the viscous stress component $\langle \tau_{tot} \rangle_{xt}(y = 0, z) = \rho_f \nu_f \partial \langle u_f \rangle_{xt} / \partial y$ alone, the mean

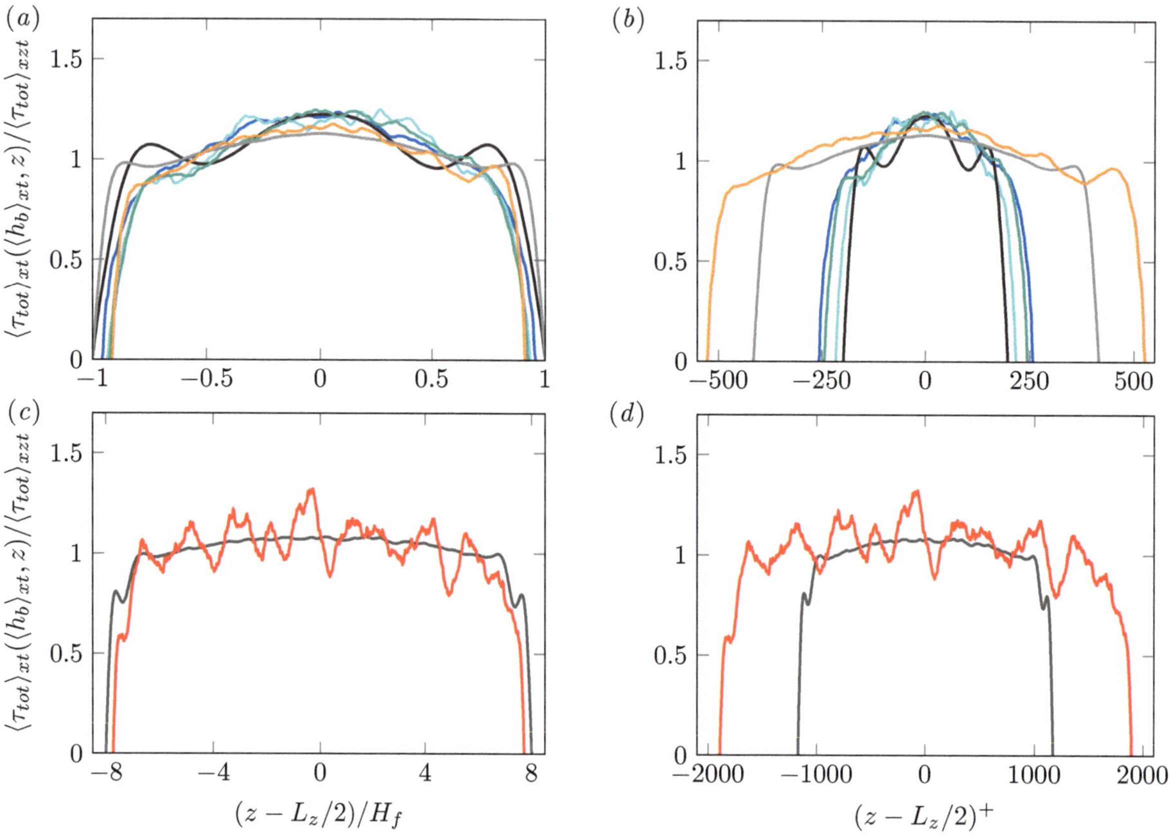

Figure 7.14: Lateral variation of the bottom wall/bed shear stress distribution $\langle\tau_{tot}\rangle_{xt}(\langle h_b\rangle_{xt}(z),z)/\langle\tau_{tot}\rangle_{xzt}$ as a function of the lateral distance to the duct bisector $(z-L_z/2)$ scaled in *(a,c)* outer and *(b,d)* inner length scales. *(a,b)* Low aspect ratio cases $AR\approx 1$: $DS250^{H2}$ (—), $DL240^{H2}$ (—), $DL250^{H2}$ (—), $DL500^{H2}$ (—), $DL200^{H2}_{smooth}$ (—) and $DL400^{H2}_{smooth}$ (—). *(c,d)* High aspect ratio cases $AR\approx 8$: $DL250^{H16}$ (—), $DL150^{H16}_{smooth}$ (—). All particle-laden datasets have been smoothed with a moving average filter using a filter width of $1.5D$.

shear stress along the sediment bed in the particle-laden cases comprises in general three contributions: viscous and turbulent terms as well as stresses arising from particle-fluid interaction. The general definition of the total shear stress then reads

$$\langle\tau_{tot}\rangle_{xt}(y,z)=\rho_f\nu_f\frac{\partial\langle u_f\rangle_{xt}}{\partial y}-\rho_f\langle u'_f v'_f\rangle_{xt}+\int_y^{L_y}\langle f_x\rangle_{xt}(s,z)\mathrm{d}s. \tag{7.11}$$

Equation (7.11) represents the equivalent of relation (6.39) in the open channel configuration. The main difference between both formulations is that in the open duct case, the second term on the RHS of equation (7.11) comprises all turbulence-related contributions, while in the open channel case two different terms (turbulent and dispersive stresses) appear in this context due to the different decomposition of the velocity field into mean and fluctuating parts in the latter case.

The smooth-wall low aspect ratio cases provided in figure 7.14*(a,b)* show a nearly symmetric wall shear stress profile, featuring a distinct local maximum and a neighbouring local minimum near both sidewalls, which are indicative of the preferential positions of the first high- and low-speed buffer layer streaks near the two sidewalls, respectively (Pinelli et al., 2010). The buffer layer streaks are more or less locked in a distance of roughly $50\delta_\nu$ (high-speed streaks) and $100\delta_\nu$-$125\delta_\nu$ (low-speed streaks) from the sidewalls due to the geometrical constraints in the solid-solid corner (Sakai, 2016). With increasing Reynolds number and thus increasing relative width L_z^+ of the duct, the duct can accommodate a larger number of individual buffer layer streaks in the cross-section and those in the centre are not subject to the geometrical restrictions felt by their counterparts near the sidewalls. The consequence is that in the duct centre, high- and low-speed streaks can appear at any position with the same probability and contributions of both types effectively cancel in the long-time average, such that the profile clearly flattens for increasing Reynolds numbers in regions away from the sidewalls (Pinelli et al., 2010; Sakai, 2016).

This peculiar structure of the mean wall shear stress is only partly visible in the particle-laden cases, even though the general profile shape and the maximum amplitude compared to the transverse mean are quite comparable. In case $DS250^{H2}$, for instance, the wall shear stress does not attain any local extrema near the sidewalls but shows an essentially monotonic increase from the sidewalls towards the duct centre, whereas case $DL500^{H2}$ develops at least in the vicinity of the sidewall at $z = L_z$ a local maximum and minimum. However, these local extrema are located at a distance of approximately $80\delta_\nu$ ($0.16H_f$) and $145\delta_\nu$ ($0.30H_f$) to the sidewall, respectively, and are thus somewhat further away from the sidewalls than what is seen and expected in the smooth-wall situation. In this regard, it appears noteworthy that the local minimum $145\delta_\nu$ ($0.30H_f$) away from the sidewall essentially collapses with the location of the highest sediment ridge in that case, explaining the low erosion activity in this region that favours the formation of a sediment ridge. The remaining cases $DL240^{H2}$ and $DL250^{H2}$ possess no pronounced local extrema near the sidewalls either, but in both cases a narrow plateau-like region can be identified close to the sidewall at $z = 0$.

One way to interpret the absence of clearly developed shear stress peaks close to the sidewalls of the particle-laden cases is that the regular organisation of the buffer layer streaks and vortices in these regions is somewhat disturbed by the active sediment transport into the corner regions, such that their signature in the mean shear stress profile is weakened or even lost. In section 7.3.3, it will indeed be shown that the individual quasi-streamwise vortices arrange differently in the vicinity of the mobile sediment beds in the corner regions compared to the smooth-wall case.

The corresponding large aspect ratio simulations are shown in figure 7.14(*c*,*d*). In this context, let us highlight that in terms of the buffer layer structures, an increase of the duct width L_z/H_f has a similar effect as an increase of the Reynolds number Re_τ, as both lead to a rise of L_z^+. Consequently, on average, a larger number of individual buffer layer streaks fits into the cross-section of the duct, from which those near the sidewalls are again more constrained in their lateral mobility than those in the duct core and thus leave a visible footprint in the wall shear stress profile (Sakai, 2016). The shear stress profile of case $DL150^{H16}_{smooth}$ hence shows distinct local maxima and minima only near both sidewalls, as opposed to the central part of the duct which reveals a rather homogeneous distribution of the wall shear stress. By contrast, case $DL250^{H16}$ is characterised by a sinusoidal variation of the wall shear stress throughout the entire cross-section, anti-correlated with the mean fluid-bed interface: regions of higher (lower) shear stress match regions of sediment troughs (ridges). While a comparably alternating shear stress is not observed in the long-time statistics of case $DL150^{H16}_{smooth}$, though, the instantaneous shear stress indeed exhibits similar lateral undulations (plots not shown). It is widely accepted that this lateral variation of the instantaneous (or short-time averaged) bed shear stress is the main reason for the evolution of sediment ridges and troughs in high aspect ratio ducts (Nezu, 2005), and the explanations in the previous chapter 6 have underlined that such a lateral variation of the bed shear stress caused by the presence of large-scale velocity streaks indeed leads to the formation of sediment ridges in the open channel cases.

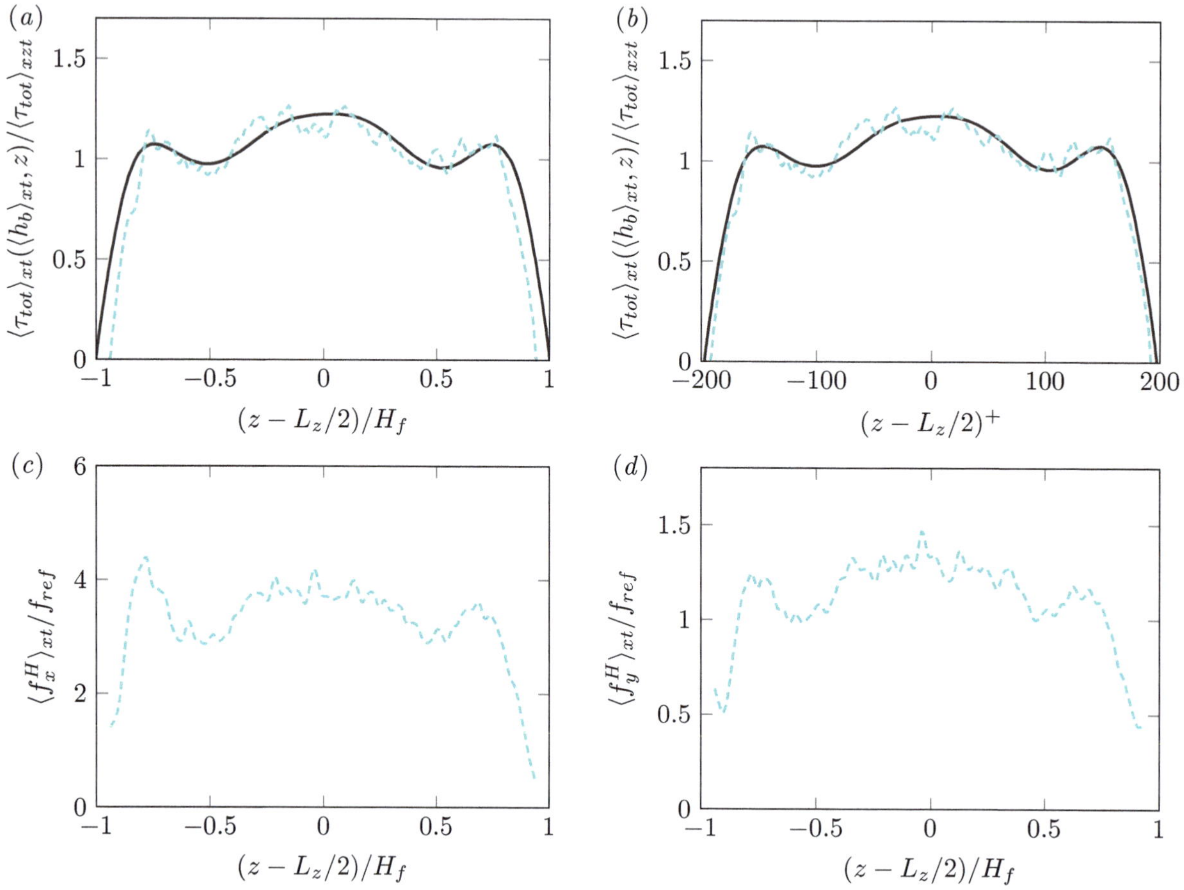

Figure 7.15: (*a,b*) Lateral profile of the mean bed/bottom shear stress $\langle\tau_{tot}\rangle_{xt}(\langle h_b\rangle_{xt}(z),z)/\langle\tau_{tot}\rangle_{xzt}$ as a function of (*a*) the outer- and (*b*) inner-scaled lateral distance to the duct bisector $(z-L_z/2)$ over the stationary flat sediment bed of case $DL220^{H2}_{fix}$ (- - -) and the smooth wall of case $DL200^{H2}_{smooth}$ (—). (*c,d*) Lateral profiles of the mean hydrodynamic force components acting at the top of the sediment bed: (*c*) streamwise force component f_x^H/f_{ref} and (*d*) wall-normal force component f_y^H/f_{ref}, with $f_{ref}=\rho_f u_\tau^2\pi(D/2)^2$. The forces have been extracted at a wall-normal offset of $0.5D$ above the mean fluid-bed interface H_b roughly at the top of the uppermost particle layer, while $\langle\tau_{tot}\rangle_{xt}$ was evaluated at the mean interface positions $\langle h_b\rangle_{xt}(z)$. All datasets have been smoothed with a moving average filter using a filter width of $1.5D$.

The role of the bed shear stress for the local erosion rate is best illustrated when comparing the lateral variation of the former with that of the hydrodynamic forces acting on the surface particle layer of a stationary sediment bed, shown in figure 7.15 exemplary for case $DL220^{H2}_{fix}$. Note that in contrast to the mobile sediment case, the submerged particle weight F_W is of no physical relevance in the case of a stationary sediment bed. In the remainder, we therefore consider as a reference scale a pseudo-force computed based on the mean wall shear stress and the projected particle area, i.e. $f_{ref}=\rho_f u_\tau^2\pi(D/2)^2$, that is effectively independent of the gravitational acceleration and the sediment density. Interestingly, the distribution of the bed shear stress over the stationary sediment bed of case $DL220^{H2}_{fix}$ collapses quite well with the curve of the smooth wall duct case $DL200^{H2}_{smooth}$ at matching Reynolds number. This indicates that the absence of the local extrema near the sidewalls in the mobile sediment cases are indeed of physical nature, rather than being a statistical effect related to the rather short averaging time window. The alternating zones of relatively higher and lower shear stress that mark the preferred positions of high- and low-speed streaks over the sediment bed induce a similarly alternating pattern of stronger and weaker hydrodynamic forces $\langle f^H\rangle_{xt}$ exerted on the stationary sediment bed, as seen in figure 7.15(*c,d*) (cf. equation (6.21) for the definition of the Eulerian force field). Both hydrodynamic

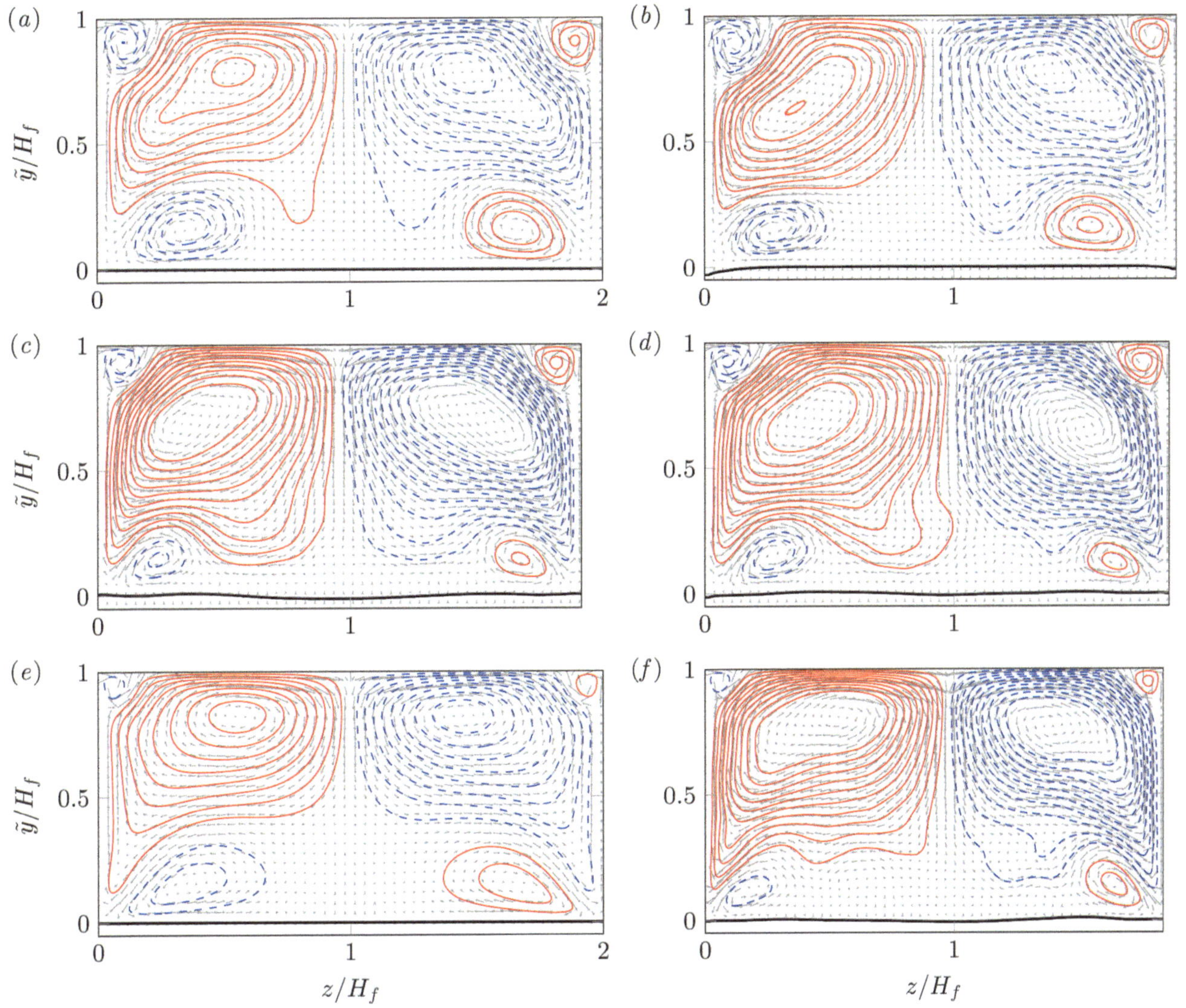

Figure 7.16: Mean secondary flow pattern $(\langle v_f \rangle_{xt}, \langle w_f \rangle_{xt})^T / u_b$ in open duct simulations with aspect ratio $AR \approx 1$ shown as vector plot together with isocontours of the secondary mean flow streamfunction $\langle \psi \rangle_{xt}/(u_b H_f)$ for cases (a) $DL200^{H2}_{smooth}$, (b) $DL220^{H2}_{fix}$, (c) $DS250^{H2}$, (d) $DL250^{H2}$, (e) $DL400^{H2}_{smooth}$ and (f) $DL500^{H2}$. Clockwise (counter-clockwise) secondary flow rotation is indicated by solid red (dashed blue) contours. Isocontours refer to values of $\langle \psi \rangle_{xt}/(u_b H_f)$ in the interval $[-0.006, 0.006]$ with an increment of $0.006/10$. The time-averaged fluid-bed interface profile is indicated by the black solid curve.

drag and lift force components are, on average, acting in a destabilising way, that is, they aim at inducing a combined downstream and upward motion of the particles. Assuming supercritical conditions $\theta_{bed} > \theta_c$, it can be easily imagined that the lateral force undulations will lead to a laterally varying erosion rate in the moment when particles are released, causing the evolution of initial ridges and troughs in the regions of lowest and highest mean hydrodynamic forces, respectively.

7.2.4 Turbulent mean secondary flow and vorticity distribution

7.2.4.1 Low aspect ratio cases

In figure 7.9 which opened the previous section, we have seen that the general mean streamwise and secondary flow pattern in the low aspect ratio duct cases is very similar over smooth walls, stationary and mobile sediment beds. Despite this qualitative similarity, though, the mean flow of the particle-laden cases quantitatively differs in several aspects from its counterpart in the smooth-wall reference simulations, as we shall see in the following.

In figure 7.16, we compare the mean cross-flow field $(\langle v_f \rangle_{xt}, \langle w_f \rangle_{xt})^T$ whose intensity and orientation are identifiable from a vector plot combined with a set of isocontours of the mean cross-flow streamfunction $\langle \psi \rangle_{xt}$. All cases exhibit the three characteristic features of the secondary mean flow pattern, that are, two *inner vortices* enclosed in the mixed corner between the free surface and the sidewalls (Grega et al., 1995), two outer-scaling *bottom vortices* that span from the wall-wall/wall-bed corner along the bottom wall/bed towards the duct mid-span and two large-scale mean *outer vortices* that occupy the remaining area of the cross-section (Nezu and Nakagawa, 1993; Sakai, 2016).

As expected, the inner vortex in the mixed corner appears to be unaffected by the presence of a mobile sediment bed and its diameter is seen to reduce with increasing Reynolds number in smooth-wall and particle-laden duct simulations. Sakai (2016) explained this inner-scaling behaviour of the inner vortices as the result of a vortex sorting process along the free surface based on the investigations of Orlandi (1990) concerning the dynamics of two-dimensional vortices in the vicinity of no-slip and free-slip boundaries. The proposed sorting mechanism basically states that instantaneous quasi-streamwise vortices are generated at the sidewalls and, depending on their sense of rotation, either move towards the duct centre forming a vortex dipole with their imaginary twin on the other side of the free surface or remain in the mixed corner. Vortices of the latter kind for which the vortex dipole would induce a motion towards the sidewall are stuck in the mixed corner and, thus, their preferential locations become visible in the mean flow pattern.

In contrast to the inner vortices, the mean bottom vortices are of somewhat different shape for the different flow configurations: In the stationary sediment bed case $DL220^{H2}_{fix}$, their size is only slightly smaller than in the smooth-bottom wall case $DL200^{H2}_{smooth}$ at essentially unchanged intensity, which is less surprising as the sediment beds considered here feature a relatively small D^+ and are hence hydraulically smooth. The bottom vortices past mobile sediment beds are, on the other hand, of markedly smaller transverse extent and are more confined to the respective wall-bed corners. As a consequence, the distance between their centre of rotation and the closest sidewall is only $(0.1\text{-}0.2)H_f$, whereas for smooth-wall open duct cases values in a range of $(0.3\text{-}0.4)H_f$ are observed (Sakai, 2016). Interestingly, the upflow region of the narrower mean bottom vortices, in which low-momentum fluid is transported away from the bed, collapses with the crests of the mean fluid-bed interface, suggesting that particles are preferentially deposited in these regions of relatively lower streamwise velocity. In most mobile sediment cases, the reduction of the bottom vortex size goes hand in hand with an expansion of the outer vortex cell towards the bed in the duct centre. Moreover, it can be seen from the spacing of the isocontours of $\langle \psi \rangle_{xt}$ that the gradient of the streamfunction is for the mobile sediment cases in some regions substantially higher than in the smooth-wall reference simulations. Recalling that $(\partial_y \langle \psi \rangle_{xt}, \partial_z \langle \psi \rangle_{xt})^T = (-\langle w_f \rangle_{xt}, \langle v_f \rangle_{xt})^T$, this indicates in turn that the cross-plane velocity components in such regions are accordingly higher: $\langle v_f \rangle_{xt}/u_b$ in the mean up- and downflow regions of the outer vortex along the sidewalls and the duct-bisector, respectively, and $\langle w_f \rangle_{xt}/u_b$ mainly along the free surface.

The mean cross-flow streamfunction $\langle\psi\rangle_{xt}$ is directly related to the mean streamwise vorticity field $\langle\omega_{f,x}\rangle_{xt}$ via the following Poisson equation subject to homogeneous Dirichlet boundary conditions, viz. (Pinelli et al., 2010; Sakai, 2016)

$$\begin{cases} \Delta_\perp \langle\psi\rangle_{xt} = -\langle\omega_{f,x}\rangle_{xt} & \text{in } \Omega_\perp \\ \langle\psi\rangle_{xt} = 0 & \text{on } \partial\Omega_\perp \end{cases}, \tag{7.12}$$

where $\Omega_\perp = [0, L_y] \times [0, L_z]$ is the cross-sectional domain, $\partial\Omega_\perp$ is the boundary enclosing the domain $\Omega_\perp$ and $\Delta_\perp = (\partial_y^2 + \partial_z^2)$ is the two-dimensional Laplacian acting only in the cross-plane. Relation (7.12) can be interpreted in such a way that $\langle\omega_{f,x}\rangle_{xt}$ is the local source term of the Poisson equation, while $\langle\psi\rangle_{xt}$ is of non-local character obtained by twofold integration of the streamwise vorticity such that $\langle\psi\rangle_{xt}$ depends on $\langle\omega_{f,x}\rangle_{xt}$ in the entire domain $\Omega_\perp$. As pointed out by Pinelli et al. (2010), this interpretation is conceptually similar to the non-local role of the pressure field in the Poisson equation (2.12) that is obtained by applying the divergence operator to the Navier-Stokes equations. In both open and closed duct flows, the non-locality of $\langle\psi\rangle_{xt}$ manifests itself in the fact that $\langle\psi\rangle_{xt}$ effectively scales in outer units, whereas $\langle\omega_{f,x}\rangle_{xt}$ scales in inner units (Pinelli et al., 2010; Sakai, 2016). The only exception of this general behaviour is the afore discussed inner vortex in the mixed corners between free surface and lateral sidewalls, for which $\langle\psi\rangle_{xt}$ shows a similar inner-scaling as $\langle\omega_{f,x}\rangle_{xt}$.

Similar to the mean secondary flow streamfunction $\langle\psi\rangle_{xt}$, the overall distribution of $\langle\omega_{f,x}\rangle_{xt}$ is qualitatively similar in all considered cases including those over a mobile sediment bed, as can be seen in figure 7.17. As reported by Sakai (2016), the regions of intense vorticity in the lower left and right quadrant of the duct tend towards the corners as the Reynolds number is increased for smooth-wall and particle-laden cases likewise. Also, the relatively wide but thin regions of high streamwise vorticity amplitude located below the free surface on either side of the duct bisector resemble each other in both flow configurations.

Despite the generally similar structure of $\langle\omega_{f,x}\rangle_{xt}$ in the upper half of the duct, the organisation of the mean vorticity near the mobile sediment bed in figures 7.17(*c,d,f*) differs from that over smooth bottom walls (figures 7.17(*a,e*)) and stationary sediment beds (figure 7.17(*b*)): In the same way as the mean bottom vortex associated with $\langle\psi\rangle_{xt}$ is seen to shrink in size for the mobile sediment cases ($DS250^{H2}$, $DL250^{H2}$ and $DL500^{H2}$), the corresponding patterns of $\langle\omega_{f,x}\rangle_{xt}$ in figure 7.17(*c,d,f*) reduce in size and are more confined to the lower corners than their counterparts over smooth walls and stationary sediment beds. Also, while the quasi-streamwise vortices in the smooth-wall cases $DL200^{H2}_{smooth}$ and $DL400^{H2}_{smooth}$ induce mirror vortex regions along the no-slip bottom walls with finite values of $\langle\omega_{f,x}\rangle_{xt}$ along the latter due to the applied boundary conditions (Orlandi, 1990), this is not the case at the top of a permeable sediment bed. In the latter case, velocity components and their gradients tend to zero when approaching the porous sediment bed and so does $\langle\omega_{f,x}\rangle_{xt}$. Instead of the mirror vortex cells at the solid walls, regions of intense vorticity arise that peak slightly above the bed and spread, in contrast to the mirror vortex cells in the smooth wall ducts, only up to the first sediment ridge. Apparently, the new 'additional' cells of strong vorticity in the vicinity of the sediment bed drive the remaining vorticity regions out, such that these latter are more strongly inclined w.r.t. the xz-plane in the mobile particle cases, as can be inferred when comparing figures 7.17(*e,f*). In section 7.3.3 below, we will show that the different structure of the mean streamwise vorticity field in the presence of a mobile sediment bed is in fact caused by a different organisation of the individual quasi-streamwise vortices in these cases. Eventually, it seems worth highlighting that the streamwise-minimal simulation $DS250^{H2}$ recovers, in

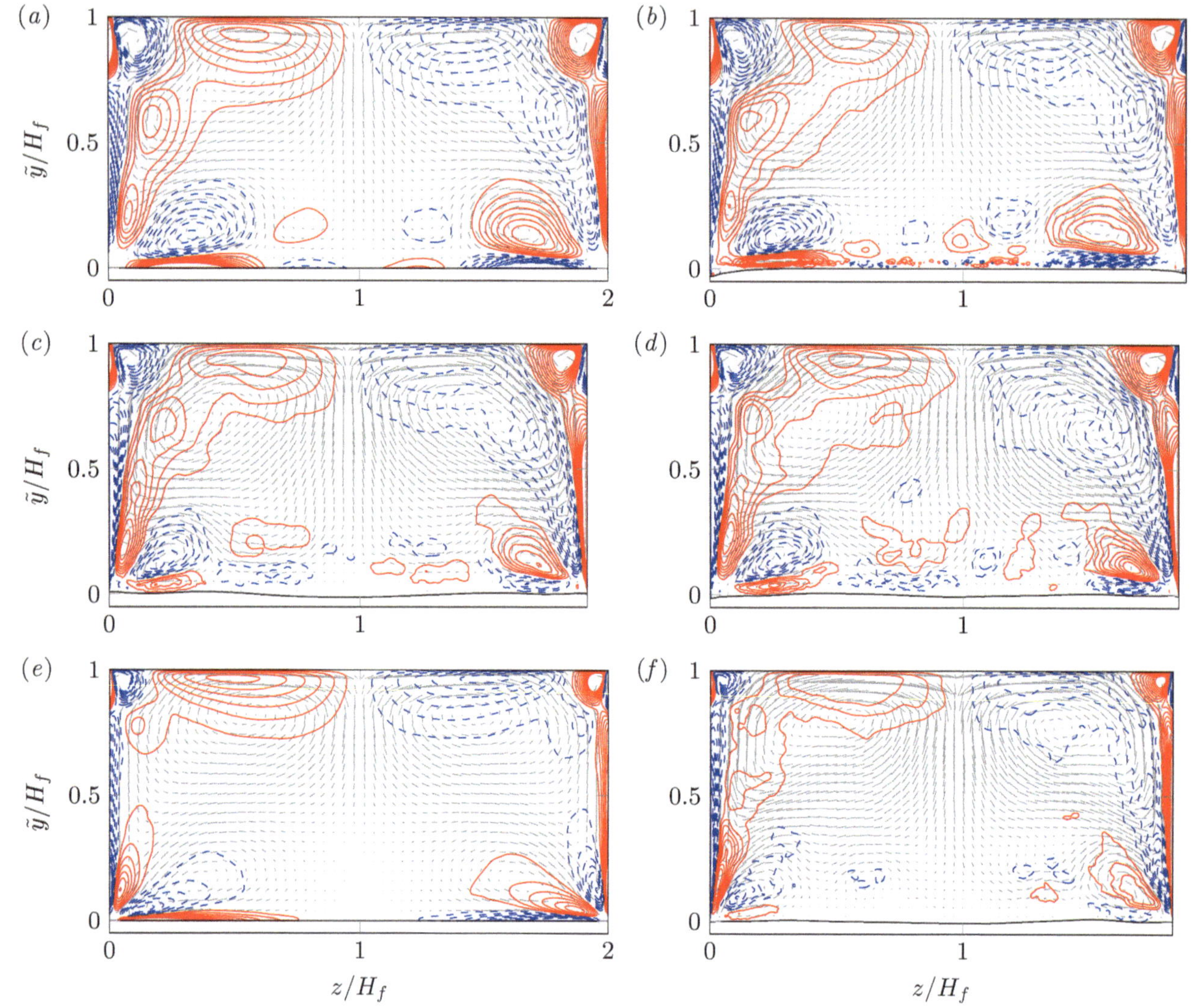

Figure 7.17: Mean streamwise vorticity $\langle\omega_{f,x}\rangle^+_{xt} = \langle\omega_{f,x}\rangle_{xt}/(u_\tau/\delta_\nu)$ for cases (*a*) $DL200^{H2}_{smooth}$, (*b*) $DL220^{H2}_{fix}$, (*c*) $DS250^{H2}$, (*d*) $DL250^{H2}$, (*e*) $DL400^{H2}_{smooth}$ and (*f*) $DL500^{H2}$. For the sake of comparison, the intensity and orientation of the mean secondary flow field $(\langle v_f\rangle_{xt}, \langle w_f\rangle_{xt})^T/u_b$ are shown as vector plot. Clockwise (counterclockwise) rotation is indicated by solid red (dashed blue) contours, while the black solid curve marks the location of the time-averaged fluid-bed interface profile. Isocontours refer to values of $\langle\omega_{f,x}\rangle^+_{xt}$ in the interval $[-0.04, 0.04]$ with an increment of $0.04/10$.

spite of its severe spatial restrictions, fairly well the mean streamfunction and vorticity patterns observed in case $DL250^{H2}$ which features a six times longer domain, in accordance with our observations concerning the mean flow $\langle u_f\rangle_{xt}$ in the previous section.

As we shall see in the following, the locally different structure of the secondary flow patterns in the mobile sediment cases cause an enhancement of the global secondary flow amplitude. For convenience, let us repeat in the following the definitions of the instantaneous and mean secondary flow intensity, $u_\perp(t)$ (cf. equation (6.37)) and $U_\perp$ (cf. equation (6.38)), respectively:

$$u_\perp(t) = \left[\frac{1}{L_z\,\langle h_f\rangle_{xz}(t)} \int_0^{L_z} \int_{\langle h_b\rangle_{xz}(t)}^{L_y} \left(\langle v_f\rangle_x^2 + \langle w_f\rangle_x^2\right)\, \mathrm{d}y\mathrm{d}z\right]^{1/2}$$

$$U_\perp = \left[\frac{1}{L_z\,H_f} \int_0^{L_z} \int_{H_b}^{L_y} E_\perp\, \mathrm{d}y\mathrm{d}z\right]^{1/2}.$$

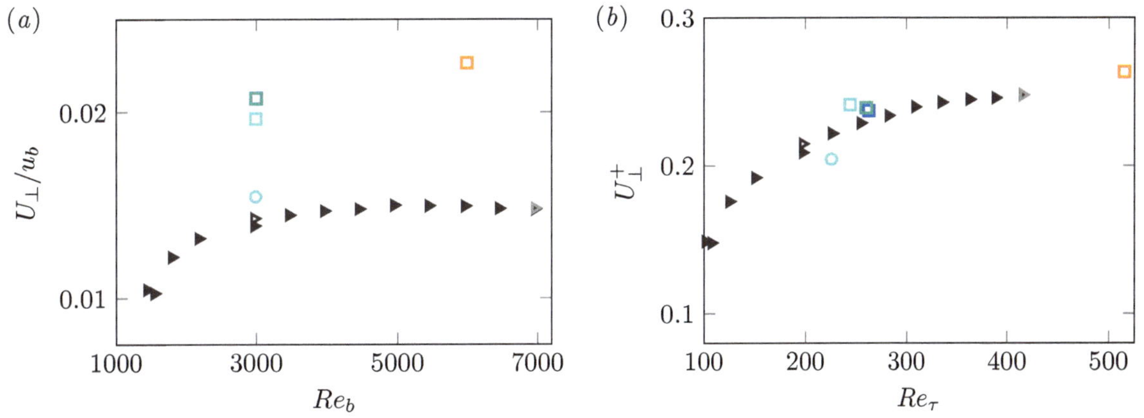

Figure 7.18: Intensity of the mean secondary flow field $U_\perp$ in open duct cases with $AR \approx 1$ scaled in (*a*) outer units as a function of Re_b and (*b*) inner units as a function of Re_τ. Symbols represent data from current open duct simulations over fixed (○) and mobile sediment beds (□), current smooth-wall open duct simulations (▷) and smooth-wall open duct data from Sakai (2016) (▶).

Here, we have introduced the mean kinetic energy of the secondary flow field as $E_\perp(y,z) = (\langle v_f \rangle_{xt}^2 + \langle w_f \rangle_{xt}^2)$. In figure 7.18, we first compare the mean secondary flow intensity $U_\perp$ in the smooth-wall duct simulations presented by Sakai (2016) with those of the current simulations. Therein, $U_\perp$ is scaled in both outer and inner units as a function of the corresponding Reynolds numbers Re_b and Re_τ, respectively. The data points from Sakai (2016) indicate that the outer-scaled mean secondary flow intensity $U_\perp/u_b$ (cf. figure 7.18(*a*)) in smooth-wall open duct flows initially increases with the bulk Reynolds number as long as the latter is of marginal to moderate size, whereas it is seen to settle at a plateau of approximately 1.5% for $Re_b \gtrsim 3500$. The same quantity scaled in inner units $U_\perp^+$ (cf. figure 7.18(*b*)) exhibits a quite similar trend, but it reaches a plateau-like level of approximately 0.25 only for the last simulations of the series at the highest values of Re_τ. This different behaviour is a consequence of the fact that Re_τ varies not exactly linearly with Re_b, as can be seen from the implicit formula (7.9) of Jones (1976) above.

For the sake of comparison, figure 7.18 contains in addition to the long-time statistics for the smooth-wall cases $DL200^{H2}_{smooth}$ and $DL400^{H2}_{smooth}$ also the values computed for the rather short time intervals of around 300 bulk time units that have been simulated in the current study at these parameter points. In both cases, a very good agreement with the corresponding data from Sakai (2016) is concluded, even though the latter statistics were collected over approximately 9000 and 4800 bulk time units, respectively. The comparison indicates that even though very long averaging intervals are usually required to obtain converged statistics fulfilling the statistical symmetries (Pinelli et al., 2010; Vinuesa et al., 2014a), the mean secondary flow intensity seems to converge relatively fast to its long-time average. Note that the influence of the averaging time interval and that of the spatial domain size on the convergence of the statistics will be discussed in the context of figure 7.20 below.

In accordance with our earlier observations that the secondary flow field over the hydraulically smooth stationary sediment bed of case $DL220^{H2}_{fix}$ agrees fairly well with that in the respective smooth-wall reference simulation $DL200^{H2}_{smooth}$, it is seen here that also the mean secondary flow intensities $U_\perp/u_b$ differ only slightly between the two cases. Considering inner-scaling, $U_\perp^+$ is somewhat lower than the corresponding smooth-wall simulations. In the mobile sediment bed simulations, in turn, the outer-scaled secondary flow intensities $U_\perp/u_b$ are markedly higher than in the smooth-wall simulations at comparable bulk Reynolds number, with $DL500^{H2}$ attaining a value that is approximately 50% higher than in the corresponding single-phase case. Interestingly, at least part of this pronounced deviation can be compensated when scaling $U_\perp$ in inner units, as the friction velocity u_τ increases with growing bed friction due to the sediment motion, whereas the bulk velocity does not depend on the specific roughness configuration in the duct. Remarkable in this context is that the secondary flow intensity in the low Shields number simulation $DL240^{H2}$ is, depending of the applied scaling, only slightly lower or even higher than the values observed in cases $DS250^{H2}$ and $DL250^{H2}$, respectively. This rather surprising finding suggests that even the weak particle transport in the former case causes a non-negligible increase of the secondary flow intensity.

Lastly, it should be stressed that the data points of cases $DS250^{H2}$ and $DL250^{H2}$ essentially collapse in both scalings, irrespective of the fact that their streamwise domain extensions differ by factor six, further supporting that the typical secondary flow pattern is correctly represented even in the streamwise minimal simulations not only qualitatively, but also quantitatively. A very similar observation was recently made by Xie et al. (2020), who studied the flow over laterally heterogeneous bottom roughness by means of direct numerical simulations in streamwise minimal domains ($L_x/\delta \approx 0.4$, with δ denoting the boundary layer thickness). The authors concluded that while the streamwise turbulent intensities were stronger than in the spatially unconstrained reference experiments, the secondary currents induced by the laterally heterogeneous bottom wall were accurately captured, which let them conjecture that the characteristic secondary flow patterns over such bottom topography cannot depend on the presence of individual long streamwise-elongated velocity structures which were naturally absent in their streamwise minimal boxes.

While the time-averaged mean secondary flow seems to be well reproduced even in the streamwise-minimal simulation concerning both general structure and amplitude, this does not hold for the instantaneous secondary flow intensity $u_\perp(t)$: As can be seen in figures 7.19(*a,b*), $u_\perp(t)$ is in case $DS250^{H2}$ almost twice as high as in most of the remaining simulations that feature streamwise domain lengths $L_x > 10H_f$. For the longer domain simulations, variations are mainly observed between smooth-wall and mobile sediment simulations (cf. figure 7.19(*a*)). A general trend here is that the latter attain, on average, somewhat higher values of $u_\perp(t)$ than the former, even though also the smooth-wall simulations intermittently attain peak values that are of comparable size as those reached in the mobile sediment cases. As for the mean secondary flow intensity $U_\perp$, we conclude that scaling $u_\perp(t)$ in u_τ compensates at least part of the deviations between smooth-wall and mobile sediment simulations (cf. figure 7.19(*b*)).

Unfortunately, $u_\perp$ as a global measure for the secondary flow intensity does not provide information in which regions of the cross-section the secondary fluid motion is enhanced. To this end, we have evaluated $u_\perp$ for the near-bed/-wall region $\tilde{y}/H_f \in [0, 0.2]$ (cf. figure 7.19(*c,d*)) and the outer flow region $\tilde{y}/H_f \in [0.2, 1]$ (cf. figure 7.19(*e,f*)) separately. The decomposition shows that, as expected, the main contribution to the secondary flow intensity comes from the large-scale outer mean vortex cells, while the secondary flow intensity in the region $\tilde{y}/H_f \in [0, 0.2]$ is markedly weaker. This highlights the fact that the pronounced deviations of $u_\perp$ that have been detected in figure 7.19(*a,b*) do not originate in the different secondary flow patterns in the near-bed/-wall region, but from the slight variations in the outer flow.

In the context of figure 7.16, it was concluded that the gradients of the streamfunction $\langle\psi\rangle_{xt}$ associated with the mean outer secondary vortices are stronger in the mobile particle simulations. Indeed, comparing the regions of intense secondary mean flow kinetic energy $E_\perp$ for both smooth-wall and mobile

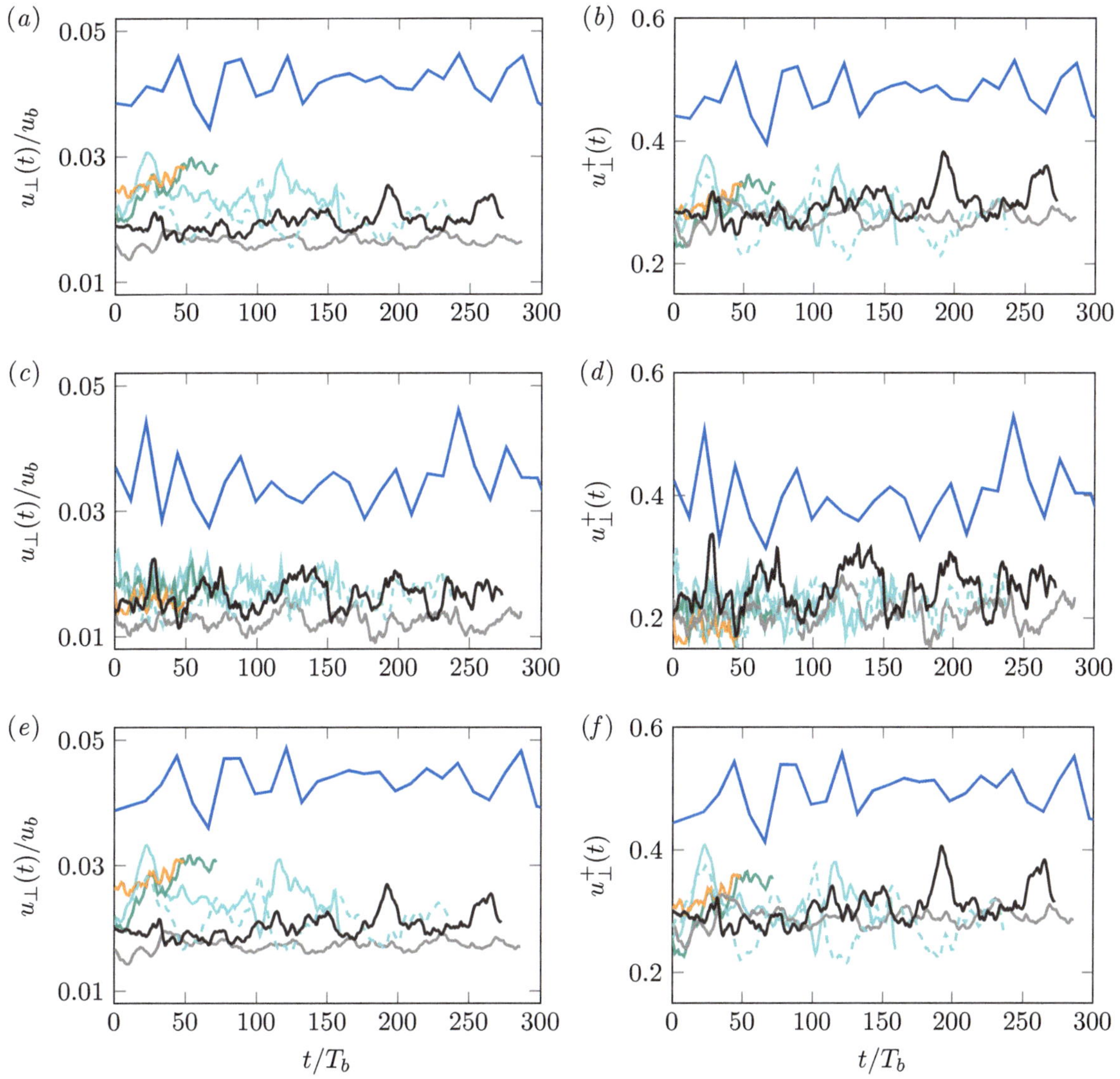

Figure 7.19: Time evolution of the instantaneous secondary flow intensity $u_\perp(t)$ in open duct flow with $AR \approx 1$ scaled in (*a,c,e*) outer and (*b,d,f*) inner units, respectively. In (*a,b*), $u_\perp(t)$ is determined by integrating the streamwise-averaged transverse velocity components over the entire mean fluid height H_f, as defined in equation (7.13). In (*c-f*), on the other hand, integration in the wall-normal direction is performed (*c,d*) over the near-bed/near-wall region $\tilde{y}/H_f \in [0, 0.2]$ and in (*e,f*) over the outer region $\tilde{y}/H_f \in [0.2, 1]$ only, respectively. Note that the black curves in all panels represent the short domain simulation of case $DL200^{H2}_{smooth}$ with $L_x = 4\pi H_f$.

sediment cases (plots omitted), it turns out that $E_\perp$ is dominated for most wall-normal distances $\tilde{y}$ by the strong upward motion $\langle v_f \rangle_{xt} > 0$ along the two sidewalls that has been seen to be higher for the mobile sediment simulations in figure 7.16. We therefore believe that the increase in both the instantaneous and mean secondary flow intensity in the mobile sediment cases is mainly due to an enhanced vertical velocity at the two lateral sidewalls and a thus intensified outer mean vortex cell.

The reason for this stronger upflow close to the sidewalls over mobile sediment beds could not yet entirely be clarified, but there exists some evidence from experiments that the outer mean secondary vortex cell is intensified in corners enclosed by a rough bottom wall/sediment bed on one side and a smooth sidewall on the other, while the bottom vortex weakens (Nezu and Nakagawa, 1993; Rodríguez and García, 2008; Albayrak and Lemmin, 2011). Note that in contrast to the current study, these experiments were performed over a fixed and hydraulically rough bottom, but we have seen earlier that the mobile sediment in the here considered cases can lead to a significant increase in friction compared

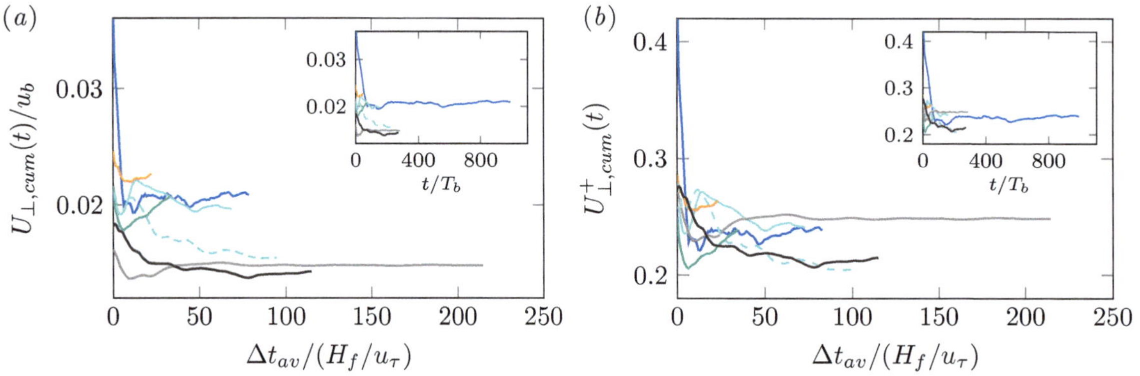

Figure 7.20: Cumulative mean of the secondary flow intensity $U_{\perp,cum}(t)$ in low aspect ratio open duct flows as a function of the averaging time interval scaled in (a) outer and (b) inner units, respectively. The length of the time window over which statistics are gathered is given in 'effective time units' $\Delta t_{av} = (t - t_0)(L_x/H_f)/(L_x/H_f)_{ref}$, with $t_0 = 0$ and $(L_x/H_f)_{ref} = 2$ for all considered cases, and is scaled in eddy turnover times. The insets show the same quantity, but as a function of the standard non-weighted time measure $(t - t_0)/T_b$, $t_0 = 0$. Note that the black curves represent again the time evolution of the short domain simulation of case $DL200^{H2}_{smooth}$ with $L_x = 4\pi H_f$.

to what would be expected based on their roughness height. In regard of the fact that the secondary flow is essentially not enhanced over the stationary bed of case $DL220^{H2}_{fix}$ for comparable averaging time intervals, these considerations suggest that it is indeed the sediment mobility that makes the difference.

A second conceivable effect that could contribute to a further enhancement of the strength of the mean outer vortex cell is also related to the mobility of the sediment bed: As was discussed in the context of figure 7.10, the mobile sediment transported along the sediment bed reduces the streamwise velocity in the vicinity of the bed, which leads to an increase of the mean shear in the outer regions required to maintain the constant mass flow rate in the duct. It is not unlikely that the modification of the primary mean flow towards an enhanced mass flow rate in the upper part of the channel results in an intensification of the wall-normal velocity.

In the current discussion, though, we should keep in mind that the time-averaging interval in the particle-laden cases is relatively short, compared with the single-phase data of Sakai (2016). In order to assess the level of statistical convergence in the current simulations, we introduce in the following an effective time averaging interval $\Delta t_{av} = (t - t_0)(L_x/H_f)/(L_x/H_f)_{ref}$ with $t_0 = 0$ that is additionally weighted by the relative streamwise domain size, similar to approaches applied, for instance, by Vinuesa et al. (2016) or Bauer et al. (2017). The advantage of considering Δt_{av} as measure for the observation time interval lies in the fact that it quantifies both the temporal and spatial ensemble size and thus allows a more objective comparison of the statistics in the current simulations, which feature a fairly wide range of domain sizes $L_x \in [1.9H_f, 8\pi H_f]$. In this measure, a single 'flow through' time interval for the longest domains with $L_x = 8\pi H_f$ is equivalent to more than 12 'flow through' periods in the shortest one for, apart from that, identical parameters. As a reference duct length, we choose $(L_x/H_f)_{ref} = 2$ which approximately matches that of the shortest considered domains, such that for case $DS250^{H2}$, Δt_{av} is almost equivalent to the standard time scale. Note that the choice of the reference length is more or less arbitrary, and an alternative physically motivated value is $(L_x/H_f)_{ref} = 6$ (Vinuesa et al., 2016), which represents the streamwise length of the minimal log-layer unit reported by Flores and Jiménez (2010) for a canonical closed channel.

In figure 7.20, we show the cumulative mean of the secondary flow intensity

$$U_{\perp,cum}(t) = \left[\frac{1}{L_z H_f} \int_0^{L_z} \int_{H_b}^{L_y} \langle\langle v_f \rangle_x \rangle^2_{[0,t]} + \langle\langle w_f \rangle_x \rangle^2_{[0,t]} \mathrm{d}y\mathrm{d}z \right]^{1/2} \tag{7.14}$$

as a function of the newly introduced effective time-averaging interval Δt_{av} normalised in terms of the eddy-turnover time H_f/u_τ and in classical time scales normalised in bulk time units. Note that in the above definition, $\langle \, \rangle_{[0,t]}$ indicates time averaging over the indicated time interval only, such that the relations $U_{\perp,cum}(0) = u_\perp(0)$ and $U_{\perp,cum}(T_{obs}) = U_\perp$ hold. Considering first the development of $U_{\perp,cum}/u_b$ in figure 7.20(*a*), it is seen that the smooth-wall and stationary bed simulations seem to converge to values that correspond to the plateau-like region visible in figure 7.18(*a*) for bulk Reynolds numbers $Re_b \gtrsim 3500$. As already seen in this latter figure, $U_\perp$ is in the two smooth-wall simulations $DL200^{H2}_{smooth}$ ($L_x = 4\pi H_f$) and $DL400^{H2}_{smooth}$ ($L_x = 8\pi H_f$) already very close to the corresponding values that Sakai (2016) observed for much longer time-averaging intervals at the same parameter point. Together with the observation that both curves in figure 7.20(*a*) settle at a more or less constant level for times $\Delta t_{av} > 50 H_f/u_\tau$ without further strong fluctuations, this implies that the statistical quantities under consideration are good approximations of the fully-converged state. The stationary bed case $DL220^{H2}_{fix}$, on the other hand, has not reached a comparable state yet, but extrapolating the general trend, one might expect that it will attain a very similar value $U_\perp$ even in a fully-converged state.

The situation is somewhat different for the mobile sediment cases, for which the effective simulation time is limited in the longer domains by the evolution of transverse bedforms, as explained earlier. Here, it is mainly for case $DS250^{H2}$ that a more or less converged state is identifiable, while the remaining cases could not be studied for sufficiently long time-intervals to observe an eventually quasi-converged state. Rather, $U_{\perp,cum}$ is seen for cases $DL250^{H2}$ and $DL500^{H2}$ to increase even further with growing averaging interval. Although it was not possible to reach a more or less converged state in all simulations, it is noticeable that the general trend of all mobile sediment cases is not directed towards the values in the smooth-wall reference simulations, but all seem to either settle or further grow, supporting the assumption that the observed differences in the secondary flow field over smooth wall and mobile sediment bed are of physical nature and would be visible also in time averages over much longer ensemble series.

7.2.4.2 High aspect ratio cases

For open duct flows with aspect ratios much larger than unity, mean secondary currents of Prandtl's second kind remain restricted to the near-sidewall region, whereas the mean flow in the duct core is essentially unidirectional (Tominaga et al., 1989; Sakai, 2016), as can be inferred from figure 7.21(*a*) that shows the mean secondary flow streamfunction and the mean streamwise vorticity for the high aspect ratio case $DL150^{H16}_{smooth}$ ($T_{obs} = 8140 T_b$). Indeed, intense regions of $\langle \psi \rangle_{xt}$ and $\langle \omega_{f,x} \rangle_{xt}$ are merely seen up to a distance of $4H_f$ from either sidewall, whereas both quantities attain negligible values in the central region of the cross-section.

Instantaneously, on the other hand, we observe a substantial cross-stream motion in form of instantaneous depth-spanning streamwise rollers with a lateral width of (1-2)H_f populating the entire cross-section, accompanied by alternating regions of intense vorticity that span up to the free surface (cf.

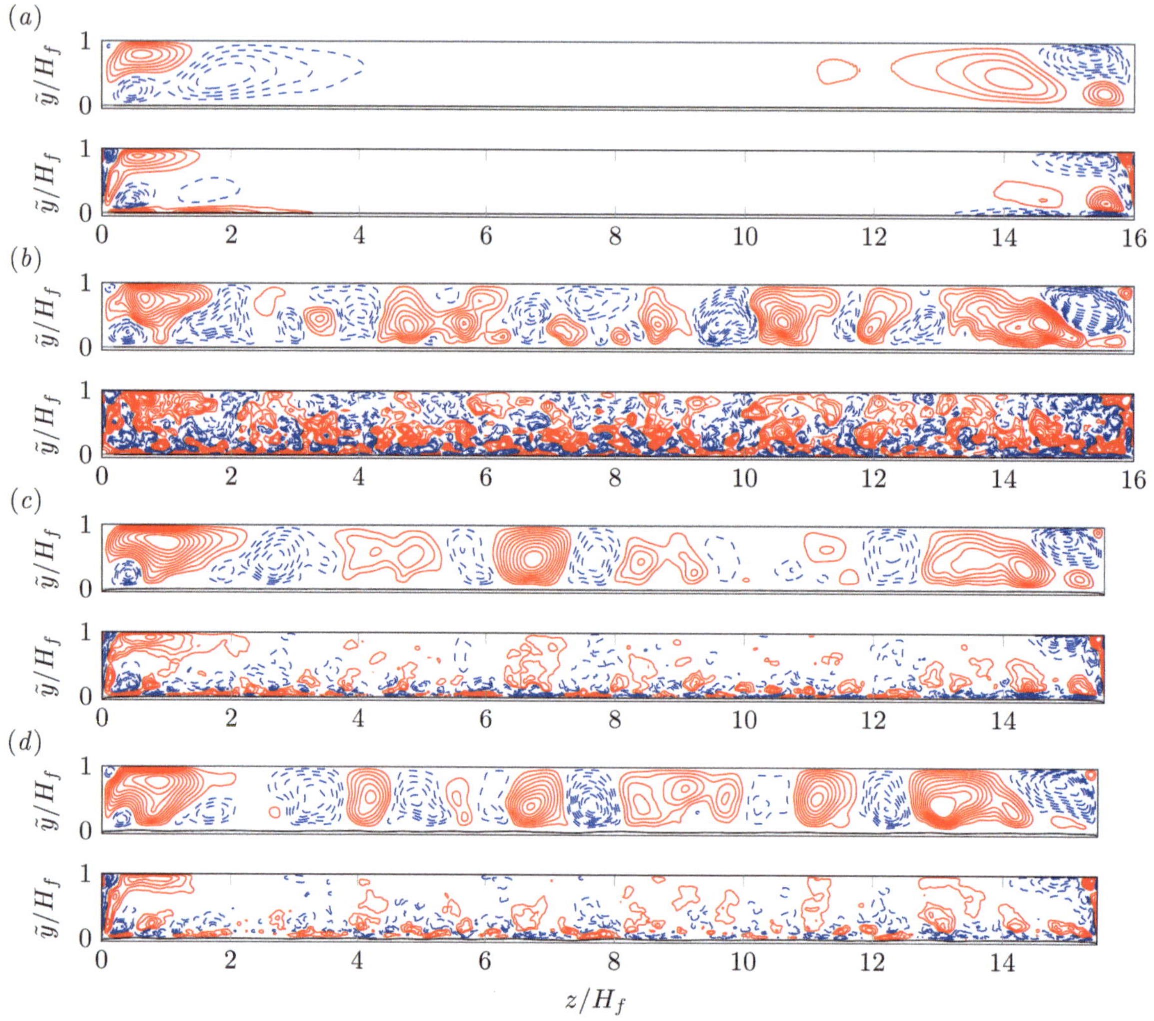

Figure 7.21: Mean secondary flow streamfunction $\langle\psi\rangle_{xt}/(u_bH_f)$ and mean streamwise vorticity $\langle\omega_{f,x}\rangle^+_{xt}$ in open duct simulations with aspect ratio $AR \approx 8$ for cases: (*a*,*b*) $DL150^{H16}_{smooth}$, (*c*) $DL210^{H16}_{fix}$ and (*d*) $DL250^{H16}$. For each case, the upper and lower figure show the mean secondary flow streamfunction and the mean streamwise vorticity, respectively. Note that panels (*a*,*b*) show for case $DL150^{H16}_{smooth}$ (*a*) the long-time statistics and (*b*) the respective quantities for a single randomly-chosen instantaneous flow field. Colour coding and isocontour values are chosen identically to those in figures 7.16 and figures 7.17, respectively.

figure 7.21(*b*)). Such 'instantaneous secondary currents' (Nezu, 2005) have been observed in many experiments (Nezu and Nakagawa, 1993; Onitsuka and Nezu, 2001) and represent most likely the footprint of large-scale streaks and Reynolds stress-carrying structures analogous to those found in open channel and other canonical wall-bounded flows (Adrian and Marusic, 2012). Let us recall that in the open channel flow cases discussed in chapter 6, these large-scale features caused the evolution of mean secondary currents even in the case of single-phase smooth-wall channels, if time averaging intervals of $\mathcal{O}(10T_b)$ were considered. The same phenomenon is visible in figure 7.21(*c*,*d*) for the high aspect ratio sediment-laden open duct simulations: Owing to their long lifetime and a relatively weak lateral mobility that both will be discussed in the subsequent section 7.2.5, the statistical footprint of these large-scale structures in form of 'instantaneous secondary currents' remains detectable for time-averaging intervals $\mathcal{O}(10T_b)$ even far away from the sidewalls.

Interestingly, the general structure of the secondary currents in the instantaneous and short-time averaged fields is qualitatively similar over all considered bed configurations, that are, the smooth bottom wall of case $DL150^{H16}_{smooth}$, the stationary sediment bed of case $DL210^{H16}_{fix}$ and the mobile sediment bed in

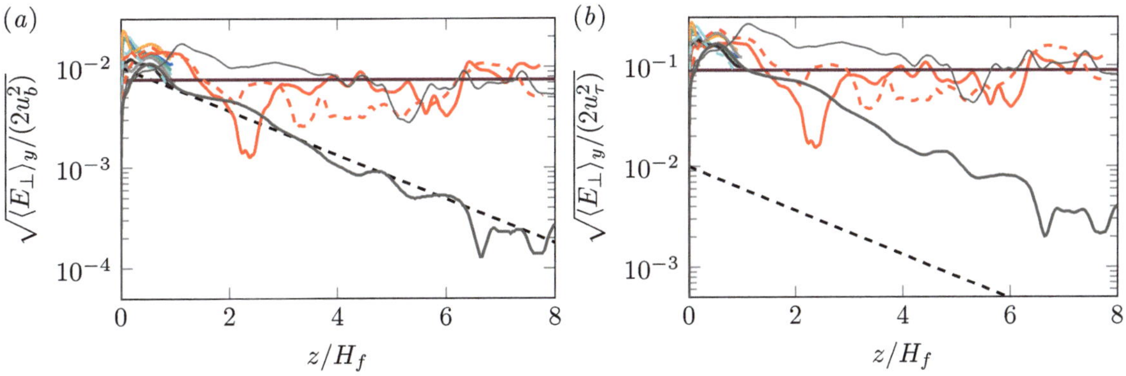

Figure 7.22: Wall-normal averaged kinetic energy of the mean secondary flow field $\langle E_\perp \rangle_y = 1/H_f \int_{H_b}^{L_y} E_\perp dy$ as a function of the distance to the lateral sidewall (*a*) scaled in outer and (*b*) inner units. Note that only the left half of each case is shown. The thick and thin solid grey lines refer to the temporal mean and the instantaneous value of $\langle E_\perp \rangle_y$ for case $DL150^{H16}_{smooth}$, from which the latter has been computed for the field shown in figure 7.21(*b*). In both figures, the horizontal dark red line marks the wall-normal and spanwise-averaged mean secondary flow kinetic energy $\langle E_\perp \rangle_{yz}$ in the channel flow reference simulation $CL250^{H3}$ and the dashed-dotted line indicates the exponential decay as $\sim \exp(-0.5z/H_f)$, proposed by Sakai (2016).

case $DL250^{H16}$. In particular, it should be remarked that the latter two cases feature an identical initial condition at $t = 0$, but while in the former case the sediment bed is kept stationary, the sediment in case $DL250^{H16}$ is allowed to move from this moment on. Despite the entirely different evolution of the sediment bed in both simulations, the mean secondary currents seen in figure 7.21(*c,d*) possess very similar shapes and lateral positions in both simulations. The latter observation strengthens our earlier conclusions that large-scale structures of this kind in sufficient distance to the sediment bed are more or less unaffected by the exact dynamics of the latter. In fact, we shall see in the subsequent section that the opposite is true: As in the open channel case, the dynamics of the large-scale streaks dictate the formation and organisation of individual sediment ridges and troughs outside the near-sidewall regions.

Just as for the high- and low-speed streaks of the buffer layer, only those large-scale structures that live in the neighbourhood of either of the two sidewalls remain visible in the long-time average, as their spatial mobility is restricted by the lateral domain boundaries and new large-scale features will always form at a very similar lateral position. In contrast to the near-sidewall structures, the large-scale structures in the duct core are not bounded and thus may appear at every position in sufficient distance to the sidewalls with the same probability. Consequently, the contributions from individual large-scale streaks in the duct centre effectively cancel out if the averaging interval is sufficiently longer than the lifetime of the individual features.

As a consequence, the mean secondary flow kinetic energy $E_\perp$ averaged over the wall-normal direction is seen in figure 7.22 to decay exponentially from the sidewall-region towards the duct centre for all long-time averaged fields. For the low aspect ratio cases with $AR \approx 1$, we observe that all simulations feature a similar decay rate of the mean secondary flow energy from the sidewalls to the duct bisector. However, the mobile sediment cases are seen to start from a higher value in the vicinity of the sidewalls, which is in line with our earlier observation that the mean wall-normal velocity along the sidewalls and thus $E_\perp$ is enhanced in the mobile sediment bed cases compared to the single-phase smooth-wall reference simulations.

For larger aspect ratios, Sakai (2016) approximated the decay rate as $\sim \exp(-0.5z/H_f)$, which is seen to match quite well the evolution of $E_\perp$ for his largest aspect ratio case $DL150^{H16}_{smooth}$. For the sake of comparison, the profile of the secondary flow kinetic energy in the instantaneous flow field previously shown in figure 7.21(*b*) was added to figure 7.22. This latter reveals no such decay of $E_\perp$, but maintains a constantly high level of kinetic energy from the sidewalls up to the duct bisector, highlighting that instantaneously large-scale structures near and far from the lateral sidewalls carry a comparable amount of secondary flow kinetic energy. This interpretation is further supported by the qualitatively similar energy distribution in the stationary and mobile bed cases $DL210^{H16}_{fix}$ and $DL250^{H16}$, respectively, which similarly lack an exponential decay when approaching the duct centre and which attain values of comparable size throughout the entire span.

In this regard, it is interesting to note that the attained level of kinetic energy in these cases is of the same order of magnitude as the mean secondary flow kinetic energy $\langle E_\perp \rangle_{yz}$ in the open channel reference case $CL250^{H3}$, agreeing with our above argumentation that the secondary currents are in both cases related to the same kind of large-scale coherent velocity structures.

7.2.5 Large-scale streak-ridge interaction

In the following, we aim to clarify how exactly these instantaneous large-scale coherent structures interact with the sediment bed under the influence of lateral sidewalls, in particular in the light of the top-down formation process that has been discovered in open channel flows in the previous chapter.

To do so, we first analyse the temporal evolution of the streamwise-averaged streamwise velocity fields provided in figure 7.23 for single- and multiphase open duct simulations as well as for the open channel reference simulation $CL250^{H3}$ at two different wall distances $\tilde{y}/H_f = 0.2$ and $\tilde{y}/H_f = 0.5$. In order to visualise the lateral variation of the streamwise velocity profiles at each height, $\langle u_f \rangle_x$ is shown relative to a reference velocity u_{ref} averaged over $z \in [0.25H_f, L_z - 0.25H_f]$ excluding only the regions in the vicinity of the two sidewalls. The choice of this width is up to a certain point arbitrary, but the visualisation appears not to be sensitive to small variations of it and the here chosen sidewall-offset $0.25H_f$ appears to well quantify the sidewall boundary layer thickness at both heights, as can be seen from the lateral velocity profiles previously shown in figure 7.8(*b,d,f*).

That way visualised, the data in figure 7.23(*f*) indicates that the cross-section of the high aspect ratio case $DL250^{H16}$ is sufficiently wide to host at $\tilde{y}/H_f = 0.5$ up to 8 individual large-scale high-speed zones with the typical lateral spacing $1H_f$-$2H_f$ of large-scale streaks in canonical wall-bounded flows (Jiménez, 2018). Comparing the positions of the high- and low-speed zones with those of the secondary currents in the short-time averaged field in figure 7.21(*d*), it is seen that the secondary rolls are, as expected, located roughly at the boundary between the high- and low-speed zones. The large-scale secondary currents rotate in such a way that they sustain the large-scale velocity streaks, moving fluid downwards inside the high-speed regions and upwards in the neighbouring low-speed zones. The low aspect ratio of the remaining cases, on the other hand, does not allow to accommodate more than a single large-scale high-speed zone that then fills the entire cross-section (cf. figures 7.23(*a-d*)). As can be seen, the central high-speed regions in these cases exhibit a relatively weak lateral meandering tendency in time due to the geometrical restrictions imposed by the domain boundaries.

The comparison underlines that, in a quite similar way as the number of buffer layer velocity streaks that fit into a duct cross-section depends on L_z^+, the amount of large-scale coherent structures that can

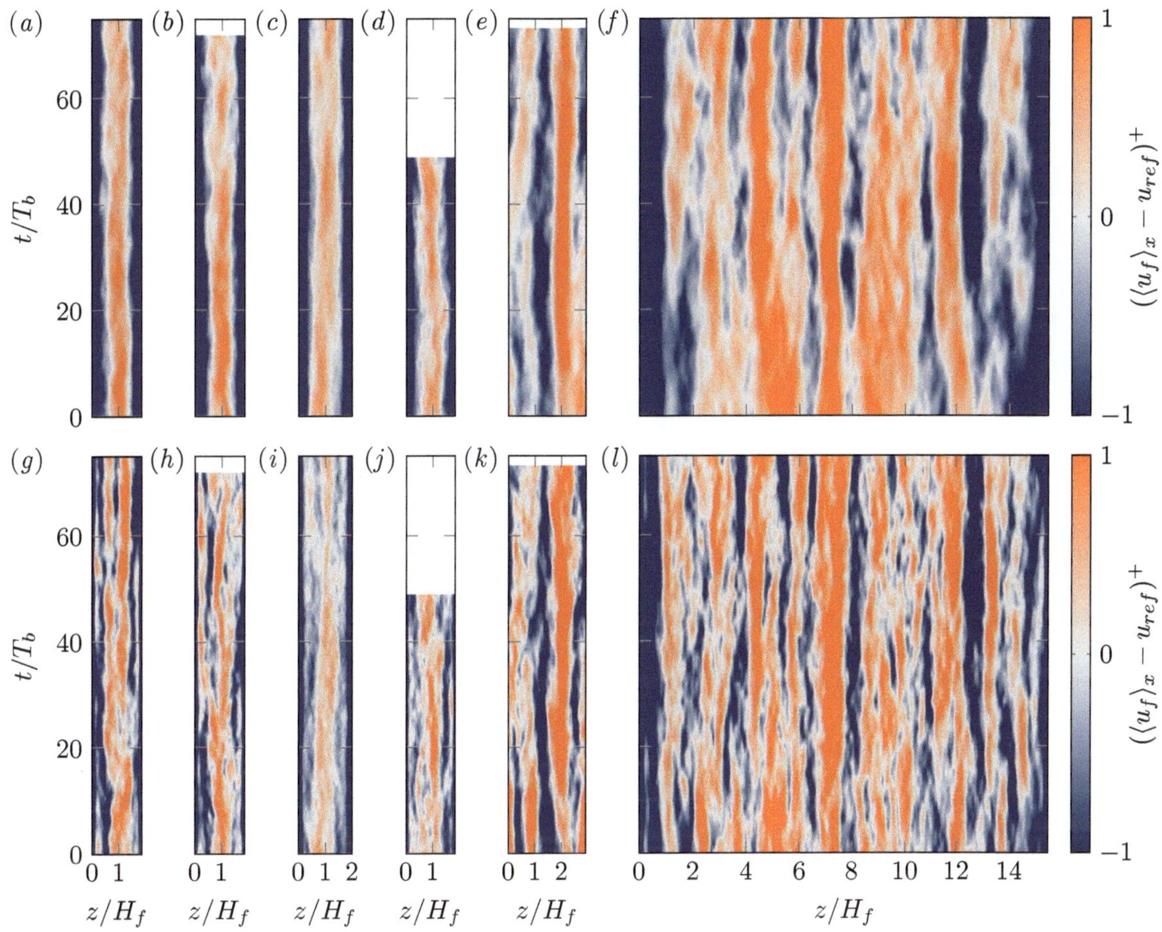

Figure 7.23: Space-time plot of the streamwise-averaged fluctuations of the streamwise velocity component $(\langle u_f \rangle_x - u_{ref})/u_\tau$ extracted at $\tilde{y}/H_f = 0.5$ (a-f) and $\tilde{y}/H_f = 0.2$ (g-l). Blue and red regions refer to streamwise-averaged low- and high-speed regions, respectively. Cases (a,g) $DL240^{H2}$, (b,h) $DL250^{H2}$, (c,i) $DL400^{H2}_{smooth}$, (d,j) $DL500^{H2}$, (e,k) $CL250^{H3}$, (f,l) $DL250^{H16}$. For open duct cases, $u_{ref}(y,t) = 1/(L_z - 2b) \int_b^{L_z - b} \langle u_f \rangle_x \mathrm{d}z$ is the mean velocity in the duct core ($b = 0.25H_f$). For the open channel case, we choose $b = 0$ such that $u_{ref}(y,t) = \langle u_f \rangle_{xz}$ and $(\langle u_f \rangle_x - u_{ref}) = u_f''$.

be accommodated in a given cross–section is a function of the outer-scaled domain width or the aspect ratio, as $L_z/H_f = 2AR$. Recalling that the mean spacing between two large-scale structure of same sign in canonical and open channel flows lies in a range $1H_f$-$2H_f$ (Flores and Jiménez, 2010; Smits et al., 2011), the mean width of a single prototype large-scale streak can be estimated as roughly $0.5H_f$-$1H_f$. An idealised symmetric streak arrangement consisting of a single large-scale low-speed streak in the duct core flanked by two large-scale high-speed streaks on either side would then feature a width of $1.5H_f$-$3H_f$. Taking moreover into account that the first large-scale streak in case $DL250^{H16}$ centres at a distance of roughly $1H_f$ from the sidewalls, the minimal duct width required to accommodate such a three-streak state can be quantified as $L_{z,min} \approx 3H_f$-$4H_f$, which is equivalent to an aspect ratio of $AR \approx 1.5$-2. Unfortunately, our database does not comprise simulations featuring an intermediate aspect ratio $AR \gtrsim 2.5$ to further test this assumption. However, observations of Nezu and Rodi (1985) and Sakai (2016) that the velocity dip in the mean velocity field $\langle u_f \rangle_{xt}$ disappears for cross-sections with $AR \gtrsim 2$-2.5 might be a hint that around this critical aspect ratio, the cross-section is now sufficiently wide to accommodate more than a single large-scale high-speed region, which are then less restricted in their mobility by the lateral domain boundaries. Interestingly, even the more or less geometrically unconstrained large-scale high- and low-speed regions in the centre of case $DL250^{H16}$ do not reveal a

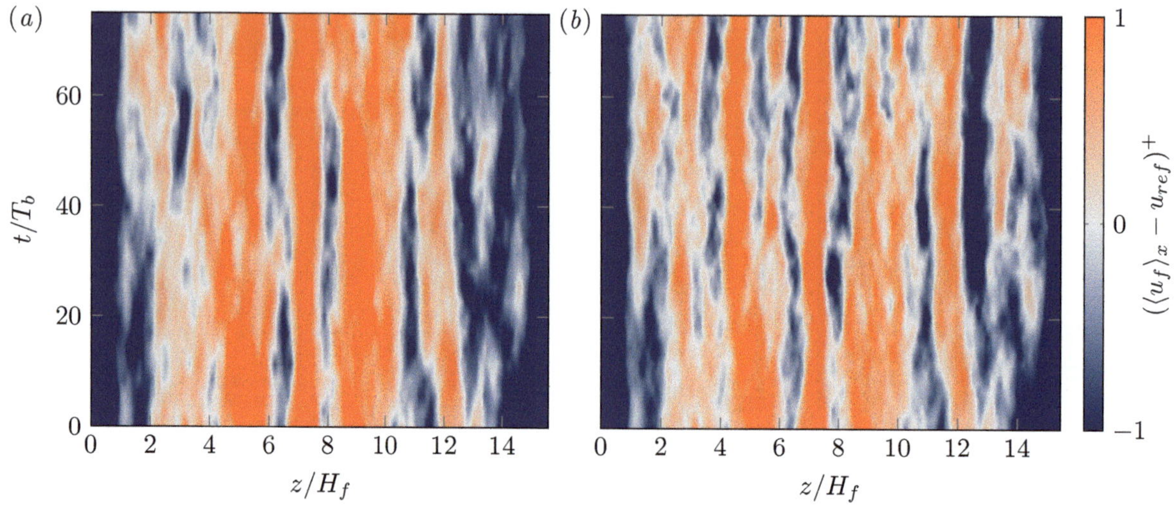

Figure 7.24: Space-time plot of the streamwise-averaged fluctuations of the streamwise velocity component $(\langle u_f\rangle_x - u_{ref})/u_\tau$ at $\tilde{y}/H_f = 0.5$ (u_{ref} as defined in figure 7.23) in high aspect ratio open duct simulations over (*a*) a stationary bed in case $DL210^{H16}_{fix}$ and (*b*) a mobile sediment bed in case $DL250^{H16}$. Note that both simulations were started from the same initial condition and are therefore identical at $t = 0$. Blue and red regions refer to streamwise-averaged low- and high-speed regions, respectively.

pronounced meandering activity in time for the here considered intervals of $\mathcal{O}(10T_b)$ length. Let us remind that the same phenomenon was also noticed for the equivalent features in the open channel cases in domains of varying size and is seen here again for the open channel reference case $CL250^{H3}$ in figures 7.23(*e*).

In figure 7.24, we compare in this context the evolution of the large-scale high- and low-speed regions at $\tilde{y}/H_f = 0.5$ in case $DL250^{H16}$ with that in the companion simulation $DL210^{H16}_{fix}$, for which the sediment was not released at $t = 0$ and instead kept stationary throughout the entire simulation. In this context, it was already at the end of the previous section in figure 7.21 noticed that many of the mean secondary currents are very similar over the stationary and the mobile sediment bed. Here, it can be seen that also the development of the large-scale high- and low-speed regions is qualitatively similar: The five most dominant high-speed regions exhibit only weak lateral mobility and reside most of the time interval at the same lateral mean position. During the first 20 bulk time units, moreover, the streamwise-averaged velocity field is essentially identical in both cases, showing that the formation of the initial sediment ridges has no particular influence on their organisation. Further progressing in time, deviations between both cases generally increase, even though individual large-scale high-speed zones such as the one located in the interval $z/H_f \in [6, 8]$ reveal even after $70T_b$ an astonishing similarity in both cases – considering that the chaotic turbulent system is strongly perturbed by the mobilised sediment and the friction is increased by about 20% compared to the flow over the stationary sediment bed. We remark that these findings are in agreement with our observations in open channel flow and further support the idea that the organisation of large-scale coherent structures in the bulk of the flow is only weakly affected by changes of the lower domain boundary and the roughness of the latter (Flores and Jiménez, 2006; Flores et al., 2007). Let us underline that the large-scale streaks in both simulations possess – on account of their size and distance to the bed – a comparably long lifetime (Jiménez, 2013a, 2018), such that their signature in the time-averaged field disappears only slowly. In fact, a similar space-time analysis combined with cumulative averaging of the flow field in case $DL150^{H16}_{smooth}$ (plots not shown) implies that even for averaging intervals of several thousand bulk time units length, a weak but detectable up- and downward-motion is found throughout the entire cross-section, reminiscent of the relatively inert large-scale velocity structures.

Having investigated the dynamics of the large-scale velocity structures, it remains to discuss how their mutual organisation eventually affects the sediment erosion activity along the sediment bed. For open channel flow, it was found that the regions of higher and weaker particle erosion correlate quite well with the lateral positions of large-scale high- and low-speed streaks in the half height of the channel. The same seems to hold in the core of wide ducts, whereas the situation in the narrow duct cases is more difficult: When comparing the streamwise-averaged flow field at the two considered heights in the low aspect ratio flows (top and bottom row of figure 7.23), it is clearly noticeable that the structures at $\tilde{y}/H_f = 0.2$ are of smaller width than those in the outer region at $\tilde{y}/H_f = 0.5$. At the same time, on account of their smaller size, the high- and low-speed regions extracted at $\tilde{y}/H_f = 0.2$ are less 'stable' than the large high-speed zones in the duct core at $\tilde{y}/H_f = 0.5$ and reveal intermittent reorganisation phases, including splitting and merging events between different high- and low-speed zones. Also, zones of relatively higher and lower velocity exhibit a much more vigorous tendency to meander in time (in the sense that their lateral mean position changes continuously) compared with the large-scale structures at $\tilde{y}/H_f = 0.5$. Remarkably, the different organisation of the streamwise velocity at $\tilde{y}/H_f = 0.2$ is not an exclusive feature of duct flow over mobile sediment beds, but also the flow over the stationary beds (plots not shown) and that in the smooth-wall case in figure 7.23(*i*) reveal permanently changing zones of relatively lower and higher streamwise velocity in this distance to the wall.

In this respect, the organisation of the streamwise velocity is qualitatively different in the doubly-periodic open channel case $CL250^{H3}$ and in the core of the wide open duct $DL250^{H16}$ shown in figures 7.23(*k*) and (*l*), respectively. In both situations, the high- and low-speed regions at $\tilde{y}/H_f = 0.2$ feature a clearly larger width than those in the narrow duct cases, quite comparable to the corresponding large-scale regions at $\tilde{y}/H_f = 0.5$ that appear at roughly the same lateral positions. This similar arrangement suggests that the large-scale velocity streaks span in these regions over the entire depth of the channel/duct in a similar way as it was seen in the previous chapter 6 for the open channel simulations. At the relatively low Reynolds numbers considered in the current study, a full scale separation between inner- and outer-scaling structures is not yet developed. In particular for the cases at $Re_b \approx 3000$, the low- and high-speed regions at $\tilde{y}/H_f = 0.2$ might thus be related to both the buffer and outer layer at the same time. However, the typical width of the individual high- and low-speed zones at $\tilde{y}/H_f = 0.2$ of case $DL500^{H2}$ (cf. figure 7.23(*j*)) is with 200 to 400 wall units (equivalent to $0.4H_f$ and $0.8H_f$) higher than the width of $\approx 50\delta_\nu$ usually reported for buffer layer streaks (Kline et al., 1967). It is therefore assumed that the observed features do not scale in wall units and hence should be detectable in similar form also at higher Reynolds numbers. In order to investigate the flow organisation in the different distances to the bottom wall in more detail, simulations comparable to the here performed ones, but at higher Reynolds numbers are essential and can be hopefully performed in the near future.

In order to eventually assess which flow structures are responsible for the formation and the evolution of the individual sediment ridges and troughs, we have collected in figure 7.25 the space-time visualisations of several of the above discussed physical quantities including the sediment bed height fluctuation h_b'' (figure 7.25(*a*)), the streamwise and wall-normal components of the velocity field $\langle u_f \rangle_x$ and $\langle v_f \rangle_x$ (figure 7.25(*b*,*c*)), respectively, as well as the streamwise particle flux $\langle q_{p,x} \rangle_x$ (figure 7.25(*d*)). The combined visualisation of these quantities reveals a strong correlation between the bed evolution,

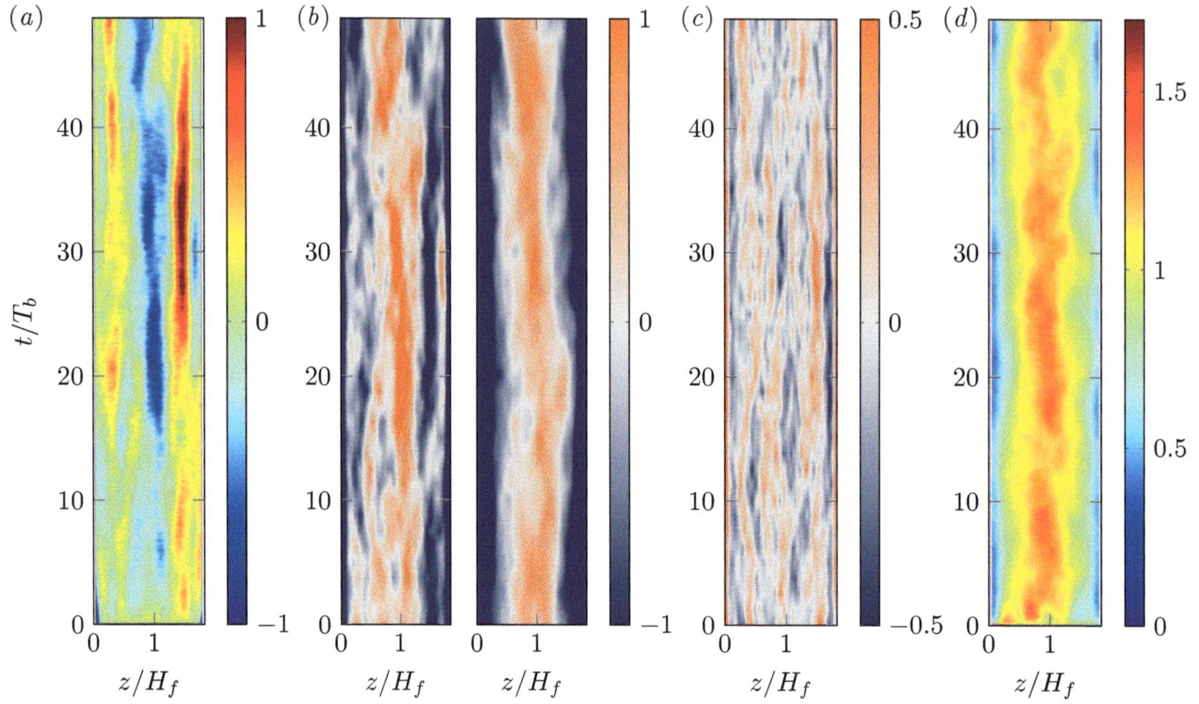

Figure 7.25: Space-time evolution of (*a*) the sediment-bed height fluctuation of the streamwise-averaged bed (identical to figure 7.3(*d*)); (*b*) the streamwise-averaged streamwise velocity fluctuation $(\langle u_f \rangle_x - u_{ref})^+$ at $\tilde{y}/H_f = 0.2$ (left) and $\tilde{y}/H_f = 0.5$ (right) (identical to figure 7.23(*d,j*)); (*c*) the streamwise-averaged wall-normal velocity $\langle v_f \rangle_x^+$ at $\tilde{y}/H_f = 0.2$; (*d*) the streamwise-averaged streamwise particle flux $\langle q_{p,x} \rangle_x(z,t)/q_{x,rms}$ with $q_{x,rms}(t) = (\langle \langle q_{p,x} \rangle_x \langle q_{p,x} \rangle_x \rangle_z)^{1/2}$ (identical to figure 7.6(*d*)). Data is visualised exemplary for case $DL500^{H2}$.

the particle flux and the streamwise velocity at $\tilde{y}/H_f = 0.2$ in a very similar way as it was observed for the open channel flows before. For instance, it is seen that the trough in the duct centre is 'carved' into the bed as a consequence of the strong high-velocity region that develops approximately 10 bulk time units after the onset of sediment erosion. On its right, on the other hand, the streamwise velocity is rather low such that sediment is predominantly deposited in this region, supporting the evolution of a local sediment ridge. Comparing the positions of troughs and ridges with the profile of the streamwise velocity in a larger distance to the bed at $\tilde{y}/H_f = 0.5$, however, the correlation is visibly much poorer, in contrast to the open channel case. This is insofar expected, as we have seen above that the large-scale organisation of the velocity field itself strongly varies along the wall-normal direction in the low aspect ratio ducts. Note that this marks an important difference to the open channel case and generally to canonical flow configurations, where, in the absence of lateral sidewalls, the largest-scale streaks spread from the centreline down to the bed maintaining an essentially constant width along most of the wall-normal direction. As a consequence, these structures are capable of inducing regions of increased bed shear stress and hence of stronger erosion that are of comparable size as their own width (Hutchins and Marusic, 2007b). In the duct flow, by contrast, regions of preferred erosion and deposition as well as the corresponding sediment troughs and ridges are seen to rather have the same width and location as the smaller high- and low-speed regions that centre at $\tilde{y}/H_f = 0.2$.

Comparing figures 7.25(*b*) and 7.25(*c*) furthermore reveals that in most of the low-speed regions at $\tilde{y}/H_f = 0.2$, fluid is preferentially moving away from the bed while the opposite is the case in regions of higher streamwise velocity. In this context, it is particularly interesting that the zones of local up- and downward motion in the vicinity of the lateral sidewalls show almost no lateral motion in the considered time interval. As a consequence, they are visible in form of the up- and downward oriented

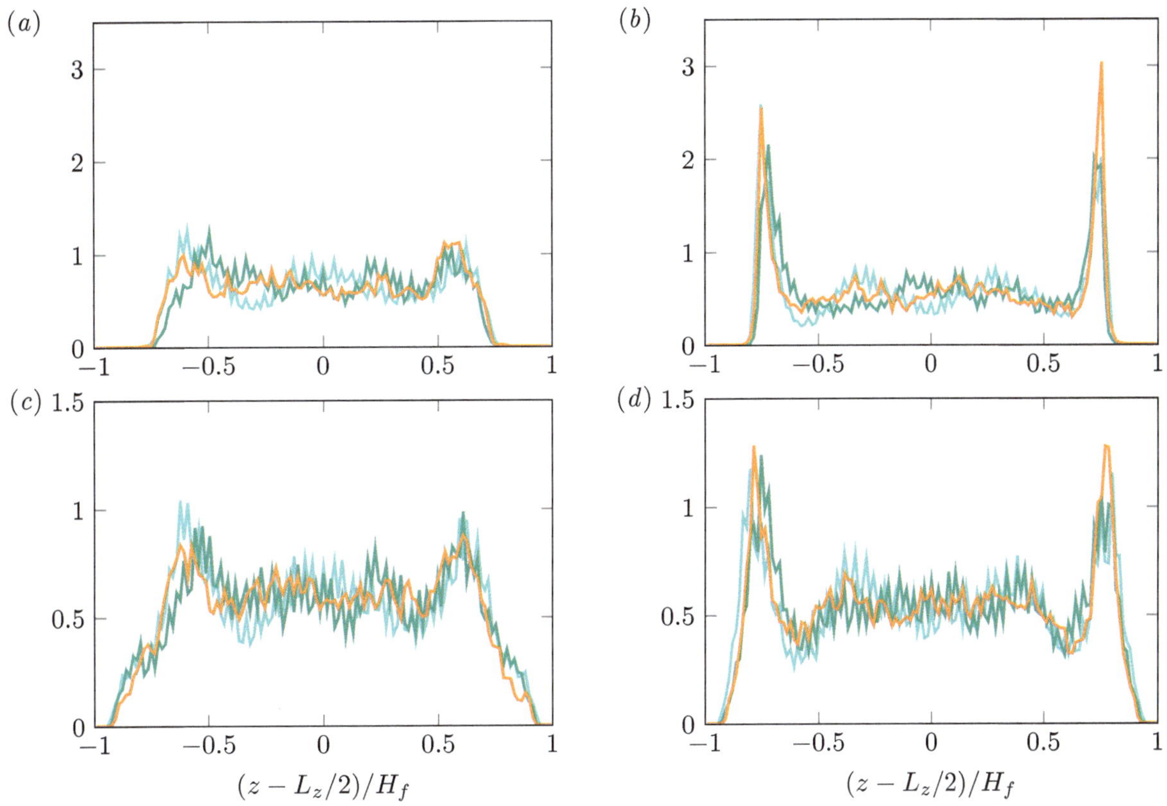

Figure 7.26: P.d.f. of the lateral position of (*a*) local maxima and (*b*) minima of the streamwise velocity as well as (*c*) local minima and (*d*) maxima of the wall-normal velocity at $\tilde{y}/H_f = 0.2$. The p.d.f. has been computed for wall-parallel planes filtered using a 2D Gaussian filter with dimensions $\Delta_x = 0.6H_f$ and $\Delta_z = 0.3H_f$, respectively. The presented datasets have been normalised such that its integral over the entire width of the duct L_z is unity. Data is shown for cases $DL240^{H2}$ (—), $DL250^{H2}$ (—) and $DL500^{H2}$ (—).

parts of the mean bottom vortex in the time-averaged secondary flow field, whereas the instantaneous local up- and downflow regions in the duct centre are meandering more strongly in time such their signature disappears in the time average. Important to pronounce is that the current observations do not imply that the wall-normal velocity organises in streamwise-elongated structures with lengths comparable to the domain length L_x. Rather, the here discussed streamwise-averaged regions of local up- and downflow represent the preferential locations of the usually much shorter instantaneous wall-normal velocity structures (Jiménez, 2018) in the cross-section – just as the corresponding regions of locally higher and lower streamwise velocity indicate the preferential positions of streamwise velocity streaks, as will be discussed in the following.

To this end, figure 7.26 shows the p.d.f. of the preferential location of local minima and maxima of the streamwise and wall-normal velocity in a distance of $\tilde{y}/H_f = 0.2$ from the mean bed height in the particle-laden narrow duct simulations. In order to remove small-scale fluctuations, the investigated xz-planes of the instantaneous velocity fields were first filtered using the two-dimensional version of the Gaussian filter introduced in equation (6.26). In order to maintain the divergence-freedom of the field, we treat the periodic boundary conditions as before, but mirror the flow field at the two solid sidewalls while this time reversing the sign of w_f in analogy to the treatment of the bottom no-slip wall proposed by Lozano-Durán et al. (2016). Thereafter, the lateral positions of the local extrema of both velocity components were determined for each streamwise location and all instantaneous flow field realisations.

The streamwise velocity profile is seen to typically attain a first maximum at a lateral distance of about $0.25H_f$ from either sidewall, a region that is at the same time most probably characterised by a marked downward fluid motion (cf. figure 7.26(*a*,*c*)). The p.d.f. of the preferential location of local streamwise velocity minima reveals less pronounced peaks, but still indicates a preferential location of regions with relatively lower streamwise velocity at a spanwise distance of approximately $0.5H_f$ from the closest sidewall. Roughly at the same location, the probability for an upward fluid motion is strongest, while that for a local minimum of the wall-normal velocity is lowest (cf. figure 7.26(*b*-*d*)). The latter zones mark the upflow regions of the mean secondary bottom vortices and are, due to the relatively lower streamwise velocity, regions of weaker sediment erosion, such that sediment ridges predominantly evolve in this part of the cross-section, as seen earlier in figure 7.16. In this context, it appears noteworthy that the curves of the different p.d.f.s more or less collapse for all three simulations despite the differing Reynolds number which is in case $DL500^{H2}$ with $Re_\tau = 515$ roughly twice as high as in the remaining simulations, suggesting that the positions of low- and high-speed regions in this distance to the bottom wall scale indeed in outer rather than in inner units, as claimed earlier. For the sake of completeness, let us furthermore remark that v_f typically attains the strongest maxima in the direct vicinity of the sidewalls, which leads in the time average to a mean upflow in the mean secondary flow field along the lateral domain boundaries. In the context of the Gaussian filtering, these near-sidewall peaks are filtered out such that their footprint is not visible in the p.d.f.s shown in figure 7.26. For the current problem of the interaction between the near-bed flow structures and the individual sediment ridges, however, these near-sidewall motions are of no direct relevance such that their absence in the filtered fields does not affect our general conclusions.

The strong correlation between the bed evolution and the organisation of the flow field even though at a lower wall-normal distance strongly suggests that the interaction of the turbulent flow with the bed is a comparable top-down mechanism as the one we have discovered in the context of the open channel case (cf. section 6.4). For the open channel case $CM850^{H3}$, the time lag between the bed evolution and the flow organisation at the channel centreline at $\tilde{y}/H_f = 0.5$ was seen to be about 10 bulk time units. Assuming that in the current open duct case $DL500^{H2}$ the speed of information propagation along the wall-normal direction is similar to the former case, one might expect a time-lag between the dynamics of the flow structures at $\tilde{y}/H_f = 0.2$ and that of the sediment bedforms of less than 5 bulk time units. In general, this relatively short time lag is hardly visible in the data, but there are few distinct instances which imply that a similar mechanism as in the open channel simulations also drives the evolution of sediment ridges in open ducts: For instance, we observe in figure 7.25 the development of a small sediment ridge in the right half of the cross-section left of the higher dominant ridge at approximately $t = 45T_b$ (cf. figure 7.25(*a*)), where the streamwise velocity is seen to be relatively low such that sediment preferentially agglomerates in this region (cf. figure 7.25(*b*)). However, the streamwise velocity in this region has reduced already a few bulk time units before at approximately $t = 41T_b$, but the trough is only slowly filled up with sediment such that the small sediment ridge develops with a small delay to the reduction of the fluid velocity.

In the previous chapter, the time lag and thus the causality between flow dynamics and bedform evolution could be unambiguously shown by evaluation of the two-point two-time cross-correlation function $\rho^t_{ua}(y, \delta t)$, as defined in equation (6.34). In the context of the open duct simulations, on the other hand, the definition in its original form does not represent an appropriate measure as in the vicinity of both sidewalls, the streamwise-averaged profiles of the sediment bed contour and that of the velocity are necessarily uncorrelated: The velocity at each height is subject to the no-slip boundary conditions,

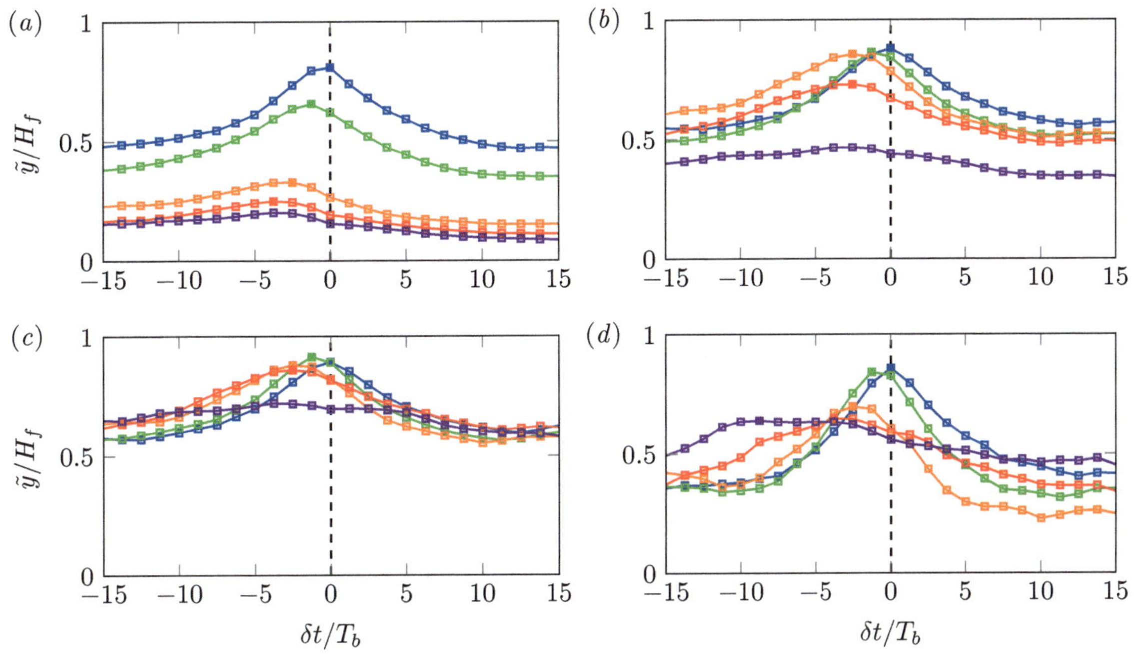

Figure 7.27: Two-point cross-time correlations $-\rho^t_{uh_b}(y,\delta t,b)$ between $(\langle u_f\rangle_x(y,z,t+\delta t)-u_{ref}(y,t+\delta t,b))$ and the sediment bed height fluctuations $h''_b(z,t)$ in case $DL500^{H2}$ computed over different lateral subintervals $[b,L_z-b]$ of the duct span and for different wall-normal positions $\tilde{y}/H_f$: (*a*) $b/H_f=0$, (*b*) $b/H_f=0.10$, (*c*) $b/H_f=0.25$ and (*d*) $b/H_f=0.50$. For each line, the horizontal distance between each two square symbols indicates the time intervals in which consecutive flow fields have been stored for the present analysis. Filled squares mark the time separation at which the maximum correlation is reached for the respective wall-normal location. The colouring indicates the wall-normal location: $y^+=10$ (—), $y^+=30$ (—), $y/H_f=0.2$ (—), $y/H_f=0.3$ (—) and $y/H_f=0.5$ (—).

whilst h''_b does in general not vanish at the sidewalls. The easiest solution therefore is to restrict the analysis, as before, to the duct centre excluding some region of width b next to the two sidewalls. An accordingly modified version of equation (6.34) for the two-point two-time cross-correlation function then reads

$$\rho^t_{ua}(y,\delta t,b)=\frac{\langle(\langle u_f\rangle_x(y,z,t+\delta t)-u_{ref}(y,t+\delta t,b))\,a''(z,t)\rangle_{[b,L_z-b]t}}{\left[\langle(\langle u_f\rangle_x(y,z,t+\delta t)-u_{ref}(y,t+\delta t,b))^2\rangle_{[b,L_z-b]t}\,\langle(a''(z,t))^2\rangle_{[b,L_z-b]t}\right]^{1/2}},\qquad(7.15)$$

where $\langle\bullet\rangle_{[b,L_z-b]t}$ indicates temporal and spanwise averaging over the interval $z\in[b,L_z-b]$ only. The duct core reference velocity u_{ref} is defined as in figure 7.23, viz.

$$u_{ref}(y,t,b)=\langle\langle u_f\rangle_x(y,z,t)\rangle_{[b,L_z-b]}=\frac{1}{L_z-2b}\int_b^{L_z-b}\langle u_f\rangle_x(y,z,t)\,\mathrm{d}z.\qquad(7.16)$$

Since the choice of the parameter b is, as stated above, more or less arbitrary, we show in figures 7.27(*a*-*d*) the two-point two-time cross-correlation function between the sediment bed evolution and that of the flow field, $-\rho^t_{uh_b}(y,\delta t,b)$, in case $DL500^{H2}$ for different values of $b\in\{0,0.1,0.25,0.5\}$ in order to assess the sensitivity of $\rho^t_{uh_b}$ to changes of b. Note that for the first two values of b, both sediment ridges and the trough in their middle are included in the analysis, whereas for a parameter value $b=0.5$, $\rho^t_{uh_b}$ basically describes the correlation between the temporal evolution of the trough and the flow field in the duct core above it.

As expected, the exact amplitude of $\rho^t_{uh_b}$ clearly depends on the chosen spatial averaging interval $[b, L_z - b]$ due to the inhomogeneity of the flow field in the lateral direction. Of higher interest is for us, however, that the peak of the correlation function continuously shifts towards earlier times (i.e. negative values of δt) when moving away from the sediment bed for all considered parameter values b. Indeed, this confirms our observations in figure 7.25 and the associated conjecture that quite similar to the 'top-down formation mechanism' seen to act in open channel flows, sediment ridge formation is also in open duct flows controlled by the dynamics of the turbulent flow, not *vice versa*.

Interestingly, a qualitative difference is seen between the correlation profiles at wall-normal locations $\tilde{y}/H_f \leq 0.3$ and those at the duct half height $\tilde{y}/H_f = 0.5$: In fact, $\rho^t_{uh_b}$ features only for locations in the lower half of the duct $\tilde{y}/H_f \leq 0.3$ a pronounced peak that is indicative of a clear causal relation between velocity field and sediment ridge organisation. For $\tilde{y}/H_f = 0.5$, as opposed to the former, the correlation profile features much weaker peaks and a slow decay with increasing values δt. This agrees well with our earlier observations that the typical width of individual velocity structures varies across the wall-normal direction of the duct and that the smaller scales closer to the bed control where sediment is predominantly eroded and deposited. One sign for an, even though weak, causal relation between the flow in the duct centre and the sediment bed evolution might be seen in figure 7.25(*d*), where $\rho^t_{uh_b}$ is determined only over the duct centre $[0.5H_f, L_z - 0.5H_f]$ and thus, for $\tilde{y}/H_f = 0.5$, basically indicates how the lateral position of the global maximum of the outer velocity profile is correlated to the minimum of the central sediment trough. For this particular relation, we obtain a time lag of about ten bulk time units quite similar to the corresponding values observed in the open channel case $CM850^{H3}$ in section 6.3.6. Let us finally mention that the correlations in the remaining considered duct flows reveal qualitatively similar characteristics, whilst the open channel reference case $CL250^{H3}$ features – as expected – a relatively high correlation between the sediment ridge patterns and the organisation of the turbulent flow field even at $\tilde{y}/H_f = 0.5$, attaining a peak at a negative time lag of around ten bulk time units, quite similar to case $CM850^{H3}$.

7.3 Organisation of coherent structures in open duct flows

The close correlation between the evolution of individual sediment ridges and the dynamics of the flow field motivates a more detailed investigation of individual coherent structures and their preferential organisation. In the previous chapter 6, we have studied the interplay between individual large-scale velocity streaks and evolving sediment ridges in the absence of lateral sidewalls. As strongly suggested by the results reported in the previous sections, the large-scale velocity structures in the core of large-aspect ratio ducts closely resemble those in laterally periodic channels, and so does the sediment bed formation process. In open ducts of low aspect ratio, on the other hand, coherent structures everywhere in the domain 'feel' the presence of the lateral sidewalls and a substantially different flow develops. In the following, we therefore restrict ourselves to the investigation of coherent structures in low aspect ratio open duct flows over both smooth bottom walls and mobile sediment beds.

While the dynamics of individual coherent structures in canonical flows have attained much attention during the last decades (Robinson, 1991; Adrian, 2007; Jiménez, 2018), the number of studies that focus on the same structures in ducts of non-circular cross-section is significantly smaller. For closed square duct flows, Uhlmann et al. (2007) and Pinelli et al. (2010) were the first to systematically investigate the role of instantaneous coherent structures for the peculiar pattern of the time-averaged turbulent mean

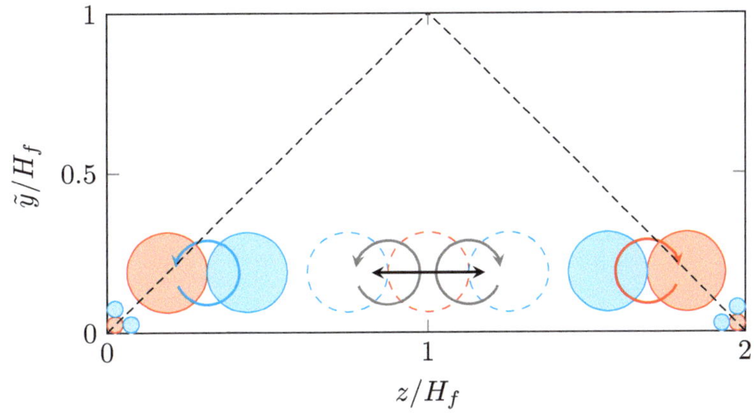

Figure 7.28: Conceptual sketch of streaks and vortices in different locations of the cross-section that contribute to the mean flow and/or the fluctuating field. Idealised clockwise (counterclockwise) rotating vortices are indicated by red (light blue) arrows, while red (light blue) circles indicate high-speed (low-speed) streaks. Structures that are constrained in their position by the domain boundaries and thus contribute to the long-time average are shown in the respective colours. Grey and dashed objects are not restricted by the domain boundaries and are therefore free to meander and to form at different lateral positions (indicated by the black double-arrow). They will not contribute to the mean flow field and will instead appear only in the fluctuating fields u'_f and $\omega'_{f,x}$, respectively.

secondary flow. For open duct flow, in turn, the work of Sakai (2016) appears to be the first and only one of its kind that focusses in detail on the importance of such structures. In both cases, i.e. open and closed duct flows, the presence of lateral sidewalls breaks the statistical lateral homogeneity of the flow such that the mean flow becomes a two-dimensional field $\langle \boldsymbol{u}_f \rangle_{xt}(y,z)$, in contrast to the one-dimensional mean flow profile in canonical channel or pipe flows. This has important implications for the analysis of coherent structures which can contribute here to both the mean and the fluctuating field, whereas in canonical flows coherent velocity structures are conventionally defined as subdomains in which the fluctuating field $\boldsymbol{u}'_f(\boldsymbol{x},t)$ exceeds a given threshold (cf. also the discussion on the implications of this conventional choice in section 5.4 of Jiménez, 2018).

As an example, let us consider the mean bottom wall shear stress in either closed or open duct flows which has been discussed earlier in this chapter (cf. figure 7.14). As outlined by Pinelli et al. (2010), the first local peaks of the wall shear stress refer to the preferential positions of the instantaneous high-speed streaks that are closest to the duct corner. These features are strongly restricted in their mobility and in their preferential location by the flanking sidewalls, such that they leave a marked imprint in the mean flow field in form of the aforementioned wall shear stress peaks. Streaks further away from the sidewalls (in terms of wall units) are not laterally bounded, and for time-averaging intervals sufficiently longer than their lifetime, contributions from streaks of different signs effectively cancel out each other in the statistical average. Consequently, these streaks can be considered as objects of the fluctuating field exclusively. It might be worth noting that this effect is certainly not limited to the buffer layer streaks, and indeed we have seen in figure 7.26 that local peaks of the streamwise velocity appear at a preferential distance to the sidewalls also in larger distances to the bottom wall, whilst they are essentially homogeneously distributed over the rest of the duct cross-section. A conceptual sketch of features that contribute to the mean field and those that are contained in the fluctuating field only is provided in figure 7.28.

In this context, let us pronounce that above considerations are under the assumption that the time-averaging interval is sufficiently longer than the lifetime of the largest structures in the domain. While this is more or less given for the smooth-wall duct reference simulations which feature observation

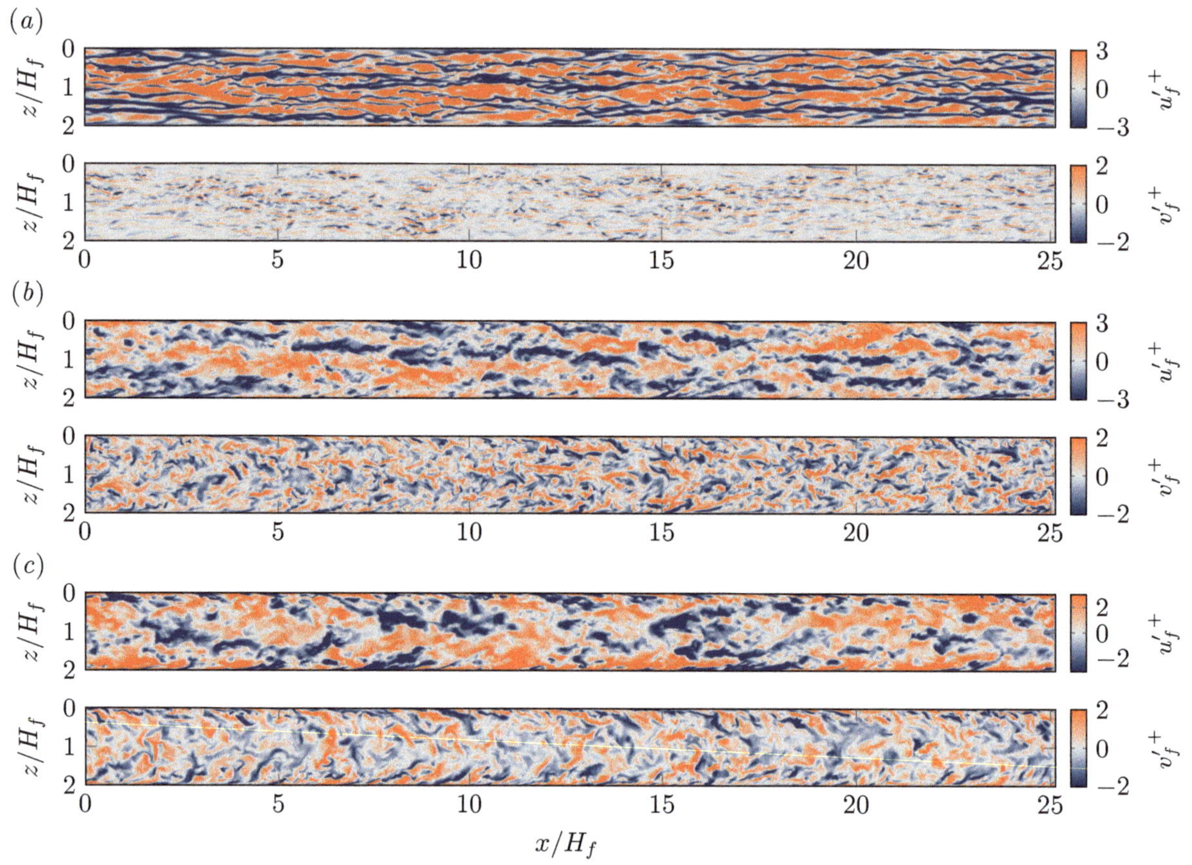

Figure 7.29: Instantaneous wall-parallel planes of the streamwise and wall-normal velocity fluctuations in case $DL400^{H2}_{smooth}$ at $t = 99.4T_b$ in different distances to the bottom wall: (*a*) $\tilde{y}^+ \approx 15$, (*b*) $\tilde{y}/H_f = 0.2$, (*c*) $\tilde{y}/H_f = 0.5$. Each panel shows two different flow field visualisations, from which the upper one represents u'_f/u_τ and the bottom one v'_f/u_τ.

intervals of $\mathcal{O}(10^3T_b)$ length, the particle-laden simulations are studied over two orders of magnitude shorter time intervals in virtue of the evolving transverse bedforms. As was discussed in the context of figure 7.21, the consequence of the shorter time interval is that the statistical footprints of individual large-scale velocity structures are still visible in the mean flow field even in the duct centre, whereas such signatures are essentially invisible for the corresponding long-time single-phase duct simulations. In the latter case, a large-scale velocity streak in the domain centre contributes only to the fluctuating field. For the subsequently considered low aspect ratio open duct cases, however, the mean flow field agrees quite well between short-time particle-laden and long-time smooth-wall simulations despite the different investigated time intervals. The reason for this is most likely that the cross-sections of these duct flows are too narrow to accommodate more than a single large-scale streak of dimensions comparable with H_f, as discussed earlier. These streaks are, in addition, strongly constrained in their lateral mobility and, thus, they contribute considerably to both the mean flow and the fluctuating field.

In the following, we now focus exclusively on those contributions of the streaks that are contained in the fluctuating velocity field. Figure 7.29 shows different horizontal xz-slices of the streamwise and wall-normal fluctuating velocity field exemplary for case $DL400^{H2}_{smooth}$, extracted from an instantaneous state of the flow field at three different wall-normal positions $\tilde{y}/H_f = 0.04$ ($\tilde{y}^+ = 15$), $\tilde{y}/H_f = 0.2$ ($\tilde{y}^+ = 83$) and $\tilde{y}/H_f = 0.5$ ($\tilde{y}^+ = 208$), respectively. Even though the friction Reynolds number is with $Re_\tau = 416$ still moderate, it is sufficiently high to allow a clear distinction between near-wall buffer layer and outer-scaling large-scale coherent structures. Figure 7.29(*a*) shows the characteristic elements of the buffer layer comprising relatively smooth streamwise-elongated velocity streaks with typical characteristic spacing $\mathcal{O}(100\delta_v)$ and shorter quasi-streamwise vortices indicated by locally high positive and negative wall-normal velocity amplitudes. While in most areas of rather low magnitude due to the underlying impermeable bottom wall, v_f attains locally higher values in particular in regions where the streaks are seen to get disordered and vorticity gains in strength, as conventionally expected in the bursting phase of the self-sustaining buffer layer cycle (Hamilton et al., 1995; Waleffe, 1997). Similar to their counterparts in the smooth channel simulations discussed in section 6.3.5, the streaks are not randomly distributed across the domain. Instead, they are seen to organise themselves in larger clusters whose dimension are comparable to the size of the larger structures further away from the wall. Indeed, it can be visually identified that the positions of these clusters correlate quite well with the larger turbulent streaks found at $\tilde{y}/H_f = 0.2$ in figure 7.29(b), taking into account that the mean speed of propagation varies with the mean shear along the wall-normal direction.

At $\tilde{y}/H_f = 0.2$, the streamwise velocity streaks reveal a 'rougher' shape and are of larger size reaching lengths of several H_f and widths up to $0.7H_f$ (approximately $300\delta_v$), which suggests that these features are fully-turbulent structures that are not part of the buffer layer anymore. Due to their larger size, the cross-section of the duct can usually accommodate only between two and four streaks of either sign. The corresponding wall-normal velocity structures are again of shorter streamwise extent such that, typically, several individual structures of positive wall-normal velocity can be associated with a single low-speed streak and *vice versa* for the high-speed streaks. In the near-sidewall region, both streaks and structures of intense wall-normal velocity are inclined to the no-slip sidewalls and thus also to the mean flow direction, just as individual coherent structures in canonical flows are inclined to the bottom wall (Jiménez, 2018). Recalling that v_f' attains in this context the role of the wall-parallel velocity component (Huser and Biringen, 1993), the observed inclination angle agrees reasonably well with values $\approx 35°$ found in two-point correlations of the spanwise velocity component of canonical flows (cf. Sillero, 2014, and the subsequent section 7.3.1). Further increasing the distance to the bottom wall, velocity structures are seen to further grow in size: At $\tilde{y}/H_f = 0.5$, i.e. half way between the bottom wall and the free surface, streamwise velocity structures reach sizes comparable to the duct width such that not more than one or two individual streaks fit into the cross-section, as is seen in figure 7.29(c). The wall-normal velocity structures are similarly larger than their counterparts closer to the wall and partly exhibit significant inclinations w.r.t. the mean flow direction, often larger than 45°.

7.3.1 Two-point velocity statistics

The seeming correlation between the locations of outer-scaling flow structures at $\tilde{y}/H_f \approx 0.2$ and the organisation of the smaller buffer layer streaks into larger clusters are in good agreement with our observation in the sediment-laden flows, as it implies that the regions of strong and weak wall shear stress and hence erosion activity are again controlled by larger outer-scaling structures from outside the buffer layer.

With this in mind, we focus in the following on the preferential organisation of velocity structures in this distance to the wall. In consequence of the two non-homogeneous spatial directions in open and closed duct flows, an analysis of the lateral organisation of individual velocity streaks based on premultiplied energy spectra as in the doubly-periodic channel is not possible. Instead, the method of choice is an analysis of the two-point correlation in physical space. To this end, we introduce the two-point velocity covariances as (Townsend, 1976; Pope, 2000)

$$R_{u_iu_i}(\delta x, y, z, \check{y}, \check{z}) = \langle u'_{fi}(x, y, z, t) u'_{fi}(x + \delta x, \check{y}, \check{z}, t) \rangle_t, \quad i = 1, 2, 3 \tag{7.17}$$

where $(\check{y}, \check{z})$ is a reference point in the yz-plane and δx is the streamwise separation length. The corresponding two-point velocity correlations are then obtained by normalising $R_{u_iu_i}$ with the respective turbulence intensities $u_{fi,rms}$, viz.

$$C_{u_iu_i}(\delta x, y, z, \check{y}, \check{z}) = \frac{R_{u_iu_i}}{u_{fi,rms}(y, z) u_{fi,rms}(\check{y}, \check{z})}, \tag{7.18}$$

such that $C_{u_iu_i}$is unity at the reference point as, by definition,

$$R_{u_iu_i}(0, \check{y}, \check{z}, \check{y}, \check{z}) = u_{fi,rms}(\check{y}, \check{z}) u_{fi,rms}(\check{y}, \check{z}). \tag{7.19}$$

As discussed in Jiménez (2018), the two-point velocity covariances and correlations are in general functions in $\mathbb{R}^6$, but their dimension reduces by one for each homogeneous direction. In regard of the streamwise periodicity of the here considered domain, the investigated velocity correlations in duct flows are therefore five-dimensional objects. For a fixed reference point $(\check{y}, \check{z})$ in the duct cross-section, however, $C_{u_iu_i}$ reduces to a three-dimensional field which can be visualised as done in figure 7.30(a), exemplary for C_{uu} and $(\check{y}, \check{z})$=$(0.2H_f, L_z/2)$ in case $DL400^{H2}_{smooth}$. Therein, red and blue surfaces refer to manifolds of constant positive and negative streamwise correlation $C_{uu} = \pm 0.1$, respectively. The correlation field features a region of high positive correlation around the reference point in the duct centre with a maximum streamwise extent of about $5H_f$, flanked by two essentially symmetric negative lobes with a somewhat shorter streamwise length. All three objects are quasi-parallel and aligned with the streamwise direction, but are inclined to the bottom wall. In addition, it is noticed that the streamwise dimension is not symmetric w.r.t. the reference point at $\delta x = 0$, but all three feature longer tails that span further in negative δx-direction than their heads do in positive δx-direction.

The length of the positive lobe is best seen in figure 7.30(b), where a cut through C_{uu} along the streamwise direction is shown for cases $DL400^{H2}_{smooth}$ and $DL500^{H2}$, respectively. Similar visualisations for the remaining velocity components, supplemented with cross-sectional views of the respective cross-correlations, follow in the remaining panels of figure 7.30. The isocontours of the different correlation functions show a qualitatively good agreement between the smooth-wall and the particle-laden case, even though the data in the latter case shows a stronger scatter and in particular the large positive object in the centre of C_{uu} is clearly shorter than its counterpart in case $DL400^{H2}_{smooth}$. It is assumed that different effects contribute to this deviation: First, let us recall that the streamwise domain extension is shorter by approximately factor two in the particle-laden case, which might affect the streamwise dimensions of the correlation objects. Second, the number of ensembles that underly the shown statistics differs markedly: while for case $DL400^{H2}_{smooth}$, 1200 individual fields are available over a time frame of $t = 286T_b$, the observation time is in case $DL500^{H2}$ with $t = 50T_b$ significantly shorter and with only 40 individual flow realisations, the available data is rather limited. Eventually, in addition to the numerical restrictions, the role of the mobile sediment bed is not entirely clear: While the investigations throughout the previous and the present chapter have shown that the large-scale structures in some distance to the bed seem to be rather unaffected by the presence of mobile particles in the near-bed

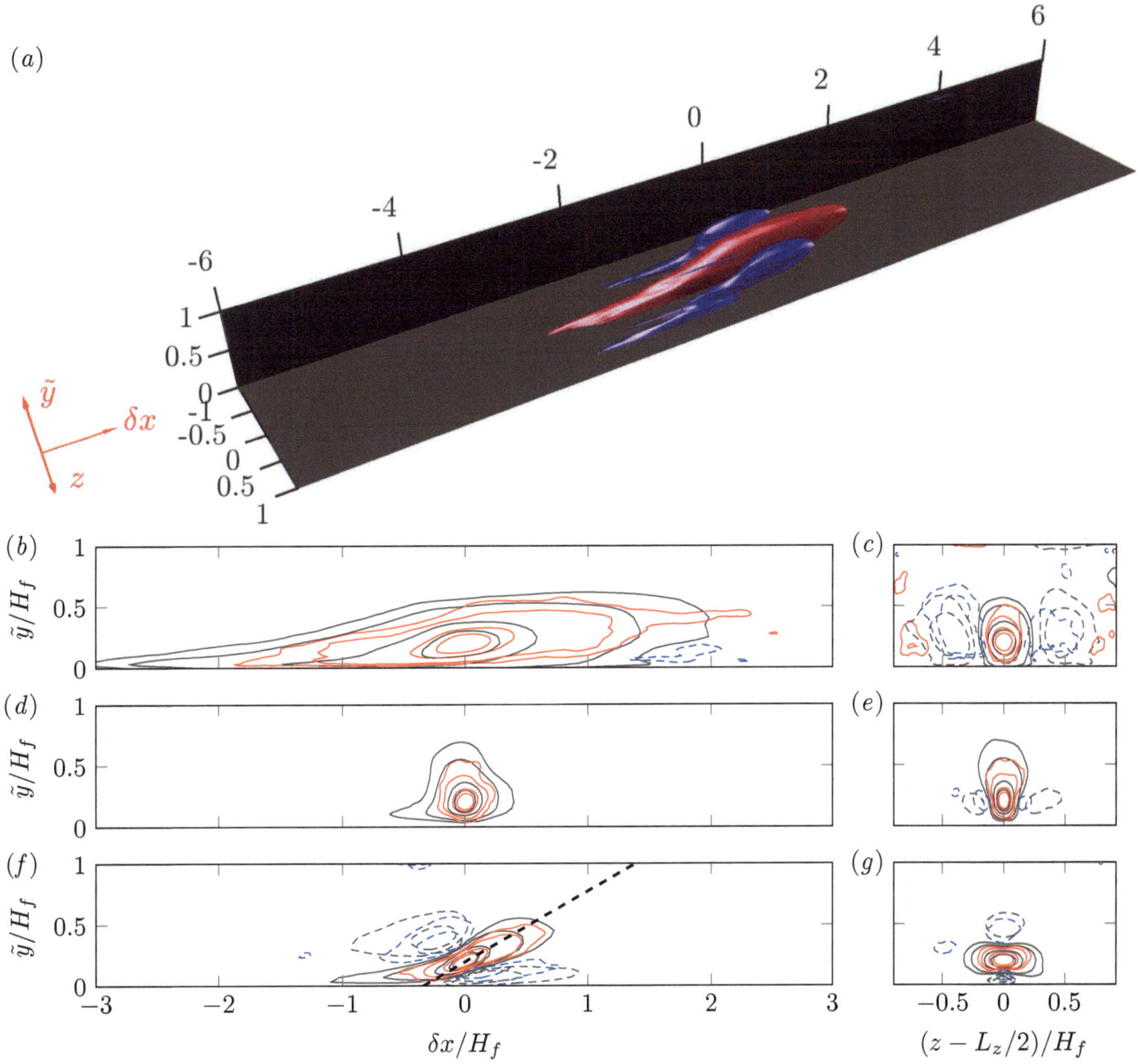

Figure 7.30: Two-point auto-correlation function of the individual velocity components in the open duct simulations $DL400^{H2}_{smooth}$ and $DL500^{H2}$ with respect to the reference point $(\check{y}, \check{z}) = (0.2H_f, L_z/2)$. *(a)* Three-dimensional visualisation of $C_{uu}(\delta x, y, \check{y}, (z-\check{z}), \check{z})$, where red (blue) iso-surfaces represent manifolds of constant $C_{uu} = 0.1$ ($C_{uu} = -0.1$). *(b,d,f)* Streamwise wall-normal xy- and *(c,e,g)* spanwise wall-normal yz-sections through $C_{u_iu_i}$, with solid (red) contours indicating a positive correlation at $C_{u_iu_i} = \{0.05, 0.1, 0.3, 0.5\}$ and dashed (blue) contours representing a negative correlation at $C_{u_iu_i} = \{-0.1, -0.05\}$. Coloured and grey lines refer to the mobile particle case $DL500^{H2}$ and the smooth-wall reference case $DL400^{H2}_{smooth}$, respectively. *(b,c)* Streamwise velocity C_{uu}, *(d,e)* wall-normal velocity C_{vv}, *(f,g)* spanwise velocity C_{ww}. The black dashed line in *(f)* indicates an inclination angle of 30° w.r.t. the bottom wall.

region, it is not unlikely that they do contribute to a shortening of the correlation tails compared to the smooth wall case near the bed, predominantly in negative δx-direction.

In contrast to the streamwise correlation function, those of the cross-stream velocity components are more localised in the streamwise direction and do not spread over streamwise extents as long as the objects associated with C_{uu} (cf. figures 7.30(*d,f*)). In particular for C_{vv}, we observe that a relevant correlation is only given in a distance of $\pm 0.5H_f$ up- and downstream of the reference point, which is in accordance with the fact that the instantaneous structures of intense wall-normal velocity seen in figure 7.29 are shorter than the streamwise velocity streaks. The spanwise correlation C_{ww}, in turn, reveals two smaller negative and a single tilted positive correlation object, from which the latter features the strongest inclination w.r.t. the bottom wall of all velocity components. The angle between this positive lobe and the bottom wall is slightly lower than 30° and thus very similar to values 30°-35° observed in canonical boundary layers (Sillero, 2014; Sillero et al., 2014) and closed channels (Jiménez, 2018). Generally, the discovered characteristic shapes of the two-point correlation objects exhibit an astonishing qualitative similarity with those observed for canonical flows at significantly higher Reynolds numbers up to $Re_\tau \approx 2000$ in the latter three studies, even though the streamwise correlation lengths are markedly different for the three cases. The strikingly similar shape of the correlation objects in boundary layer, closed channel and open duct flows supports the suggestion that not only the characteristic buffer layer structures, but also the larger outer-scaling coherent motions in duct flows share many similarities with those in canonical wall-bounded flows. A qualitative comparison with the datasets of Sillero (2014), Sillero et al. (2014) and Jiménez (2018) as well as a discussion of the varying streamwise correlation lengths in the different flow configurations is out of the scope of the current work, but would be worth performing in a follow-up study.

The cross-sectional cut through C_{uu} in figure 7.30(c) shows that the three main objects of the streamwise velocity correlation are all between $0.4H_f$ and $0.5H_f$ wide and span merely over the lower half of the duct cross-section. As such, they represent the largest objects for which more than a single representative fit into the cross-section of the duct, considering the typical aspect ratio of the streamwise velocity streaks reported by Sillero (2014). In the sense of the attached-eddy model (Townsend, 1961; Jiménez, 2012), the extracted correlation objects might refer to the family of self-similar streamwise velocity streaks attached to the bottom wall, that we will investigate in more detail in the following section. Though, it has to be kept in mind that the considered correlations are mean statistics themselves and thus have to interpreted accordingly. In particular, the shown organisation of three neighbouring velocity patterns of alternating sign should not be expected to occur in this symmetric arrangement in instantaneous flow fields. Nonetheless, the shown correlations suggest that individual structures of the given size might significantly contribute to the full correlations.

In this context, it is especially interesting to compare the spanwise organisation of the correlation objects of the different velocity components with each other. Both canonical boundary layers and closed channels feature a homogeneous transverse direction such that the two-point correlations do not depend on the physical spanwise position, but they are functions of the spanwise separation length δz only. As a consequence, the correlations shown for these two flow configurations have to be understood w.r.t. an arbitrary spanwise position in physical space, which is not necessarily the same for C_{uu}, C_{vv} and C_{ww}. In these configurations, it is therefore hardly possible to comment on the relative organisation of the different velocity components w.r.t. each other only based on the two-point statistics. In the current case, on the other hand, all three correlation functions are extracted relative to a single point of the cross-section in physical space, such that the only possible shift between the correlations is in the homogeneous streamwise direction. Comparing the lateral positioning of the objects of C_{uu} with that of the wall-normal counterparts associated with C_{vv} in figure 7.30(e), we remark that the positive wall-normal velocity correlation object similarly features two smaller zone of negative correlation on either side. Although the comparison of C_{uu} and C_{vv} neither provides information on the sign of the velocity components nor does it ensure that intense events of the two velocity components occur simultaneously (Jiménez, 2018), the results are most likely reflecting the conventionally observed upward-directed fluid motion inside instantaneous low-speed streaks and the corresponding downflow in the high-speed streaks. In this regard, objects of C_{vv} and C_{ww} in the respective locations are assumed to form closed conditional rollers between the streaks, as similarly concluded for canonical flows by Jiménez (2018).

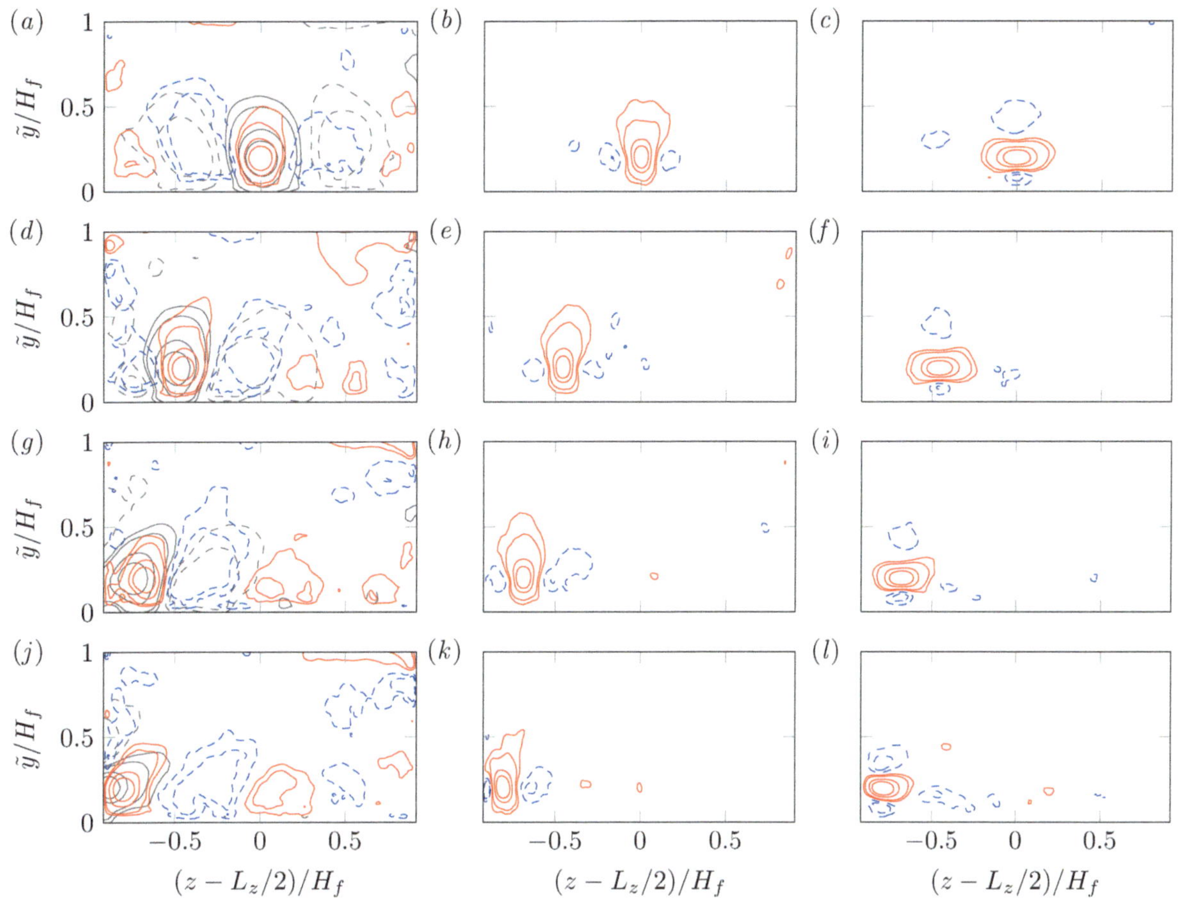

Figure 7.31: Spanwise wall-normal yz-section through the two-point auto-correlation function $C_{u_iu_i}$ of the individual velocity components in the open duct simulations $DL400^{H2}_{smooth}$ and $DL500^{H2}$ with respect to different reference points: (*a-c*) $(\check{y}, \check{z}) = (0.2H_f, L_z/2)$, (*d-f*) $(\check{y}, \check{z}) = (0.2H_f, L_z/4)$, (*g-i*) $(\check{y}, \check{z}) = (0.2H_f, L_z/8)$ and (*j-l*) $(\check{y}, \check{z}) = (0.2H_f, L_z/16)$, respectively. Solid (red) contours indicate a positive correlation at $C_{u_iu_i} = \{0.05, 0.1, 0.3, 0.5\}$, while dashed (blue) contours represent a negative correlation at $C_{u_iu_i} = \{-0.1, -0.05\}$. (*a,d,g,j*) Streamwise velocity C_{uu}, the grey contours in the background correspond to case $DL400^{H2}_{smooth}$. Note that due to the slightly different aspect ratio, the reference points do not exactly collapse. (*b,e,h,k*) Wall-normal velocity C_{vv}; (*c,f,i,l*) spanwise velocity C_{ww}.

In figure 7.31, the correlation functions $C_{u_iu_i}$ are compared for different lateral reference positions $\check{z}/L_z \in \{1/16, 1/8, 1/4, 1/2\}$ across the cross-section of case $DL500^{H2}$ for a constant reference height $\check{y}/H_f = 0.2$. For the cross-stream velocity components, the two-point correlations exhibit essentially no dependence on the spanwise position of the reference point. Even if this latter is chosen in the direct vicinity of the wall as for $\check{z} = L_z/16$ in figures 7.31(*k,l*), there is only a weak asymmetry visible. The relatively weak influence of the spanwise location is most probably a consequence of the small size of the v_f- and w_f-structures in this distance to the bottom wall. In fact, the structures essentially do not feel the presence of the lateral sidewalls unless they are located in their direct vicinity, which indeed corroborates our observations concerning the instantaneous field in figure 7.29.

The isocontours of the streamwise two-point correlations C_{uu} reveal in case $DL500^{H2}$ again a rather strong scatter as a consequence of the relatively low number of samples in the particle-laden cases. For the sake of comparison, we therefore show C_{uu} in addition in the smooth-wall reference case $DL400^{H2}_{smooth}$ (cf. figure 7.31(*a,d,g,j*)). In this context, one should note that the reference points cannot exactly collapse for both simulations as their aspect ratios slightly differ. It is nevertheless identifiable that the general shape and size of the positive correlation object changes little with the lateral variation of the reference point. The characteristics of the neighbouring negative correlation cells, on the other hand, indeed depend on the spanwise position: If the positive lobe associated with the reference point is located in sufficient distance to the sidewall, it is flanked by negative correlation objects on either side which are essentially of same shape and size as the positive object itself. Choosing a reference point closer to the wall, the correlation pattern becomes single-sided in the sense that there is not enough space between the positive object and the sidewall to accommodate another streak of this size. While the positive correlation object in the particle-laden case is always accompanied by at least one second object of opposite sign no matter how close the reference point approaches the wall, this is not the case in the smooth-wall simulation (cf. figure 7.31(*j*)). In the latter case, we generally observe a faster decay of the two-point correlation in the spanwise direction, while for case $DL500^{H2}$ C_{uu} is of non-trivial amplitude even in significant distance to the reference point, which is another indication that the statistics in the latter case are less converged than those in the smooth-wall case.

7.3.2 Streamwise velocity streaks

The organisation of the two-point correlation functions implies that instantaneous streamwise velocity streaks with a width of $(0.4\text{-}0.5)H_f$ are a key feature of the turbulent flow in the investigated wall-parallel layer $\tilde{y}/H_f = 0.2$. However, two-point correlations themselves are mean flow statistics, such that they represent the average organisation of the flow, but their meaningfulness concerning the instantaneous structure of the turbulent flow field is limited. In the following, we will therefore extract fully three-dimensional instantaneous streamwise velocity streaks from the individual flow fields and study their preferential organisation in the duct cross-section depending on their size.

To do so, we make use of an extraction technique that was first used by Moisy and Jiménez (2004) to detect regions of strong vorticity and dissipation in isotropic turbulence and by Del Álamo et al. (2006) to identify self-similar vortex cluster in a turbulent closed channel flow. Ever since, it has been applied to detect coherent structures associated with different physical quantities such as velocity fluctuations or Reynolds stresses in a variety of turbulent shear flows, including channels (Lozano-Durán et al., 2012; Bae and Lee, 2021), boundary layers (Sillero, 2014), free shear (Dong et al., 2017) and closed duct flows (Atzori et al., 2021). The basic idea is to define an instantaneous structure as a connected subdomain of Ω_f, for which a physical quantity of interest a fulfils a given condition, for instance, that it exceeds a given threshold. Here, connectivity is understood along the orthogonal lines of the Cartesian computational grid, i.e. a given grid point that fulfils the condition is connected to a neighbouring structure if at least one of its six orthogonal neighbours is an element of this structure.

Surely, the technique depends on the adopted threshold and a reasonable choice of the latter is crucial to obtain physically meaningful results. In the extreme cases of a very low or high threshold, the technique can be assumed to detect either a single very large object as the condition is fulfilled essentially everywhere in the domain or only few very small features, as the condition is fulfilled almost nowhere. Of interest is therefore a threshold somewhere in between these two extremes, which is in most of the above listed studies chosen based on the results of a 'percolation analysis' (Moisy and Jiménez, 2004), in which the extraction of structures is performed repeatedly while slowly increasing the threshold. An optimal choice of the threshold is then typically defined around what is called the 'percolation crisis', a range of thresholds for which the ratio between the volume of the largest extracted structure and the total extracted volume abruptly decreases, while the number of extracted structures is close to its maximum. Needless to say, such percolation analyses are relatively expensive from a computational point of view as a large amount of data has to be analysed again and again. That is why in most of

Figure 7.32: Visualisation of instantaneous streamwise high- and low-speed streaks in case $DL500^{H2}$ ($t/T_b = 23$), as seen from the top of the open duct. High- and low-speed streaks were detected as connected regions for which $u'^{+}_{f}(x,t) > 2$ and $u'^{+}_{f}(x,t) < -2$, respectively, with flow from left to right. High-speed streaks are indicated by reddish colours, while the surface of low-speed streak structures is coloured in blue. In both cases, brighter colours indicate a larger distance to the bottom wall and thus to the sediment bed.

the above studies, extra simulations at lower resolution are performed only for the purpose of defining the optimal threshold (Sillero, 2014), or the percolation analysis is performed on a subdomain of the total computational domain only (Del Álamo et al., 2006; Lozano-Durán et al., 2012). Performing this relatively expensive and time-consuming analysis for the current open duct simulations is out of the scope of the current work. However, in the past years percolation analyses have been performed for all physical quantities that we will focus on in the following such that we choose the same thresholds adopted in these studies, also to allow for a comparison of our data with that in previous investigations of other flow configurations such as channel and developing boundary layer flows.

In the current section, we will first study the organisation of streamwise velocity streaks in the duct cross-section. In this context, we define high- and low-speed streaks as connected regions for which $u'^{+}_{f}(x,t) > 2$ and $u'^{+}_{f}(x,t) < -2$, respectively, using the velocity threshold detected by means of a percolation analysis by Sillero (2014) in turbulent boundary layers. As an example, figure 7.32 shows the so extracted high- and low-speed streaks for an instantaneous flow realisation of case $DL500^{H2}$. In agreement with the procedure described in Sillero (2014), we evaluate statistics only for structures that possess a volume larger than $(30\delta_\nu)^3$ to avoid resolution issues in the framework of the discrete numerical grid. For the same reason, we additionally exclude all structures with streamwise extensions $\widetilde{\Delta x} \leq 3\Delta x$ from the analysis (the definition of $\widetilde{\Delta x}$ follows below). Note that in terms of wall-bounded turbulent structures, this does not pose a series restriction on the analysis as even the quasi-streamwise vortices of the buffer layer are typically larger than the adopted thresholds.

An overview of the available number of individual flow fields and the simulated time intervals is given in table 7.3 for all simulations considered in this section and the appendices of this work. Also, it is shown which of the performed coherent structure detection techniques that will be used in the remainder of this section are applied to which simulation. Due to the relatively short observation interval and the consequently small number of available fields, statistics on three-dimensional flow structures remain relatively poor for the particle-laden case. For the subsequent study, we therefore restrict ourselves to a discussion of the results obtained for the smooth-wall duct case $DL400^{H2}_{smooth}$, for which a sufficiently large ensemble of 1200 individual flow fields could be collected. For the particle-laden cases, on the other hand, we performed an alternative streak eduction analysis per cross-section following the concept first presented by Nakatsuji (2012) for closed duct and later applied to open duct flows by Sakai (2016). Details on the extraction procedure and a comparison of the streak organisation in particle-laden and single-phase smooth-wall simulations are given in appendix D.

Case	Re_τ	N_{fields}	$\frac{tu_b}{H_f}$	$\frac{tu_\tau}{H_f}$	$\frac{tu_\tau}{\delta_v}$	u'_{f2D}	$\omega_{f,x2D}$	u'_{f3D}	$\mathcal{D}_{3D}$	ψ_{2D}
$DL240^{H2}$	244	153	159	13	3157	×	×	–	–	–
$DL220^{H2}_{fix}$	226	225	237	18	4037	×	×	–	–	–
$DL250^{H2}$	259	71	72	6	1617	×	×	–	–	–
$DL500^{H2}$	515	40	49	4	2168	×	×	–	–	–
$DL200^{H2}_{smooth}$	198	1000	767	51	10 073	×	×	–	–	–
$DL400^{H2}_{smooth}$	416	1200	286	17	7100	×	–	×	×	×

Table 7.3: Overview of available datasets for the coherent structure eduction in open duct flow simulations. In the course of this analysis, coherent structures are extracted from N_{fields} individual flow field realisations in a total observation time t that is presented in terms of bulk, eddy turnover and viscous time units, respectively. Except for the statistics in case $DL200^{H2}_{smooth}$ which are taken from Sakai (2016), all statistics have been computed in the course of this work. In the remainder of this section, different coherent structure eduction techniques will be discussed, including eduction of streak centres in the cross-plane (u'_{f2D}, appendix D), individual vortex cores in the cross-plane ($\omega_{f,x2D}$, section 7.3.3), three-dimensional streamwise velocity streaks (u'_{f3D}, section 7.3.2), three-dimensional vortex clusters ($\mathcal{D}_{3D}$, section 7.3.4) and two-dimensional streamwise rolls (ψ_{2D}, appendix E). Crosses in the corresponding columns indicate in which cases the respective coherent structure eduction technique was applied.

In the following, for each three-dimensional streak a mean streamwise velocity fluctuation is determined as the mean value over its volume $\mathcal{V}_u$, viz.

$$u'_{\mathcal{V}} = \frac{\int_{\mathcal{V}_u} u'_f(\mathbf{x},t)\mathrm{d}v}{\int_{\mathcal{V}_u} \mathrm{d}v}. \tag{7.20}$$

Dimensions of the streaks are quantified based on the size of their bounding box, which represents the smallest cuboid with edges parallel to the three Cartesian unit vectors that encloses the entire volume $\mathcal{V}_u$. The minimal and maximal coordinates of this bounding box in the ith direction are $x_{i,min}$ and $x_{i,max}$, respectively, such that the edge length of the bounding box in the respective direction accordingly is $\widetilde{\Delta x}_i = x_{i,max} - x_{i,min}$ ($i \in \{1,2,3\}$). In the remainder of this study, we associate each individual structure with its centre of gravity $\mathbf{x}_c = (x_c, y_c, z_c)^T$, where

$$x_{c,i} = \frac{\int_{\mathcal{V}_u} x_i \mathrm{d}v}{\int_{\mathcal{V}_u} \mathrm{d}v} \quad \forall i \in \{1,2,3\}. \tag{7.21}$$

Figure 7.33 shows the p.d.f. of the preferential locations of streamwise velocity streaks of different size in the duct cross-plane for high- and low-speed streaks separately. As a characteristic cross-sectional length scale of the individual streaks, we choose the edge length of an equivalent square with identical cross-sectional area as the bounding box, viz. $\widetilde{l}_{yz} = (\widetilde{\Delta y}\widetilde{\Delta z})^{1/2}$. Velocity streaks of different size are seen to cluster in different regions of the cross-section: The smallest structures with a characteristic length $\widetilde{l}^+_{yz} \in [25, 50)$ shown in figures 7.33(*a,b*) are the inner-scaling streaks inside the buffer layer, and thus they are found almost exclusively in the direct vicinity of the three solid walls. The locally highest values of the p.d.f. are in this context attained in the lower two solid-solid corners as well as in the mixed corners enclosed by the sidewalls and the free surface, respectively. Let us remark that low-speed streaks are seen to preferentially reside directly inside the two solid-solid corners, whereas

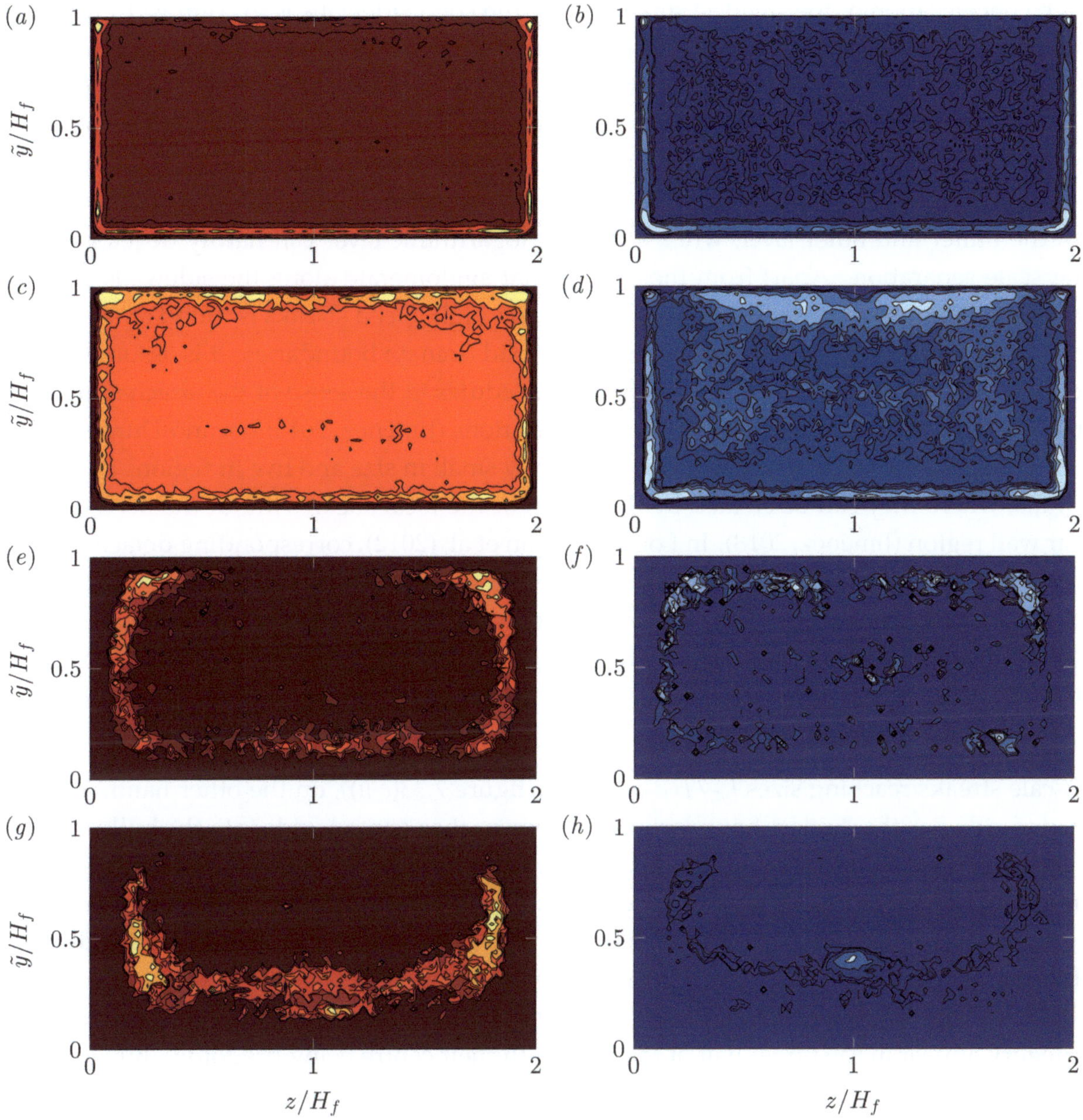

Figure 7.33: P.d.f.s of the preferential positioning of instantaneous three-dimensional streamwise velocity streaks in the cross-plane of case $DL400^{H2}_{smooth}$. Individual streaks are associated with their centre of gravity (y_c, z_c). *(a,c,e,g)* High- and *(b,d,f,h)* low-speed streaks were identified as connected regions of high $(u'^{+}_{f} > 2)$ and low streamwise velocity fluctuations $(u'^{+}_{f} < -2)$ (Sillero, 2014). The streaks are further classified according to the cross-sectional size of their bounding box, with a characteristic length scale defined as $\widetilde{l}_{yz} = (\widetilde{\Delta y}\widetilde{\Delta z})^{1/2}$: *(a,b)* $25 < \widetilde{l}^{+}_{yz} \leq 50$; *(c,d)* $50 < \widetilde{l}^{+}_{yz} \wedge \widetilde{l}_{yz}/H_f \leq 0.5$; *(e,f)* $0.5 < \widetilde{l}_{yz}/H_f \leq 0.75$; *(g,h)* $0.75 < \widetilde{l}_{yz}/H_f$. Contour lines contain 10, 30, 50, 70 and 90% of the total mass of the respective p.d.f.s.

the peaks of the p.d.f. associated with the high-speed streaks are attained in a small distance to the corner along each of the enclosing solid walls. From the investigation of the mean shear stress and the findings of Pinelli et al. (2010) and Sakai (2016), on the other hand, it is known that the first velocity streak next to the corner is indeed a high-speed streak. This apparent contradiction originates in the above discussed averaging effect: The high-speed streaks in the corner are more or less locked in their position by the presence of the two solid walls, such that their signature is visible in the mean flow field $\langle u_f \rangle_{xt}$ as a locally higher streamwise velocity. As a consequence, the velocity fluctuations considered here have to be understood w.r.t. the mean flow signature of the high-speed streaks.

The high concentration of the smallest streaks along the solid sidewalls is in contrast to the relatively low probability of small streaks residing along the free surface. This changes for the larger structures of size $\widetilde{l}_{yz} \in [50\delta_\nu, 0.5H_f)$, whose p.d.f.s are presented in figure 7.33(*c,d*): Streaks of this size seem to populate both the near-wall regions and the vicinity of the free surface, the latter revealing the highest probability for the occurrence of high-speed streaks in the entire cross-section. Note that due to the rather moderate Reynolds number, this intermediate interval features structures that can be part of both the buffer and outer layer, while a distinct logarithmic layer can hardly develop due to the missing scale separation. Apart from the streaks that agglomerate along the sidewalls and the free surface, a small but non-negligible number of low- and high-speed streaks is found at all positions in the cross-section, also in quite large distance to the solid domain boundaries. It is assumed that these structures belong to a family of detached structures/eddies in the sense of the attached-eddy model (Townsend, 1961, 1976) and its more recent interpretations (Jiménez, 2012; Lozano-Durán et al., 2012; Jiménez, 2018). Such detached structures are generally small in size and live in regions away from the solid walls, where they can be created by the action of the local shear without the need to ever visit the near wall region (Jiménez, 2018). In Lozano-Durán et al. (2012), corresponding detached Reynolds stress carrying structures are reported to represent background turbulent fluctuations that, on average, do not significantly contribute to the overall Reynolds stress. In a similar way, the here discovered detached velocity streaks are assumed to belong to the same kind of background turbulence. They are, in particular, of no direct relevance for the erosion and transport of sediment grains along the sediment bed as they are not in contact with the latter.

Large-scale streaks reaching sizes $\widetilde{l}_{yz}/H_f > 0.5$ (cf. figure 7.33(*e-h*)), on the other hand, are typically attached to either of the domain boundaries, from where they spread wide into the bulk flow (Sillero, 2014) and thus undoubtedly scale with the outer length scale H_f. Their centres are hence seen to be preferentially located at distances of around $(0.2\text{-}0.3)H_f$ to the closest part of the domain boundary, extending from there towards the solid walls or the free surface on the one hand and the duct core on the other hand. Comparing the p.d.f.s for streaks with sizes $0.5 < \widetilde{l}_{yz}/H_f \leq 0.75$ in figure 7.33(*e,f*) and those with $\widetilde{l}_{yz}/H_f > 0.75$ in figure 7.33(*g,h*) which each other, it is seen that only streaks of the first category appear in the upper half of the duct with their centre being predominantly located close to the upper left and right corner of the cross-section. The largest high-speed structures for which $\widetilde{l}_{yz}/H_f > 0.75$, on the other hand, centre merely in three distinct regions in some distance to the free surface, from which two are located at a lateral distance of $(0.2\text{-}0.3)H_f$ from either sidewall and a third one is found approximately at the duct bisector, $\check{y}/H_f = 0.2$ above the bottom wall. While the former two regions are located in a zone of mean secondary upflow and they consequently spread over a wider wall-normal range, the latter zone lies below the mean secondary downflow and is more confined to the bottom wall. For the low-speed streaks, a qualitatively similar distribution is visible, even though a much larger fraction of the total mass of the p.d.f. is concentrated in a narrow region in the duct centre. The reason for this peculiar distribution of the probability density for large-scale low-speed streaks is not completely clear and it would require longer time series to verify whether this effect is of statistical or physical nature, as the current simulation time is with approximately $286T_b$ or 17 eddy turnover times still relatively short compared to the lifetime of the largest streaks. Eventually, it should be pointed out that the preferential location of the larger high-speed streaks with $0.5 < \widetilde{l}_{yz}/H_f$ agrees reasonably well with the general organisation of the two-point correlation functions in figures 7.30 and 7.31, respectively. Choosing, for instance, the reference point at $(\check{y},\check{z})$=$(0.2H_f, L_z/2)$, around which the high-speed streak p.d.f. in figure 7.33(*g*) exhibits a local maximum. Then, the neighbouring zones in which less large high-speed streaks have been detected collapse with the regions of strongest anti-correlation, being then representative of local low-speed streaks.

Figures D.3 and D.4 in appendix D show a comparable set of p.d.f.s for the streak organisation in the here considered case $DL400^{H2}_{smooth}$ as well as for the corresponding particle-laden case $DL500^{H2}$, respectively, computed based on the alternative two-dimensional extraction technique of Nakatsuji (2012) and Sakai (2016) that was mentioned earlier. In the latter procedure, regions of high and low streamwise velocity fluctuations are detected in each two-dimensional yz-plane of the flow field separately, and statistics are computed based on the size and preferential organisation of these two-dimensional structures (cf. appendix D for a detailed overview of the methodology). The different procedure based on which the streaks are extracted from the individual flow realisations and the deviating definition of the characteristic length scale adopted therein do not allow a direct quantitative comparison of the two datasets, but we observe a qualitative agreement of both statistics concerning the regions in which small- and large-scale structures typically cluster. Also, the results in appendix D show – within the bounds of the available data in the particle-laden cases – that the general organisation of streamwise high- and low-speed streaks in the cross-section is not strongly altered by the presence of a mobile sediment bed and the thereon emerging ridges.

In the following, we further investigate the size and aspect ratio of the detected streaks. To this end, figure 7.34(*a*) provides the joint p.d.f. of the minimum (y^+_{min}) and maximum distance (y^+_{max}) of a given streak to the bottom wall. In agreement with earlier studies in canonical channel and boundary layer flows considering different types of coherent structures (Del Álamo et al., 2006; Lozano-Durán et al., 2012; Sillero, 2014), the streaks can be classified as either attached to the bottom wall or detached from it. Following the aforementioned studies, we categorise a structure as attached to the bottom wall if $y^+_{min} \leq 20$, and as detached otherwise. Here, we do not further distinguish between structures that are attached or detached to one of the sidewalls or the free surface as done by Atzori et al. (2021) for closed duct flows, as our main interest is in those structures that are attached to the bottom wall and that, thus, influence the wall shear stress as well as the near-wall flow organisation. By definition, the attached structures base inside the buffer layer and part of them can spread far into the outer layer, while others, the actual buffer layer streaks, do not. Note that in some studies, the former larger-scale structures are separately studied as 'tall-attached structures' (Sillero, 2014), but we will forego such additional distinction for the moment. In figure 7.34(*a*), attached structures correspond to the data on the left of the vertical dashed line, whereas the contributions to the joint p.d.f. on the right are related to the detached streak family. Members of the latter are seen to be limited in their wall normal extent, revealing for all distances to the wall essentially the same range of heights $\widetilde{\Delta y}$ that are seldom larger than $100\delta_\nu$ ($0.24H_f$).

The mass distribution of the joint p.d.f. in the (y_{min}, y_{max})-plane of the here considered open duct flow is very similar to those found in canonical flows, with one remarkable difference: In contrast to closed channels where $y^+ = Re_\tau$ marks the channel centreline, here $y^+ = Re_\tau$ refers to the wall-normal location of the free surface, to which structures can again be attached. Indeed, figure 7.34(*a*) shows a second horizontal band of locally increased p.d.f. values, that are related to surface attached streaks with $y^+_{max} \approx Re_\tau$. Apparently, some of these surface attached structures span more than 200 wall units towards the duct centre. As the focus of this study is the influence of coherent structures on the erosion and transport of sediment and the related evolution of sediment bedforms, unfortunately, a more elaborate investigation of the velocity and vortex structures in the vicinity of the free surface as

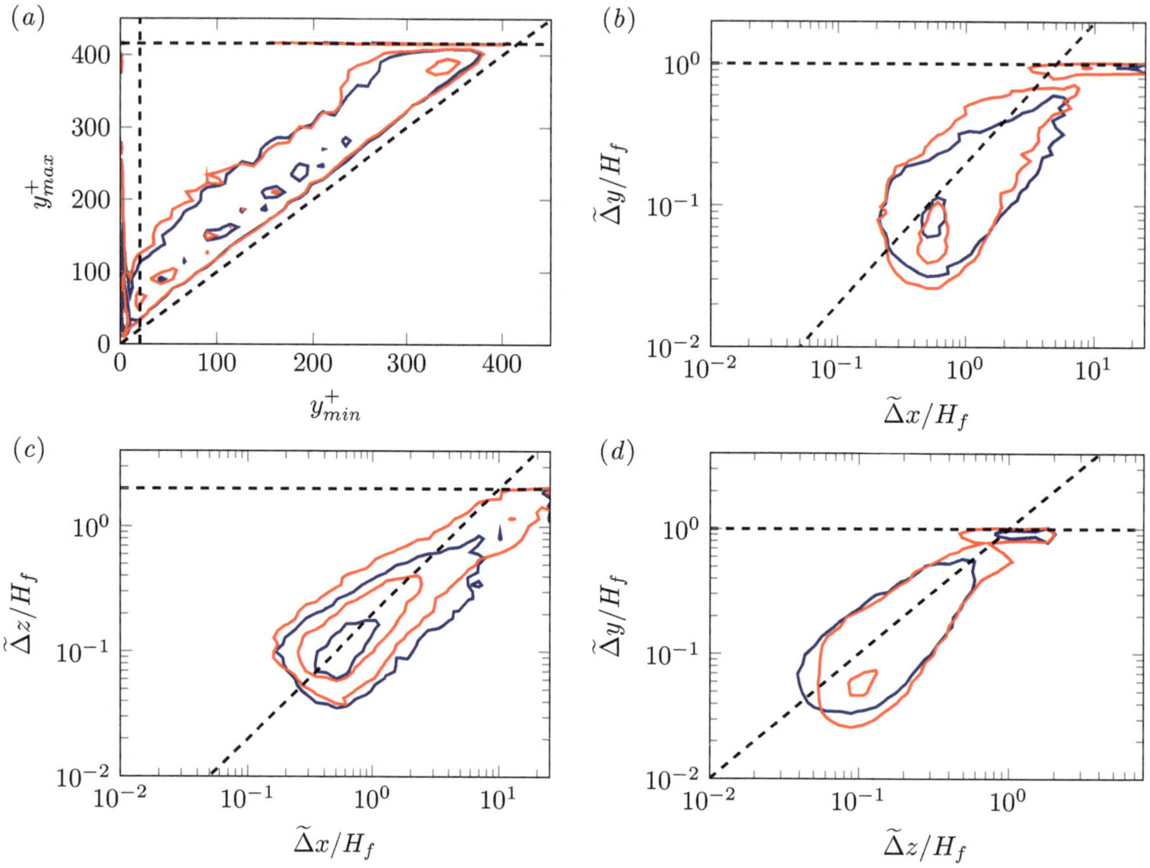

Figure 7.34: Joint p.d.f. of the axis-parallel extensions of individual wall-attached streamwise velocity streaks in case $DL400^{H2}_{smooth}$, measured in terms of their circumscribing bounding box. (*a*) Joint p.d.f. of the minimum (y_{min}) and maximum wall-normal distance (y_{max}) of the structure to the bottom wall. The vertical and horizontal dashed lines mark $y^+_{min} = 20$ and $y^+_{max} = Re_\tau$, respectively. (*b*) Joint p.d.f. of the streamwise ($\widetilde{\Delta x}$) and wall-normal dimension ($\widetilde{\Delta y}$) of the circumscribing bounding box of streaks that are attached to the bottom wall (i.e. $y^+_{min} \leq 20$); (*c*) same for the streamwise ($\widetilde{\Delta x}$) and spanwise dimension ($\widetilde{\Delta z}$) and (*d*) the spanwise ($\widetilde{\Delta z}$) and wall-normal dimension ($\widetilde{\Delta y}$). Horizontal dashed lines indicate the domain sizes in the respective directions, i.e. H_f and L_z, while the diagonal lines mark: (*a*) $y_{min} = y_{max}$, (*b*) $\widetilde{\Delta x} = 5\widetilde{\Delta y}$, (*c*) $\widetilde{\Delta x} = 5\widetilde{\Delta z}$, (*d*) $\widetilde{\Delta z} = \widetilde{\Delta y}$. Red (high-speed streaks) and blue (low-speed streaks) contours contain 20 and 90% of the total mass of the joint p.d.f.s.

in Nagaosa (1999) for open channel flow is beyond the scope of this study, but represents a valuable task for a future study.

For canonical boundary layer and closed channel flows, Sillero (2014) and Jiménez (2018) report that the bottom wall-attached streamwise velocity streaks are self-similar throughout the logarithmic layer, featuring a constant aspect ratio of approximately $\widetilde{\Delta x}/\widetilde{\Delta y}/\widetilde{\Delta z} = 5/1/1$. Even though the here considered Reynolds number is too low to exhibit a distinct log-layer, the joint p.d.f.s of the two-dimensional aspect ratios $\widetilde{\Delta x}/\widetilde{\Delta y}$, $\widetilde{\Delta x}/\widetilde{\Delta z}$ and $\widetilde{\Delta z}/\widetilde{\Delta y}$ shown in figures 7.34(*b-d*) follow quite well the predicted aspect ratio for both low- and high-speed streaks for most scales. Similar to Sillero (2014) and Jiménez (2018), however, we also observe an 'overhanging' tail at the upper end of all three p.d.f.s, which is indicative of very long streamwise-elongated streaks that reach streamwise dimensions of $\widetilde{\Delta x} = \mathcal{O}(10H_f)$. Clearly, the aspect ratio $5/1/1$ cannot be maintained for streaks of that length as the cross-stream dimensions are, in contrast to the streamwise extent, geometrically restricted by the presence of the lateral sidewalls and can thus not grow any further. The determined cross-sectional dimensions of the largest streaks are in excellent agreement with those estimated by Sekimoto (2011) for closed square duct flows based on two-point correlations similar to those evaluated in the previous section. Similar to our observations, he concludes that the surrounding (no-slip) boundaries hinder the streaks in their growth, causing their maximum width to saturate at about $\widetilde{\Delta z}/H \approx 1.0$-$1.2$ (H indicating the half height of the closed duct).

Note that some of the very long streaks even span the entire domain length, and thus have to be interpreted as infinitely long modes. As discussed by Lozano-Durán et al. (2012), structures with streamwise length $\mathcal{O}(10H_f)$ most likely correspond to what is conventionally termed as 'global modes' (Del Álamo and Jiménez, 2003), 'superstructures' or 'very-large scale motions' (VLSMs) (Adrian, 2007; Smits et al., 2011). In consequence, these features do not belong to the family of self-similar streaks that includes all scales from the buffer layer streaks to the outer-scaling large-scale streaks/motions with streamwise lengths $\mathcal{O}(H_f)$, as argued by Jiménez (2013a).

7.3.3 Quasi-streamwise vortices

In contrast to the self-similar velocity streaks, quasi-streamwise vortices do not scale self-similarly across the log-layer. Instead, these structures are more or less of the same scale close to and far away from the solid wall (Jiménez, 2012). In section 7.2.4, the relation between the mean streamwise vorticity field $\langle \omega_{f,x} \rangle_{xt}$ and the mean secondary flow streamfunction $\langle \psi \rangle_{xt}$ was discussed for smooth-wall and particle-laden flows. It was observed that the organisation of both quantities in the lower half of the duct is modified for all simulations that feature a mobile sediment bed, while it is essentially identical for simulations over smooth walls and stationary sediment beds in the considered roughness regime. This qualitative difference leads to the question, how the mean organisation of the streamwise vorticity is altered by the presence of mobile particles. As shown by Uhlmann et al. (2007), Pinelli et al. (2010) and Sakai (2016) for closed and open duct flows, the mean streamwise vorticity field $\langle \omega_{f,x} \rangle_{xt}$ in fact represents the preferential organisation of instantaneous quasi-streamwise vortices in the cross-plane. In order to better understand the influence of a mobile sediment bed on the preferential location of these instantaneous quasi-streamwise vortices, we will identify in the following the instantaneous position of the vortex centres in each cross-plane. The here applied method was first presented by Uhlmann et al. (2007) and bases on a criterion proposed by Kida and Miura (1998), according to which vortex cores are represented by local minima of the pressure field in the yz-plane, at which the local two-dimensional velocity gradient tensor $\nabla_\perp (v_f, w_f)^T$ fulfils a 'swirl-condition' $\mathcal{D}_\perp < 0$. In the latter expression, $\mathcal{D}_\perp$ is the two-dimensional discriminant of the velocity gradient tensor (i.e. of its characteristic polynomial), defined as

$$\mathcal{D}_\perp = \frac{1}{4}\left(\frac{\partial v_f}{\partial y} - \frac{\partial w_f}{\partial z}\right)^2 + \frac{\partial v_f}{\partial z}\frac{\partial w_f}{\partial y}. \tag{7.22}$$

In this context, a negative value of $\mathcal{D}_\perp$ refers to an elliptic arrangement of the streamlines around the given point, which indicates that the latter lies inside a region of vortical motion. Note that, in this context, $\mathcal{D}_\perp$ can be understood as the two-dimensional analogue of the three-dimensional discriminant $\mathcal{D}$ of the full velocity gradient tensor $\nabla \boldsymbol{u}_f$ that was introduced in equation 6.35 in the previous chapter. As in Uhlmann et al. (2007), relation (7.22) is evaluated for all local pressure minima detected in each cross-section of an instantaneous field. In the subsequent analysis, only those points are included for which $\mathcal{D}_\perp < 0$ holds and the local streamwise vorticity attains non-negligible values $|\omega_{f,x}| > 0.01 \max_{y,z}(|\langle \omega_{f,x} \rangle_{xt}|)$.

Figure 7.35 provides the corresponding p.d.f.s of the vortex centre locations in the duct cross-section for the flow over a smooth wall (case $DL200^{H2}_{smooth}$) and a mobile sediment bed (case $DL250^{H2}$), further

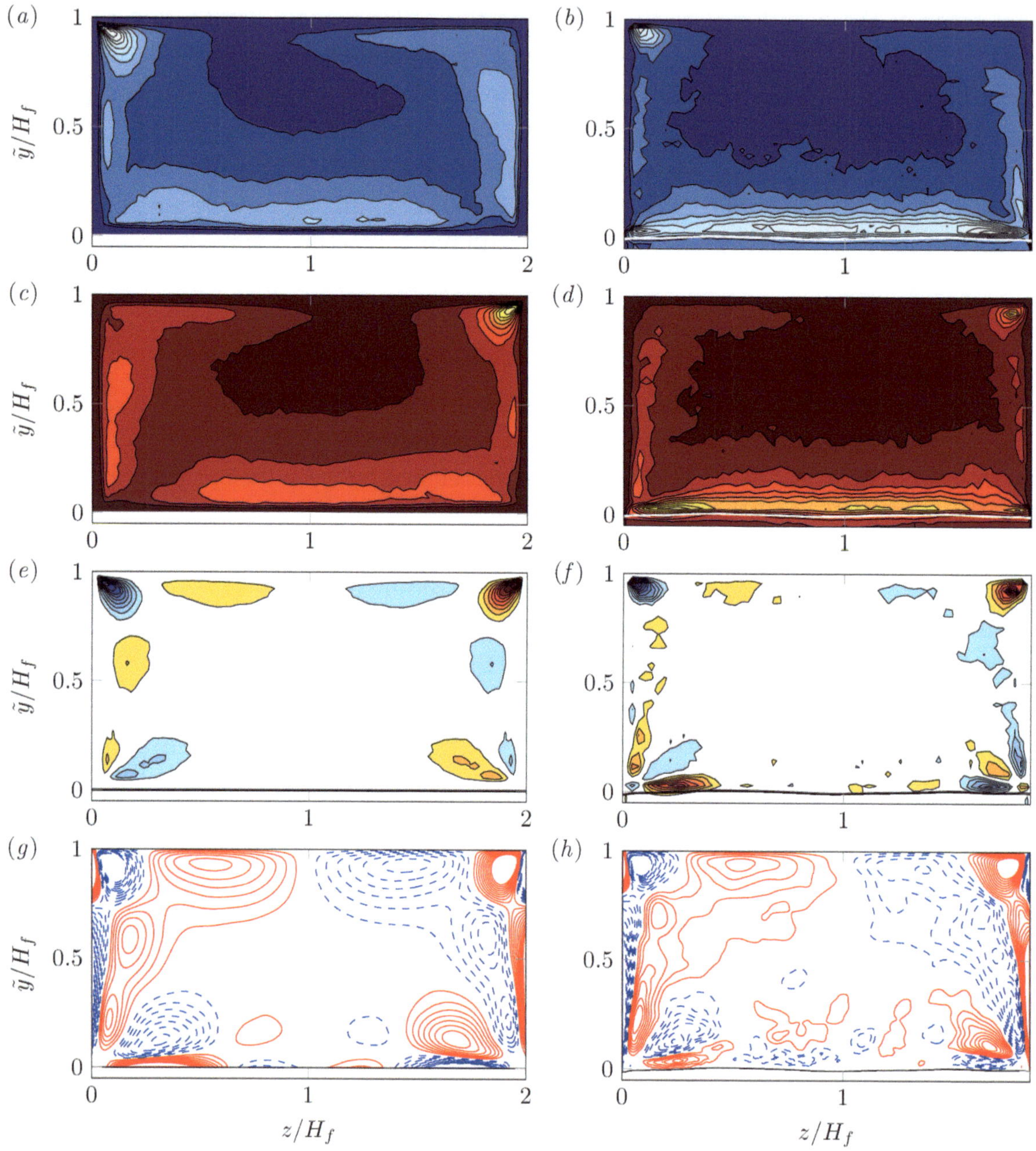

Figure 7.35: P.d.f. of the preferential positioning of quasi-streamwise vortices in the cross-plane of (*a,c,e,g*) the smooth-wall case $DL200^{H2}_{smooth}$ and (*b,d,f,h*) the mobile sediment bed case $DL250^{H2}$. Vortex cores were identified as local minima of the pressure field in each cross-section that additionally fulfil the swirl condition $\mathcal{D}_\perp < 0$ and for which $|\omega_{f,x}|$ is larger than 1% of the maximum of $|\langle\omega_{f,x}\rangle_{xt}|$. Each p.d.f. is normalised with its global maximum. (*a,b*) vortices with negative sense of rotation ($\omega_{f,x} < 0$), isocontours are $0.1(0.1)0.9$; (*c,d*) vortices with positive sense of rotation ($\omega_{f,x} > 0$), isocontours are $0.1(0.1)0.9$; (*e,f*) difference between the p.d.f.s for vortices featuring positive and negative vorticity, isocontours are $-0.9(0.1)0.9$. In case $DL200^{H2}_{smooth}$, the shown data was additionally symmetrised about the duct bisector taking care of the reversed sign. (*g,h*) Mean streamwise vorticity $\langle\omega_{f,x}\rangle_{xt}$ (red: positive, blue: negative) as in figure 7.17, repeated for convenience. In the individual panels, solid white and black lines indicate the location of the mean fluid-bed interface.

distinguished based on the sign of the local streamwise vorticity $\omega_{f,x}$. For the sake of comparison, the presented data is supplemented with a visualisation of the respective mean streamwise vorticity field $\langle \omega_{f,x} \rangle_{xt}$, which has been presented earlier in this chapter (cf. figure 7.17). The overall distribution of clockwise and counterclockwise rotating quasi-streamwise vortices across the cross-section shows a reasonable agreement between both simulations in all regions except for the near-bed zone (cf. figure 7.35(*a-d*)). In particular, the preferential agglomeration of vortices with a common sense of rotation in the mixed corners due to the vortex sorting process proposed by Sakai (2016) is clearly identifiable in both simulations.

The good match of the p.d.f.s in most regions of the cross-section is expected, since we have seen earlier that the mobile sediment bed has little influence on the dynamics of individual turbulent structures except for those in its direct vicinity. In the fluid layer above the mean fluid-bed interface, on the other hand, significantly more individual vortex cores are counted than in the corresponding region over a smooth wall, such that the local probability for the occurrence of a vortex core in the vicinity of the sediment bed is comparable with that in the mixed corners. The larger number of extracted vortex cores along the sediment bed is most likely related to vortex shedding processes in the wake of individual sediment grains, located near by or directly at the top of the sediment bed (Zeng et al., 2010; Li et al., 2019). Even though large in number, these individual vortices are induced by quasi-randomly occurring events and thus appear with more or less the same probability along the cross-section for clockwise and counterclockwise rotation. In the unconditioned average, thus, contributions from these structures cancel out (cf. figure 7.35(*f*)) such that there is no net mean vorticity seen in the corresponding $\langle \omega_{f,x} \rangle_{xt}$ field (cf. figure 7.35(*h*)). For the smooth wall case, a lower number of individual vortices is measured in the duct centre compared to the particle-laden case, but the effect is similar: In sufficient distance to the sidewalls, vortices with opposite sense of rotation occur with the same probability at all positions (cf. figure 7.35(*e*)), such that there is essentially no net mean streamwise vorticity $\langle \omega_{f,x} \rangle_{xt}$ in these regions.

As pointed out earlier, the situation is different in the proximity of the sidewalls, where the positions of quasi-streamwise vortices with a given sense of rotation are geometrically restricted, such that these structures can occur only in distinct regions. For the smooth-wall case, figure 7.35(*e*) reveals that each of the two lower solid-solid corners contains two regions with an excess of either clockwise or counterclockwise rotating vortices. This peculiar arrangement eventually leads to the characteristic patterns of the mean streamwise vorticity $\langle \omega_{f,x} \rangle_{xt}$ in these regions, visualised in figure 7.35(*g*) and described in detail by Sakai (2016). The mean mirror vorticity cells along the bottom wall, on the other hand, are induced by quasi-streamwise vortices approaching the impermeable no-slip wall in consequence of the there applied boundary conditions (Orlandi, 1990, and references therein). Indeed, figures 7.35(*a,c*) underline that the impermeable no-slip bottom wall does not allow vortices to come very close to the wall.

Corresponding regions for the particle-laden case, in which vortices with a certain sense of rotation cluster close to the sidewalls, are visible in figure 7.35(*f*), but these zones are of smaller size and span less wide into the duct than in the smooth-wall case. Most strikingly is, however, the appearance of a third zone of strong net vortical activity in the particle-laden case, located directly above the sediment bed. While the no-slip condition along the walls in the single-phase case prevents the agglomeration of vortices in the direct vicinity of the walls, such regions of intense vorticity can indeed form near a mobile and permeable lower boundary, explaining the presence of intense mean vorticity zones $\langle \omega_{f,x} \rangle_{xt}$

in figure 7.35(*h*). Further, a comparison of figures 7.35(*b,d*) exhibits that vortices of both sense of rotation are detected in these near-bed regions of intense vortical activity, but the ones supporting the mean transport of fluid into the corner along the bed are clearly dominant. Their dominance is, however, restricted to the narrow region between the sidewalls and the first sediment ridge crest, where particle transport is comparably weak and, thus, particle-induced vortices form less frequently. The different organisation of the quasi-streamwise vortices in the lower left and right corners of the duct is of direct relevance for the structure of the mean secondary flow $\langle\psi\rangle_{xt}$ near the bed in form of the mean secondary bottom vortices, which have been seen to be more restricted to the near-sidewall region in the mobile sediment cases (cf. section 7.2.4). As both are connected via the Poisson equation (7.12), a modification in the organisation of the quasi-streamwise vortices immediately affects the structure of $\langle\psi\rangle_{xt}$ as well.

7.3.4 Instantaneous vortex clusters and mean secondary flow

The small-scale quasi-streamwise vortices discussed in the previous section naturally scale in wall units. As a consequence, the mean vorticity patterns seen in figures 7.35(*g,h*) move towards the respective corners as the Reynolds number is increased, whereas the outer-scaling mean secondary flow streamfunction $\langle\psi\rangle_{xt}$ changes little for non-marginal Reynolds numbers (Sakai, 2016). As was pointed out by Pinelli et al. (2010), the different scaling behaviour of streamwise vorticity and mean secondary flow is indeed compatible with the physical interpretation of the Poisson equation (7.12), highlighting the analogy with the Poisson equation for the pressure. In equation (7.12), $\langle\omega_{f,x}\rangle_{xt}$ represents a local source term that scales in inner units, whereas the mean secondary flow streamfunction $\langle\psi\rangle_{xt}$ is of non-local character and depends on $\langle\omega_{f,x}\rangle_{xt}$ everywhere in the domain by the action of the two-dimensional Laplace operator $\Delta_\perp$. While the characteristic patterns of $\langle\omega_{f,x}\rangle_{xt}$ are undoubtedly related to the preferential quasi-streamwise vortices, the question remains whether it is possible to associate instantaneous vortical structures of larger scales up to H_f with the mean secondary flow pattern in a similar way, as was speculated by Pinelli et al. (2010).

One candidate for such larger-scale vortical structures are the self-similar attached vortex clusters investigated by Del Álamo et al. (2006) in canonical closed channels, which represent concatenations of small-scale vortices. Structures of this kind can be extracted from the flow field analogously to the way streamwise velocity streaks were identified in section 7.3.2. Based on the results of the percolation analysis of Del Álamo et al. (2006), we define vortex clusters as connected regions for which the discriminant of the velocity gradient tensor $\mathcal{D}(\boldsymbol{x},t)$ defined in equation (6.35) exceeds 2% of the local r.m.s. value $\mathcal{D}_{rms}(y,z) = \langle\mathcal{D}'\mathcal{D}'\rangle_{xt}^{1/2}$. In figure 7.36, the extracted vortex clusters are visualised for an instantaneous state of the flow field of case $DL500^{H2}$. In the following, we however restrict ourselves to the analysis of the vortex clusters in the smooth wall case $DL400^{H2}_{smooth}$, for which the number of available flow field realisations is large enough to obtain meaningful statistics.

In analogy to figure 7.33, figure 7.37 provides p.d.f.s of the preferential location of the vortex cluster centres x_c in the cross-section, classified based on their characteristic length scale $\widetilde{l}_{yz}$ and the sign of the mean vorticity of each cluster defined as the average over the cluster volume $\mathcal{V}_\mathcal{D}$ as

$$\omega_{x,\mathcal{V}} = \frac{\int_{\mathcal{V}_\mathcal{D}} \omega_{f,x}(\boldsymbol{x},t)\mathrm{d}v}{\int_{\mathcal{V}_\mathcal{D}} \mathrm{d}v}. \tag{7.23}$$

Figure 7.36: Visualisation of instantaneous vortex clusters in case $DL500^{H2}$ ($t/T_b = 23$) as seen from the top of the open duct. The structures were detected as connected regions for which the discriminant of the velocity gradient tensor, $\mathcal{D}$, is larger than 2% of its local r.m.s. value. In contrast to the extraction procedure outlined in the text, we here use for visualisation purposes only the instantaneous r.m.s. value $\langle(\mathcal{D} - \langle\mathcal{D}\rangle_x)^2\rangle_x^{1/2}$ rather than $\mathcal{D}_{rms}$, as statistics of $\mathcal{D}$ have not been computed for the entire time series in this case. Colouring of the individual structures represents their distance to the bottom wall and thus the sediment bed, with brighter green clusters being located closer to the free surface than dark green objects. Flow is from left to right.

In addition, figure 7.38 shows the difference between the p.d.f.s associated with clusters of positive and negative mean vorticity $\omega_{x,\mathcal{V}}$ for each length scale range, that way highlighting regions of the cross-section in which either of the two types of vortex clusters is dominant.

Most of the small-scale clusters with $\widetilde{l}_{yz} = \mathcal{O}(10\delta_v)$ seen in figures 7.37(*a-d*) and figures 7.38(*a,b*) can be assumed to consist of a single quasi-streamwise vortex. Indeed, their spatial organisation follows more or less that of the inner-scaling vortices discussed in the previous section and consequently also the mean vorticity distribution $\langle\omega_{f,x}\rangle_{xt}$ (cf. figure 7.38(*e*)), with a markedly high concentration of individual structures enclosed in the two mixed corners and along the free surface itself. A comparison with the individual structures displayed in figure 7.36 reveals that the vortex clusters directly beneath the free surface are indeed of rather small size and more or less parallel to the free surface, either aligned with the mean flow or inclined to it if they are additionally attached to one of the sidewalls. Note that surface-parallel vortices of different scales have been observed to be of high relevance for mass and momentum transport in the vicinity of the free surface (Nagaosa, 1999; Pinelli et al., 2022), and thus a more detailed future analysis of their dynamics under the influence of near lateral sidewalls could be of large interest in the context of heat and gas transfer in the vicinity of and across the air-water interface.

On the other hand, vortex clusters much larger than single quasi-streamwise vortices are found in some distance to the solid walls and the free surface, reaching characteristic length scales of $\widetilde{l}_{yz} > 0.5H_f$. Structures with $0.5 < \widetilde{l}_{yz}/H_f \leq 0.75$ for which the corresponding p.d.f.s are presented in figures 7.37(*e,f*) and figure 7.38(*c*) predominantly occur in a horizontal layer approximately $0.2H_f$ below the free surface, separated depending on the sign of $\omega_{x,\mathcal{V}}$: Clockwise and counterclockwise rotating structures reside primarily in the left and right half of the cross-section, respectively. This distance to the free surface roughly corresponds to the height of the outer mean secondary flow cells' centre, and the sense of rotation of these large-scale clusters favours the rotating motion of the mean secondary currents (cf. figure 7.38(*f*)). In spite of their larger size and their accordingly larger distance to the free surface, the vortex clusters in this layer are assumed to be still attached to the former and thus, there is no reason why the vortex sorting process proposed by Sakai (2016) should not apply to them in a similar way as it does to the small-scale quasi-streamwise vortices. The latter would explain the clear separation of clockwise and counterclockwise rotating structures to the respective sides of the cross-section. Even though the highest probability for the occurrence of these larger clusters is found in the near-surface region, structures of similar size also appear in the lower half of the duct, in particular inside the mean secondary bottom vortex cell (cf. figure 7.38(*f*)). Again, clusters that possess the same sense of rotation as the mean secondary motion dominate the p.d.f.s.

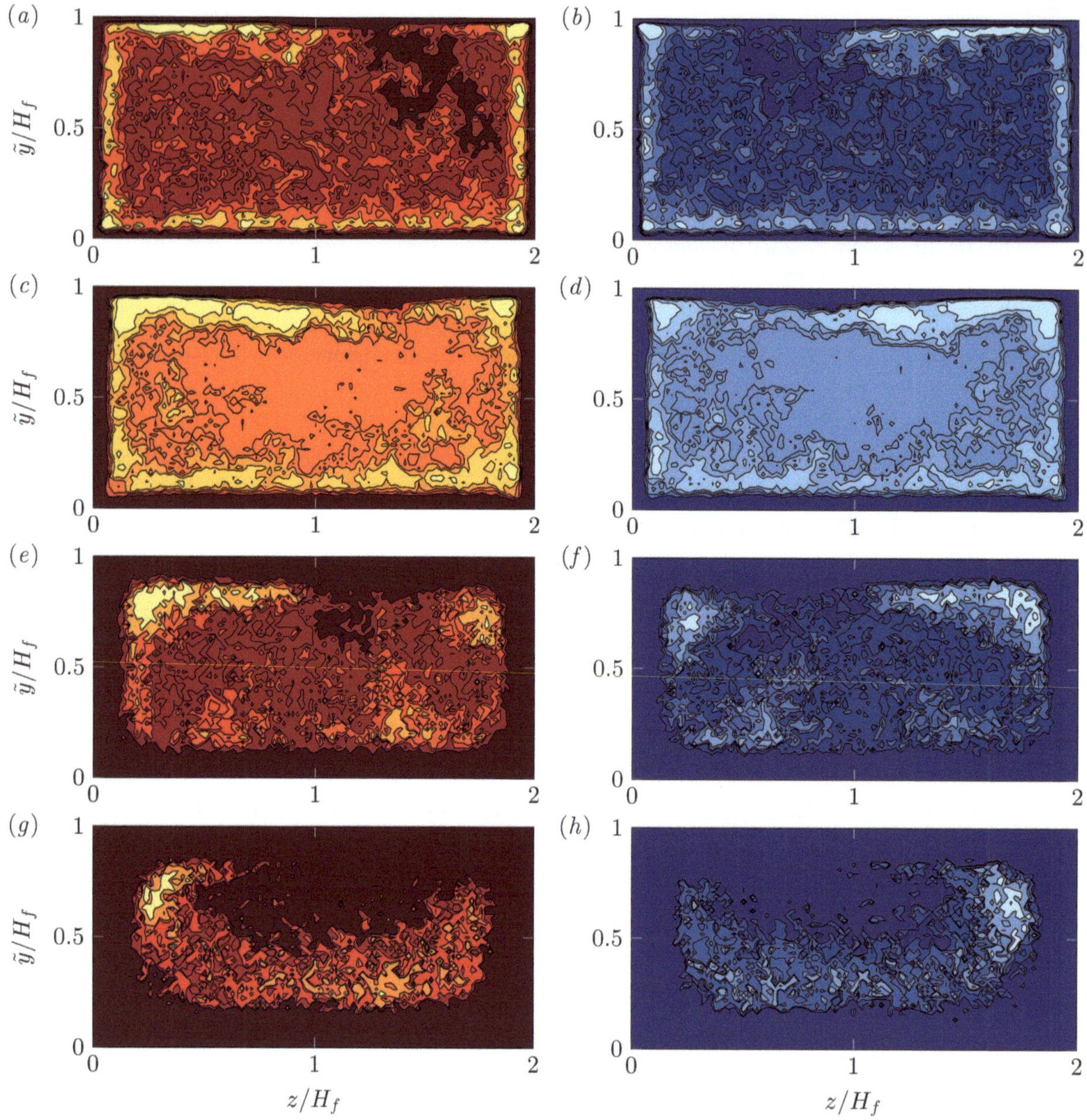

Figure 7.37: P.d.f.s of the preferential positioning of instantaneous three-dimensional vortex clusters in the cross-plane of case $DL400^{H2}_{smooth}$. Individual clusters are associated with their centre of gravity (y_c, z_c). (*a,c,e,g*) Vortex cluster with a positive $(\omega_{x,\mathcal{V}} > 0)$ and (*b,d,f,h*) negative mean sense of rotation $(\omega_{x,\mathcal{V}} < 0)$. Vortex cluster were identified as connected regions for which the discriminant of the velocity gradient tensor, $\mathcal{D}$, exceeds a value of $0.02\mathcal{D}_{rms}(y,z)$ (Del Álamo et al., 2006). The vortex clusters are further classified according to the cross-sectional size of their bounding box: (*a,b*) $25 < \tilde{l}^+_{yz} \leq 50$; (*c,d*) $50 < \tilde{l}^+_{yz} \wedge \tilde{l}_{yz}/H_f \leq 0.5$; (*e,f*) $0.5 < \tilde{l}_{yz}/H_f \leq 0.75$; (*g,h*) $0.75 < \tilde{l}_{yz}/H_f$. Contour lines contain 10, 30, 50, 70 and 90% of the total mass of the respective p.d.f.s.

The even larger structures with characteristic lengths $\tilde{l}_{yz} > 0.75H_f$ reveal a qualitatively similar distribution across the cross-section, with clockwise rotating clusters dominating the upper left and lower right quadrant of the plane and *vice versa* for the counterclockwise circulating structures, as can be seen in figures 7.37(*g,h*) and 7.38(*d*). The vortex clusters in the upper half of the duct are not randomly distributed over the cross-section, but their centres of gravity are seen to agglomerate in two rather narrow regions around the points $(0.35H_f, L_y - 0.35H_f)$ and $(L_z - 0.35H_f, L_y - 0.35H_f)$, respectively. Recalling that the mean extent of these structures is with $\tilde{l}_{yz} > 0.75H_f$ comparable to the half duct width, it is quite conceivable that these structures are, owing to their size, more or less locked in their spatial position by the lateral sidewalls and the free surface to which they are attached. The detection

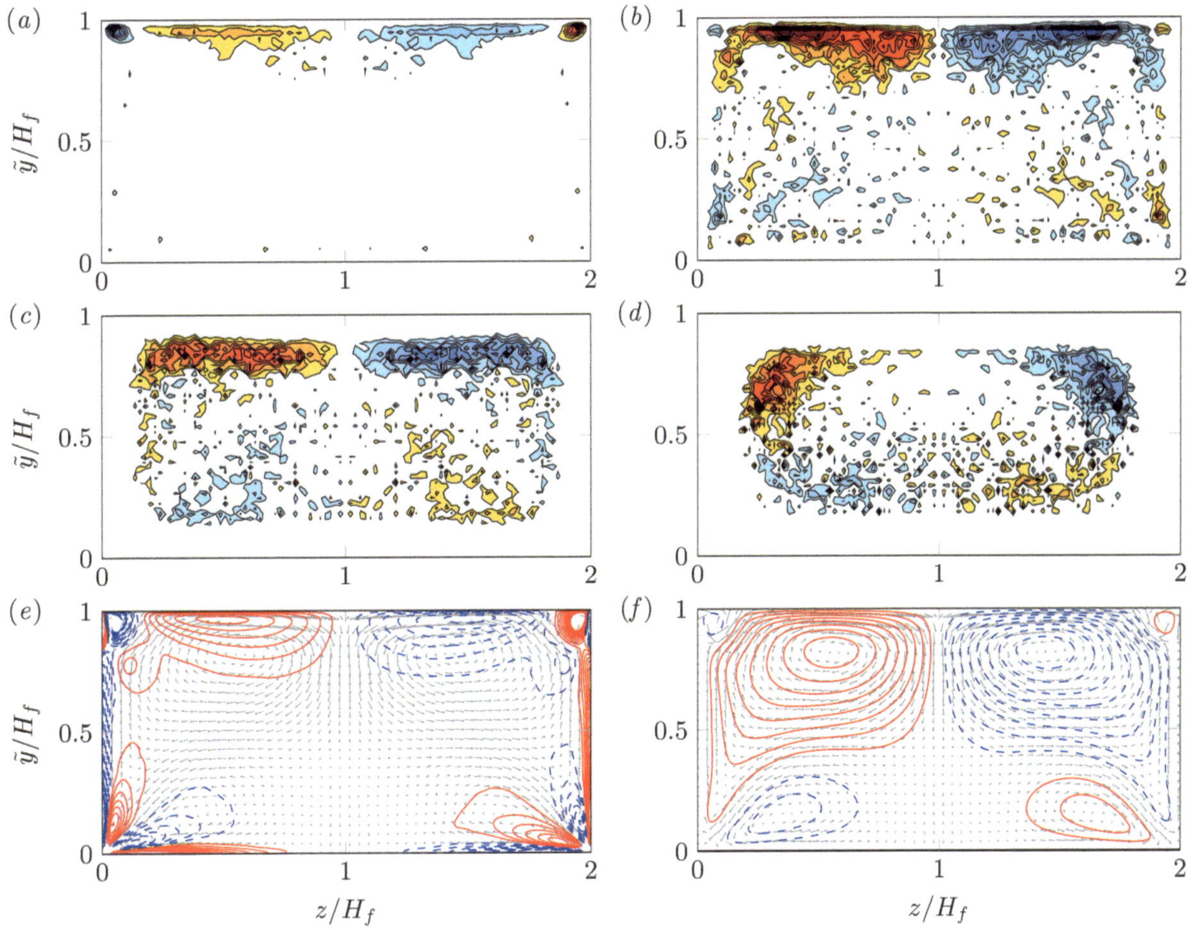

Figure 7.38: Differences between p.d.f.s of the preferential positioning of instantaneous three-dimensional vortex clusters with positive and negative mean vorticity $\omega_{x,\mathcal{V}}$ in the cross-plane of the open duct simulation $DL400^{H2}_{smooth}$ which are shown in figure 7.37. The here presented data are additionally symmetrised about the duct bisector taking care of the reversed sign. The p.d.f.s are normalised with their maximum absolute value and isocontours refer to values $-0.9(0.1)0.9$, with reddish (blueish) regions indicating local dominance of structures with $\omega_{x,\mathcal{V}} > 0$ ($\omega_{x,\mathcal{V}} < 0$). (*a*) $25 < \tilde{l}^+_{yz} \leq 50$; (*b*) $50 < \tilde{l}^+_{yz} \wedge \tilde{l}_{yz}/H_f \leq 0.5$; (*c*) $0.5 < \tilde{l}_{yz}/H_f \leq 0.75$; (*d*) $0.75 < \tilde{l}_{yz}/H_f$. In (*e*) and (*f*), the mean streamwise vorticity $\langle\omega_{f,x}\rangle_{xt}$ and the mean secondary flow streamfunction $\langle\psi\rangle_{xt}$ are repeated for convenience, respectively, with red (blue) contours indicating clockwise (counterclockwise) rotation.

of the large-scale vortex clusters highlights that small-scale vortices do not only preferentially reside in distinct regions of the cross-section making them visible in the time-averaged vorticity field $\langle\omega_{f,x}\rangle_{xt}$, but they also aggregate instantaneously in form of these large-scale clusters that feature a mean rotation themselves.

Similar to the three-dimensional velocity streaks, the here discussed vortex clusters can be classified into three different families depending on their minimal (y^+_{min}) and maximal distance (y^+_{max}) to the bottom wall: The corresponding p.d.f. provided in figure 7.39(*a*) indicates that a first set of clusters is attached to the bottom wall (i.e. $y^+_{min} < 20$), a second one is attached to the free surface (i.e. $(Re_\tau - y^+_{max}) < 20$) and the remaining structures are detached from both. While the mean height of the detached structures does essentially not vary with the wall-normal coordinate being limited to about $\widetilde{\Delta y}^+ \approx 150$, the attached clusters partly extend over more than $\widetilde{\Delta y} \approx 0.75 H_f$ ($312\delta_\nu$), in agreement with the largest structures seen to contribute to the p.d.f.s in figures 7.37 and 7.38.

The aspect ratios of the vortex clusters attached to the bottom wall and those connected to the free surface are depicted in figure 7.39(*b*-*d*) in terms of the joint p.d.f.s of the bounding box edge lengths. In

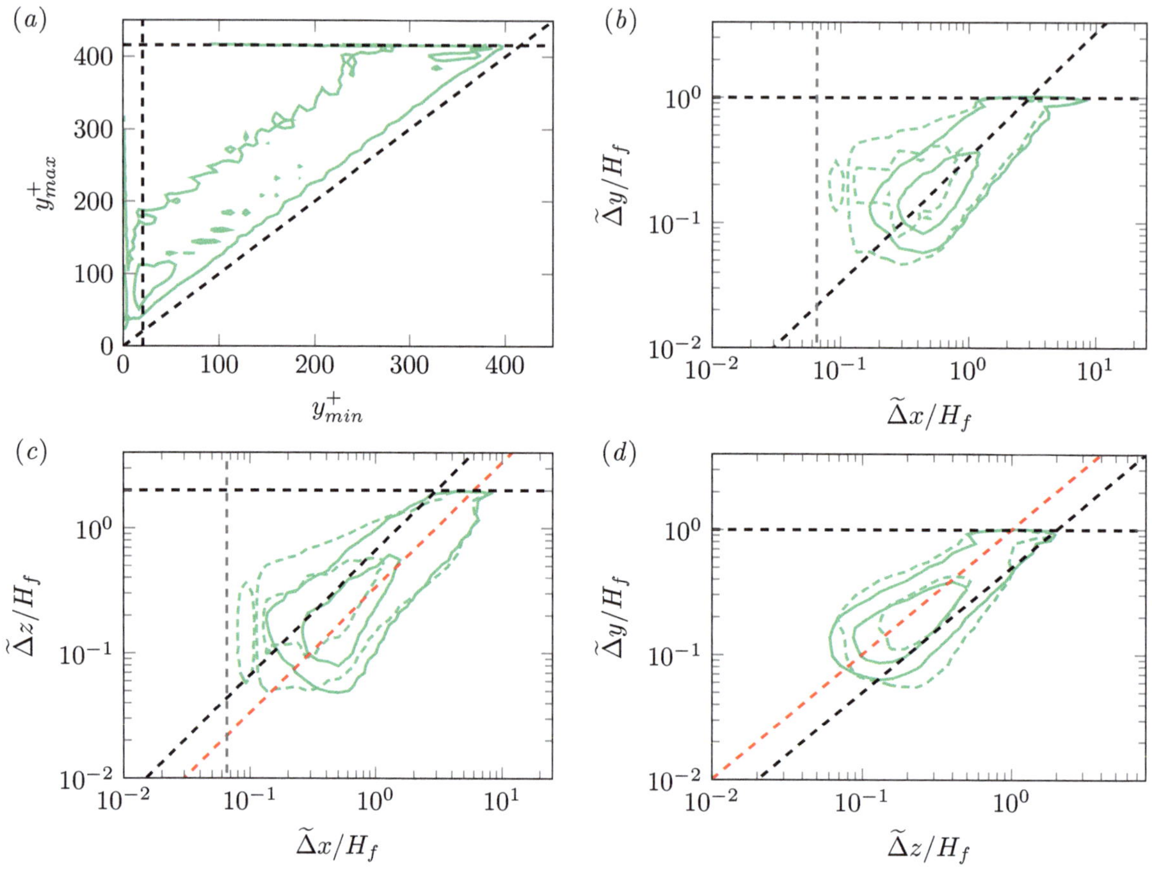

Figure 7.39: Joint p.d.f. of the axis-parallel extensions of individual wall-attached and surface-attached vortex clusters in the open duct simulation $DL400^{H2}_{smooth}$, measured in terms of their circumscribing bounding box. (*a*) Joint p.d.f. of the minimum (y_{min}) and maximum wall-normal distance (y_{max}) of the structures to the bottom wall. The vertical and horizontal dashed lines mark $y^+_{min} = 20$ and $y^+_{max} = Re_\tau$, respectively. (*b*) Joint p.d.f. of the streamwise ($\widetilde{\Delta x}$) and wall-normal dimension ($\widetilde{\Delta y}$) of the circumscribing bounding box of vortex cluster attached to the bottom wall ($y^+_{min} < 20$, —) or the free surface ($(Re_\tau - y^+_{max}) < 20$, - - -); (*c*) same for the streamwise ($\widetilde{\Delta x}$) and spanwise dimension ($\widetilde{\Delta z}$) and (*d*) the spanwise ($\widetilde{\Delta z}$) and wall-normal dimension ($\widetilde{\Delta y}$). Horizontal dashed lines indicate the domain sizes in the respective directions H_f and L_z, while the diagonal lines mark: (*a*) $y_{min} = y_{max}$; (*b*) $\widetilde{\Delta x} = 3\widetilde{\Delta y}$ (black); (*c*) $\widetilde{\Delta x} = 1.5\widetilde{\Delta z}$ (black), $\widetilde{\Delta x} = 3\widetilde{\Delta y}$ (red); (*d*) $\widetilde{\Delta z} = 2\widetilde{\Delta y}$ (black), $\widetilde{\Delta z} = \widetilde{\Delta y}$ (red). In (*b,c*), the vertical grey dashed line marks $\widetilde{\Delta x} = 3\Delta x$ which is the minimal streamwise length that structures have to possess to be included in the present analysis. In each panel, contours contain 20 and 90% of the total mass of the joint p.d.f.s.

agreement with the findings of Del Álamo et al. (2006) and Lozano-Durán et al. (2012) in closed channel flow, the bottom-attached vortex clusters exhibit a comparable self-similar behaviour as the streamwise velocity streaks. However, they feature in the self-similar regime with $\widetilde{\Delta x}/\widetilde{\Delta y}/\widetilde{\Delta z} = 3/1/1$ a slightly different aspect ratio than the one found by the aforementioned authors in closed channel flow, that is, $\widetilde{\Delta x}/\widetilde{\Delta y}/\widetilde{\Delta z} = 3/1/2$. The apparently different aspect ratio should be interpreted with caution, as we here do not separate between the structures in the centre of the duct and those in the near-sidewall regions, as was done for instance by Atzori et al. (2021) for closed ducts. It is imageable that the clusters located near the sidewalls feature a 'flipped' aspect ratio compared with that in the duct core which feel the presence of the lateral sidewalls only in form of the modified mean shear, with the consequence that the structures reveal – on average – a symmetric cross-sectional aspect ratio $\widetilde{\Delta y}/\widetilde{\Delta z} = 1/1$. Indeed, to clarify the possible effect of sidewalls on the shape and aspect ratio of individual large-scale structures, further investigations are required which are beyond the scope of the current study and are thus left for future studies.

The p.d.f.s of the surface-attached clusters in figure 7.39(*b-d*), in turn, show two different regimes: On the one hand, structures shorter than approximately $\widetilde{\Delta x} \approx 0.2H_f$ ($83\delta_v$) reveal essentially no self-similarity and feature more or less constant cross-sectional dimensions for varying streamwise extensions. Indeed, checking the spatial distribution of clusters with a given streamwise length (plots not shown) reveals that these very short structures live almost exclusively in the direct vicinity of the free surface, where the local flow is fairly different from that in the vicinity of an impermeable no-slip wall. With increasing streamwise length $\widetilde{\Delta x} > 0.2H_f$, on the other hand, the aspect ratio of the surface-attached structures approaches that of the bottom wall-attached structures. Note that this is not unexpected, as the dimensions of the largest structures are assumed to be essentially controlled by the outer geometry of the domain, rather than by the specific boundary conditions on the boundaries they are connected to (Jiménez, 2018). The question is where these structures are actually born: In contrast to the near-wall region, the mean shear and with it the turbulent kinetic energy production along the free surface is fairly low, but the mean shear is the main source of energy for turbulent coherent structures and it is closely linked to the self-similar log-layer cascade (Jiménez, 2012). In agreement with Sakai (2016), it is assumed that the larger structures that are attached to the free surface are actually formed along one of the solid walls as members of the self-similar family of attached cluster. At some point, these structures detach from the walls and move upwards towards the free surface. Alternatively, when born at one of the lateral sidewalls, vortex clusters might also be attached to both a sidewall and the free surface at the same time.

Note that an alternative attempt to locate large-scale rotating structures was made by Nakatsuji (2012) and Kawahara et al. (2012a) based on a poloidal-toroidal decomposition of the instantaneous velocity field in square duct flows. Since both publications are, except for the abstracts, written in Japanese, we will briefly summarise the main findings in the following. The mathematical background of the underlying poloidal-toroidal decomposition is discussed in appendix E, appropriate boundary conditions for the open duct case are proposed and an application of their methodology to the current dataset is presented.

In complete agreement with the here observed vortex clusters, Kawahara et al. (2012a) found two families of large-scale rotations, from which the first one consists of self-similar outer-scaling structures that are attached to the duct walls. The second one features both smaller and larger structures which are, however, detached from the solid walls and scale, depending on their size, either in inner or outer units. The preferential location of the circulating structures depends on their size, with larger structures found further away from the walls than smaller ones. The authors argue that the mean secondary flow pattern is the collective effect of the individual small and large circulating structures and their particular mutual organisation in the duct cross-section. Eventually, Kawahara et al. (2012a) pronounce the close similarity of the structures detected in their study with the vortex clusters investigated in Del Álamo et al. (2006). Indeed, when comparing the data presented in figure 7.37 with the preferential organisation of the streamwise rolls shown in figure E.1 in appendix E, it is seen that all but the largest rolls organise in a very similar way as the vortex clusters discussed above. The largest streamwise rolls concentrate somewhat more in the bottom half of the duct cross-section. Note that for consistency with the original work of Kawahara et al. (2012a), we have for now classified the streamwise rolls extracted in appendix E based on the same length scale and with the same definition of the roll centre as in their study, which are not exactly the same as the definitions considered here for the vortex clusters. So, further investigations are necessary to directly compare the results. Nonetheless, the good agreement between the mutual organisation of small- and large-scale streamwise rolls on the one hand and that

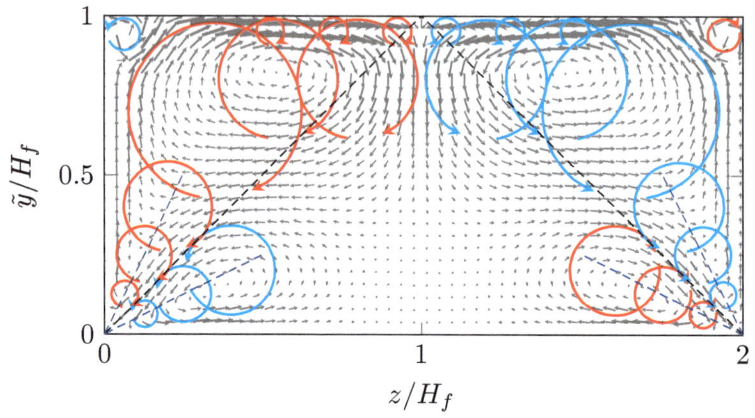

Figure 7.40: Conceptual sketch of the preferential organisation of small and large vortex clusters in the duct cross-section together with the mean secondary flow patterns. Clockwise (counterclockwise) rotating vortex clusters are indicated by red (light blue) colour. The vector field visualisation in the background represents the mean secondary flow field $(\langle v_f \rangle_{xt}/u_b, \langle w_f \rangle_{xt}/u_b)^T$ in case $DL400^{H2}_{smooth}$.

of the vortex clusters on the other hand strengthen the assumption of Kawahara et al. (2012a) that both structures might be related to each other.

In analogy to a visualisation presented in Nakatsuji (2012) for closed duct flow, figure 7.40 shows a conceptual sketch of the preferential organisation of small and large vortex clusters in the open duct cross-section, created based on the spatial organisation of small- and large-scale vortex clusters observed in figures 7.37 and 7.38. The visualisation is supplemented with a vector field representation of the mean secondary flow field in case $DL500^{H2}$. Based on the observations of Kawahara et al. (2012a) and the here discussed dataset, we believe that the mean secondary flow in the shown open duct flow represents the collective statistical footprint of a large number of instantaneous vortex clusters of different size and scale. Owing to their size, these structures are not randomly distributed over the duct cross-section: While small-scale clusters are found all along the walls and the free surface where they are indicative of the quasi-streamwise buffer layer vortices, large-scale clusters are controlled and bounded in their cross-sectional size by the domain boundaries. The largest structures that feature cross-sectional dimensions of comparable size as the mean fluid height H_f are strongly restricted in their mobility, and they are seen to centre predominantly in a relatively narrow region in the upper left and right duct quadrant, in some distance to the mixed corners. Along the free surface, in turn, small- and large-scale structures that are essentially attached to the free surface are seen to agglomerate either on the left or on the right of the duct bisector depending on their sense of rotation and the thereby induced lateral direction of motion, most likely as a result of the sorting process proposed by Sakai (2016). The so arranged vortex clusters naturally induce, by virtue of their mean rotation, a secondary flow in the cross-section of the duct that strikingly resembles the mean secondary flow, as can be inferred from the idealised conditional rollers provided in figure 7.40.

7.4 Summary and conclusion

The current chapter was dedicated to the interaction between turbulent coherent structures and sediment ridges in domains that are laterally bounded by impermeable sidewalls. For this purpose, the database on turbulent flows in infinitely wide open channel flows that was build up in the previous chapter was extended by a series of 13 open duct simulations with finite aspect ratio, over a range of friction Reynolds numbers $140 \leq Re_\tau \leq 515$. This new simulation series comprises simulations with three different bed/bottom configurations, including the flow over a mobile sediment bed, that over a stationary porous sediment bed as well as four simulations of single-phase smooth-wall open duct flow. Part of the latter smooth-wall reference data were kindly provided by Y. Sakai (Sakai, 2016). Individual simulations were performed in a wide range of different domain sizes, from streamwise-minimal narrow duct cases with streamwise and spanwise domain size $L_x \times L_z \approx 2H_f \times 2H_f$ to simulations in relatively long domains ($10 \lesssim L_x/H_f \lesssim 26$) and either low ($AR \approx 1$) or high aspect ratios ($AR \approx 8$). The variation of the lateral domain size offers the possibility to investigate sediment ridge formation both in the vicinity as well as in some distance to the laterally bounding sidewalls and to compare the results with our observations in spanwise homogeneous open channel flows. The Shields number θ_{bed}, on the other hand, was varied between fairly low values only slightly above the threshold for the onset of sediment erosion and relatively high values $\theta_{bed} \approx 0.22$, the latter indicating the development of an intense bedload transport layer. Purpose of this parameter variation was primarily to investigate whether the evolution of transverse ripple-like sediment bedforms can be delayed, as the appearance of these bed features strongly limits the time window during which sediment ridge dynamics can be studied exclusively. Indeed, the growth rate of the amplitude associated with initial ripple-like features was for the lowest Shields number about 50% lower than in a simulation with $\theta_{bed} \approx 0.15$ and otherwise essentially identical parameters.

The newly created database corroborates observations in laboratory experiments that sediment ridges evolve quasi-simultaneously at varying positions of the duct span, readily covering the entire cross-section of the duct (Nezu and Nakagawa, 1989). This is in line with our investigations in open channel flows in the previous chapter, where it was shown that sediment ridge formation is essentially independent of the presence of lateral sidewalls. Nonetheless, the formation of sediment ridges near lateral sidewalls and in some distance to the latter was seen to differ qualitatively. Most narrow ducts with an aspect ratio $AR \approx 1$ were seen to typically accommodate a single trough roughly in the duct centre enclosed by two sediment ridges, which are seen to form about $0.3H_f$-$0.5H_f$ away from the lateral sidewalls. The specific bedform arrangement is in good accordance with those observed in the experiments conducted by McLelland et al. (1999) in a slightly wider duct ($AR = 1.5$) with weakly bimodal sediment. The here observed bedforms are, on average, of smaller amplitude than their counterparts investigated in open channel flows or in the core of large aspect ratio ducts with $AR \approx 8$. In the latter case, the first sediment ridges evolve at a slightly larger distance of $0.5H_f$-$0.8H_f$ to the sidewalls and are relatively weaker in amplitude than their counterparts in the duct core, where the bedforms reveal essentially the same characteristics as those studied in infinitely wide channels. Especially, the mean lateral spacing between two ridges agrees fairly well with the values observed in open channel flow simulations as well as laboratory flume experiments (Wolman and Brush, 1961; McLelland et al., 1999) and the most-amplified wavelength of $1.3H_f$ in the linear stability analysis of Colombini (1993) that we have analysed in a previous chapter.

The qualitatively different properties of sediment ridges in narrow open duct flows compared to the open channel cases was found to be a consequence of the different organisation of the turbulent flow field near sidewalls and in sufficient distance to them. In fact, the core regions of sufficiently wide open ducts accommodate essentially the same alternating large-scale high- and low-speed streaks with average widths of $(1\text{-}2)H_f$ as in canonical or open channel flows (Jiménez, 2013a). Similar to the latter case, these large-scale structures are elongated in the streamwise direction and span over the entire depth of the clear fluid region, thereby inducing regions of strong and weak erosion activity along the sediment bed in which troughs and sediment ridges develop with time, respectively. In virtue of their long lifetime and their relatively weak lateral mobility, such large-scale velocity streaks and the associated Reynolds stress-carrying structures leave an imprint in the short-time averaged flow field ($\mathcal{O}(10T_b)$) in form of large-scale depth-spanning secondary rollers that populate the entire duct cross-section. Such 'instantaneous secondary currents' (Nezu, 2005) have been observed earlier in experimental studies and were, in agreement with our observations here, suspected to interact with a mobile sediment bed to form sediment ridges and troughs, respectively. Again, these features are the largest analogue of the conditional quasi-streamwise rollers of Lozano-Durán et al. (2012) and Jiménez (2018) and they occur over smooth walls, stationary and mobile sediment beds in essentially the same form.

For sufficiently longer averaging intervals of $T_{obs} \geq \mathcal{O}(10^3 T_b)$ length, mean secondary currents in some distance $z \gtrsim 4H_f$ to the sidewalls disappear, as the instantaneous large-scale velocity structures appear over time at different positions of the cross-section and contributions from structures with different sign cancel in the averaged framework. Their counterparts in the near-sidewall regions, on the other hand, are restricted in both their mobility and their preferential location, appearing in distinct lateral positions such that they remain visible even in the long-time statistics. For sediment-laden cases, a clear prognosis on the long-time evolution cannot be given based on the current database, as the ridges are typically displaced by transverse ripple-like bedforms after approximately 100-200 bulk time units. Therefore, it is not clear whether or not the sediment ridges indeed possess the potential to 'lock' or stabilise instantaneous large-scale velocity structures in their lateral position, as is often suggested based on experimental observations (Nezu and Nakagawa, 1993). At least in the initial phase of pattern formation considered in this study, the sediment ridges seemed to follow the evolution of the turbulent flow field rather than the other way round, what might however be related to the adopted Shields numbers. Indeed, the mobility of individual sediment ridges was observed to reduce with decreasing Shields number, and so it is not unlikely that a stabilisation of the large-scale structures by the developed ridges similar to the situation over fixed inhomogeneous bottom walls (Kevin et al., 2019a) can occur as long as the Shields number is sufficiently low. This latter assumption is also in line with the fact that sediment ridges are mostly observed in experiments with limited sediment transport availability, such that larger bedforms cannot develop (Wolman and Brush, 1961; Kleinhans et al., 2002). Future numerical simulations comparable to the here conducted that additionally sweep a range of different Shields numbers could clarify this open question.

In contrast to the situation in sufficient distance to lateral sidewalls, sediment ridges that evolve in the vicinity of solid sidewalls necessarily 'feel' the presence of the impermeable domain boundaries, especially in form of a clearly modified organisation of the flow field: The narrow domain width can accommodate only a single large-scale high-speed region in its centre, and it was observed that the peculiar domain geometry does not allow this structure to extend down to the bed in a similar way as the large-scale structures in the open channel do. Instead, it was seen that velocity structures of lower width that centre at approximately $\tilde{y}/H_f \approx 0.2$ control the regions of weak and strong erosion along the sediment bed, which in turn leads to the development of local sediment ridges and troughs. In analogy to the open channel case, the causal relation between the dynamics of these smaller structures and the organisation of sediment ridges and troughs along the sediment bed was shown based on two-point two-time correlations. These latter imply a time lag of several bulk time units length that it takes for the sediment bed to adapt to changes of the flow organisation at $\tilde{y}/H_f \approx 0.2$. Note that this time lag is accordingly smaller than the one observed for the open channel case, as in the latter we measured the time lag w.r.t. the flow field evolution in the centreline at $\tilde{y}/H_f \approx 0.5$.

The presence of a mobile sediment bed was seen to affect predominantly the flow field in its direct vicinity, whereas characteristic large-scale structures are seen to be rather weakly influenced, in agreement with the corresponding observations in the sediment-laden open channel cases or canonical closed channel flows past fully-rough walls (Flores and Jiménez, 2006; Flores et al., 2007). In a similar way as for the open channel case, particles that move inside the bedload layer above the sediment bed substantially reduce the mean velocity in the near-bed region while, simultaneously, the bottom friction is markedly enhanced compared to the corresponding smooth-wall reference cases at matching bulk Reynolds numbers Re_b: In case $DL500^{H2}$, for instance, Re_τ is about 38% higher than the value detected in the smooth-wall single-phase simulation at identical Re_b. In a similar way as in the corresponding open channel simulations, the turbulence activity in the bedload layer is damped and the characteristic peak of the streamwise turbulence intensity that is indicative of the buffer-layer regeneration cycle (Hamilton et al., 1995) is moved away from the bed, while its amplitude is clearly reduced. However, in contrast to the higher Reynolds number open channel case $CM850^{H3}$ discussed in the previous chapter, the relative particle size D^+ and the intensity of the bedload transport layer seem to be in all simulations low enough such that the buffer layer can remain intact.

The general pattern of both the mean secondary flow streamfunction $\langle\psi\rangle_{xt}$ and the streamwise vorticity field $\langle\omega_{f,x}\rangle_{xt}$ were seen to be qualitatively similar over all three considered bottom wall/bed configurations in some distance to the bed. In the lower left and right corners of the duct, however, the there residing mean secondary bottom vortex is seen to shrink as sediment is mobilised and the vorticity organisation in these corners is accordingly modified. Nonetheless, the cross-plane averaged mean secondary flow intensity is higher over the mobile sediment beds, which is related to an intensification of the outer mean secondary flow cells in the upper half of the duct due to a stronger mean upflow along the two sidewalls. Noteworthy, the secondary flow intensity over the stationary sediment bed is not significantly higher than the value in the single-phase case, highlighting that the origin of the secondary flow intensification is indeed the particle mobility. The exact mechanism that causes this enhancement of the outer secondary flow cell over mobile sediment beds cannot be fully clarified based on the current database, but it is assumed that a stronger redistribution of momentum in the cross-plane occurs due to the difference in friction between the two smooth sidewalls on the one hand and the mobile sediment bed on the other hand, as was suspected in a number of experimental studies for fully-rough beds (Nezu and Nakagawa, 1993; Rodríguez and García, 2008; Albayrak and Lemmin, 2011). To the best of the author's knowledge, no high-fidelity numerical study exists that investigates the exact role of bottom and/or sidewall roughness on the strength of the mean bottom and outer secondary flow cell, which could shed some light on this discussion. It would be therefore of high interest to perform simulations of this kind in the near future to clarify the impact of boundary roughness on the secondary flow pattern. In order to be able to compare the results with the above listed experiments, it appears suitable to perform such simulations over similar fixed roughness elements (e.g. particles) with a roughness height large enough to ensure fully-rough conditions, in analogy to the investigations of Mazzuoli and Uhlmann (2017) in open channel flows over fully-rough bottom walls.

Similar to the mean field, the preferential organisation of individual instantaneous coherent structures is seen to be essentially unaffected by the underlying mobile sediment bed unless the structures are located in its direct vicinity. The characteristics of the large-scale high- and low-speed streaks that dictate the bed organisation in sufficient distance to the sidewalls have been discussed in the previous chapter and much is known about them from studies in canonical wall-bounded flows, that is why we focused in a second part of the chapter on the peculiar organisation of the coherent structures in narrow open duct flows. In this context, one major focus was set on the organisation of the fluctuating velocity field $\boldsymbol{u}'_f(\boldsymbol{x},t)$ in the cross-section. To this end, two-point velocity correlations were analysed for flows over smooth walls and mobile sediment beds, focussing primarily on the organisation of the flow at a wall-normal distance $\tilde{y}/H_f \approx 0.2$ above the bed/wall, as the flow organisation in this layer had been previously found to play a major role in the formation process of sediment ridges. Indeed, the evaluated two-point correlations reveal elongated streamwise velocity objects of laterally alternating sign of comparable width $(0.4\text{-}0.5)H_f$ as the high- and low-speed regions that have been seen to influence the sediment bed evolution. The corresponding wall-normal correlation objects are clearly shorter, but reveal a similar laterally alternating sign which leads to the conjecture that these pairs of streamwise and wall-normal correlation patterns represent high- and low-speed streaks with associated down- and upflow regions, respectively. Interestingly, correlation objects of all velocity components bear an astonishing qualitative and in part also quantitative similarity with those investigated by Sillero (2014) and Jiménez (2018) in developing boundary layer and closed channel flow, respectively. This provides some evidence that not only the buffer layer structures studied by Pinelli et al. (2010), but also larger-scale velocity structures in duct flows feature important similarities with their counterparts in canonical flows.

This hypothesis was further strengthened by analysing the preferential organisation and size distribution of individual three-dimensional streamwise velocity structures, using the extraction procedure proposed by Moisy and Jiménez (2004) and Del Álamo et al. (2006). In complete agreement with an analogous investigation of velocity streaks in turbulent boundary layer flow by Sillero (2014) at however much higher Reynolds number, high- and low-speed streaks could be classified into a group of self-similar structures that are attached to the bottom wall and a second family of structures that are detached from it. In the special case of the open duct flow, a third group of structures was identified that is attached to the free surface which is naturally not existing in boundary layer flows. The bottom-wall attached objects scale self-similarly with their distance to the bottom, maintaining an aspect ratio of roughly $5/1/1$ in the streamwise, wall-normal and spanwise direction that is measured in terms of the structure enclosing cuboidal bounding box, in very good agreement with the results of Sillero (2014). The smallest structures of the self-similar family essentially correspond to the buffer layer streaks, while the largest objects feature dimensions comparable to the clear fluid height H_f. Additionally, a two-dimensional streak eduction study was performed following the approach proposed by Nakatsuji (2012) and adopted to open duct flows by Sakai (2016), which turned out to be better suited for the sediment-laden cases due to the clearly smaller number of available flow realisations in these cases, as opposed to the considered smooth-wall simulations. Both methods agreed in that streaks of different cross-sectional scale preferentially agglomerate in different regions of the domain for single-phase and particle-laden cases likewise, with the small buffer layer structures residing predominantly near the three solid sidewalls. Larger structures, on the other hand, were seen to centre preferentially in some distance to the walls and the free surface in virtue of their larger size, revealing some similarity with the earlier discussed organisation of the two-point correlations.

Perhaps most strongly affected from the mobile sediment were the statistics of the preferential positioning of quasi-streamwise vortices, that were extracted from the fields using a technique first used by Uhlmann et al. (2007) in the context of closed ducts based on the swirl-condition criterion of Kida and Miura (1998). As for the streaks, the organisation of small-scale vortices was in most areas of the cross-section similar for smooth-wall and mobile sediment beds. In the direct vicinity of the sediment bed, a substantial amount of individual vortex cores were counted, most likely related to individual vortices that develop in the context of vortex shedding in the wake of the bed and mobile sediment grains. In consequence, these vortices are essentially randomly distributed and thus cancel when averaging in time, such that they do not contribute to the mean vorticity field $\langle\omega_{f,x}\rangle_{xt}$. In the lower left and right corners, however, individual vortices were seen to organise in such a way that they induce a mean flow into the corner along the sediment bed. The resulting mean vorticity cell is naturally absent in the single-phase case, where the no-slip conditions do not allow vortices to come so close to the wall. Recalling that $\langle\omega_{f,x}\rangle_{xt}$ and $\langle\psi\rangle_{xt}$ are directly linked by a Poisson equation, it was concluded that the different organisation of the quasi-streamwise vortices in the corner regions affect not only the mean vorticity distribution, but also indirectly that of the mean secondary flow streamfunction, that way causing the observed modification of the bottom secondary flow cell.

In a final section, the instantaneous spatial aggregation of small-scale vortices in form of larger-scale vortex clusters and their relation to the mean secondary flow were studied. As for the streamwise velocity streaks, these vortex cluster can be classified into families of bottom-wall or surface-attached structures on the one hand and objects that are detached from both on the other hand. In agreement with the observations of Del Álamo et al. (2006), the bottom-wall attached clusters scale again self-similarly with their distance to the wall. While the smallest representatives of this family essentially represent a single quasi-streamwise vortex, the largest clusters reach dimensions comparable to H_f and thus take up a significant fraction of the cross-section. As for the velocity streaks, vortex clusters preferentially agglomerate in different regions of the cross-section depending on their size: The smallest clusters primarily concentrate in the vicinity of the domain boundaries, whereas the largest are typically located in the upper left and right quadrant roughly $0.35H_f$ away from both the sidewalls and the free surface. The organisation of the vortex clusters of different size and sense of rotation reveals an astonishing similarity with the general secondary flow pattern. In this context, Kawahara et al. (2012a) investigated large-scale circulating quasi-streamwise rolls in square duct flows based on a poloidal-toroidal decomposition of the flow field and conjectured that these features are related to the here studied vortex clusters. We have applied their methodology to the current open duct case, which revealed that except for the largest rolls in the domain, the vortex clusters and the streamwise rolls organise indeed in a very similar way in the cross-section. Based on the conclusions of Kawahara et al. (2012a) and the current investigations of the self-similar vortex clusters and rolls, we believe that the mean secondary flow could be the collective statistical footprint of the individual vortex clusters or large-scale rolls. In order to verify this hypothesis, additional efforts are necessary and a more detailed analysis of both types of structures in closed as well as open channel flows should be performed. In this regard, the analysis could be ideally applied to datasets at higher Reynolds numbers that allow for a proper scale separation between buffer layer and outer-scaling structures.

Even though turbulent flows in ducts of square or rectangular cross-section have raised much more attention since the pioneering works of Uhlmann et al. (2007) and Pinelli et al. (2010) and higher Reynolds number simulations up to $Re_\tau \approx 1000$ were indeed performed (Modesti et al., 2018), there are only few studies that really focus on the characteristics and dynamics of the individual structures (Atzori et al., 2021). In the case of open duct flows, as stated earlier, the database of Sakai (2016) represents, to the best of the author's knowledge, in fact the only set of fully-resolved simulations at sufficient size to cover also the largest emerging flow structures. In the same way, the current simulations seem to be the first of its kind that allow to study the evolution of sediment bedforms under 'laboratory conditions', i.e. in domains of finite aspect ratio analogously to experimental flumes. In this regard, simulations of the current setup at desirably higher Reynolds numbers will hopefully soon provide the opportunity for a direct comparison between carefully planned laboratory and numerical experiments. It is in this context also worth considering to perform simulations featuring bimodal sediment mixtures, that can additionally enhance the development of sediment ridges in form of sediment sorting processes (Colombini and Parker, 1995; McLelland et al., 1999).

Chapter 8

Transition from sediment ridges to initial ripples – an outlook

In the previous chapters, our goal was to investigate the relation between large-scale turbulent structures and a mobile sediment bed that causes the evolution of sediment ridges. In this context, it was repeatedly mentioned that the time window during which sediment ridges can be studied exclusively is limited by the appearance of larger amplitude ripple-like bedforms for $t \gtrsim 100T_b$. This last chapter is devoted to the transformation of the ridge-covered sediment bed into a ripple-dominated one and should be understood as an outlook on future studies that are planned to investigate this transition phase.

To this end, we have analysed the bed evolution in the large-domain open channel ($CL250^{H16}$) and open duct simulations ($DL250^{H16}$) well over the earlier studied time interval by letting the bed evolve over 500 bulk time units (cf. tables 8.1 and 8.2 for the respective parameters determined over the longer time interval).

An instantaneous state of the sediment bed and the overlying flow field in a later stage of bedform evolution is visualised in figure 8.1, wherein the transverse bedforms are clearly seen to dominate the bed surface. A cross-section of the flow field indicating instantaneous high- and low-speed regions at the downstream end of the periodic domain highlights that the general lateral organisation of the flow field in alternating large-scale high- and low-speed streaks is maintained even over developed transverse bedforms. In the near-bed region, on the other hand, we expect a significant altering of the flow, for instance, downstream of the ripple crests, where the flow detaches from the bed and a recirculation zone develops (Kidanemariam et al., 2022). In this regard, the here shown patterns belong to the class of 'vortex dunes' (Ouriemi et al., 2009). Note that the terms ripples and dunes typically refer to bedforms with a much longer evolution time than the patterns that arise in the current simulations, and so the here discussed transverse bedforms should be understood as 'initial ripples' (Langlois and Valance, 2007; Kidanemariam and Uhlmann, 2017). For the sake of simplicity, we nevertheless term these earliest evolving transverse bedforms as 'ripples' in the remainder, keeping in mind that the term refers to this specific type of early-stage bedforms.

Figure 8.1 also shows a 'pseudo free surface' of the open duct, visualising the local amplitude of the pressure fluctuations along the strictly flat free-slip plane of the open duct. Even though the currently used computational method does not allow for a deformation of the free surface, it is that way possible to estimate how a freely-deformable surface would react to the pressure fluctuations in the domain. Interestingly, the pressure fluctuations are seen to be correlated with the evolving bedforms in that they would cause an upwelling of a deformable free surface downstream of the bedform crest over

Figure 8.1: Three-dimensional visualisation of an instantaneous state of the sediment bed and the flow field in case $DL250^{H16}$ in the ripple-dominated time period. Flow is from bottom left to top right. As before, the particle colour ranges from dark to light brown with increasing distance to the bottom wall. The local turbulence activity is indicated by individual vortices, visualised as isosurfaces of constant negative value of λ_2 (second eigenvalue of the Hessian of the pressure, cf. Jeong and Hussain (1995)). White and black regions in the cross-plane at the downstream end of the periodic domain indicate high- and low-speed regions of the streamwise velocity (u'_f). The curved manifold at the top of the domain indicates pressure fluctuations along the free surface, which was kept flat in the simulation and thus represents a 'pseudo-deformable fluid-air interface'. Vertical deformations with dark (light) blue colour indicate low (high) values of the pressure field p_f at $\tilde{y} = H_f$. For the sake of visualisation, the second sidewall in the foreground has been removed.

the recirculation zone. Here, the streamlines expand and so-called 'kolk-boil vortices' are generated which are typically assumed to detach from the sediment bed contour and to afterwards propagate towards the free surface (Nezu and Nakagawa, 1993, and references therein). More detailed investigations of the free-surface dynamics cannot be performed at this point due to the strictly flat free surface in our simulations, but it strongly motivates to perform such analysis using appropriate numerical tools that are able to capture the deformation of the duct free surface. In this field, some progress is recently observed in that nowadays fully-resolved numerical studies of turbulent flows over complex bottom topographies and their interaction with a deformable free surface can be performed (e.g. Kidanemariam and Marusic, 2020).

In the following, we now focus on the bedform dynamics in the phase subsequent to the ridge-dominated time-interval. Figure 8.2(a) shows the time evolution of the streamwise- and spanwise-averaged r.m.s. of the sediment bed height fluctuations for the open channel and duct cases, respectively, which we typically consider as a measure for the mean amplitude of streamwise-aligned sediment ridges and transverse-oriented ripples, respectively. For the sake of comparison, the presented datasets are supplemented with corresponding curves for a narrow open channel case from Kidanemariam and Uhlmann (2017) (case $H12^1$ in their notation) with streamwise and spanwise domain periods $L_x/H_f \approx 12$ and $L_z/H_f \approx 3$, respectively, under otherwise comparable physical conditions. Interestingly, the time evolution of $\sigma_{h,x}$ is markedly different in the wide open channel case $CL250^{H16}$ compared to the narrow channel $H12^1$ and the wide open duct case $DL250^{H16}$, respectively. The strongest increase is observed in the narrow open channel case $H12^1$, where we observe an initial exponential growth almost from the very beginning of particle erosion – an effect that has been studied in detail by Kidanemariam and Uhlmann (2017). The wide open duct case $DL250^{H16}$ reveals a comparable growth rate, but the onset of the phase of strong increase is shifted in time by about 50 bulk time units. The wide open channel case $CL250^{H16}$, finally, exhibits a much slower increase of $\sigma_{h,x}$ at least for the first 400 bulk time units of the simulation. Only then, a stronger increase of the mean bedform amplitude is observed. In this phase, the remaining two simulations seem to have already reached a quasi-stationary plateau regime.

	Case	Re_b	Re_τ	ρ_p/ρ_f	Ga	D^+	H_f/D	H_b/D	H_f/H_b	θ_{bed}
—	$CL250^{H16}$	3012	283.67	2.5	28.37	10.68	26.57	11.83	2.25	0.14
—	$DL250^{H16}$	2998	286.72	2.5	28.37	10.77	26.62	11.78	2.26	0.14

Table 8.1: Physical parameters of long-time ripple-featuring open channel and high aspect ratio duct flow simulations. The presented parameters are defined as described in sections 6.1 and 7.1, respectively.

	Case	$[L_x \times L_z]/H_f$	$[L_x \times L_z]/D$	$D/\Delta x$	Δy^+	N_p	T_{obs}/T_b
—	$CL250^{H16}$	11.56×15.42	307.2×409.6	10	1.07	1 406 983	503
—	$DL250^{H16}$	11.54×15.39	307.2×409.6	10	1.08	1 402 849	499

Table 8.2: Numerical parameters of long-time ripple-featuring open channel and high aspect ratio duct flow simulations. The shown parameters follow the definitions in sections 6.1 and 7.1, respectively.

The strikingly different evolution of the amplitude of the spanwise-averaged sediment bed motivates a more detailed analysis of the bedform development in the three cases. To this end, figure 8.2(*b*) shows the time evolution of the two-dimensional r.m.s. of the sediment bed height fluctuations that is determined for the non-averaged bed in analogy to equation (6.7), viz.

$$\sigma_{h,2D}(t) = \left[\langle (h_b(x,z,t) - \langle h_b \rangle_{xz}(t))^2 \rangle_{xz}\right]^{1/2} . \tag{8.1}$$

In Scherer et al. (2020), this alternative measure was used to identify the growth of bedforms that are arbitrarily oriented w.r.t. the mean flow direction (e.g. bedforms that span diagonally across the domain) and thus might disappear in the spanwise average. Indeed, the time evolution of $\sigma_{h,2D}$ is seen to be very similar for the three cases in figure 8.2(*b*), implying that the bed instability mechanism and the associated growth rate is comparable in narrow and wide domains. However, the narrow width of the channel case $H12^1$ artificially forces the bedforms to be quasi two-dimensional, i.e. more or less independent of the spanwise direction. The wide domains, on the other hand, allow for the development of initial bed perturbations that represent nuclei for the later evolving ripples at different locations in the domain, which can be shifted to one another in both the streamwise and spanwise direction. In such cases, contributions from different local sediment patches can cancel in the spanwise average and thus do not contribute to $\sigma_{h,x}$.

This effect is discernible in figure 8.3, which shows instantaneous states of the extracted fluid-bed interface at different instances of the bedform development $t/T_b \in \{50, 150, 400, 500\}$ for the wide open channel and duct simulations. During the early phase of sediment bed evolution at $t/T_b = 50$ (cf. figures 8.3(*a,b*)), the bed is entirely covered by streamwise-elongated sediment ridges that reveal only a weak streamwise modulation. Already 100 bulk time units later (cf. figures 8.3(*c,d*)), a dominant spanwise organisation of the sediment bed is clearly visible superimposed on streamwise sediment 'lines', being reminiscent of the original sediment ridges. However, the local sediment agglomerations are only partly connected and some of them feature a three-dimensional shape comparable to small-scale versions of three-dimensional barchan dunes (Franklin and Charru, 2011). The described evolution is in line with the observations of Kleinhans et al. (2002) according to which "gradual transitions exist from sand ribbons to barchans, and from barchans to fully developed dunes" with increasing particle

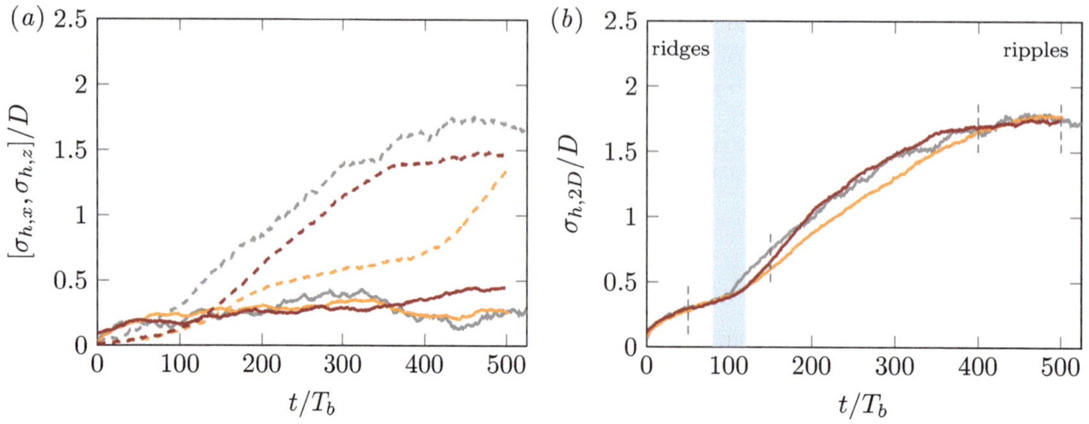

Figure 8.2: Time evolution of the mean bedform amplitude in the long-time ripple-featuring open channel and high aspect ratio duct flow simulations $CL250^{H16}$ (—) and $DL250^{H16}$ (—), respectively. (*a*) One-dimensional root mean square of the fluctuation of the streamwise-averaged ($\sigma_{h,z}/D$,—) and spanwise-averaged fluid-bed interface ($\sigma_{h,x}/D$,---). (*b*) Two-dimensional root mean square of the fluctuation of the non-averaged fluid-bed interface $\sigma_{h,2D}/D$ (cf. definition (8.1)). The same quantities have been also evaluated for a narrow domain open channel case $H12^1$ from Kidanemariam and Uhlmann (2017) (—), which features except for the domain width $L_z/H_f \approx 3$ comparable parameters: $Re_\tau \approx 250$, $L_x/H_f \approx 12$ and $H_f/D \approx 26$. In (*b*), additional vertical dashed lines refer to the times at which the instantaneous snapshots of the fluid-bed interface have been extracted in the following figure 8.3.

transport intensity. In good agreement with the evolution process described therein, we observe that the initially localised sediment seeds connect with each other after a while (cf. figures 8.3(*e*,*f*)), forming spanwise-elongated sediment bedforms that span the entire cross-section of almost $16H_f$ width. The synchronisation of individual bedforms is most likely associated with varying pattern heights, as it is well known that the streamwise propagation velocity of sediment patterns is inversely proportional to their size (Kidanemariam and Uhlmann, 2017). Hence, small patterns travel faster than larger ones such that they can 'catch up' and merge with the latter (Zgheib et al., 2018a,b).

However, there is a remarkable difference between the channel and the duct case in that the crest lines of the bedforms in the latter case are more or less aligned with the spanwise direction, such that the local bed profile at a given position z is essentially the same as in the spanwise average. In the open channel, on the other hand, the crest lines of the ripple-like patterns oscillate w.r.t. the lateral direction such that spanwise averaging partially leads to a cancelling of troughs and crests at specific streamwise positions. Time-resolved visualisations of the bedform evolution in the open channel case $CL250^{H16}$ further strengthen the conclusion that it is predominantly due to this 'cancelling effect' in the spanwise averaging that $\sigma_{h,x}$ features a markedly lower growth rate compared to that in the open duct case $DL250^{H16}$. In this regard, it is worth noting that the lateral synchronisation of the individual local sediment patterns occurs much earlier for the open duct case than for the open channel simulation, which motivates the question whether lateral sidewalls have a 'stabilising' effect on the formation of transverse sediment ripples. To clarify this point, more simulations have to be performed to allow for ensemble averages that will show whether or not there is such an effect that is of statistical relevance.

In agreement with our earlier observations in chapter 7, it is seen that the lower mass flow rate in the near-sidewall region of the open duct case $DL250^{H16}$ leads to a rather weak streamwise particle transport in the corner regions. While the time period investigated in the previous chapter was too short to observe how the bed-sidewall corner is filled up with sediment, it is now seen that the streamwise-averaged bed profile $h_b''(z,t)$ even attains its global maxima at the sidewalls. In figures 8.3(*e*-*h*), the sediment agglomerations in the near-sidewall regions are seen to be still connected to transverse-bedforms

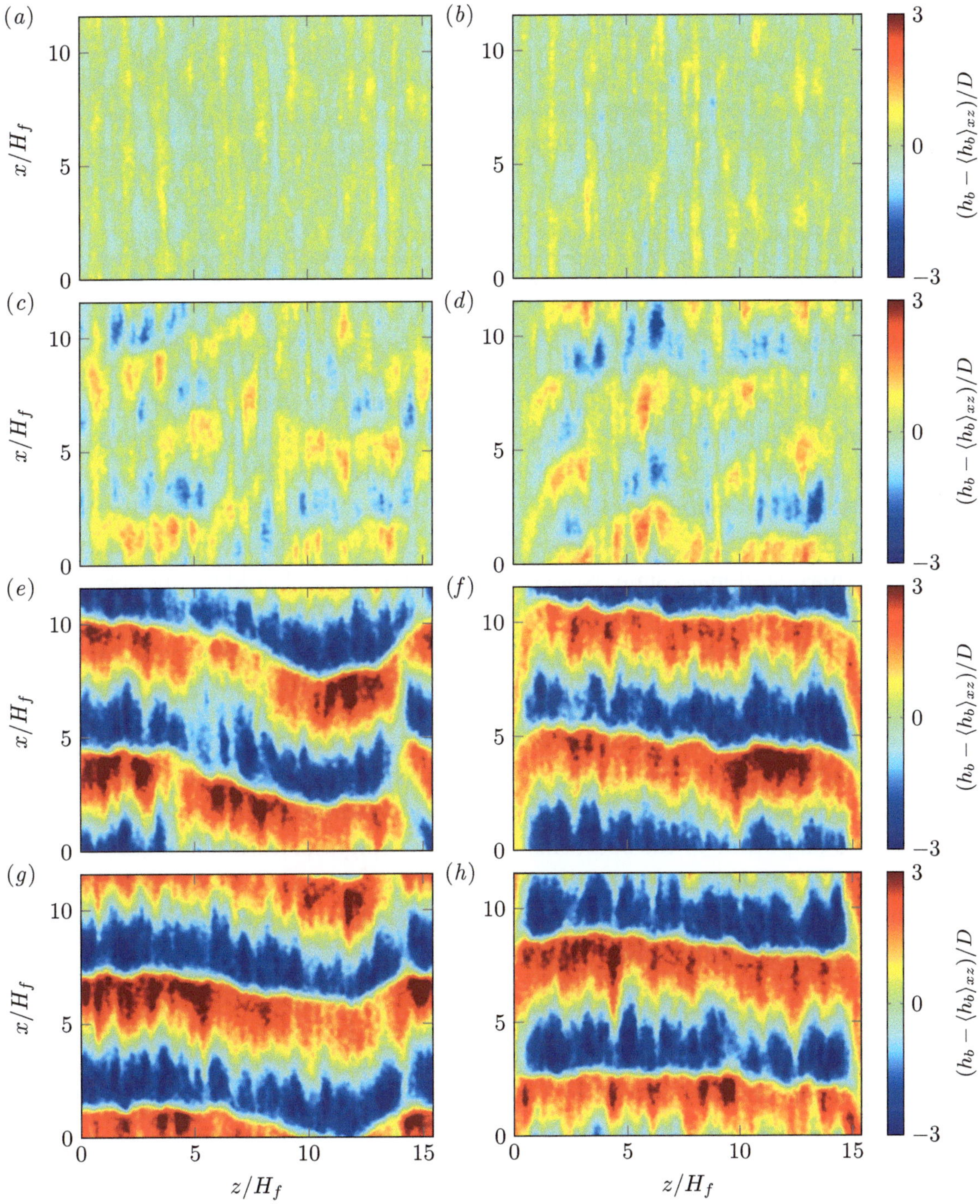

Figure 8.3: Instantaneous snapshots of the fluid-bed interface as seen from the top in the long-time ripple-featuring open channel and high aspect ratio duct flow simulations at *(a,b)* $t = 50T_b$, *(c,d)* $t = 150T_b$, *(e,f)* $t = 400T_b$ and *(g,h)* $t = 500T_b$. Flow is from bottom to top in each panel. *(a,c,e,g)* $CL250^{H16}$ (—), *(b,d,f,h)* $DL250^{H16}$ (—).

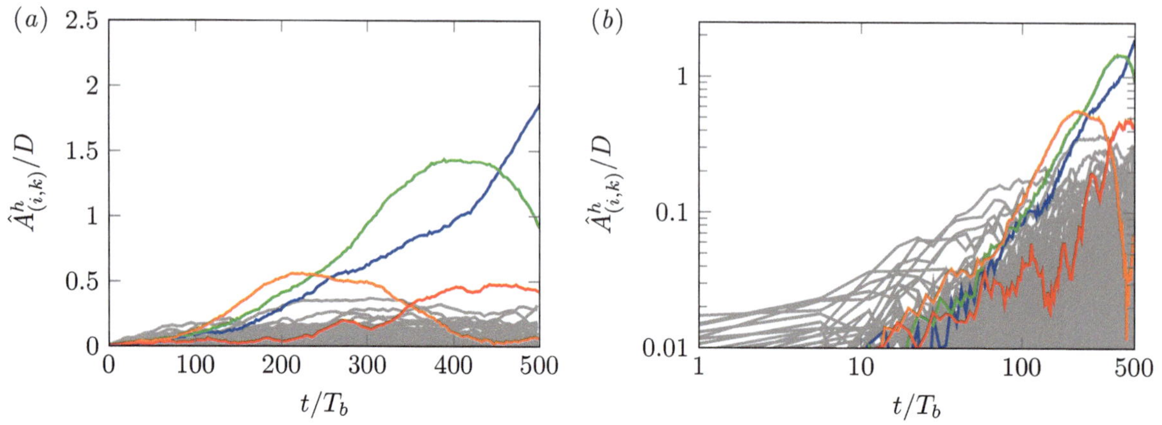

Figure 8.4: Long-time evolution of individual Fourier modes of the single-sided two-dimensional amplitude spectra of the fluid-bed interface in the large-domain open channel flow simulation $CL250^{H16}$ in (*a*) linear and (*b*) logarithmic scaling. Dominant modes which exceed a value of $0.4D$ at least once during the time interval are highlighted with the following colours: $\hat{A}^h_{(2,0)}/D$ (—), $\hat{A}^h_{(2,1)}/D$ (—), $\hat{A}^h_{(3,0)}/D$ (—) and $\hat{A}^h_{(4,1)}/D$ (—). Here, the notation $\hat{A}^h_{(i,k)}$ refers to the single-sided amplitude of the (i,k)th harmonic w.r.t. the computational domain (L_x, L_z).

in the duct core that have already travelled downstream, resembling in that way a kind of 'shear layer' along the sidewalls. Finally, figures 8.3(*g*,*h*) indicate that 500 bulk time units after the onset of sediment erosion, the transverse bedforms in the open channel case $CL250^{H16}$ have almost reached an essentially spanwise independent state. In the open duct case $DL250^{H16}$, on the other hand, the general structure of the fluid-bed interface is almost the same as at $t = 400T_b$, implying that this case has already reached a quasi-stationary state. Note that quasi-stationary should be understood in the context of the time scale and phase of pattern evolution on which our study focuses. For much longer time intervals and sufficiently large domains, on the other hand, the patterns are expected to further grow towards larger-scale bedforms (Coleman and Melville, 1996).

The transition of the ridge-covered sediment bed to one that is dominated by ripple-like patterns in the doubly-periodic channel can be conveniently quantified in spectral space, as can be seen in figure 8.4. Having exploited the analogy of the bed undulations to a two-dimensional wave earlier when defining wavelength and amplitude of the bedforms, it is now straightforward to extend the study to the Fourier transform of the fluid-bed interface fluctuations, viz. $\hat{h}_b(k_x, k_z, t) = \mathcal{F}(h_b(x,z,t) - \langle h_b \rangle_{xz}(t))$. We compute the single-sided amplitude spectra for the physically relevant non-negative wavenumbers $k_x, k_z \geq 0$ as (Kidanemariam, 2016; Scherer et al., 2020)

$$\hat{A}^h(k_x,k_z,t) = 2|\hat{h}_b(k_x,k_z,t)| = 2\left[\hat{h}_b(k_x,k_z,t)\hat{h}_b^*(k_x,k_z,t)\right]^{1/2}. \tag{8.2}$$

In the remainder of this chapter, we will make use of the short-hand notation $\hat{A}^h_{(i,k)}$ to refer to the single-sided amplitude of the (i,k)th harmonic w.r.t. the computational domain (L_x, L_z).

Figure 8.4 shows the time evolution of the first 15 streamwise and spanwise harmonics as well as the 'mixed' modes with $i, k \neq 0$ for case $CL250^{H16}$, such that the shortest spanwise wavelength considered therein is close to unity, while the shortest streamwise wavelength is somewhat lower. Only four out of 225 considered modes exceed an amplitude of $0.4D$ at least once during the simulation interval (coloured in figure 8.4). These four modes are related to the second to forth streamwise harmonic and are thus associated with transverse ripple-like bedforms. In figure 8.4(*b*), it can be seen that the four 'dominant' modes initially feature an essentially exponential increase in agreement with earlier observations in linear stability analyses and numerical simulations (Colombini and Stocchino, 2011; Kidanemariam and Uhlmann, 2017). They become relevant during the ripple-dominated phase $t \gtrsim 150T_b$, while their amplitude is negligible in the initial 100 bulk time units during which the pure spanwise harmonics $\hat{A}^h_{(i=0,k)}$ related to sediment ridges still dominate the spectra, even though at relatively low amplitude $\mathcal{O}(0.1D)$. Three of the four coloured modes dominate the spectra for a while, and the sequence at which these modes rise and fall is directly correlated with the different phases of bedform evolution observed in figures 8.3(*a,c,e,g*): Between $t = 100T_b$ and $t = 200T_b$, the domain accommodates three transverse patterns at shorter streamwise wavelength, followed by a period between $t = 250T_b$ and $t = 450T_b$ during which a 'mixed' mode $(2,1)$ is dominant. This mode is indicative of two larger transverse bedforms which are however oscillating in the lateral direction, in complete agreement with the shape of the two sediment patterns in figure 8.3(*e*). At the end of the observation interval, the crests of the ripple-like bedforms in figure 8.3(*g*) are more or less aligned with the z-direction, accordingly the spanwise constant mode $(2,0)$ now dominates the spectra.

A detailed analysis of the flow modulation around individual, partly three-dimensional initial bedforms is outside the scope of the current study. But the here generated datasets will allow to perform exactly these kinds of studies in the future by offering the 'full' flow field and sediment bed information in sufficiently large domains to investigate how 'vortex ripples' and the associated shear layers evolve. The results of such studies are expected to be of high interest for hydraulic and environmental engineers likewise.

Let us finally comment on a conceivable mechanism in the consequence of which sediment ridges transform into transverse ripple-like bedforms. The minimal streamwise wavelength of these initial sediment ripples was recently quantified to be about $\lambda_{crit} \approx 80D$ in the current parameter regime (Kidanemariam and Uhlmann, 2017; Scherer et al., 2020), which is in excellent agreement with the current observation that initially three distinct transverse patterns form on the sediment bed ($\lambda_{h,x}/D \approx 100$). In Scherer et al. (2020), a mechanism was proposed following the conceptual ideas of Coleman and Nikora (2009) according to which a quasi-randomly generated sediment seed induces a downstream wake in which more quiescent conditions lead to a considerably weaker erosion activity. Consequently, downstream initial seeds in this wake lack a sufficient sediment supply from upstream to grow further, such that the next 'amplified crest' can develop only at a certain downstream distance to the initial seed where the wake effect has sufficiently weakened. The collective effect of these processes is a relatively regular streamwise spacing of initial ripple-like bedforms that is of the order of the streamwise extension of the seed-wake. The here newly created dataset allows to study the proposed mechanism in a more or less spatially unconstrained environment.

Figure 8.5 shows an exemplary time series of instantaneous snapshots of the fluid-bed interface similar to figures 8.3(*a,c,e,g*), but now focussing on a single ridge that is initially located at a lateral position of $z \approx 3.75H_f$ in the open channel case $CL250^{H16}$. Initially ($t/T_b = 25$), the sediment ridge spans essentially the entire streamwise domain length L_x/H_f, revealing two regions of minimally higher amplitude at approximately $x \approx 4H_f$ and $x \approx 10H_f$. In spite of their weak amplitude, the two initial 'perturbations' of the streamwise sediment ridge are seen to significantly grow in time over the following 125 bulk time units, during which both are seen to propagate downstream at an almost constant propagation speed of $\mathcal{O}(0.01u_b)$ or $\mathcal{O}(0.1u_\tau)$, as expected slightly faster than the larger bedforms that

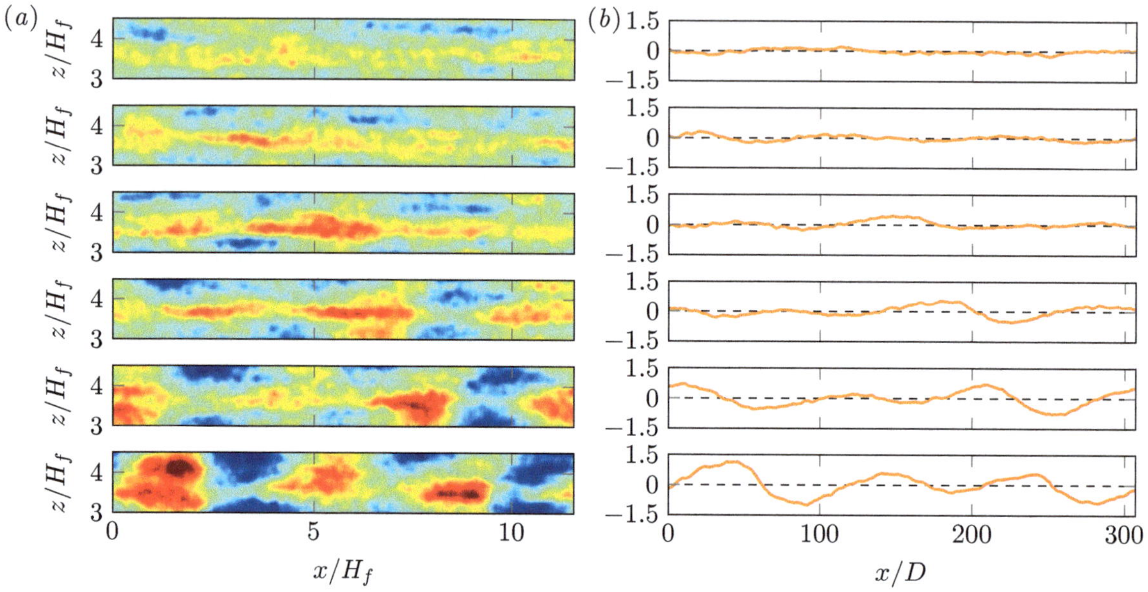

Figure 8.5: Time series of snapshots visualising how a single sediment ridge perturbation can act as initial seed for transverse bedform development in the long-time open channel flow simulation $CL250^{H16}$. (*a*) Instantaneous snapshots of a subsection of the fluid-bed interface as seen from the top of the channel are presented at a sequence of times $t/T_b = \{25, 50, 75, 100, 125, 150\}$ (from top to bottom). In each panel, the flow is from left to right. The colour ranges from blue (troughs) to red (crests) in the interval $(h_b - \langle h_b \rangle_{xz}) \in [-1.5D, 1.5D]$. (*b*) Mean fluid-bed interface contour $\langle h_b - \langle h_b \rangle_{xz} \rangle_z / D$, computed as an average over the spanwise subsection of the fluid-bed interface shown in the corresponding panels in (*a*) only.

are seen at later times of these simulations. Simultaneously, a deepening of the sediment bed on the leeward side of these seeds is observed which eventually causes the initial ridge to break into a sequence of localised sediment patches. These latter are then seen to connect with neighbouring seeds to form larger transverse-aligned sediment bed structures. The observed 'instability' of the streamwise sediment ridge further strengthens the above proposed formation mechanism in that local quasi-randomly occurring perturbations of the sediment bed can lead to the evolution of transverse bedforms at a preferential streamwise spacing.

To conclude, let us remark that the current discussion of the sediment bed transformation is of rather phenomenological nature and more comprehensive investigations of an ensemble of individual ridges are necessary to assess whether or not the outlined process is of statistical relevance for the formation of transverse sediment patterns. However, the present results strongly suggest that the mechanistic process conceived by Scherer et al. (2020) based on the results of Coleman and Nikora (2009) can indeed explain the development of transverse ripple-like patterns in spatially unconstrained domains.

Chapter 9

Summary and conclusion

Sediment bedforms are ubiquitous in many natural ecosystems, ranging from massive sand dunes down to small sand seeds in subaqueous environments such as rivers or estuaries. In this context, streamwise-aligned subaqueous sediment ridges represent a peculiar class of sediment bedforms that are closely coupled to the structure and organisation of the turbulent flow above them, and they typically appear side by side with large-scale secondary currents of Prandtl's second kind. Understanding the formation of these sediment ridges, the development of a mean secondary flow and the relation of both to individual turbulent structures is a formidable challenge and in spite of almost a century of intensive scientific work, many questions remain unanswered. The aim of the current work is to contribute to a better understanding of the complex interactions between the chaotic turbulent flow and a mobile sediment bed that eventually cause the formation of both secondary currents and sediment ridges. The novelty of the present work lies in the use of direct numerical simulations featuring up to 1.4 million fully-resolved mobile particles in order to study the mutual interaction between a mobile sediment bed and individual flow structures from first principles. The current work is, to the best of our knowledge, the first of its kind in providing high-fidelity data of the formation of sediment ridges at that level of detail. The newly generated database has allowed us to study instantaneous coherent velocity structures embedded within the turbulent flow field and their impact on the bed topography in open channel and duct flows. In this regard, a 'top-down mechanism' was detected in which large-scale coherent structures initiate the formation of long streamwise-aligned sediment ridges. A detailed summary of the most important findings in each of the four chapters is given in the following, together with an outlook on future works for which the herein presented results might represent a good starting point.

9.1 Linear instability of a turbulent flow over initial sediment ridges

Practically the only extensive theoretical work on the evolution of sediment ridges and the development of the associated secondary currents was presented roughly 30 years ago by Colombini (1993). The author performed a linear stability analysis of a coupled fluid-sediment bed system in an infinitely wide open channel, in which a turbulent base velocity profile is perturbed by a sinusoidal undulation of the sediment bed with an infinitesimal amplitude. The study was the first to prove that both sediment ridges and secondary currents can arise only due to an interaction of a turbulent channel flow with a slightly perturbed sediment bed. In particular, pre-existing secondary currents induced by lateral sidewalls are not required for this mechanism. In the current work, we first revisited and implemented the model of Colombini (1993) in its original formulation, using a standard Chebyshev-tau collocation method to solve the linearised system of equations. After a validation of the current method against the original results, the model results were investigated in more detail than it was possible in the original work, especially in light of recent numerical and theoretical studies on the organisation of secondary flows over non-homogeneous bottom walls. It turned out that the model – even though originally formulated for the specific case of streamwise-aligned sediment ridges – predicts quite well the overall organisation of secondary currents in the cross-plane of these more general type of flow configurations. Specifically, the spanwise wavelength at which the secondary flow kinetic energy attains its maximum and the secondary currents become 'space-filling' is captured, and so is the tendency of secondary currents to concentrate to the near-bed region when the spanwise wavenumber tends to infinity. Next, a comparison with a theoretical model recently proposed by Zampino et al. (2022) that predominantly aims to predict the mean secondary flow patterns over non-homogeneous bottom walls showed very good agreement with the model of Colombini (1993) revisited here, despite the use of a different turbulence closure. The comparison indicates that the predictions obtained with the original model of Colombini (1993) are of physical nature and are not sensitive to the specific turbulence model, as long as it is elaborate enough to allow for the necessary anisotropy of the Reynolds stress tensor, as otherwise no secondary currents can evolve.

A second part of the chapter was devoted to the analysis of the original model and to the question of how the most amplified lateral wavelength $\lambda_{h,z} \approx 1.3H_f$ depends on the different ingredients of the model. It turned out that the bed modulation acts in a similar way as a stationary external force field to the linearised stationary Navier-Stokes operator, which is required to maintain the lateral instability that manifests itself in large-scale depth-spanning secondary currents. In order to further scrutinise this observation, the original theoretical model was modified by replacing the bottom curvature that induces the lateral perturbation by an arbitrary external force field. To be able to compare the results with theoretical investigations of the linear processes in canonical turbulent channel flows, we furthermore added viscous terms to the linear operator that were neglected in the original model formulation. In the remainder, different external forcing schemes and the corresponding induced flow fields were investigated, in comparison with studies on transient growth mechanisms (Butler and Farrell, 1993; Del Álamo and Jiménez, 2006; Pujals et al., 2009) and the externally forced Orr-Sommerfeld problem (Hwang and Cossu, 2010b). Even though the current model is independent of the streamwise direction and formulated in a stationary way, it was able to reproduce the formation of streamwise velocity streaks in the buffer layer for the respective wavenumber and under suitable external forcing, in line with the less simplified linearised model developed by Hwang and Cossu (2010b). Interestingly, most intense large-scale velocity streaks and secondary currents arose at essentially the same characteristic lateral wavelength $\lambda_{h,z} \approx 1.3H_f$ as in the original model for all forcing schemes; including one scheme that exclusively forced the buffer layer region. This striking observation underlines that the choice of the most-amplified lateral wavelength in the model is mainly due to the structure of the linearised Navier-Stokes operator, rather than depending on the specific type of external forcing. The findings thus imply that a perturbed bottom wall is only one conceivable way to trigger the discussed lateral instability. In particular, other laterally-varying external forcing types at a suitable lateral wavelength $\lambda_{h,z} \approx 1.3H_f$ are assumed to equally lead to the evolution of sediment ridges and mean secondary currents.

9.2 Turbulent large-scale streaks and sediment ridges in open channel flow

In contrast to the idealised setting in the linear models discussed before, a fully-turbulent flow is strongly non-linear and of intermittent character. However, it is nowadays well established that a variety of recurrent coherent structures exist in wall-bounded turbulence that are of crucial importance for the physical processes in such flows. Of particular relevance for sediment transport are laterally alternating large-scale high- and low-speed streaks of the streamwise velocity which represent an intermittent finite-amplitude modulation of the turbulent mean flow profile (Smits et al., 2011; Jiménez, 2018). In order to study the role of these flow structures for the sediment ridge evolution, a series of direct numerical simulations of doubly-periodic open channel flow has been performed in the course of this chapter. The series comprises simulations of both single-phase smooth-wall open channel flows as well as multiphase simulations over thick mobile sediment beds at varying friction Reynolds numbers $200 \lesssim Re_\tau \lesssim 830$. Remarkably, it was observed that the large-scale velocity structures centring in the channel bulk are essentially the same in both flow configurations in what concerns shape, organisation and energy contribution. Conversely, this striking similarity strongly suggests that the presence of a mobile sediment bed has a rather weak influence on the turbulent large-scale structures, which is in line with corresponding observations over fully-rough walls (Townsend, 1976; Jiménez, 2004a; Flores et al., 2007).

Turbulent large-scale structures in canonical wall-bounded flows are well known to penetrate deep into the near-wall region, where they are able to alter the organisation of the small-scale near-wall flow structures and to induce outer-scaling regions of intense high and low wall shear stress. The same observation was made in the context of the flow over mobile sediment beds, where the large-scale velocity streaks generated laterally alternating zones of high and low bed shear stress, which directly implies an inhomogeneous erosion activity along the lateral channel direction. After short time intervals of $\mathcal{O}(10)$ bulk time units, this laterally varying erosion rate led to the formation of sediment ridges below the large-scale low-speed streaks where erosion is reduced. Troughs developed accordingly below the corresponding large-scale high-speed streaks.

That the bed organisation in the context of this 'top-down mechanism' is indeed controlled from the bulk of the channel by these large-scale streaks was shown by means of two-point two-time correlations. These implied a strong correlation between the formation of sediment ridges at a given time and the organisation of the turbulent large-scale structures at an earlier time (about ten bulk time units for the investigated case $CM850^{H3}$), revealing the causality between flow structures and the sediment patterns. In the context of the 'top-down mechanism', the sediment ridges naturally 'inherit' the characteristic spanwise spacing of the large-scale structures of $\lambda_z/H_f \approx$ 1-2 (Jiménez, 2013a), values that are in complete agreement with those measured for sediment ridges in laboratory experiments (Wolman and Brush, 1961; McLelland et al., 1999).

Finally, the loop back to the secondary currents that typically flank sediment ridges on either side was closed by showing that this mean secondary fluid motion is basically the statistical footprint of the well-organised large-scale Reynolds stress-carrying structures. In turn, the associated laterally varying regions of lower and higher mean streamwise velocity in the up- and downflow regions, respectively, are the signature of the large-scale velocity streaks and their preferential alignment. In particular, when averaged over intermediate time intervals of $\mathcal{O}(10)$ bulk time units length, secondary currents evolved even in the laterally homogeneous smooth wall open channel featuring a similar size, shape and strength as their counterparts in the particle-laden cases. These observations eventually strengthen the conclusion that both secondary currents and sediment ridges originate in the same turbulent large-scale flow structures, with the secondary currents representing an analogue to the conditional rollers discussed by Lozano-Durán et al. (2012) and Jiménez (2013a, 2018).

The observed 'top-down mechanism' is in accordance with the conceptual model on turbulent structures of different type and scale in turbulent wall-bounded shear flows proposed by Jiménez (2018, and references therein) and further supported by Zhou et al. (2022). Therein, turbulent coherent structures are assumed to form at arbitrary distances to the solid walls in virtue of the local mean shear, rather than being generated exclusively in the vicinity of the bottom wall. Our simulations show indeed that large-scale turbulent streaks intermittently break up in analogy to the log-layer bursting of Flores and Jiménez (2010) and afterwards form again in the channel bulk in significant distance to the sediment bed. With time, they extend over the entire channel depth, which is again in agreement with the aforementioned model, wherein for the Reynolds stresses, on average, a preferred direction of information propagation towards the wall was established. In contrast, the 'top-down mechanism' conceptually differs from the linearised model of Colombini (1993) in that a finite-amplitude modulation of the mean velocity profile in form of the large-scale streaks induces sediment ridge formation, whereas in the theoretical model an initial infinitesimal bed perturbation triggers a spanwise variation of the velocity field ('bottom-up mechanism'). However, even in the model of Colombini (1993) the lateral spacing of the sediment bedforms is chosen by the linearised Navier-Stokes operator, as it represents the lateral wavelength for which the flow field exhibits the strongest response (i.e. the maximum lateral bed shear stress) to the bed formation. This eventually explains the good match of the lateral ridge-spacing in both cases, despite the conceptually different formation mechanisms.

The data analysed in this chapter contains for the first time a simulation of sediment transport at a friction Reynolds number high enough to provide a reasonable scale separation between the typical size of buffer layer structures and that of the large-scale structures in the bulk. On the other hand, execution of simulations at moderate Reynolds numbers $Re_\tau \approx 250$ are today relatively affordable even if they include $\mathcal{O}(10^5)$ fully-resolved particles. We are thus now in a position to perform numerical experiments in the style of Jiménez and Moin (1991), Jiménez and Pinelli (1999) or Kwon and Jiménez (2021), wherein the conditions of the simulations are intentionally modified to exclude, isolate or deactivate specific parts or scales of the flow field and check the influence of this modifications on the remaining flow. In the context of sediment transport, one might, for instance, artificially damp the outer large-scale flow structures to 'deactivate' them such that they cannot contribute to the sediment erosion and transport. An alternative way of studying the interaction of sediment with turbulent structures at a very specific scale is by means of multiphase simulations of invariant solutions such as travelling waves or periodic orbits. Pestana et al. (2020) recently used the upper branch solution of Nagata (1990) to investigate the dynamics of a small number of particles in such systems, allowing them to reproduce the typical behaviour of particles in fully-turbulent flows to propagate into the low-speed streaks of the buffer layer. One of the advantages of such simulations is that they consist exclusively of 'exact coherent structures' (Waleffe, 2003) whose dynamics are known a priori (except for particle-induced modulations), which greatly simplifies the analysis of causal connections between individual flow structures and the particle dynamics.

9.3 Coherent structures and sediment ridges in open duct flow

Even though lateral sidewalls as they appear in man-made canals or in form of river side-banks in natural flows have been shown to be not required for the evolution of sediment ridges, the presence of lateral domain boundaries nonetheless affects the afore discussed formation process. The subsequent chapter was therefore dedicated to the analysis of sediment ridge formation in narrow and wide open duct flows. For this purpose, a series of direct numerical simulations of sediment bed evolution in open duct flows was performed for low and high aspect ratios of $AR \approx 1$ and $AR \approx 8$ ($AR = L_z/(2H_f)$), respectively. The simulations cover a range of friction Reynolds numbers $150 \lesssim Re_\tau \lesssim 500$ and were supplemented with a number of single-phase smooth-wall open duct flows, that were partly performed in the course of this work. The remaining single-phase datasets and long-time statistics of those were kindly provided by Y. Sakai (Sakai, 2016).

The formation of sediment ridges turns out to be somewhat different in narrow open duct flows, where the influence of the lateral sidewalls on the flow is omnipresent, and in the core of wide open duct flows, where the impact of the boundary is much weaker. The flow organisation in the core of high-aspect ratio open ducts thus closely resembled that in open channel flows, including the appearance of large-scale velocity streaks and Reynolds stress-carrying structures at essentially the same preferential lateral spacing of $\lambda_z/H_f \approx$ 1-2 as in canonical wall-bounded flows. As a consequence, sediment ridges and troughs were seen to form in virtue of the same 'top-down mechanism' as in open channel cases, caused by large-scale velocity streaks that span the entire channel depth and thus induce a laterally varying bed shear stress. Considering time averaging intervals of $\mathcal{O}(10)$ bulk time units length, the entire cross-section of the single-phase and particle-laden wide ducts were covered by depth-spanning counterrotating large-scale secondary currents. These features thus represent the 'instantaneous secondary currents' that were postulated by Nezu (2005) to play a role in sediment ridge formation. Only for sufficiently long time intervals of at least $\mathcal{O}(10^3)$ bulk time units length, the mean secondary currents were seen to disappear in the duct core due to a cancelling of structures of different sign in the long-time average. The ridge-featuring cases, on the other hand, could not be executed over comparably long time intervals which would be however necessary to give a conclusive answer on whether or not ridges are able to stabilise secondary currents, as was argued in some experimental studies (Nezu, 2005).

Sediment ridges in narrow open duct flows feature – on average – a lower amplitude than those in the core of wide ducts. In these cases, the small duct width typically allows to accommodate two ridges and a single trough in between, roughly located at the duct bisector where the mean secondary flow transports high momentum fluid down to the bed. The flow structures responsible for the development of these ridges cannot be the same as in the open channel case, since the domain is too narrow to host more than a single large-scale high-speed region. In fact, it was seen that the bed development is in these narrow cases predominantly controlled by smaller, but nevertheless outer-scaling velocity structures, which centre roughly $0.2H_f$ above the sediment bed. The causal connection between these flow structures and the sediment ridges and troughs was shown in a similar way as for the open channel case based on two-point two-time correlations, which similarly revealed a time shift at which the sediment bed evolution lags the dynamics of the turbulent velocity streaks.

The presence of a mobile sediment bed was seen to alter the mean secondary flow patterns in the duct cross-section. On the one hand, the mean bottom vortices in the corners between sidewalls and sediment bed were seen to shrink, being more confined to the corner regions for sediment-laden flows than for smooth-wall ducts. On the other hand, the mean secondary flow intensity averaged over the entire cross-section seemed to increase over the mobile sediment bed due a strengthening of the outer mean secondary vortex cells. Even though this point cannot be finally clarified based on the current datasets alone due to the relatively short averaging time interval, we nevertheless suspect that the intensification of the mean secondary flow is a consequence of a stronger redistribution of momentum across the cross-section due to the different friction along the smooth sidewall and the mobile sediment bed, in agreement with findings of various experimental investigations (Nezu and Nakagawa, 1993; Rodríguez and García, 2008; Albayrak and Lemmin, 2011). The question on the influence of sidewall and bottom roughness on the secondary flow organisation is here of particular interest, but to the best of the author's knowledge no detailed numerical investigations of this problem exist up to the present day. On the other hand, channel flow over a fully rough wall covered by fully-resolved spherical roughness elements was recently studied by Mazzuoli and Uhlmann (2017), and a similar scenario with immobile spheres attached to the bottom wall, the sidewalls or both is also conceivable for the here considered open ducts in future studies.

Having seen that the flow organisation responsible for the sediment ridge evolution in narrow domains differed clearly from that in laterally more or less unconstrained regions of the flow, a special focus was placed on the preferential organisation of instantaneous coherent structures in the remainder of the chapter. To this end, two-point velocity correlations were analysed in order to get an impression on the average organisation of large-scale flow structures. The correlation patterns exhibit a surprisingly good qualitative similarity to their counterparts in canonical wall-bounded flows (Sillero et al., 2014; Jiménez, 2018). Also, the characteristic structures of the correlation field in the lower half of the duct cross-section revealed comparable sizes as the flow structures seen to be responsible for the sediment ridge formation. In addition to these two-point statistics, instantaneous small- and large-scale streamwise velocity streaks were extracted from the available flow realisations using two different approaches, one originally proposed by Moisy and Jiménez (2004) to detect three-dimensional coherent structures and one used by Nakatsuji (2012) and Sakai (2016) to extract high- and low-speed regions in instantaneous cross-sections of closed and open duct flows, respectively. Both agree in the conclusion that streamwise velocity streaks of different size preferentially reside in different regions of the domain for single-phase and particle-laden simulations likewise. In agreement with studies in turbulent channels and boundary layers (Sillero, 2014; Jiménez, 2018), the streaks could be classified into a self-similar family of wall-attached structures (Townsend, 1976; Jiménez, 2012) and a group of detached objects. In addition, a third group was seen to be attached to the free surface. While the shortest representatives of the self-similar family are the small-scale buffer layer streaks, the largest ones feature dimensions comparable to the mean fluid height H_f.

As opposed to large-scale streaks, the preferential locations of small-scale quasi-streamwise vortices appearing in the near-bed and near-sidewall region were seen to clearly differ in the presence of a mobile particle bed. On the one hand, more vortices were generated along the bed in the wakes of individual particles, but these are rather randomly spaced and cancel each other such that they do not contribute to the mean vorticity field $\langle \omega_{f,x} \rangle_{xt}$. In the corners between the sidewalls and the sediment bed where sediment transport is rather weak, on the other hand, the preferred organisation of the quasi-streamwise vortices was seen to differ from that in the smooth wall case due to the absence of a nearby impermeable bottom wall. This deviation is most likely responsible for the modified mean secondary bottom vortex, which is in turn connected to the local vorticity field by a Poisson equation.

In the final section of the chapter, agglomerations of individual smaller vortices in form of small- and large-scale vortex clusters were scrutinised, in a similar way as in the studies by Del Álamo et al. (2006) and Lozano-Durán et al. (2012) on closed channel flows. In accordance with their findings, also the vortex clusters could be classified into a self-similar cascade of structures attached to the bottom wall and a second group of detached clusters, from which some where however attached to the free surface. These latter are of large interest especially in the context of interfacial mass transfer (Nagaosa, 1999; Pinelli et al., 2022) and a more detailed investigation of these structures could be of high relevance for the understanding of gas exchange processes across the fluid-air interface. The vortex clusters themselves span a range of sizes, from the smallest clusters that basically consist of single buffer layer vortices to structures whose dimensions attain values comparable to H_f and thus take up a significant fraction of the cross–section. The particular organisation of these small- and large-scale structures within the cross-section together with their sense of rotation revealed an astonishing similarity with the mean secondary flow field. These observations corroborate those of Kawahara et al. (2012b) who investigated large-scale rotating motions in closed ducts by means of a poloidal-toroidal decomposition and concluded that these features are closely related to both the mean secondary flow and the vortex clusters studied in the current work. It was therefore claimed that the mean secondary flow could be the collective statistical footprint of the small- and large-scale vortex clusters. In the current work, the analysis of Kawahara et al. (2012b) was repeated for the here considered open duct flows and similar conclusions could be drawn. In the near future, further investigations both in open and closed duct flows are planned in order to clarify this suspected relation. In this context, it might be also worth studying the dynamics of the extracted streaks and vortex clusters in greater detail, following a similar approach as in Lozano-Durán and Jiménez (2014b) or in Sakai (2016). In particular, such analysis would tackle the question where in the cross-section coherent structures are born, whether they are advected throughout the cross-section and where they eventually 'die'.

Generally, it would be desirable to extend the current database of open duct flows both over smooth walls and mobile sediment beds, which is today still limited to relatively low Reynolds numbers $Re_\tau \lesssim 500$. To allow for a clear scale separation of at least one order of magnitude between the buffer layer and the outer length scales, Reynolds numbers of $Re_\tau \geq 1000$ are required such that $H_f^+/100 \gtrsim 10$. Recently, this level was reached for direct numerical simulations in closed ducts (Pirozzoli et al., 2018), but for open duct flows such datasets are still lacking.

9.4 Transition from sediment ridges to initial ripples

While the main objective of the previous chapters was to contribute to the understanding of the basic formation processes due to which sediment ridges arise, in the final chapter we aimed to provide a prospect on later stages of the bed evolution during which sediment ridges are concurring and interacting with more dominant transverse bedforms. Specifically, we showed how the ridge-covered sediment bed gradually transforms into a ripple-featuring sediment bed in the largest domains with streamwise and spanwise dimensions $L_x/H_f \approx 12$ and $L_z/H_f \approx 16$, respectively. Three main stages of the sediment bed evolution can be distinguished in this context, the first of which is characterised by a set of streamwise-elongated sediment ridges at rather low amplitude. In the third phase, on the other hand, the domain accommodates two ripple-like transverse bedforms at clearly higher amplitude whose crest-lines are essentially parallel to the lateral direction. The phase in between these two states is a transient period, during which small sediment seeds were observed to appear on the sediment ridges that grow and propagate downstream, while a trough region develops in their lee. This evolution of transverse bedforms starting from initially localised sediment seeds is in good agreement with recently proposed formation mechanisms for this kind of bedforms (Kleinhans et al., 2002; Coleman and Nikora, 2009; Scherer et al., 2020). Eventually, the sediment ridge breaks into several of these seeds, which were seen to synchronise with their lateral neighbours after a while to connect to wider bedforms that finally span over the entire channel or duct width. However, the transformation process itself was observed to happen faster for the duct flow, while the bed in the channel flow remained over a longer period in an intermediate state during which the transverse bedforms still exhibited a lateral variation.

A full analysis of this later stages of bedform evolution would have taken us too far from the actual topic of this work, i.e. the evolution of sediment ridges and their relation to turbulent coherent structures, but the datasets provided in this final chapter represent an interesting point of departure for future studies on transverse pattern formation. In particular, it will be an interesting task to scrutinise in detail how the near-bed flow field changes with rising amplitude of the transverse bedforms with a focus on the flow separation that occurs as soon as the pattern crests reach a sufficiently large height.

Appendix A

Derivation of the perturbed system for the linear stability analysis

A.1 Derivation of the perturbed linearised equations

A.1.1 Governing equations in the new non-orthogonal coordinate system

Substituting the transformed derivatives (5.18) into the reduced non-dimensional system (5.2) and multiplying by $H(\zeta)$, we eventually obtain the transformed governing equations of fluid motion as

$$H(\zeta)\frac{\partial U}{\partial t} + [V - m(\zeta)W]\frac{\partial U}{\partial \eta} + H(\zeta)W\frac{\partial U}{\partial \zeta} = \frac{\partial \tau_{xy}}{\partial \eta} - m(\zeta)\frac{\partial \tau_{xz}}{\partial \eta} + H(\zeta)\frac{\partial \tau_{xz}}{\partial \zeta} + H(\zeta)\frac{S}{Fr^2} \tag{A.1a}$$

$$H(\zeta)\frac{\partial V}{\partial t} + [V - m(\zeta)W]\frac{\partial V}{\partial \eta} + H(\zeta)W\frac{\partial V}{\partial \zeta} = \frac{\partial \tau_{yy}}{\partial \eta} - m(\zeta)\frac{\partial \tau_{yz}}{\partial \eta} + H(\zeta)\frac{\partial \tau_{yz}}{\partial \zeta} - \frac{\partial P}{\partial \eta} \tag{A.1b}$$

$$\begin{aligned} H(\zeta)\frac{\partial W}{\partial t} + [V - m(\zeta)W]\frac{\partial W}{\partial \eta} + H(\zeta)W\frac{\partial W}{\partial \zeta} &= \frac{\partial \tau_{yz}}{\partial \eta} - m(\zeta)\frac{\partial \tau_{zz}}{\partial \eta} + H(\zeta)\frac{\partial \tau_{zz}}{\partial \zeta} \\ &\quad + m(\zeta)\frac{\partial P}{\partial \eta} - H(\zeta)\frac{\partial P}{\partial \zeta} \end{aligned} \tag{A.1c}$$

$$\frac{\partial V}{\partial \eta} - m(\zeta)\frac{\partial W}{\partial \eta} + H(\zeta)\frac{\partial W}{\partial \zeta} = 0. \tag{A.1d}$$

For the sake of readability, we have therein introduced the notation

$$m(\zeta) = \left[\eta\frac{\mathrm{d}H}{\mathrm{d}\zeta} + \frac{\mathrm{d}B}{\mathrm{d}\zeta}\right] = -\left[h_1\,\eta + 1\right]\,\epsilon\exp(\sigma T)\alpha\sin(\alpha\zeta).$$

A.1.2 Perturbed mean rate of strain tensor

Inserting the perturbation ansatz functions (5.14) for the velocity gradients in D_{ij}, the six independent entries of the mean rate of strain tensor are obtained as

$$\left\{\begin{array}{lcl} D_{xx} & = & 0 \\ D_{xy} & = & \frac{1}{2}\frac{dU_0}{dy} + \frac{1}{2}\epsilon\frac{du_1}{dy}\ \exp(\sigma T)\cos(\alpha z) \\ D_{xz} & = & -\frac{1}{2}\epsilon u_1 \alpha\ \exp(\sigma T)\sin(\alpha z) \\ D_{yy} & = & \epsilon\frac{dv_1}{dy}\ \exp(\sigma T)\cos(\alpha z) \\ D_{yz} & = & -\frac{1}{2}\epsilon\left[\frac{dw_1}{dy} + v_1\right]\alpha\ \exp(\sigma T)\sin(\alpha z) \\ D_{zz} & = & -\epsilon w_1 \alpha^2\ \exp(\sigma T)\cos(\alpha z). \end{array}\right. \tag{A.2}$$

It can then directly be seen that the quadratic terms $D_{im}D_{mj}$ and $D_{mn}D_{mn}$ in equation (5.9) are of order $\mathcal{O}(\epsilon^0)$ or $\mathcal{O}(\epsilon^1)$ if and only if one of the two matrix entries is D_{xy}. Mapping D_{ij} into the new coordinate system (η, ζ) using the non-orthogonal coordinate transformation introduced in equations (5.16), (5.17b) and (5.18), the components of the mean rate of strain tensor D_{ij} transform into

$$\left\{\begin{array}{lcl} D_{xx} & = & 0 \\ D_{xy} & = & \frac{1}{2H(\zeta)}\frac{dU_0}{d\eta} + \frac{1}{2H(\zeta)}\epsilon\frac{du_1}{d\eta}\ \exp(\sigma T)\cos(\alpha\zeta) \\ D_{xz} & = & -\frac{m(\zeta)}{2H(\zeta)}\frac{dU_0}{d\eta} - \frac{m(\zeta)}{2H(\zeta)}\epsilon\frac{du_1}{d\eta}\ \exp(\sigma T)\cos(\alpha\zeta) - \frac{1}{2}\epsilon\ u_1\alpha\ \exp(\sigma T)\sin(\alpha\zeta) \\ D_{yy} & = & \frac{1}{H(\zeta)}\epsilon\frac{dv_1}{d\eta}\ \exp(\sigma T)\cos(\alpha\zeta) \\ D_{yz} & = & -\frac{m(\zeta)}{2H(\zeta)}\epsilon\ \frac{dv_1}{d\eta}\ \exp(\sigma T)\cos(\alpha\zeta) - \frac{1}{2}\epsilon\ \left[\frac{1}{H(\zeta)}\frac{dw_1}{d\eta} + v_1\right]\alpha\ \exp(\sigma T)\sin(\alpha z) \\ D_{zz} & = & \frac{m(\zeta)}{H(\zeta)}\epsilon\ \frac{dw_1}{d\eta}\alpha\ \exp(\sigma T)\sin(\alpha\zeta) - \epsilon\ w_1\alpha^2\ \exp(\sigma T)\cos(\alpha\zeta). \end{array}\right. \tag{A.3}$$

A.1.3 Perturbed non-linear Reynolds stresses

The resulting entries of the linear turbulent shear stress tensor $\boldsymbol{\tau}^L$ then read

$$\begin{cases} \tau_{xy}^L &= \dfrac{\nu_{t0}}{H(\zeta)} U_0' + \epsilon \dfrac{\nu_{t0}}{H(\zeta)} \left[2u' + U_0' h_1\right] \exp(\sigma T)\cos(\alpha\zeta) + \mathcal{O}(\epsilon^2) \\ \tau_{xz}^L &= -\dfrac{m(\zeta)}{H(\zeta)} \nu_{t0} U_0' - \epsilon \left[\dfrac{m(\zeta)}{H(\zeta)} \nu_{t0} \left(2u' + U_0' h_1\right)\right] \exp(\sigma T)\cos(\alpha\zeta) \\ & \quad -\epsilon \left[\nu_{t0} u_1 \alpha\right] \exp(\sigma T)\sin(\alpha\zeta) + \mathcal{O}(\epsilon^2) \\ \tau_{yy}^L &= -\dfrac{2}{3}k + \epsilon \left[\dfrac{2}{H(\zeta)} \nu_{t0} v_1'\right] \exp(\sigma T)\cos(\alpha\zeta) + \mathcal{O}(\epsilon^2) \\ \tau_{yz}^L &= -\epsilon \left[\dfrac{m(\zeta)}{H(\zeta)} \nu_{t0} v_1'\right] \exp(\sigma T)\cos(\alpha\zeta) \\ & \quad -\epsilon \left[v_1 \nu_{t0} \alpha + \dfrac{1}{H(\zeta)} \nu_{t0} w_1' \alpha\right] \exp(\sigma T)\sin(\alpha\zeta) + \mathcal{O}(\epsilon^2) \\ \tau_{zz}^L &= -\dfrac{2}{3}k + \epsilon \left[\dfrac{2m(\zeta)}{H(\zeta)} \nu_{t0} w_1' \alpha\right] \exp(\sigma T)\sin(\alpha\zeta) - \epsilon \left[2\nu_{t0} w_1 \alpha^2\right] \exp(\sigma T)\cos(\alpha\zeta) \\ & \quad + \mathcal{O}(\epsilon^2). \end{cases}$$

In the equations for τ_{xy}^L and τ_{xz}^L, we have made use of the relation $\nu = \nu_{t0}\left(h_1 + u_1'/U_0'\right)$ (equation (5.7a, b) in Colombini, 1993) which originates in the adopted mixing length hypothesis. Similarly, one derives the non-linear contribution of the Reynolds stress tensor, $\boldsymbol{\tau}^{NL}$, arising from the turbulence model presented by Speziale (1987). Keeping only contributions of order $\mathcal{O}(\epsilon^0)$ or $\mathcal{O}(\epsilon^1)$ therein, the individual entries read:

$$\begin{cases} \tau_{xy}^{NL} &= \epsilon \dfrac{1}{H(\zeta)^2} \left[\left(\dfrac{1}{2}C_D - \dfrac{3}{2}C_E\right) l_0^2 U_0' v_1' + \dfrac{1}{2} l_0^2 U_0'' v_1\right] \exp(\sigma T)\cos(\alpha\zeta) + \mathcal{O}(\epsilon^2) \\ \tau_{xz}^{NL} &= \epsilon \dfrac{1}{H(\zeta)} \left[-\left(\dfrac{1}{4}C_D - C_E\right) l_0^2 U_0' w_1' \alpha - \left(\dfrac{1}{4}C_D - \dfrac{1}{2}C_E\right) l_0^2 U_0' v_1 \alpha\right] \exp(\sigma T)\sin(\alpha\zeta) \\ & \quad + \mathcal{O}(\epsilon^2) \\ \tau_{yy}^{NL} &= \epsilon \dfrac{1}{H(\zeta)^2} \left[\left(\dfrac{1}{6}C_D + \dfrac{2}{3}C_E\right) l_0^2 U_0' u_1' + \left(\dfrac{1}{6}C_D + \dfrac{2}{3}C_E\right) l_0^2 (U_0')^2 h_1\right] \exp(\sigma T)\cos(\alpha\zeta) \\ & \quad + \mathcal{O}(\epsilon^2) \\ \tau_{yz}^{NL} &= -\dfrac{m(\zeta)}{H(\zeta)^2} \left[\dfrac{1}{4}C_D (U_0')^2 l_0^2\right] \\ & \quad -\epsilon \dfrac{1}{H(\zeta)} \left[\dfrac{1}{4}C_D U_0' u_1 \alpha l_0^2\right] \exp(\sigma T)\sin(\alpha\zeta) \\ & \quad + \mathcal{O}(\epsilon^2) \\ \tau_{zz}^{NL} &= -\dfrac{1}{H(\zeta)^2} \left[\left(\dfrac{1}{6}C_D - \dfrac{1}{3}C_E\right) (U_0')^2 l_0^2\right] \\ & \quad +\epsilon \dfrac{1}{H(\zeta)^2} \left[\left(\dfrac{1}{3}C_D - \dfrac{2}{3}C_E\right) l_0^2 U_0' u_1' - \left(\dfrac{1}{3}C_D - \dfrac{2}{3}C_E\right) (U_0')^2 l_0^2 h_1\right] \exp(\sigma T)\cos(\alpha\zeta) \\ & \quad + \mathcal{O}(\epsilon^2) \end{cases}$$

A.1.4 Perturbed linear system

The equations for the perturbation amplitudes u_1, v_1, w_1 and p_1 are found when collecting all terms of $\mathcal{O}(\epsilon^1)$ in the transformed equations (A.1) after having inserted the above derived expressions for the Reynolds stress components. The resulting linear system can be written in matrix-vector form as

$$A\boldsymbol{q} = \begin{pmatrix} a_{11} & a_{12} & a_{13} & a_{14} \\ a_{21} & a_{22} & a_{23} & a_{24} \\ a_{31} & a_{32} & a_{33} & a_{34} \\ a_{41} & a_{42} & a_{43} & a_{44} \end{pmatrix} \begin{pmatrix} u_1 \\ v_1 \\ w_1 \\ p_1 \end{pmatrix} = h_1\boldsymbol{B} + \boldsymbol{R}, \tag{A.4}$$

where the individual entries of the matrix A are defined as:

$$\begin{cases} a_{11} &= 2\left(v_{t0}\mathrm{d}^2/\mathrm{d}\eta^2 + v_{t0}'\mathrm{d}/\mathrm{d}\eta\right) - \alpha^2 v_{t0} \\ a_{12} &= -\left[U_0' + \left(\frac{1}{4}C_D - \frac{1}{2}C_E\right)\alpha^2 l_0^2 U_0' - C_E l_0 l_0' U_0'' - \frac{1}{2}C_E l_0^2 U_0'''\right] \\ a_{13} &= \left(\frac{1}{4}C_D - \frac{1}{2}C_E\right)\alpha^2 l_0^2 U_0'\mathrm{d}/\mathrm{d}\eta + \left[\left(\frac{1}{2}C_D - \frac{3}{2}C_E\right)\alpha^2\left(2l_0 l_0' U_0' + l_0^2 U_0''\right) + \frac{1}{2}C_E\alpha^2 l_0^2 U_0''\right] \\ a_{14} &= 0 \\ a_{21} &= \left(\frac{1}{6}C_D + \frac{2}{3}C_E\right)\left[l_0^2 U_0'\mathrm{d}^2/\mathrm{d}\eta^2 + \left(2l_0 l_0' U_0' + l_0^2 U_0''\right)\mathrm{d}/\mathrm{d}\eta\right] - \frac{1}{4}C_D\alpha^2 l_0^2 U_0' \\ a_{22} &= -\alpha^2 v_{t0}, \quad a_{23} = \alpha^2 v_{t0}\mathrm{d}/\mathrm{d}\eta + 2\alpha^2 v_{t0}', \quad a_{24} = -\mathrm{d}/\mathrm{d}\eta \\ a_{31} &= -\left(\frac{1}{12}C_D - \frac{2}{3}C_E\right) l_0^2 U_0'\mathrm{d}/\mathrm{d}\eta + \frac{1}{4}C_D\left(2l_0 l_0' U_0' + l_0^2 U_0''\right) \\ a_{32} &= v_{t0}' \\ a_{33} &= 1\left(v_{t0}\mathrm{d}^2/\mathrm{d}\eta^2 + v_{t0}'\mathrm{d}/\mathrm{d}\eta\right) - \alpha^2 v_{t0} \\ a_{34} &= -1 \\ a_{41} &= 0, \quad a_{42} = \mathrm{d}/\mathrm{d}\eta, \quad a_{43} = -\alpha^2, \quad a_{44} = 0. \end{cases} \tag{A.5}$$

Note that it is not explicitly written in Colombini (1993) that the pre-factor in a_{11} differs from that in a_{33}, but it was verified in the course of the present derivations that this is indeed the case. The vectors $\boldsymbol{B}$ and $\boldsymbol{R}$ that form the RHS of equation (A.4) read

$$\boldsymbol{B} = \begin{pmatrix} -C_0 - \alpha^2 v_{t0} U_0' \eta \\ -\frac{1}{4}C_D\alpha^2 l_0^2 (U_0')^2 \eta \\ \frac{1}{4}C_D l_0^2 (U_0')^2 \\ 0 \end{pmatrix}, \quad \boldsymbol{R} = \begin{pmatrix} -\alpha^2 v_{t0} U_0' \\ -\frac{1}{4}C_D\alpha^2 l_0^2 (U_0')^2 \\ 0 \\ 0 \end{pmatrix}. \tag{A.6}$$

Note that in the equations presented in appendix A of Colombini (1993), a minus sign (highlighted in red) is missing in the second entry of $\boldsymbol{B}$ and $\boldsymbol{R}$, as can be verified by either deriving the equations following the procedure described above or by comparison with the respective equations provided in Colombini and Parker (1995). The current problem is a special case of the more general situation described therein.

For the perturbed flow equations, no-slip boundary conditions are imposed at the reference level η_0 (i.e. $u_1(\eta_0) = v_1(\eta_0) = w_1(\eta_0) = 0$), while at the free surface a solution has to fulfil the following boundary conditions:

$$\begin{aligned}
\tau_{yz}|_{\eta=1} = 0 \implies 0 &= \nu_{t0}\left(v_1 + w_1'\right) + \frac{1}{4} C_D l_0^2 U_0' u_1 \\
\tau_{xy}|_{\eta=1} = 0 \implies 0 &= 2\nu_{t0} u_1' + \left(\frac{1}{2} C_D - \frac{3}{2} C_E\right) l_0^2 U_0' \alpha^2 w_1 + \frac{1}{2} C_E l_0^2 U_0'' v_1 \\
(P + \tau_{yy})|_{\eta=1} = 0 \implies 0 &= -p_1 + \frac{h_1}{Fr^2} + 2\nu_{t0}\alpha^2 w_1 + \left(\frac{1}{6} C_D + \frac{2}{3} C_E\right) l_0^2 U_0' u_1' \\
V|_{\eta=1} = 0 \implies 0 &= v_1.
\end{aligned} \tag{A.7}$$

While the first three equations represent dynamical boundary conditions, the last line indicates an impermeability condition for the free surface, i.e. it ensures that there is no mass flux across the interface.

A.1.5 Perturbed sediment bed continuity equation

In a similar way as for the reduced RANS system (5.2), the bed continuity equation (5.4) can be rewritten in the context of the new non-orthogonal coordinate system $(\eta, \zeta)^T$ as

$$\frac{\partial B}{\partial t} = -Q_{ref}\left(-\frac{m(\zeta)}{H}\frac{\partial Q}{\partial \eta} + \frac{\partial Q}{\partial \zeta}\right). \tag{A.8}$$

Inserting the empirical relation of Meyer-Peter and Müller (1948) (i.e. relation (5.5b)) into equation (A.8) to link the spanwise sediment flux to the fluid shear stress, we obtain

$$\frac{\partial B}{\partial (tQ_{ref})} = -\left(-\frac{m(\zeta)}{H}\frac{\partial}{\partial \eta} + \frac{\partial}{\partial \zeta}\right)\left[\Phi(\theta)\left(\frac{\tau_t}{\tau} - \frac{0.3}{\theta^{1/2}}\left(-\frac{m(\zeta)}{H}\frac{\partial B}{\partial \eta} + \frac{\partial B}{\partial \zeta}\right)\right)\right]. \tag{A.9}$$

Recalling the following expressions for the derivatives of the sediment bed height $B(\zeta)$

$$\begin{aligned}
\frac{\partial B}{\partial (t\, Q_{ref})} &= \frac{\partial B}{\partial T} = \epsilon\sigma \exp(\sigma T)\cos(\alpha\zeta) \\
\frac{\partial B}{\partial \eta} &= 0 \\
\frac{\partial B}{\partial \zeta} &= -\epsilon\, \alpha\, \exp(\sigma T)\sin(\alpha\zeta) \\
\frac{\partial^2 B}{\partial \zeta^2} &= -\epsilon\, \alpha^2\, \exp(\sigma T)\cos(\alpha\zeta)
\end{aligned}$$

and focussing on terms of $\mathcal{O}(\epsilon^1)$ only, equation (A.9) can be further simplified, viz.

$$\frac{1}{\Phi(\theta_0)}\epsilon\, \sigma\, \exp(\sigma T)\cos(\alpha\zeta) = -\left[\frac{\partial}{\partial \zeta}\left(\frac{\tau_t}{\tau}\right)\right]_{\eta_0} - \epsilon\, \alpha^2 \frac{0.3}{\theta^{1/2}} \exp(\sigma T)\cos(\alpha\zeta). \tag{A.11}$$

Substituting the first term on the RHS which contains the ratio between tangential perturbed shear stress $\tau_t = W'(\eta)$ and the amplitude of the bottom shear stress in the base flow $\tau = U_0'(\eta)$, viz.

$$\begin{aligned}\left[\frac{\partial}{\partial\zeta}\left(\frac{\tau_t}{\tau}\right)\right]_{\eta_0} &= \left[\frac{\partial}{\partial\zeta}\left(\frac{\epsilon\, w_1'\alpha\, \exp(\sigma T)\sin(\alpha\zeta)}{U_0'}\right)\right]_{\eta_0} \\ &= \left[-\epsilon\,\frac{w_1'\alpha^2}{U_0'}\,\exp(\sigma T)\cos(\alpha\zeta)\right]_{\eta_0},\end{aligned} \tag{A.12}$$

we finally arrive at a dispersion relation for the growth rate σ as

$$\frac{\sigma}{\Phi(\theta_0)\,\alpha^2} = \left[\frac{w_1'}{U_0'}\right]_{\eta_0} - \frac{0.3}{\theta_0^{1/2}}. \tag{A.13}$$

A.2 Discrete problem

The continuous problem consisting of the base flow equations (5.25) and the perturbed equations (A.4) is in the following discretised on a set of $N_\eta + 1$ Chebyshev-Gauss-Lobatto collocation points following the definitions in section 5.2.

For the base flow equations (5.25), the discrete system reads

$$(\boldsymbol{D}_\eta \nu_{t0N} \boldsymbol{D}_\eta + \mathbf{diag}(\nu_{t0N})\boldsymbol{D}_{\eta\eta})U_{0N} = -C_0\mathbf{1}_{(N_\eta+1)\times 1}, \tag{A.14}$$

introducing the notations $\mathbf{1}_{m\times n}$ for a $m \times n$ matrix whose entries are all equal to unity and $\mathbf{diag}(a_n)$ for a $n \times n$ diagonal matrix whose main diagonal is filled with the n entries of vector a_n. By integration of the second equation in (5.25), we arrive at an algebraic expression for the linear profile of the discrete pressure

$$(P_{0N})_k = P_0|_{\eta=1} - \left(\frac{1}{6}C_d + \frac{2}{3}C_e\right)\frac{C_0}{2}(1-\eta_k). \tag{A.15}$$

It should be noted that U_{0N} and P_{0N} are $(N_\eta + 1) \times 1$ vectors that contain the unknowns evaluated at each collocation point.
The discrete version of the perturbed system (A.4) can be written in the following matrix-vector form:

$$A\boldsymbol{q}_N = \begin{pmatrix}(a_{11})_N & (a_{12})_N & (a_{13})_N & (a_{14})_N \\ (a_{21})_N & (a_{22})_N & (a_{23})_N & (a_{24})_N \\ (a_{31})_N & (a_{32})_N & (a_{33})_N & (a_{34})_N \\ (a_{41})_N & (a_{42})_N & (a_{43})_N & (a_{44})_N\end{pmatrix}\begin{pmatrix}u_N \\ v_N \\ w_N \\ p_N\end{pmatrix} = h_1\boldsymbol{B}_N + \boldsymbol{R}_N. \tag{A.16}$$

Here, u_N, v_N, w_N, p_N are vectors of size $(N_\eta + 1) \times 1$ that contain the unknowns evaluated at each of the collocation points, e.g. $[u_N]_k = \hat{u}_1(\eta_k)$. The discrete counterpart to the continuous vector $\boldsymbol{q}$ contains the four sub-vectors, i.e.

$$\begin{aligned}\boldsymbol{q}_N &= (u_N, v_N, w_N, p_N)^T \\ &= \left(\hat{u}_1(\eta_0)\ldots\hat{u}_1(\eta_N), \hat{v}_1(\eta_0)\ldots\hat{v}_1(\eta_N), \hat{w}_1(\eta_0)\ldots\hat{w}_1(\eta_N), \hat{p}_1(\eta_0)\ldots\hat{p}_1(\eta_N)\right)^T,\end{aligned} \tag{A.17}$$

and is thus of size $4(N_\eta+1)\times 1$, just as the right hand side vectors $\boldsymbol{B}_N$ and $\boldsymbol{R}_N$. Similarly, $(a_{ij})_N$ represents a submatrix of the discrete linear differential operator of size $(N_\eta+1)\times(N_\eta+1)$ in which the derivatives w.r.t. η are expressed using the afore defined collocation derivative matrices:

$$
\begin{aligned}
(a_{11})_N &= 2\,\mathbf{diag}(\nu_{t0N})\boldsymbol{D}_{\eta\eta} + 2\,\mathbf{diag}(\boldsymbol{D}_\eta\nu_{t0N})\boldsymbol{D}_\eta - \alpha^2\,\mathbf{diag}(\nu_{t0N})\\
(a_{12})_N &= -\Big[\mathbf{diag}(\boldsymbol{D}_\eta U_0) + \left(\tfrac{1}{4}C_D - \tfrac{1}{2}C_E\right)\alpha^2\,\mathbf{diag}(l_{0N})^2\,\mathbf{diag}(\boldsymbol{D}_\eta U_{0N})\\
&\quad -C_E\,\mathbf{diag}(l_{0N})\,\mathbf{diag}(\boldsymbol{D}_\eta l_{0N})\,\mathbf{diag}(\boldsymbol{D}_{\eta\eta}U_{0N}) - \tfrac{1}{2}C_E\,\mathbf{diag}(l_{0N})^2\,\mathbf{diag}(\boldsymbol{D}_{\eta\eta\eta}U_{0N})\Big]\\
(a_{13})_N &= \left(\tfrac{1}{4}C_D - \tfrac{1}{2}C_E\right)\alpha^2\,\mathbf{diag}(l_{0N})^2\,\mathbf{diag}(\boldsymbol{D}_\eta U_{0N})\boldsymbol{D}_\eta\\
&\quad +\Big[\left(\tfrac{1}{2}C_D - \tfrac{3}{2}C_E\right)\alpha^2\left(2\,\mathbf{diag}(l_{0N})\,\mathbf{diag}(\boldsymbol{D}_\eta l_{0N})\,\mathbf{diag}(\boldsymbol{D}_\eta U_{0N})\right.\\
&\quad \left.+\,\mathbf{diag}(l_{0N})^2\,\mathbf{diag}(\boldsymbol{D}_{\eta\eta}U_{0N})\right) + \tfrac{1}{2}C_E\alpha^2\,\mathbf{diag}(l_{0N})^2\,\mathbf{diag}(\boldsymbol{D}_{\eta\eta}U_{0N})\Big]\\
(a_{14})_N &= \mathbf{0}_{(N_\eta+1)\times(N_\eta+1)}\\
(a_{21})_N &= \left(\tfrac{1}{6}C_D + \tfrac{2}{3}C_E\right)\Big[\mathbf{diag}(l_{0N})^2\,\mathbf{diag}(\boldsymbol{D}_\eta U_{0N})\boldsymbol{D}_{\eta\eta}\\
&\quad +\left(2\,\mathbf{diag}(l_{0N})\,\mathbf{diag}(\boldsymbol{D}_\eta l_{0N})\,\mathbf{diag}(\boldsymbol{D}_\eta U_{0N}) + \mathbf{diag}(l_{0N})^2\,\mathbf{diag}(\boldsymbol{D}_{\eta\eta}U_{0N})\right)\boldsymbol{D}_\eta\Big]\\
&\quad -\tfrac{1}{4}C_D\alpha^2\,\mathbf{diag}(l_{0N})^2\,\mathbf{diag}(\boldsymbol{D}_\eta U_{0N})\\
(a_{22})_N &= -\alpha^2\,\mathbf{diag}(\nu_{t0N})\\
(a_{23})_N &= \alpha^2\,\mathbf{diag}(\nu_{t0N})\boldsymbol{D}_\eta + 2\alpha^2\,\mathbf{diag}(\boldsymbol{D}_\eta\nu_{t0N})\\
(a_{24})_N &= -\boldsymbol{D}_\eta\\
(a_{31})_N &= -\left(\tfrac{1}{12}C_D - \tfrac{2}{3}C_E\right)\mathbf{diag}(l_{0N})^2\,\mathbf{diag}(\boldsymbol{D}_\eta U_{0N})\boldsymbol{D}_\eta\\
&\quad +\tfrac{1}{4}C_D\left(2\,\mathbf{diag}(l_{0N})\,\mathbf{diag}(\boldsymbol{D}_\eta l_{0N})\,\mathbf{diag}(\boldsymbol{D}_\eta U_{0N}) + \mathbf{diag}(l_{0N})^2\,\mathbf{diag}(\boldsymbol{D}_{\eta\eta}U_{0N})\right)\\
(a_{32})_N &= \mathbf{diag}(\boldsymbol{D}_\eta\nu_{t0N})\\
(a_{33})_N &= 1\,\mathbf{diag}(\nu_{t0N})\boldsymbol{D}_{\eta\eta} + 1\,\mathbf{diag}(\boldsymbol{D}_\eta\nu_{t0N})\boldsymbol{D}_\eta - \alpha^2\,\mathbf{diag}(\nu_{t0N})\\
(a_{34})_N &= -\mathbf{I}_{(N_\eta+1)\times(N_\eta+1)}\\
(a_{41})_N &= \mathbf{0}_{(N_\eta+1)\times(N_\eta+1)},\quad (a_{42})_N = \boldsymbol{D}_\eta\\
(a_{43})_N &= -\alpha^2\mathbf{I}_{(N_\eta+1)\times(N_\eta+1)},\quad (a_{44})_N = \mathbf{0}_{(N_\eta+1)\times(N_\eta+1)}.
\end{aligned}
\tag{A.18}
$$

Here, we have used the matrix definitions $\mathbf{0}_{m\times n}$ for a zero matrix of dimensions $m\times n$ and $\mathbf{I}_{n\times n}$ for the $n\times n$ identity matrix. The vectors $\boldsymbol{B}_N$ and $\boldsymbol{R}_N$ are defined as (again highlighting the minus sign that was missing in Colombini (1993) by red colour)

$$
\boldsymbol{B}_N = \begin{pmatrix}
-C_0\mathbf{1}_{(N_\eta+1)\times 1} - \alpha^2\,\mathbf{diag}(\nu_{t0N})\,\mathbf{diag}(\boldsymbol{D}_\eta U_{0N})\eta_N\\
-\dfrac{1}{4}C_D\alpha^2\,\mathbf{diag}(l_{0N})^2\,\mathbf{diag}(\boldsymbol{D}_\eta U_{0N})^2\eta_N\\
\dfrac{1}{4}C_D\,\mathbf{diag}(l_{0N})^2\,\mathbf{diag}(\boldsymbol{D}_\eta U_{0N})^2\mathbf{1}_{(N_\eta+1)\times 1}\\
\mathbf{0}_{(N_\eta+1)\times 1}
\end{pmatrix}
\tag{A.19}
$$

and

$$\boldsymbol{R}_N = \begin{pmatrix} -\alpha^2 \,\mathbf{diag}(\nu_{t0N})\,\mathbf{diag}(\boldsymbol{D}_\eta U_{0N})\mathbf{1}_{(N_\eta+1)\times 1} \\ -\frac{1}{4}C_D\alpha^2\,\mathbf{diag}(l_{0N})^2\,\mathbf{diag}(\boldsymbol{D}_\eta U_{0N})^2\mathbf{1}_{(N_\eta+1)\times 1} \\ \mathbf{0}_{(N_\eta+1)\times 1} \\ \mathbf{0}_{(N_\eta+1)\times 1} \end{pmatrix}, \tag{A.20}$$

respectively. It should be noted that the component-wise product of two vectors (also known as 'Hadamard product') is here established as the product of the diagonal matrices of the two vectors. In matlab, there is an alternative way to come to the same result, i.e. by taking the `.*` product between two vectors. Usually, both procedures lead to identical results in matlab, but this is not true in the case of a singularity in one of the two vectors, as it appears in the present case for $l_0'(1)$. In this case, only the `.*` product ensures that the singularity remains in the same vector entry also for the product, while the matrix resulting from the product of the diagonalised vectors becomes singular. For this reason, above equations are implemented in matlab making use of the `.*` product.

A.2.1 Discrete boundary conditions

In the framework of the here chosen Chebyshev-tau method, the discrete linearised boundary conditions are directly written into the discrete linear operator matrix A_N in the rows related to $\eta_N(1)$ (i.e. $k=0$) and $\eta_N(N_\eta+1)$ (i.e. $k=N_\eta$). Thus, for the no-slip condition, $[A_N]_{ii}=1$ and $[\boldsymbol{B}_N\cdot h_1+\boldsymbol{R}_N]_i=0$ for $i=\{1,(N_\eta+1)+1,2(N_\eta+1)+1\}$. For the free surface conditions, on the other hand, we have to replace the entire row of the matrix for $i=\{(N_\eta+1),2(N_\eta+1),3(N_\eta+1),4(N_\eta+1)\}$, respectively, by the discretised version of the boundary conditions:

$$
\begin{aligned}
&\underline{\text{for}\quad i=N_\eta+1, i_{loc}=N_\eta+1:}\\
&\left[\frac{1}{4}C_D l_{0N}^2(i_{loc})\boldsymbol{D}_\eta(i_{loc},\bullet)U_{0N}\right]u_N(i_{loc})+\left[\nu_{t0N}(i_{loc})\right]v_N(i_{loc})\\
&+\left[\nu_{t0N}(i_{loc})\boldsymbol{D}_\eta(i_{loc},\bullet)\right]w_N+[0]\,p_N(i_{loc})\\
&=[0]\,h_1+[0]\\
&\underline{\text{for}\quad i=2(N_\eta+1), i_{loc}=N_\eta+1:}\\
&\left[2\nu_{t0N}(i_{loc})\boldsymbol{D}_\eta(i_{loc},\bullet)\right]u_N+\left[\frac{1}{2}C_E l_{0N}^2(i_{loc})\boldsymbol{D}_{\eta\eta}(i_{loc},\bullet)U_{0N}\right]v_N(i_{loc})\\
&+\left[\left(\frac{1}{2}C_D-\frac{3}{2}C_E\right)l_{0N}^2(i_{loc})\boldsymbol{D}_\eta(i_{loc},\bullet)U_{0N}\alpha^2\right]w_N(i_{loc})+[0]\,p_N(i_{loc})\\
&=[0]\,h_1+[0]\\
&\underline{\text{for}\quad i=3(N_\eta+1), i_{loc}=N_\eta+1:}\\
&\left[\left(\frac{1}{6}C_D+\frac{2}{3}C_E\right)l_{0N}^2(i_{loc})\boldsymbol{D}_\eta(i_{loc},\bullet)U_{0N}\boldsymbol{D}_\eta(i_{loc},\bullet)\right]u_N+[0]\,v_N(i_{loc})\\
&+\left[2\nu_{t0N}(i_{loc})\alpha^2\right]w_N(i_{loc})+[-1]\,p_N(i_{loc})\\
&=\left[-\frac{1}{Fr^2}\right]h_1+[0]\\
&\underline{\text{for}\quad i=4(N_\eta+1), i_{loc}=N_\eta+1:}\\
&[0]\,u_N(i_{loc})+[1]\,v_N(i_{loc})+[0]\,w_N(i_{loc})+[0]\,p_N(i_{loc})\\
&=[0]\,h_1+[0]\,,
\end{aligned}
\tag{A.21}
$$

where i_{loc} is the local row-index in each of the $(N_\eta+1)\times(N_\eta+1)$ sub-matrices $(a_{ij})_N$. A $\bullet$ as matrix index indicates that the entire index range is used here, so the notation $\boldsymbol{D}_\eta(i,\bullet)$ stands for the ith row of the derivative matrix. In case of De Vriend (1977)'s model, however, $l_0(\eta=1)=\nu_{t0}(\eta=1)=0$ and thus, the first two equations reduce to the trivial statement $0=0$. To avoid a singular discrete linear differential operator A_N, we set homogeneous Neumann boundary conditions at the free surface for u_1, w_1, p_1 instead.

A.2.2 Discrete linear system

In case the solution of the linear system should be solved for the case that h_1 is a free additional parameter, we reformulate problem (A.16) by adding h_1 to the vector of unknowns q_N:

$$\widetilde{A}_N \widetilde{q}_N = \begin{pmatrix} (a_{11})_N & (a_{12})_N & (a_{13})_N & (a_{14})_N & -(\boldsymbol{B}_N)_1 \\ (a_{21})_N & (a_{22})_N & (a_{23})_N & (a_{24})_N & -(\boldsymbol{B}_N)_2 \\ (a_{31})_N & (a_{32})_N & (a_{33})_N & (a_{34})_N & -(\boldsymbol{B}_N)_3 \\ (a_{41})_N & (a_{42})_N & (a_{43})_N & (a_{44})_N & -(\boldsymbol{B}_N)_4 \\ \multicolumn{5}{c}{\text{top b.c. in z-momentum}} \end{pmatrix} \begin{pmatrix} u_N \\ v_N \\ w_N \\ p_N \\ h_1 \end{pmatrix} = \begin{pmatrix} (\boldsymbol{R}_N)_1 \\ (\boldsymbol{R}_N)_2 \\ (\boldsymbol{R}_N)_3 \\ (\boldsymbol{R}_N)_4 \\ 0 \end{pmatrix}. \tag{A.22}$$

Here, we have used the free surface boundary condition that results from the momentum equation in the z-direction as constraint for the parameter h_1 to give the system a unique solution. The remaining boundary conditions at the bottom and at the free surface are directly imposed into the matrix by replacing the respective rows in the modified system matrix and vectors.

A.2.3 Discrete dispersion relation

As a consequence of the assumptions concerning the different time scales of flow field dynamics and sediment bed evolution, the perturbed equations for fluid and sediment bed are only 'one-way coupled', i.e. the linearised equations of fluid motion are completely decoupled from the sediment bed continuity equation and can be solved independently of the latter. The discretised sediment bed continuity equation which has the role of a dispersion relation in the given problem, on the other hand, depends on the perturbed and base flow via the term $[w_1'/U_0']_{\eta_0}$. Thus, the growth rate σ can be computed *a posteriori* from the discretised version of equation (A.13) that reads

$$\frac{\sigma}{\Phi(\theta_0)\,\alpha^2} = \left[\frac{\hat{w}_1'(1)}{U_0'^N(1)}\right] - \frac{c}{\theta_0^{1/2}}. \tag{A.23}$$

A.3 Finite-Reynolds number horizontal channel

In the case of finite Reynolds number flows in which viscous effects are not neglected, the boundary conditions at the free surface change slightly compared to those used in case of an infinitely large Reynolds number in equations (A.7). For the perturbed flow equations, no-slip boundary conditions are imposed at the reference level which is for the smooth wall case $\eta_0 = 0$ (i.e. $u_1(\eta_0) = v_1(\eta_0) =$

$w_1(\eta_0) = 0$), while at the free surface an admissible solution has to fulfil the following boundary conditions:

$$
\begin{aligned}
(\tau_{yz} + \tau^v_{yz})|_{\eta=1} = 0 \implies 0 &= \left(\nu_{t0} + \frac{1}{Re_b}\right)(v_1 + w_1') + \frac{1}{4}C_D l_0^2 U_0' u_1 \\
(\tau_{xy} + \tau^v_{xy})|_{\eta=1} = 0 \implies 0 &= \left(2\nu_{t0} + \frac{1}{Re_b}\right) u_1' + \left(\frac{1}{2}C_D - \frac{3}{2}C_E\right) l_0^2 U_0' \alpha^2 w_1 + \frac{1}{2} C_E l_0^2 U_0'' v_1 \\
(P + \tau_{yy} + \tau^v_{yy})|_{\eta=1} = 0 \implies 0 &= -p_1 + 2\left(\nu_{t0} + \frac{1}{Re_b}\right)\alpha^2 w_1 + \left(\frac{1}{6}C_D + \frac{2}{3}C_E\right) l_0^2 U_0' u_1' \\
V|_{\eta=1} = 0 \implies 0 &= v_1 .
\end{aligned}
\tag{A.24}
$$

Here, $\tau^v_{ij} = (\partial_i U_j + \partial_j U_i)/Re_b$ are the components of the non-dimensional Newtonian viscous stress tensor. In practice, we use again homogeneous Neumann boundary conditions at the free surface for u_1, w_1, p_1 instead.

Appendix B

Definition of discrete averaging operators

B.1 Wall-parallel plane-averaged flow fields

The following definitions of the discrete fluid phase averaging operators are introduced in analogy to the definitions in Kidanemariam et al. (2013) and Kidanemariam (2016). First, let us introduce a fluid indicator function $I_f(\boldsymbol{x}, t)$ which provides information on whether a given position $\boldsymbol{x} \in \Omega$ is at time t occupied by the fluid phase, viz.

$$I_f(\boldsymbol{x}, t) = \begin{cases} 1 & \text{if } \boldsymbol{x} \in \Omega_f(t) \\ 0 & \text{else} \end{cases} . \tag{B.1}$$

Further, we denote the number of grid points in the streamwise, wall-normal and spanwise direction with N_x, N_y and N_z, respectively, for a number of N_t available flow field realisations over the simulation interval. Discrete grid positions will be written in the following as $x_i = i\Delta x$, $y_j = j\Delta y$ and $z_k = k\Delta z$ and can be collected in a triple vector $\boldsymbol{x}_{ijk} = (x_i, y_j, z_k)$. Discrete time instances $t_m = m\Delta t$ are introduced accordingly. The total number of grid points occupied by the fluid phase in a given wall-parallel plane at wall-normal position y_j is expressed by the following sums

$$n_{xz}(y_j, t_m) = \sum_{i=0}^{N_x-1} \sum_{k=0}^{N_z-1} I_f(\boldsymbol{x}_{ijk}, t_m) \tag{B.2a}$$

$$n_{xzt}(y_j) = \sum_{i=0}^{N_x-1} \sum_{k=0}^{N_z-1} \sum_{m=0}^{N_t-1} I_f(\boldsymbol{x}_{ijk}, t_m), \tag{B.2b}$$

where n_{xzt} indicates the total number of samples in the fluid phase for the entire simulation interval. The instantaneous wall-plane average and the average over a given wall-parallel plane and time for an arbitrary Eulerian vector field $\boldsymbol{\xi}(\boldsymbol{x}, t)$ are then introduced as

$$\langle \xi \rangle_{xz}(y_j, t_m) = \frac{1}{n_{xz}(y_j, t_m)} \sum_{i=0}^{N_x-1} \sum_{k=0}^{N_z-1} \xi(\boldsymbol{x}_{ijk}, t_m) I_f(\boldsymbol{x}_{ijk}, t_m) \tag{B.3a}$$

$$\langle \xi \rangle_{xzt}(y_j) = \frac{1}{n_{xzt}(y_j)} \sum_{i=0}^{N_x-1} \sum_{k=0}^{N_z-1} \sum_{m=0}^{N_t-1} \xi(\boldsymbol{x}_{ijk}, t_m) I_f(\boldsymbol{x}_{ijk}, t_m). \tag{B.3b}$$

B.2 Streamwise-averaged flow fields

In an analogous way, the number of grid points occupied by the fluid phase at a point $(y_j, z_k)^T$ in the cross-plane of the domain accumulated over the streamwise direction (and time) is

$$n_x(y_j, z_k, t_m) = \sum_{i=0}^{N_x-1} I_f(\boldsymbol{x}_{ijk}, t_m) \tag{B.4a}$$

$$n_{xt}(y_j, z_k) = \sum_{i=0}^{N_x-1} \sum_{m=0}^{N_t-1} I_f(\boldsymbol{x}_{ijk}, t_m). \tag{B.4b}$$

Based on these expressions, the instantaneous streamwise average and the average over the streamwise direction and time can be defined as

$$\langle \xi \rangle_x(y_j, z_k, t_m) = \frac{1}{n_x(y_j, z_k, t_m)} \sum_{i=0}^{N_x-1} \xi(\boldsymbol{x}_{ijk}, t_m) I_f(\boldsymbol{x}_{ijk}, t_m) \tag{B.5a}$$

$$\langle \xi \rangle_{xt}(y_j, z_k) = \frac{1}{n_{xt}(y_j, z_k)} \sum_{i=0}^{N_x-1} \sum_{m=0}^{N_t-1} \xi(\boldsymbol{x}_{ijk}, t_m) I_f(\boldsymbol{x}_{ijk}, t_m). \tag{B.5b}$$

Note that analogous definitions can be obtained for the spanwise-averaging operators $\langle \bullet \rangle_z$ and $\langle \bullet \rangle_{zt}$ for the fluid phase, respectively, and defining a particle indicator function as $1 - I_f(\boldsymbol{x}_{ijk}, t_m)$ leads to accordingly-defined averaging operators of Eulerian fields associated with the dispersed solid phase.

Appendix C

Validation of the open duct boundary conditions in the IBM-code

C.1 Simulation parameters

In the current validation study, the in-house multiphase finite-difference code is used to simulate single-phase flow in an open duct configuration at moderate Reynolds number $Re_b = 2195\text{-}2225$, which corresponds to a friction Reynolds number $Re_\tau = 150\text{-}154$. Two finite-difference simulations have been performed, one on a finer grid ($\Delta x^+ = 0.80$, statistics accumulated over $1.41 \cdot 10^5$ time steps) and one on a coarser grid ($\Delta x^+ = 1.17$, statistics accumulated over $1.60 \cdot 10^5$ time steps). For open duct flow with the same aspect ratio and a very similar Reynolds number of $Re_b = 2205$ ($Re_\tau = 150$), a dataset computed with a pseudo-spectral method is available (Sakai, 2016). The results of this latter work are used in the following to validate the newly implemented open duct boundary conditions in the finite-difference code. The relevant physical and numerical parameters of all three simulations are presented in table C.1.

To limit the computational expenses of this validation study, the streamwise extension of the domain is with $L_x/H_f = 12$ not long enough to ensure a decorrelation of the two-point velocity correlations, for which a two times longer domain as in the spectral simulation would be necessary (Sakai, 2016). Also, due to the significantly higher computational cost of the finite-difference simulations, the simulated time interval was limited to approximately 1070 bulk time units, while in the pseudo-spectral case statistics could be gathered over a roughly ten times longer time interval. The following comparison should therefore be interpreted in view of the statistical error that comes with the shorter domain length and observation interval in the finite-difference simulations.

To increase the ensemble size in all three cases, we exploit in the following the statistical symmetry in a fully-developed open duct flow w.r.t. the duct bisector $z = L_z/2$, that read

$\langle u_f \rangle_{xt}(e)$	$\langle v_f \rangle_{xt}(e)$	$\langle w_f \rangle_{xt}(o)$
$\langle u'_f u'_f \rangle_{xt}(e)$	$\langle v'_f v'_f \rangle_{xt}(e)$	$\langle w'_f w'_f \rangle_{xt}(e)$
$\langle u'_f v'_f \rangle_{xt}(e)$	$\langle u'_f w'_f \rangle_{xt}(o)$	$\langle v'_f w'_f \rangle_{xt}(o)$.

	Case	Re_b	Re_τ	L_x/H_f	AR	$N_x \times N_y \times N_z$	Δx^+	$\max\{\Delta y^+, \Delta z^+\}$	t_{stat}/T_b
—	*FD1*	2225	154	12.0	1.0	$2304 \times 193 \times 385$	0.80	0.80	1069
—	*FD2*	2196	150	12.0	1.0	$1536 \times 129 \times 257$	1.17	1.17	1068
—	*Spec*	2205	150	8π	1.0	$256 \times 97 \times 193$	14.8	2.46	12 940

Table C.1: Physical and numerical parameters in the single-phase open duct flow simulations conducted to validate the boundary conditions in the finite-difference IBM. Cases *FD1* ($1.41 \cdot 10^5$ samples) and *FD2* ($1.60 \cdot 10^5$ samples) have been computed with the finite-difference code, while the simulation *Spec* is from Sakai (2016), using a pseudo-spectral method. *FD1* and *FD2* differ only in the number of grid points per unit width, i.e. the grid in case *FD1* is finer by a factor 1.5 with respect to *FD2* in each direction. N_i is the number of grid points in the ith coordinate direction, while t_{stat} is the time period over which statistics have been accumulated.

The notation is such that (o) and (e) stand for an odd and even parity in the spanwise direction w.r.t. the duct bisector, respectively. We can take advantage of these symmetry properties when investigating the statistics in the following by averaging all field variables over the left and the right half of the duct (taking into account the possibly changing sign for the odd symmetries).

C.2 Pointwise comparison of velocity field and Reynolds stresses

To allow for pointwise comparison of the finite-difference results with the data obtained by the pseudo-spectral code, the results obtained with the latter method have been linearly interpolated to the equidistant finite-difference pressure grid, which collapses in the cross-stream plane with the grid points of the streamwise velocity component. Note that linear interpolation might bring an additional interpolation error into play, which should be taken into account when interpreting the following results.

In table C.2, the root mean square error (RMSE) of the finite difference quantities w.r.t. the pseudo-spectral results is provided. The RMSE has been computed for the mean velocity fields as

$$\varepsilon_{RMSE}(\langle u_f \rangle_{xt}) = \frac{1}{\max\left(\left|\frac{u_{j,k}}{u_b}\right|\right)^{Spec}} \cdot \left[\frac{1}{N_y N_z} \sum_{j=0}^{N_y-1} \sum_{k=0}^{N_z-1} \left(\left(\frac{u_{j,k}}{u_b}\right)^{FD} - \left(\frac{u_{j,k}}{u_b}\right)^{Spec}\right)^2\right]^{\frac{1}{2}}, \tag{C.1}$$

where N_i is the number of grid points of the finite-difference pressure grid in the ith direction and $u_{j,k} = \langle u_f \rangle_{xt}(y_j, z_k)$. For the computation of the RMSE of the Reynolds stresses, the bulk velocity u_b in equation (C.1) has been replaced by the turbulent kinetic energy averaged over the cross-plane, viz. $\langle k \rangle_{yzt} = \langle (\langle u'_f u'_f \rangle_{xt} + \langle v'_f v'_f \rangle_{xt} + \langle w'_f w'_f \rangle_{xt}) \rangle_{yz}/2$. In the case of the streamwise vorticity component $\langle \omega_{f,x} \rangle_{xt}$, the field is normalised by (u_b/H_f). The RMSE has been computed for both the symmetrised and the non-symmetrised statistics. It is seen that the deviation is for all variables at most of order $\mathcal{O}(1\%)$. Taking into account the statistical error as well as that originating in the linear interpolation, the finite-difference results show a reasonably good agreement with the pseudo-spectral datasets.

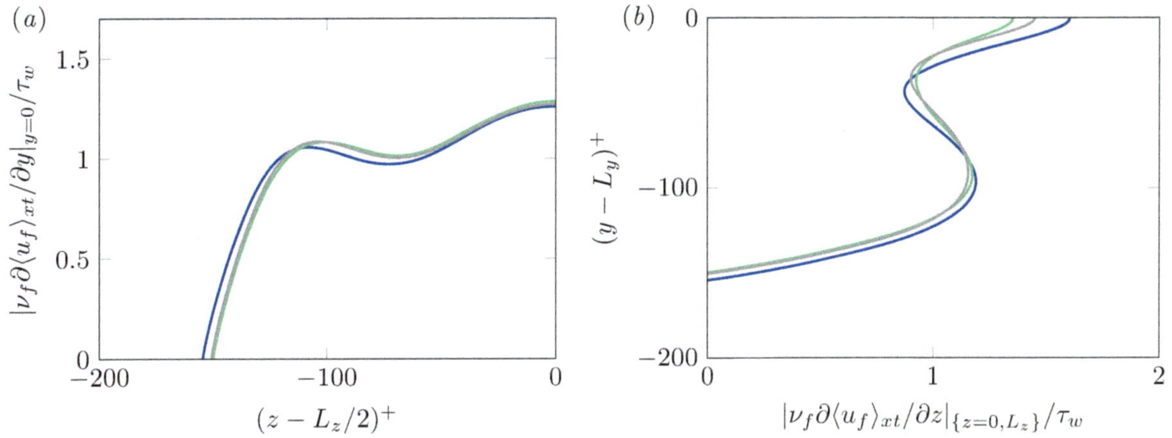

Figure C.1: Variation of (*a*) the bottom wall and (*b*) sidewall shear stress distribution in the finite-difference and pseudo-spectral open duct flow simulations, conducted to validate the boundary conditions in the finite-difference IBM. Results are symmetrised about the duct bisector, thus only the left half of the bottom wall shear stress profile and only one curve for the sidewall shear stress are shown. (—,—) finite-difference simulations *FD*1 and *FD*2; (—) pseudo-spectral simulations.

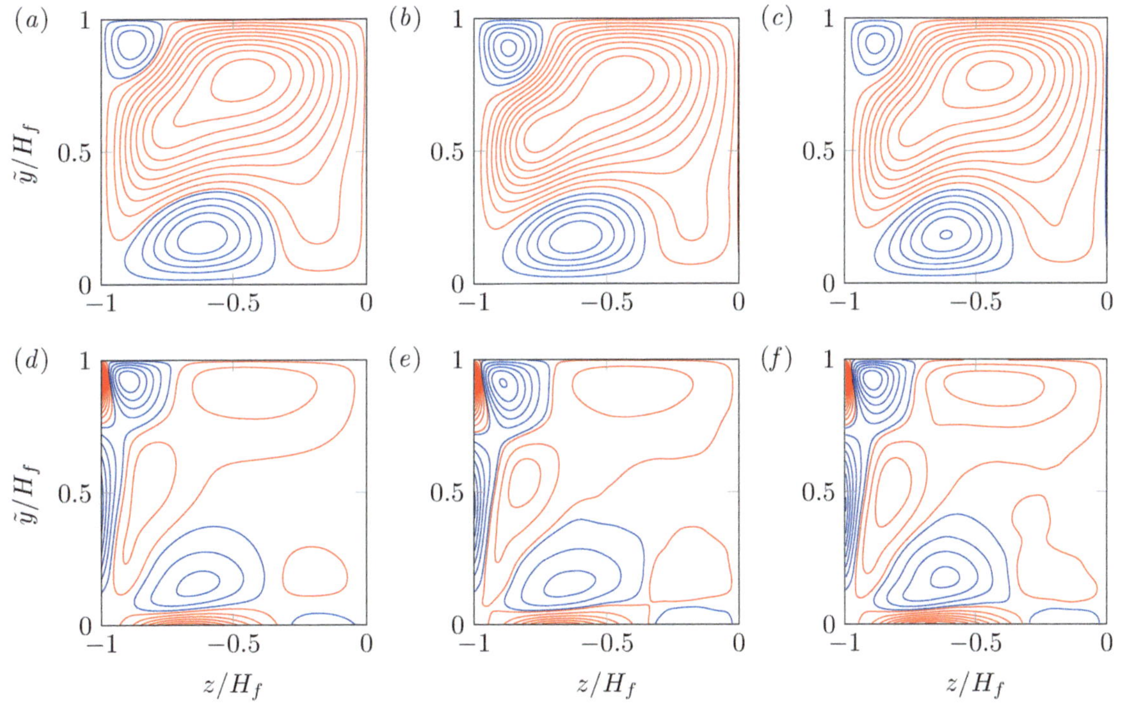

Figure C.2: (*a-c*) Mean secondary flow streamfunction $\langle \psi \rangle_{xt} / (u_b H_f)$ and (*d-f*) mean streamwise vorticity $\langle \omega_{f,x} \rangle_{xt}^+$ in the finite-difference and pseudo-spectral open duct flow simulations, conducted to validate the boundary conditions in the finite-difference IBM. (*a,d*) pseudo-spectral simulation, (*c,e*) finite-difference simulation *FD*1, (*d,f*) finite-difference simulation *FD*2. Results are symmetrised about the duct bisector, thus only the left half of the duct cross-section is shown. Clockwise (counterclockwise) secondary flow and vorticity rotation is indicated by red (blue) isolines. The shown contours indicate 20 equally spaced subintervals of the interval $[-\max_{y,z} |a|, \max_{y,z} |a|]$ with $a \in \{\langle \psi \rangle_{xt}, \langle \omega_{f,x} \rangle_{xt}\}$.

	$FD1$ ($\Delta x^+ = 0.80$)		$FD2$ ($\Delta x^+ = 1.17$)	
	ε_{RMSE}	$\varepsilon_{RMSE,sym}$	ε_{RMSE}	$\varepsilon_{RMSE,sym}$
$\langle u_f \rangle_{xt}$	0.015063	0.011831	0.009850	0.004714
$\langle v_f \rangle_{xt}$	0.099569	0.070640	0.082093	0.034196
$\langle w_f \rangle_{xt}$	0.077027	0.056956	0.059063	0.025583
$\langle u'_f u'_f \rangle_{xt}$	0.023123	0.013583	0.032820	0.018377
$\langle v'_f v'_f \rangle_{xt}$	0.049985	0.037907	0.042106	0.029909
$\langle w'_f w'_f \rangle_{xt}$	0.031975	0.025697	0.032684	0.025059
$\langle u'_f v'_f \rangle_{xt}$	0.078483	0.054857	0.065519	0.033934
$\langle u'_f w'_f \rangle_{xt}$	0.042477	0.027355	0.049159	0.021112
$\langle \omega_{f,x} \rangle_{xt}$	0.043553	0.024142	0.033570	0.012907

Table C.2: Root mean square error of the mean velocities and Reynolds stresses of the finite-difference simulations w.r.t the pseudo-spectral results, conducted to validate the boundary conditions in the finite-difference IBM. The RMSE has been computed for the whole flow field (ε_{RMSE}) as well as for the half duct statistics, in which the symmetry is used to double the number of samples for the statistics ($\varepsilon_{RMSE,sym}$), using expressions analogue to equation (C.1). Note that for the Reynolds stresses, the mean turbulent kinetic energy $\langle k \rangle_{yzt}$ is used for normalisation instead of the bulk velocity in the above equation, whereas in the case of the streamwise vorticity component, the scale (u_b/H_f) is used.

C.3 Wall-shear stress

Figure C.1 shows the wall-shear stress distribution along the bottom and sidewalls, normalised with the mean wall shear stress τ_w that was determined as the average along the entire perimeter. It can be seen that the finite-difference simulations correctly reproduce the general shape of the shear stress distribution obtained with the pseudo-spectral code along bottom and sidewalls, with some small deviations of 0.1-0.2τ_w which are partly because the Reynolds number is not exactly the same in the three simulations.

C.4 Mean secondary flow and streamwise vorticity

Figure C.2 shows contours of constant mean secondary streamfunction $\langle \psi \rangle_{xt}$ and mean streamwise vorticity $\langle \omega_{f,x} \rangle_{xt}$ for the finite-difference and pseudo-spectral simulations, respectively. All three cases reveal very similar patterns of both $\langle \psi \rangle_{xt}$ and $\langle \omega_{f,x} \rangle_{xt}$, even though the exact contours still differ somewhat. The slight deviations are assumed to be of statistical nature, recalling that the statistics in the pseudo-spectral simulation were gathered over a ten times longer averaging interval.

Appendix D

Two-dimensional streak eduction in open duct flows

For the sake of comparison with the three-dimensional streak detection procedure described in section 7.3.2, we have performed a second streak eduction study in which connected subdomains are identified in each cross-section of the discrete velocity field which are characterised by a high positive or negative streamwise velocity fluctuation. The methodology in the here discussed form was first used by Nakatsuji (2012) to detect streamwise streaks in closed square duct flows and later adopted to open duct flows by Sakai (2016).

D.1 Analysis of the preferential organisation of streamwise velocity streaks

In a similar way as in the detection method of quasi-streamwise vortex cores presented in Uhlmann et al. (2007) (cf. also section 7.3.3), we seek for connected regions for which the normalised streamwise velocity fluctuations exceed a given threshold u_{th}, viz.

$$u'_f/\sigma_u > u_{th} \quad \text{and} \quad u'_f/\sigma_u < -u_{th} \tag{D.1}$$

for high- and low-speed streaks, respectively. The velocity fluctuations u'_f are therein normalised by the root mean square of the velocity fluctuations averaged over the clear fluid region, $\sigma_u = \sqrt{\langle u'_f u'_f \rangle_{\mathcal{V}_f}}$, where the volume-averaging operator reads

$$\langle \bullet \rangle_{\mathcal{V}_f} = \frac{1}{L_x \langle h_b \rangle_{xz}(t) L_z} \int_0^{L_z} \int_{\langle h_b \rangle_x(z,t)}^{L_y} \int_0^{L_x} (\,\bullet\,)\; I_f(\boldsymbol{x},t)\; \mathrm{d}x\mathrm{d}y\mathrm{d}z. \tag{D.2}$$

In order to include only the fluid domain in the analysis, we use a fluid-phase indicator function $I_f(\boldsymbol{x},t)$ that attains a value of unity if the point $\boldsymbol{x}$ lies at time t in the fluid domain Ω_f and zero if not (cf. definition (B.1) in appendix B). Each detected high- and low-speed streak region is in the following associated with a single point in the cross-section with wall-normal and spanwise coordinates

$$y_{c,2D} = \frac{\int_{\mathcal{A}_s} u'_f\, y\, \mathrm{d}y\mathrm{d}z}{\int_{\mathcal{A}_s} u'_f \mathrm{d}y\mathrm{d}z} \quad \text{and} \quad z_{c,2D} = \frac{\int_{\mathcal{A}_s} u'_f\, z\, \mathrm{d}y\mathrm{d}z}{\int_{\mathcal{A}_s} u'_f \mathrm{d}y\mathrm{d}z}, \tag{D.3}$$

respectively, which represent the centre of gravity of the streaks' cross-sectional area $\mathcal{A}_s$, additionally weighted with the local velocity fluctuations. The cross-sectional area is then also considered to define a characteristic cross-sectional length scale of the streaks as $l_s = \sqrt{\mathcal{A}_s}$.

As in the case of the three-dimensional streak structures extracted in section 7.3.2, the choice of an appropriate velocity threshold u_{th} is crucial in the extraction procedure to obtain meaningful physical results. In analogy to the percolation analysis that is conventionally used to determine such a suitable threshold for the three-dimensional extraction process (Moisy and Jiménez, 2004), Nakatsuji (2012) and Sakai (2016) varied the velocity threshold in a range $0.25 \leq u_{th} \leq 2$ and investigated the characteristic length scale and the number of extracted structures for each of the values. We have applied the same thresholds to the current simulations as well, but for the sake of brevity we will restrict ourselves in the following on the results for a single threshold value $u_{th} = 0.5$.

Figures D.1 and D.2 show the p.d.f.s of the preferential location of the streak centre points $(y_{c,2D}, z_{c,2D})^T$ in the cross-plane of the smooth wall case $DL200^{H2}_{smooth}$ and the mobile bed simulation $DL250^{H2}$, respectively. Note that the former dataset is based on the data in Appendix B, Figure B.4 of Sakai (2016) and has been replotted here for the sake of comparison with the sediment-laden case. The p.d.f.s are further conditioned on the sign of the streaks and their characteristic length scale l_s, based on which they are classified into four categories: (i) $l_s \leq 25\delta_\nu$, (ii) $25\delta_\nu < l_s \leq 50\delta_\nu$, (iii) $50\delta_\nu < l_s < 0.5H_f$ and (iv) $0.5H_f < l_s$.

Figures D.3 and D.4 show the analogous p.d.f.s for the high Reynolds number smooth-wall case $DL400^{H2}_{smooth}$ and the corresponding mobile sediment case $DL500^{H2}$, respectively.

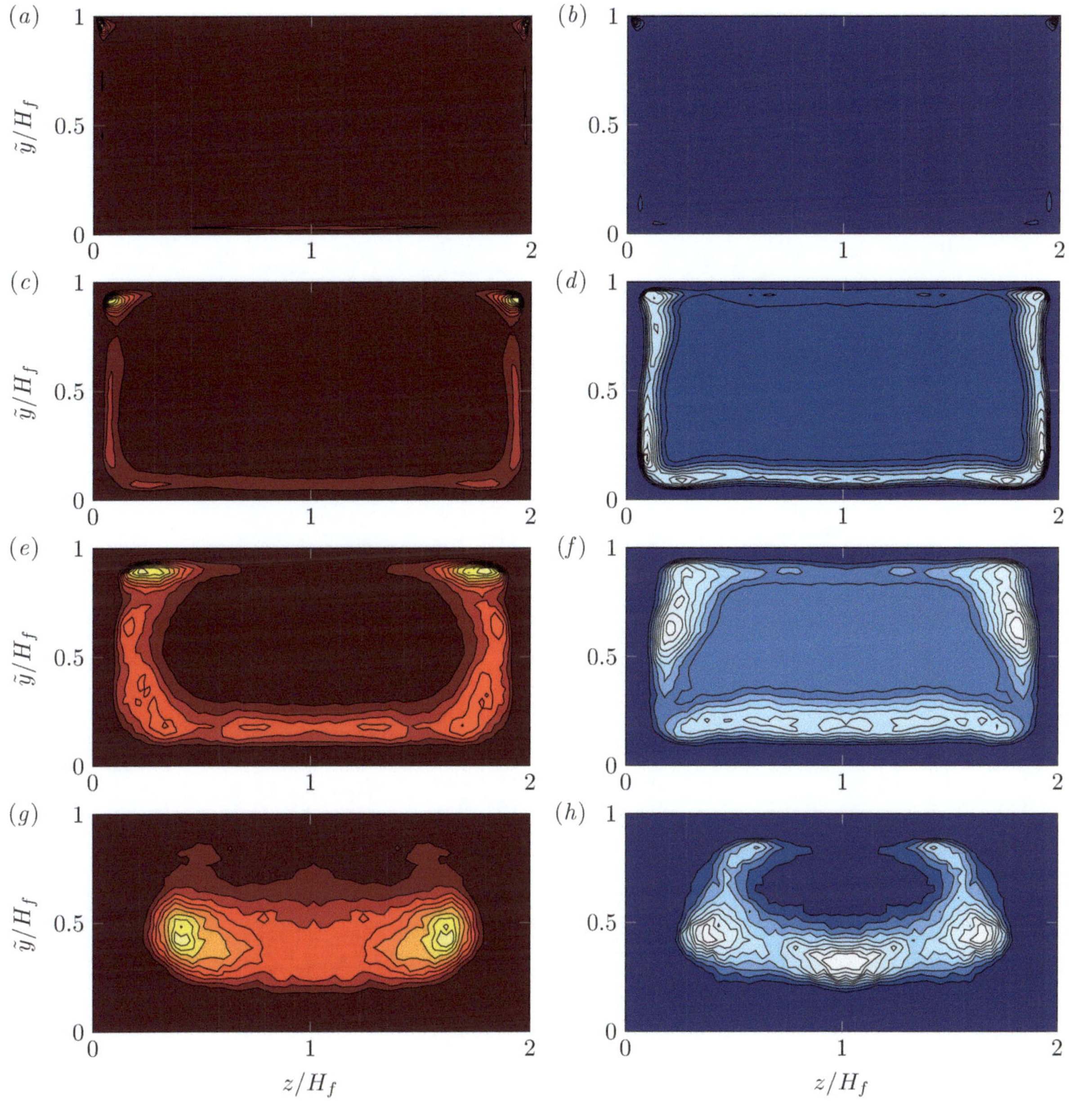

Figure D.1: P.d.f.s of the preferential positioning of instantaneous two-dimensional streamwise velocity streaks in the cross-plane of the open duct simulation $DL200^{H2}_{smooth}$. Individual streaks are associated with their weighted centre of gravity $(y_{c,2D}, z_{c,2D})$. (*a,c,e,g*) High- and (*b,d,f,h*) low-speed streaks were identified as two-dimensional connected regions of high ($u'_f/\sigma_u > 0.5$) and low streamwise velocity fluctuations ($u'_f/\sigma_u < -0.5$) in each cross-plane. The streaks are further classified according to their characteristic cross-sectional length scale l_s: (*a,b*) $l_s \leq 25\delta_v$; (*c,d*) $25\delta_v < l_s \leq 50\delta_v$; (*e,f*) $50\delta_v < l_s \leq 0.5H_f$; (*g,h*) $0.5H_f < l_s$. Each p.d.f. is normalised by its global maximum and contours represent 0.1(0.1)0.9 of this maximum. Note that in this specific case, statistics are additionally symmetrised over the left and right half of the duct cross-section.

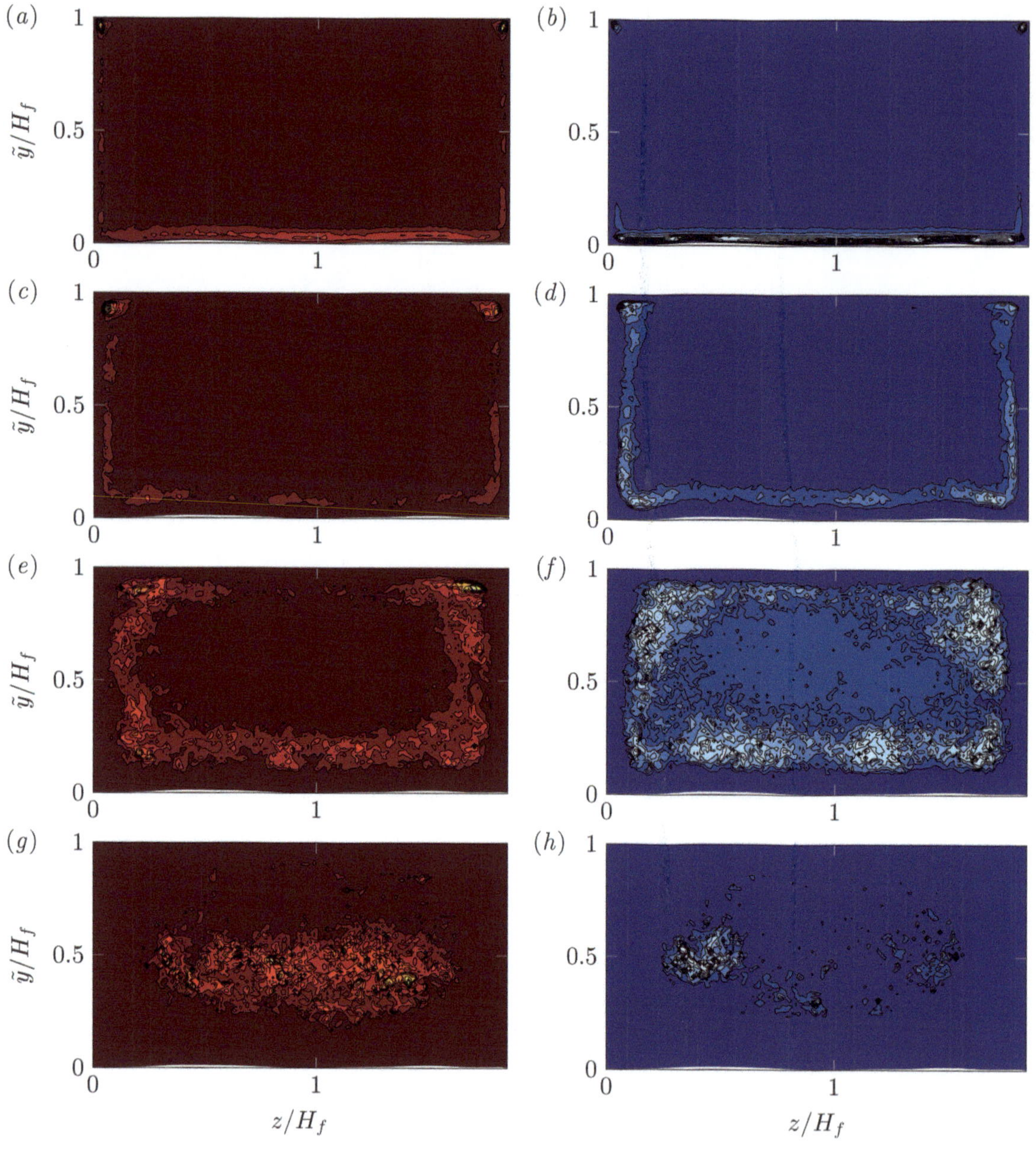

Figure D.2: P.d.f.s of the preferential positioning of instantaneous two-dimensional streamwise velocity streaks in the cross-plane of the open duct simulation $DL250^{H2}$. Individual streaks are associated with their weighted centre of gravity $(y_{c,2D}, z_{c,2D})$. *(a,c,e,g)* High- and *(b,d,f,h)* low-speed streaks were identified as two-dimensional connected regions of high $(u'_f/\sigma_u > 0.5)$ and low streamwise velocity fluctuations $(u'_f/\sigma_u < -0.5)$ in each cross-plane. The streaks are further classified according to their characteristic cross-sectional length scale l_s: *(a,b)* $l_s \leq 25\delta_\nu$; *(c,d)* $25\delta_\nu < l_s \leq 50\delta_\nu$; *(e,f)* $50\delta_\nu < l_s \leq 0.5H_f$; *(g,h)* $0.5H_f < l_s$. Each p.d.f. is normalised by its global maximum and contours represent 0.1(0.1)0.9 of this maximum. The mean fluid bed interface is represented by a white solid line.

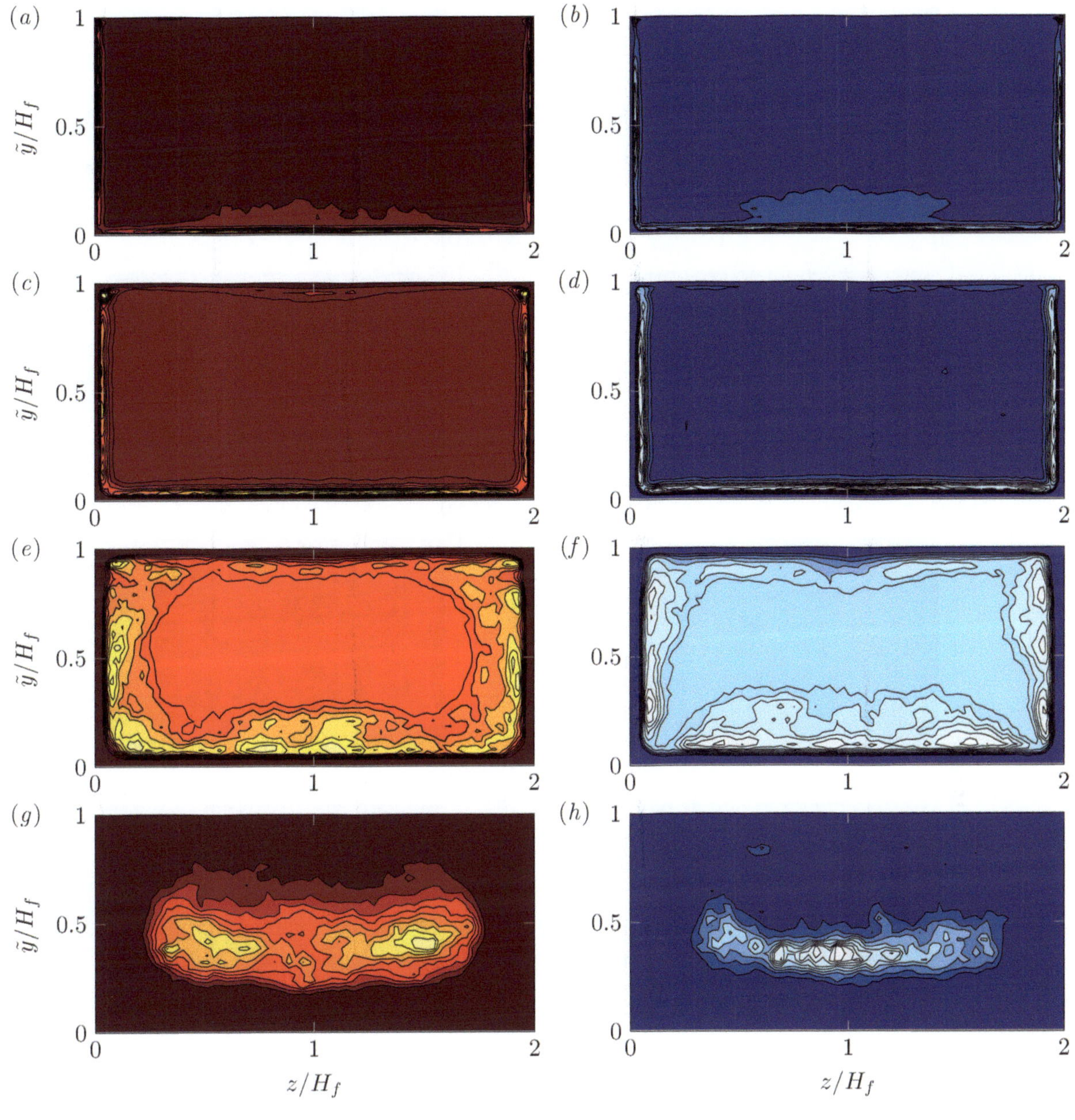

Figure D.3: P.d.f.s of the preferential positioning of instantaneous two-dimensional streamwise velocity streaks in the cross-plane of the open duct simulation $DL400^{H2}_{smooth}$. Individual streaks are associated with their weighted centre of gravity $(y_{c,2D}, z_{c,2D})$. (*a,c,e,g*) High- and (*b,d,f,h*) low-speed streaks were identified as two-dimensional connected regions of high ($u'_f/\sigma_u > 0.5$) and low streamwise velocity fluctuations ($u'_f/\sigma_u < -0.5$) in each cross-plane. The streaks are further classified according to their characteristic cross-sectional length scale l_s: (*a,b*) $l_s \leq 25\delta_\nu$; (*c,d*) $25\delta_\nu < l_s \leq 50\delta_\nu$; (*e,f*) $50\delta_\nu < l_s \leq 0.5H_f$; (*g,h*) $0.5H_f < l_s$. Each p.d.f. is normalised by its global maximum and contours represent 0.1(0.1)0.9 of this maximum.

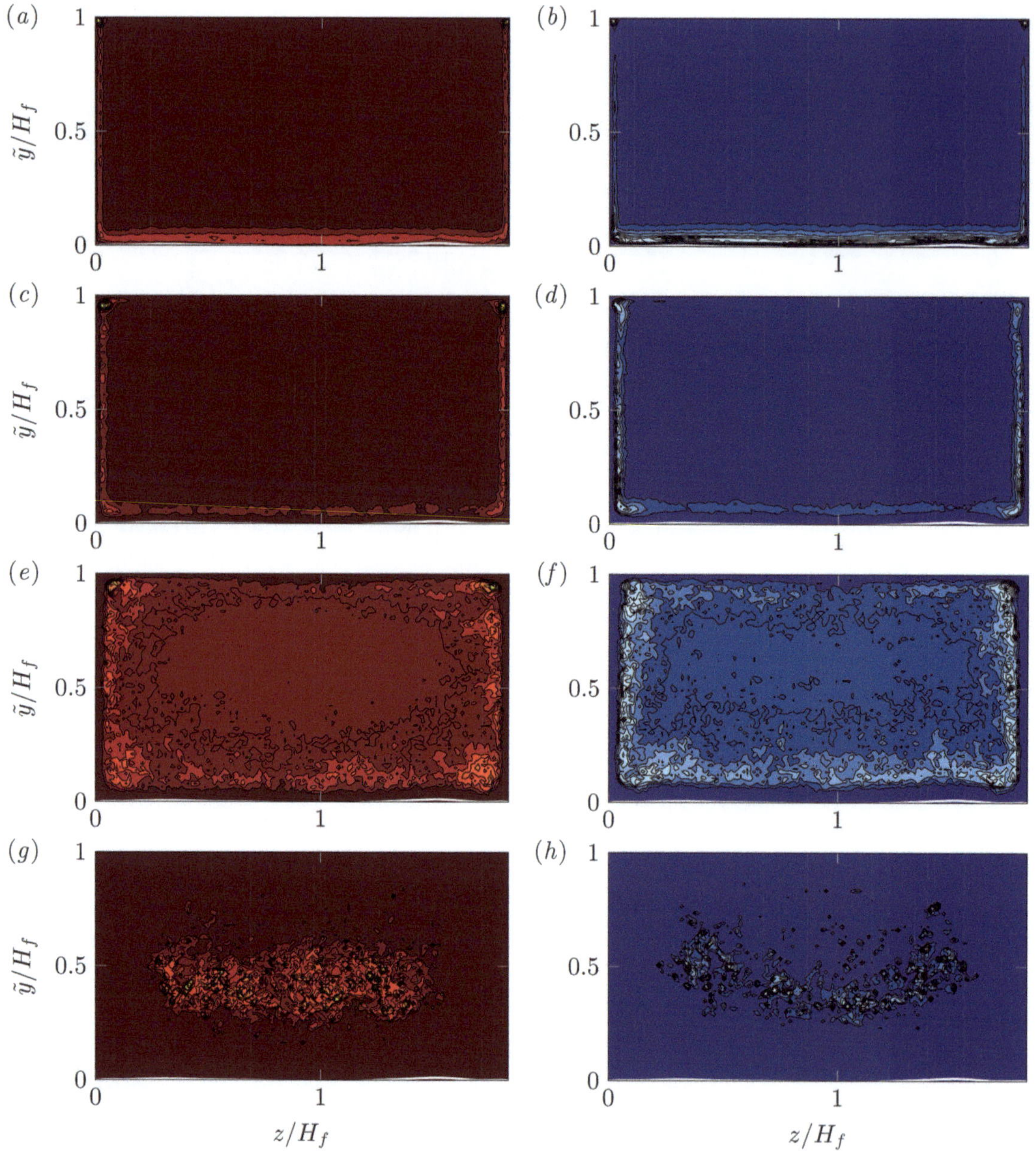

Figure D.4: P.d.f.s of the preferential positioning of instantaneous two-dimensional streamwise velocity streaks in the cross-plane of the open duct simulation $DL500^{H2}$. Individual streaks are associated with their weighted centre of gravity $(y_{c,2D}, z_{c,2D})$. (a,c,e,g) High- and (b,d,f,h) low-speed streaks were identified as two-dimensional connected regions of high ($u'_f/\sigma_u > 0.5$) and low streamwise velocity fluctuations ($u'_f/\sigma_u < -0.5$) in each cross-plane. The streaks are further classified according to their characteristic cross-sectional length scale l_s: (a,b) $l_s \leq 25\delta_\nu$; (c,d) $25\delta_\nu < l_s \leq 50\delta_\nu$; (e,f) $50\delta_\nu < l_s \leq 0.5H_f$; (g,h) $0.5H_f < l_s$. Each p.d.f. is normalised by its global maximum and contours represent 0.1(0.1)0.9 of this maximum. The mean fluid bed interface is represented by a white solid line.

Appendix E

Two-dimensional streamwise roll eduction in open duct flows

In the following, we discuss the poloidal-toroidal decomposition used by Kawahara et al. (2012a) and Nakatsuji (2012) to identify instantaneous small- and large-scale rotating motions in the cross-section of closed square ducts at varying Reynolds numbers. Note that the basic extraction procedure is similar to the one used by the same authors to detect individual velocity streaks that was outlined in the previous section appendix D.

E.1 Poloidal-toroidal decomposition in open duct flows

Let Ω_f be a bounded and simply connected domain and $\boldsymbol{u}_f : \mathbb{R}^3 \times \mathcal{I} \to \mathbb{R}^3$, $\boldsymbol{u}_f \in C^1(\Omega_f) \cap C^0(\overline{\Omega}_f)$ be a solenoidal vector field defined on this domain. Then, $\exists\, \phi, \psi : \mathbb{R}^3 \times \mathcal{I} \to \mathbb{R}$ s.t. (Warner, 1972; Chandrasekhar, 1981; Marqués, 1990)

$$\boldsymbol{u}_f = \nabla \times (\psi \boldsymbol{e}) + \nabla \times \nabla \times (\phi \boldsymbol{e}), \tag{E.1}$$

where $\boldsymbol{e}$ denotes a unit vector of arbitrary direction. The two scalar fields ϕ and ψ are typically referred to as poloidal and toroidal potentials, respectively. Note that for a Cartesian basis, there might be additional terms appearing on the RHS of relation (E.1) related to the mean flow in the directions perpendicular to $\boldsymbol{e}$ depending on the particular choice of the latter (Nagata, 1990; Clever and Busse, 1997; Waleffe, 2003). In contrast to the latter studies, however, Kawahara et al. (2012b) and Nakatsuji (2012) choose a unit vector $\boldsymbol{e} = (1, 0, 0)^T$ that is parallel to the streamwise direction such that the toroidal potential ψ is associated with fluid motions in the duct cross-plane. In consequence, there is no global mean flow in the directions orthogonal to $\boldsymbol{e}$ in this case due to the impermeability of the surrounding walls and no additional terms arise in equation (E.1). For the here considered choice of the unit vector, the velocity field can be rewritten in terms of the two potentials, viz. (Nakatsuji, 2012)

$$\boldsymbol{u}_f = \begin{pmatrix} -\Delta_\perp \phi \\ \partial_x \partial_y \phi + \partial_z \psi \\ \partial_x \partial_z \phi - \partial_y \psi \end{pmatrix}. \tag{E.2}$$

Cross-differentiation and subtraction of the spanwise and wall-normal velocity components then directly lead to a Poisson equation for the toroidal potential $\psi(x,t)$, with the streamwise vorticity field $\omega_{f,x}(x,t) = \partial_y w_f - \partial_z v_f$ as source term quite similar to equation (7.12):

$$\begin{cases} \Delta_\perp \psi = -\omega_{f,x} & \text{in } \Omega_\perp \\ \psi = 0 & \text{on } \partial\Omega_\perp^D \\ \dfrac{\partial \psi}{\partial n} = 0 & \text{on } \partial\Omega_\perp^N \end{cases}, \tag{E.3}$$

where n is a unit vector normal to the boundary at each $x \in \Omega_\perp$ oriented such that it is pointing into the domain Ω_f. In a similar way, the streamwise flow field represents the source term of a Poisson equation for the poloidal potential $\phi(x,t)$, viz.

$$\begin{cases} \Delta_\perp \phi = -u_f & \text{in } \Omega_\perp \\ \phi = 0 & \text{on } \partial\Omega_\perp^D \\ \dfrac{\partial \phi}{\partial n} = 0 & \text{on } \partial\Omega_\perp^N \end{cases}, \tag{E.4}$$

In both cases, $\partial\Omega_\perp^D$ and $\partial\Omega_\perp^N$ refer to the parts of the domain boundary at which homogeneous Dirichlet and Neumann boundary conditions are applied, respectively, such that $\partial\Omega_\perp \equiv \partial\Omega_\perp^D \cup \partial\Omega_\perp^N$ and $\emptyset = \partial\Omega_\perp^D \cap \partial\Omega_\perp^N$. Kawahara et al. (2012a) apply homogeneous Dirichlet conditions for ϕ and homogeneous Neumann conditions for ψ in the closed duct, supplied with a zero mean condition for the latter to give the pure Neumann problem a unique solution. For the closed duct, this choice of the boundary conditions for the two potentials ensures the no-slip and impermeability conditions along the four solid walls for all velocity components, as shown by Nakatsuji (2012). Note that when averaging equation (E.3) along the streamwise direction and time, we recover the Poisson equation (7.12) for $\langle\psi\rangle_{xt}$. It can be shown that only in this case, the boundary conditions for all velocity components in both open and closed duct flows are fulfilled even for the pure homogeneous Dirichlet problem for $\langle\psi\rangle_{xt}$, which were chosen to solve equation (7.12) (G. Kawahara, personal communication).

In analogy to the study of Kawahara et al. (2012a), we first derive in the following the correct set of boundary conditions for both potentials at the four domain boundaries of an open duct, starting with the bottom wall at $y = 0$. Following Kawahara et al. (2012a), we impose at the solid wall homogeneous Dirichlet boundary conditions for ϕ and homogeneous Neumann conditions for ψ, viz.

$$\phi = 0 \implies \frac{\partial \phi}{\partial z} = \frac{\partial \phi}{\partial x} = 0; \quad \frac{\partial \psi}{\partial y} = 0 \quad \text{at } y = 0,\ \forall z \in [0, L_z]. \tag{E.5}$$

The potential ϕ is determined by solving the Poisson equation (E.4), the source term of which is $-u_f$ which vanishes at the wall due to the imposed no-slip condition. So, the boundary condition for u_f is directly fulfilled. Considering next the wall-parallel velocity component w_f, it is seen that it vanishes identically at $y = 0$ due to the chosen boundary conditions, viz.

$$w_f = \left(\frac{\partial^2 \phi}{\partial x \partial z} - \frac{\partial \psi}{\partial y} \right) = 0. \tag{E.6}$$

Next we consider the streamwise vorticity $\omega_{f,x}$ at the bottom wall, which reduces due to the impermeability condition $v_f = \partial_z v_f = 0$ at $y = 0$ to

$$\omega_{f,x} = \frac{\partial w_f}{\partial y} - \frac{\partial v_f}{\partial z} = \frac{\partial w_f}{\partial y}. \tag{E.7}$$

Recalling that the streamwise vorticity represents the RHS of the Poisson equation (E.3) for the potential ψ, we obtain

$$\begin{aligned}
\frac{\partial^2 \psi}{\partial y^2} + \frac{\partial^2 \psi}{\partial z^2} &= -\frac{\partial w_f}{\partial y} \\
&= -\frac{\partial}{\partial y}\left(\frac{\partial^2 \phi}{\partial x \partial z} - \frac{\partial \psi}{\partial y}\right) \\
\Longleftrightarrow 0 &= \frac{\partial}{\partial z}\left(\frac{\partial^2 \phi}{\partial x \partial y} + \frac{\partial \psi}{\partial z}\right) = \frac{\partial v_f}{\partial z}.
\end{aligned} \tag{E.8}$$

The last equality indicates that v_f is constant along the bottom wall. Note that this information was first brought into the system by $\omega_{f,x}$ as the RHS of the Poisson equation.

In the following, we move on to the lateral sidewalls at $z = \{0, L_z\}$, where the boundary conditions are chosen as

$$\phi = 0 \Longrightarrow \frac{\partial \phi}{\partial y} = \frac{\partial \phi}{\partial x} = 0; \quad \frac{\partial \psi}{\partial z} = 0 \quad \text{at } z = \{0, L_z\} \ \forall y \in [0, L_y]. \tag{E.9}$$

As before, the potential ϕ is determined by solving the Poisson equation (E.4) with the RHS $-u_f$, which vanishes also at the sidewalls due to the imposed no-slip condition. The boundary condition for u_f is therefore again directly fulfilled. Considering next the wall-parallel velocity component v_f, it is seen that this latter vanishes identically at $z = \{0, L_z\}$ due to the applied boundary conditions, viz.

$$v_f = \left(\frac{\partial^2 \phi}{\partial x \partial y} + \frac{\partial \psi}{\partial z}\right) = 0. \tag{E.10}$$

Again, the streamwise vorticity $\omega_{f,x}$ at the sidewalls then reduces due to the impermeability constraint $w_f = \partial_y w_f = 0$ at $z = \{0, L_z\}$ to

$$\omega_{f,x} = \frac{\partial w_f}{\partial y} - \frac{\partial v_f}{\partial z} = -\frac{\partial v_f}{\partial z}. \tag{E.11}$$

With this as RHS, the Poisson equation (E.3) for the potential ψ reads

$$\begin{aligned}
\frac{\partial^2 \psi}{\partial y^2} + \frac{\partial^2 \psi}{\partial z^2} &= \frac{\partial v_f}{\partial z} \\
&= \frac{\partial}{\partial z}\left(\frac{\partial^2 \phi}{\partial x \partial y} + \frac{\partial \psi}{\partial z}\right) \\
\Longleftrightarrow 0 &= \frac{\partial}{\partial y}\left(\frac{\partial^2 \phi}{\partial x \partial z} - \frac{\partial \psi}{\partial y}\right) = \frac{\partial w_f}{\partial y},
\end{aligned} \tag{E.12}$$

which implies that w_f is constant along both sidewalls. Considering the corner points at which sidewalls and bottom wall meet, we conclude that w_f is constant along both sidewalls but $w_f = 0$ along the bottom wall, such that $w_f = \text{const.} = 0$ follows for the sidewalls. In a similar way, we show that $v_f = 0$ at the sidewalls and $v_f = \text{const.}$ at the bottom wall imply that $v_f = 0$ at all three solid walls. Under the given boundary conditions, the no-slip and impermeability conditions at all three walls are fulfilled.

Eventually, let us discuss the boundary conditions at the top of the domain $y = H_f$, where free-slip conditions are imposed for the surface-parallel components ($\partial_y u_f = \partial_y w_f = 0$) and an impermeability condition has to be fulfilled for the surface-normal component $v_f = 0$. Simply applying, as before, homogeneous Dirichlet boundary conditions for ϕ and homogeneous Neumann boundary conditions for ψ wrongly implies a no-slip rather than a shear-free boundary condition for w_f along the free surface $y = H_f$, viz.

$$w_f = \left(\frac{\partial^2 \phi}{\partial x \partial z} - \frac{\partial \psi}{\partial y} \right) = 0. \tag{E.13}$$

Instead, we 'reverse' the boundary conditions at the free surface, choosing homogeneous Dirichlet boundary conditions for ψ and homogeneous Neumann boundary conditions for ϕ, viz.

$$\psi = 0 \implies \frac{\partial^n \psi}{\partial^n z} = \frac{\partial^n \psi}{\partial^n x} = 0; \quad \frac{\partial \phi}{\partial y} = 0 \quad \text{at } y = H_f, \ \forall z \in [0, L_z]. \tag{E.14}$$

As for the remaining boundaries, u_f fulfils by definition the boundary condition at the free surface and as it is the RHS of the Poisson equation in relation (E.4), this latter automatically fulfils the condition $\partial_y u_f = 0$ at the free surface. Next, considering the transverse velocity component w_f, we obtain due the boundary conditions for its surface-normal derivative

$$\frac{\partial w_f}{\partial y} = \frac{\partial}{\partial y} \left(\frac{\partial^2 \phi}{\partial x \partial z} - \frac{\partial \psi}{\partial y} \right) = \frac{\partial^2}{\partial x \partial z} \frac{\partial \phi}{\partial y} - \frac{\partial^2 \psi}{\partial y^2} = -\frac{\partial^2 \psi}{\partial y^2}. \tag{E.15}$$

It turns out that the streamwise vorticity $\omega_{f,x}$ vanishes at the free surface due to the free-slip and impermeability conditions, which imply that $\partial_y w_f = \partial_z v_f = \omega_{f,x} = 0$ at $y = H_f$. As a consequence, the Poisson equation (E.3) for the potential ψ reduces to the following condition

$$\begin{aligned} 0 &= \frac{\partial^2 \psi}{\partial y^2} + \frac{\partial^2 \psi}{\partial z^2} \\ \implies \quad -\frac{\partial^2 \psi}{\partial y^2} &= \frac{\partial w_f}{\partial y} = \frac{\partial^2 \psi}{\partial z^2} = 0, \end{aligned} \tag{E.16}$$

where for the last equality we have used the Dirichlet boundary condition $\psi = \partial_{zz}\psi = 0$. It remains to show the impermeability condition for v_f, which again directly follows from the imposed boundary conditions, viz.

$$v_f = \left(\frac{\partial^2 \phi}{\partial x \partial y} + \frac{\partial \psi}{\partial z} \right) = \frac{\partial}{\partial x} \left(\frac{\partial \phi}{\partial y} \right) + \frac{\partial \psi}{\partial z} = 0. \tag{E.17}$$

We conclude that the impermeability and free-slip conditions are fulfilled for the chosen boundary condition at the free surface.

E.2 Analysis of the preferential organisation of streamwise rolls

Based on the afore discussed mathematical decomposition of the flow field subject to the appropriate boundary conditions at the no-slip and free-slip boundaries, the toroidal velocity potential ψ can be

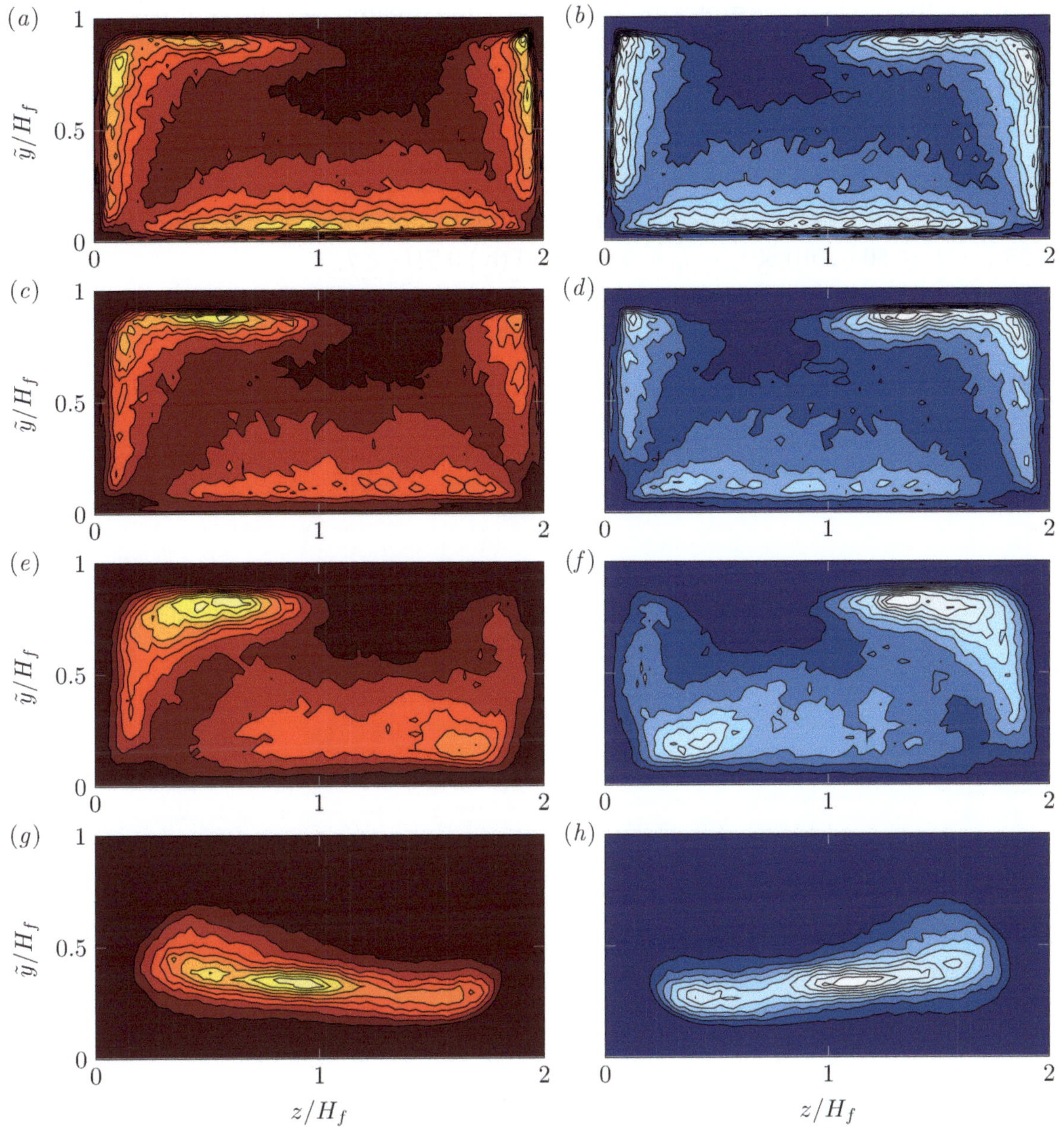

Figure E.1: P.d.f.s of the preferential positioning of instantaneous two-dimensional streamwise rolls in the cross-plane of the open duct simulation $DL400^{H2}_{smooth}$. Individual rolls are associated with their weighted centre of gravity $(y_{c,2D}, z_{c,2D})$. *(a,c,e,g)* Clockwise- and *(b,d,f,h)* counterclockwise rotating rolls were identified as two-dimensional connected regions of high $(\psi/\sigma_\psi > 1.0)$ and low values of the toroidal potential $(\psi/\sigma_\psi < -1.0)$ in each cross-plane. The streamwise rolls are further classified according to their characteristic cross-sectional length scale l_s: *(a,b)* $l_s \leq 25\delta_\nu$; *(c,d)* $25\delta_\nu < l_s \leq 50\delta_\nu$; *(e,f)* $50\delta_\nu < l_s \leq 0.5H_f$; *(g,h)* $0.5H_f < l_s$. Each p.d.f. is normalised by its global maximum and contours represent 0.1(0.1)0.9 of this maximum.

computed in analogy to the streak detection process outlined in appendix D. In the following, we then seek for connected regions for which the normalised potential ψ exceeds a given threshold ψ_{th}, viz.

$$\psi/\sigma_\psi > \psi_{th} \quad \text{and} \quad \psi/\sigma_\psi < -\psi_{th} \tag{E.18}$$

for clockwise and counterclockwise rotating motions, respectively, where the r.m.s. of ψ over the fluid dominated volume reads $\sigma_\psi = \sqrt{\langle\psi\psi\rangle_{\mathcal{V}_f}}$ using the averaging operator $\langle\bullet\rangle_{\mathcal{V}_f}$ introduced in equation (D.2).

Figure E.1 shows the p.d.f.s of the preferential location of the streamwise rolls' centre points $(y_{c,2D}, z_{c,2D})^T$ in the cross-plane of the smooth wall case $DL400^{H2}_{smooth}$ for the same threshold $\psi_{th} = 1.0$ as in Kawahara et al. (2012a). Note that the coordinates $y_{c,2D}$ and $z_{c,2D}$ are defined in analogy to equations D.3, using the local values of ψ as weights.

In analogy to appendix D, the p.d.f.s are further conditioned on the sign of the rolls' sense of rotation and their characteristic length scale l_s, based on which they are classified into four categories: (i) $l_s \leq 25\delta_v$, (ii) $25\delta_v < l_s \leq 50\delta_v$, (iii) $50\delta_v < l_s < 0.5H_f$ and (iv) $0.5H_f < l_s$.

Bibliography

H. Abe, R. A. Antonia, and S. Toh. Large-scale structures in a turbulent channel flow with a minimal streamwise flow unit. *J. Fluid Mech.*, 850:733–768, 2018. (Cited on p. 87).

R. J. Adrian. Hairpin vortex organization in wall turbulence. *Phys. Fluids*, 19(4):041301, 2007. (Cited on pp. 22, 23, 145, 200, 215, and 297).

R. J. Adrian and I. Marusic. Coherent structures in flow over hydraulic engineering surfaces. *J. Hydr. Res.*, 50(5):451–464, 2012. (Cited on pp. 2, 31, 36, 39, and 190).

R. J. Adrian, C. D. Meinhart, and C. D. Tomkins. Vortex organization in the outer region of the turbulent boundary layer. *J. Fluid Mech.*, 422:1–54, 2000. (Cited on pp. 22, 23, and 145).

I. Albayrak and U. Lemmin. Secondary currents and corresponding surface velocity patterns in a turbulent open-channel flow over a rough bed. *J. Hydr. Eng.*, 137(11):1318–1334, 2011. (Cited on pp. 187, 227, and 244).

J. R. L. Allen. The nature and origin of bed-form hierarchies. *Sedimentology*, 10(3):161–182, 1968. (Cited on p. 91).

M. Amir, V. I. Nikora, and M. T. Stewart. Pressure forces on sediment particles in turbulent open-channel flow: a laboratory study. *J. Fluid Mech.*, 757:458–497, 2014. (Cited on p. 106).

W. Anderson. Non-periodic phase-space trajectories of roughness-driven secondary flows in high-Re_τ boundary layers and channels. *J. Fluid Mech.*, 869:27–84, 2019. (Cited on p. 34).

W. Anderson, J. M. Barros, K. T. Christensen, and A. Awasthi. Numerical and experimental study of mechanisms responsible for turbulent secondary flows in boundary layer flows over spanwise heterogeneous roughness. *J. Fluid Mech.*, 768:316–347, 2015. (Cited on p. 33).

M. Atzori, R. Vinuesa, A. Lozano-Durán, and P. Schlatter. Characterization of turbulent coherent structures in square duct flow. In *J. Physics: Conference Series*, volume 1001, page 012008. IOP Publishing, 2018. doi: 10.1088/1742-6596/1001/1/012008. (Cited on p. 30).

M. Atzori, R. Vinuesa, A. Lozano-Durán, and P. Schlatter. Intense Reynolds-stress events in turbulent ducts. *Int. J. Heat Fluid Flow*, 89:108802, 2021. (Cited on pp. 30, 208, 213, 222, and 229).

D. Auerbach, P. Cvitanović, J. Eckmann, G. Gunaratne, and I. Procaccia. Exploring chaotic motion through periodic orbits. *Phys. Rev. Lett.*, 58:2387–2389, 1987. (Cited on p. 21).

S. Azimi, O. Ashtari, and T. M. Schneider. Constructing periodic orbits of high-dimensional chaotic systems by an adjoint-based variational method. *Phys. Rev. E*, 105:014217, 2022. (Cited on p. 21).

H. J. Bae and M. Lee. Life cycle of streaks in the buffer layer of wall-bounded turbulence. *Phys. Rev. Fluids*, 6:064603, 2021. (Cited on pp. 124 and 208).

F. Bagherimiyab and U. Lemmin. Large-scale coherent flow structures in rough-bed open-channel flow observed in fluctuations of three-dimensional velocity, skin friction and bed pressure. *J. Hydr. Res.*, 56(6):806–824, 2018. (Cited on pp. 37 and 114).

R. A. Bagnold. The flow of cohesionless grains in fluids. *Philos. Trans. R. Soc. A*, 249(964):235–297, 1956. (Cited on p. 88).

S. Balachandar and J. K. Eaton. Turbulent dispersed multiphase flow. *Ann. Rev. Fluid Mech.*, 42:111–133, 2010. (Cited on p. 42).

G. I. Barenblatt and A. J. Chorin. New perspectives in turbulence: Scaling laws, asymptotics, and intermittency. *SIAM review*, 40(2):265–291, 1998. (Cited on p. 15).

D. Barkley. Theoretical perspective on the route to turbulence in a pipe. *J. Fluid Mech.*, 803:P1, 2016. (Cited on p. 28).

J. M. Barros and K. T. Christensen. Observations of turbulent secondary flows in a rough-wall boundary layer. *J. Fluid Mech.*, 748:R1, 2014. (Cited on p. 34).

G. K. Batchelor and A. A. Townsend. The nature of turbulent motion at large wave-numbers. *Proc. R. Soc. Lond. A*, 199(1057):238–255, 1949. (Cited on p. 15).

C. Bauer. Direct numerical simulation of turbulent open channel flow. M.Sc. thesis, Institute for Hydromechanics, Karlsruhe Institute of Technology (KIT), 2015. (Cited on p. 48).

C. Bauer, D. Feldmann, and C. Wagner. On the convergence and scaling of high-order statistical moments in turbulent pipe flow using direct numerical simulations. *Phys. Fluids*, 29(12):125105, 2017. (Cited on p. 188).

C. Bauer, Y. Sakai, and M. Uhlmann. DNS of plane open channel flow: Scaling near the free-slip boundary and evidence of very-large-scale motions. *In preparation*, 2022. (Cited on pp. 31, 48, 87, and 110).

M. Bernardini, S. Pirozzoli, and P. Orlandi. Velocity statistics in turbulent channel flow up to $Re_\tau = 4000$. *J. Fluid Mech.*, 742:171–191, 2014. (Cited on p. 22).

S. Bertin and H. Friedrich. Effects of sand addition and bed flushing on gravel bed surface microtopography and roughness. *Water Resour. Res.*, 55(10):8076–8095, 2019. (Cited on p. 38).

H. M. Blackburn, N. N. Mansour, and B. J. Cantwell. Topology of fine-scale motions in turbulent channel flow. *J. Fluid Mech.*, 310:269–292, 1996. (Cited on p. 126).

P. Blondeaux and G. Seminara. A unified bar-bend theory of river meanders. *J. Fluid Mech.*, 157:449–470, 1985. (Cited on pp. 54 and 98).

J. Boussinesq. *Essai sur la théorie des eaux courantes*. Mémoires présentés par divers savants à l'Académie des Sciences. Paris, 1877. (Cited on p. 13).

J. P. Boyd. *Chebyshev and Fourier spectral methods*. Dover Publications (Reprint of Springer, Berlin, 1989), 2001. (Cited on pp. 61 and 62).

P. Bradshaw. Turbulent secondary flows. *Ann. Rev. Fluid Mech.*, 19(1):53–74, 1987. (Cited on p. 26).

M. Brin and G. Stuck. *Introduction to dynamical systems*. Cambridge Univ. Press, 1st edition, 2002. doi: 10.1017/CBO9780511755316. (Cited on p. 19).

E. Brundrett and W. D. Baines. The production and diffusion of vorticity in duct flow. *J. Fluid Mech.*, 19(3):375–394, 1964. (Cited on p. 26).

E. Buckingham. On physically similar systems; illustrations of the use of dimensional equations. *Phys. Rev.*, 4:345–376, 1914. (Cited on p. 84).

A. Busse and T. O. Jelly. Influence of surface anisotropy on turbulent flow over irregular roughness. *Flow, Turb. Combust.*, 104(2):331–354, 2020. (Cited on p. 132).

K. M. Butler and B. F. Farrell. Three-dimensional optimal perturbations in viscous shear flow. *Phys. Fluids A: Fluid Dynamics*, 4(8):1637–1650, 1992. (Cited on p. 68).

K. M. Butler and B. F. Farrell. Optimal perturbations and streak spacing in wall-bounded turbulent shear flow. *Phys. Fluids A: Fluid Dynamics*, 5(3):774–777, 1993. (Cited on pp. 68, 76, 79, and 240).

S. M. Cameron, V. I. Nikora, and S. E. Coleman. Double-averaged velocity and stress distributions for hydraulically-smooth and transitionally-rough turbulent flows. *Acta Geophysica*, 56(3):642–653, 2008. (Cited on p. 147).

S. M. Cameron, V. I. Nikora, and M. T. Stewart. Very-large-scale motions in rough-bed open-channel flow. *J. Fluid Mech.*, 814:416–429, 2017. (Cited on p. 31).

S. M. Cameron, V. I. Nikora, and M. J. Witz. Entrainment of sediment particles by very-large-scale motions. *J. Fluid Mech.*, 888:A7, 2020. (Cited on pp. 31, 37, and 100).

C. Camporeale, F. Cannamela, C. Canuto, and C. Manes. Stability analysis of open-channel flows with secondary currents. *J. Fluid Mech.*, 927:A32, 2021. (Cited on pp. 31, 77, and 79).

C. Canuto, M. Y. Hussaini, A. Quarteroni, and T. A. Zang. *Spectral methods: Fundamentals in single domains*. Springer, 1st edition, 2006. doi: 10.1007/978-3-540-30726-6. (Cited on pp. 49 and 62).

H. J. Casey. *Über Geschiebebewegung*, volume 19. Eigenverlag der Preussischen Versuchsanstalt für Wasserbau und Schiffbau, Berlin, 1935. (Cited on pp. 2, 35, and 158).

R. Cess. A survey of the literature on heat transfer in turbulent tube flow. Technical Report 8-0529-R24, Westinghouse Research Rep, 1958. (Cited on pp. xxviii, 70, and 71).

C. Chan-Braun. *Turbulent open channel flow, sediment erosion and sediment transport*. PhD thesis, Institute for Hydromechanics, Karlsruhe Institute of Technology (KIT), 2012. doi: 10.5445/KSP/1000029253. (Cited on p. 103).

C. Chan-Braun, M. García-Villalba, and M. Uhlmann. Force and torque acting on particles in a transitionally rough open-channel flow. *J. Fluid Mech.*, 684:441–474, 2011. (Cited on pp. 83, 99, 100, 105, 106, and 107).

G. J. Chandler and R. R. Kerswell. Invariant recurrent solutions embedded in a turbulent two-dimensional kolmogorov flow. *J. Fluid Mech.*, 722:554–595, 2013. (Cited on p. 21).

S. Chandrasekhar. *Hydrodynamic and hydromagnetic stability*. Dover Publications (Reprint of Oxford University Press, 1961), 1981. (Cited on p. 271).

F. Chiodi, P. Claudin, and B. Andreotti. A two-phase flow model of sediment transport: transition from bedload to suspended load. *J. Fluid Mech.*, 755:561–581, 2014. (Cited on pp. 91 and 97).

M. S. Chong, A. E. Perry, and B. J. Cantwell. A general classification of three-dimensional flow fields. *Phys. Fluids A: Fluid Dynamics*, 2(5):765–777, 1990. (Cited on p. 126).

M. S. Chong, J. Soria, A. E. Perry, J. Chacin, B. J. Cantwell, and Y. Na. Turbulence structures of wall-bounded shear flows found using DNS data. *J. Fluid Mech.*, 357:225–247, 1998. (Cited on pp. 126 and 127).

A. J. Chorin and J. E. Marsden. *A mathematical introduction to fluid mechanics*. Texts in applied mathematics. Springer, 3rd edition, 2000. doi: 10.1007/978-1-4612-0883-9. (Cited on p. 5).

D. Chung, J. P. Monty, and N. Hutchins. Similarity and structure of wall turbulence with lateral wall shear stress variations. *J. Fluid Mech.*, 847:591–613, 2018. (Cited on pp. 34, 66, and 78).

R. M. Clever and F. H. Busse. Tertiary and quaternary solutions for plane Couette flow. *J. Fluid Mech.*, 344:137–153, 1997. (Cited on pp. 21 and 271).

R. Clift, J. Grace, and M. Weber. *Bubbles, drops and particles*. Academic Press, 1978. (Cited on pp. 5 and 42).

S. E. Coleman and B. W. Melville. Initiation of bed forms on a flat sand bed. *J. Hydr. Eng.*, 122(6): 301–310, 1996. (Cited on p. 236).

S. E. Coleman and V. I. Nikora. Bed and flow dynamics leading to sediment-wave initiation. *Water Resour. Res.*, 45(4), 2009. (Cited on pp. 164, 237, 238, and 246).

S. E. Coleman and V. I. Nikora. Fluvial dunes: initiation, characterization, flow structure. *Earth Surface Processes and Landforms*, 36(1):39–57, 2011. (Cited on p. 90).

M. Colombini. Turbulence-driven secondary flows and formation of sand ridges. *J. Fluid Mech.*, 254: 701–719, 1993. (Cited on pp. 2, 3, 4, 36, 51, 52, 53, 54, 55, 56, 58, 61, 62, 63, 66, 67, 68, 69, 73, 76, 77, 78, 79, 82, 94, 95, 98, 106, 141, 143, 144, 146, 163, 165, 225, 239, 240, 242, 249, 250, 253, and 297).

M. Colombini. Formation of sand ridges: an instability mechanism. In J. F. A. Sleath, R. D. Rajaona, and M. Belorgey, editors, *Sediment transport mechanisms in coastal environments and rivers - Euromech*

310. World Scientific, 1994. doi: 10.1142/9789814533966. (Cited on p. 63).

M. Colombini. Revisiting the linear theory of sand dune formation. *J. Fluid Mech.*, 502:1–16, 2004. (Cited on pp. 58 and 106).

M. Colombini and G. Parker. Longitudinal streaks. *J. Fluid Mech.*, 304:161–183, 1995. (Cited on pp. 230 and 250).

M. Colombini and A. Stocchino. Ripple and dune formation in rivers. *J. Fluid Mech.*, 673:121–131, 2011. (Cited on p. 237).

M. Colombini and A. Stocchino. Three-dimensional river bed forms. *J. Fluid Mech.*, 695:63–80, 2012. (Cited on pp. 54, 63, 98, and 99).

M. Colombini, G. Seminara, and M. Tubino. Finite-amplitude alternate bars. *J. Fluid Mech.*, 181:213–232, 1987. (Cited on pp. 54, 58, and 98).

C. Cossu and Y. Hwang. Self-sustaining processes at all scales in wall-bounded turbulent shear flows. *Philos. Trans. R. Soc. A*, 375(2089):20160088, 2017. (Cited on pp. 24, 35, and 145).

R. Courant, K. Friedrichs, and H. Lewy. Über die partiellen Differenzengleichungen der mathematischen Physik. *Math. Annalen*, 100(1):32–74, 1928. (Cited on p. 45).

T. J. Craft, B. E. Launder, and K. Suga. Development and application of a cubic eddy-viscosity model of turbulence. *Int. J. Heat Fluid Flow*, 17(2):108–115, 1996. (Cited on p. 55).

C. T. Crowe, T. R. Troutt, and J. N. Chung. Numerical models for two-phase turbulent flows. *Ann. Rev. Fluid Mech.*, 28(1):11–43, 1996. (Cited on p. 43).

C. T. Crowe, M. Sommerfeld, and Y. Tsuji. *Multiphase flows with particles and droplets*. CRC Press, New York, 1998. (Cited on pp. 45 and 47).

J. K. Culbertson. Evidence of secondary circulation in an alluvial channel. *US Geological Survey Professional Paper*, pages D214–D216, 1967. (Cited on p. 35).

P. A. Cundall and O. D. L. Strack. A discrete numerical model for granular assemblies. *Géotechnique*, 29(1):47–65, 1979. (Cited on p. 45).

P. Cvitanović. Recurrent flows: the clockwork behind turbulence. *J. Fluid Mech.*, 726:1–4, 2013. (Cited on p. 21).

P. Cvitanović, R. Artuso, R. Mainieri, G. Tanner, G. Vattay, N. Whelan, and A. Wirzba. Chaos: Classical and Quantum, 2022. URL `https://chaosbook.org`. (Cited on p. 21).

H. Darcy. *Recherches expérimentales relatives au mouvement de l'eau dans les tuyaux*, volume 1. Mallet-Bachelier, Paris, 1857. (Cited on p. 13).

P. A. Davidson, Y. Kaneda, K. Moffatt, and K. R. Sreenivasan. *A voyage through turbulence*. Cambridge University Press, 1st edition, 2011. doi: 10.1017/CBO9781139018241. (Cited on p. 13).

H. J. De Vriend. A mathematical model of steady flow in curved shallow channels. *J. Hydr. Res.*, 15(1): 37–54, 1977. (Cited on pp. 58, 71, and 255).

J. C. Del Álamo and J. Jiménez. Spectra of the very large anisotropic scales in turbulent channels. *Phys. Fluids*, 15(6):L41–L44, 2003. (Cited on pp. 31, 110, 115, 147, and 215).

J. C. Del Álamo and J. Jiménez. Linear energy amplification in turbulent channels. *J. Fluid Mech.*, 559: 205–213, 2006. (Cited on pp. 68, 70, 76, 79, 144, and 240).

J. C. Del Álamo and J. Jiménez. Estimation of turbulent convection velocities and corrections to taylor's approximation. *J. Fluid Mech.*, 640:5–26, 2009. (Cited on p. 116).

J. C. Del Álamo, J. Jiménez, P. Zandonade, and R. D. Moser. Self-similar vortex clusters in the turbulent logarithmic region. *J. Fluid Mech.*, 561:329–358, 2006. (Cited on pp. 24, 25, 35, 113, 114, 126, 127, 145, 148, 151, 208, 209, 213, 218, 220, 222, 223, 228, 229, and 245).

A. O. Demuren and W. Rodi. Calculation of turbulence-driven secondary motion in non-circular ducts. *J. Fluid Mech.*, 140:189–222, 1984. (Cited on pp. 1, 27, 54, and 55).

S. Dong, A. Lozano-Durán, A. Sekimoto, and J. Jiménez. Coherent structures in statistically stationary homogeneous shear turbulence. *J. Fluid Mech.*, 816:167–208, 2017. (Cited on p. 208).

W. Dörfler. Lecture notes 'Numerical methods in fluid mechanics' (unpublished). Institute for Applied and Numerical Mathematics, Karlsruhe Institute of Technology (KIT), 2019. (Cited on p. 5).

P. G. Drazin and W. H. Reid. *Hydrodynamic stability*. Cambridge monographs on mechanics and applied mathematics. Cambridge University Press, 10th edition, 1981. doi: 10.1017/CBO9780511616938. (Cited on pp. 55, 67, and 68).

Y. Duan, Q. Chen, D. Li, and Q. Zhong. Contributions of very large-scale motions to turbulence statistics in open channel flows. *J. Fluid Mech.*, 892:A3, 2020. (Cited on p. 31).

R. Dugas. *A history of mechanics*. Dover Publications (Reprint of Editions du Griffon, Neuchâtel, Switzerland, 1955), 2012. (Cited on p. 13).

B. Eckhardt, H. Faisst, A. Schmiegel, and T. M. Schneider. Dynamical systems and the transition to turbulence in linearly stable shear flows. *Philos. Trans. R. Soc. A*, 366(1868):1297–1315, 2008. (Cited on p. 21).

H. Edelsbrunner and E. P. Mücke. Three-dimensional alpha shapes. *ACM Transactions on Graphics (TOG)*, 13(1):43–72, 1994. (Cited on p. 89).

H. A. Einstein. *The bed-load function for sediment transportation in open channel flows*, volume 1026. US Government Printing Office, 1950. (Cited on p. 58).

S. Elghobashi. On predicting particle-laden turbulent flows. *Appl. Sci. Res.*, 52(4):309–329, 1994. (Cited on p. 43).

F. Engelund. The motion of sediment particles on an inclined bed. *Progress Rep.*, 53, 1981. (Cited on pp. 54 and 98).

F. Engelund and E. Hansen. A monograph on sediment transport in alluvial streams. *Technical University of Denmark, Copenhagen*, 1967. (Cited on p. 58).

F. M. Exner. Über die Wechselwirkung zwischen Wasser und Geschiebe in Flüssen. *Akad. Wiss. Wien Math. Naturwiss. Klasse*, 134(2a):165–204, 1925. (Cited on p. 53).

E. Fadlun, R. Verzicco, P. Orlandi, and J. Mohd-Yusof. Combined immersed-boundary finite-difference methods for three-dimensional complex flow simulations. *J. Comput. Phys.*, 161(1):35–60, 2000. (Cited on p. 44).

M. Farano, S. Cherubini, J.-C. Robinet, P. De Palma, and T. M. Schneider. Computing heteroclinic orbits using adjoint-based methods. *J. Fluid Mech.*, 858:R3, 2019. (Cited on p. 21).

O. Flores and J. Jiménez. Effect of wall-boundary disturbances on turbulent channel flows. *J. Fluid Mech.*, 566:357–376, 2006. (Cited on pp. 105, 108, 146, 194, and 227).

O. Flores and J. Jiménez. Hierarchy of minimal flow units in the logarithmic layer. *Phys. Fluids*, 22(7): 071704, 2010. (Cited on pp. 25, 35, 87, 112, 121, 141, 145, 147, 188, 193, and 242).

O. Flores, J. Jiménez, and J. C. Del Álamo. Vorticity organization in the outer layer of turbulent channels with disturbed walls. *J. Fluid Mech.*, 591:145–154, 2007. (Cited on pp. 113, 145, 146, 194, 227, and 241).

A. Fourriere, P. Claudin, and B. Andreotti. Bedforms in a turbulent stream: formation of ripples by primary linear instability and of dunes by nonlinear pattern coarsening. *J. Fluid Mech.*, 649:287–328, 2010. (Cited on p. 106).

E. d. M. Franklin and F. Charru. Subaqueous barchan dunes in turbulent shear flow. Part 1. dune motion. *J. Fluid Mech.*, 675:199–222, 2011. (Cited on pp. 84 and 233).

B. Ganapathisubramani, E. K. Longmire, and I. Marusic. Characteristics of vortex packets in turbulent boundary layers. *J. Fluid Mech.*, 478:35–46, 2003. (Cited on p. 23).

S. Gavrilakis. Numerical simulation of low-Reynolds-number turbulent flow through a straight square duct. *J. Fluid Mech.*, 244:101–129, 1992. (Cited on pp. 26, 27, and 172).

S. Gavrilakis. Post-transitional periodic flow in a straight square duct. *J. Fluid Mech.*, 859:731–753, 2019. (Cited on p. 27).

F. B. Gessner. The origin of secondary flow in turbulent flow along a corner. *J. Fluid Mech.*, 58(1):1–25, 1973. (Cited on p. 26).

B. Ghidersa and J. Dušek. Breaking of axisymmetry and onset of unsteadiness in the wake of a sphere. *J. Fluid Mech.*, 423:33–69, 2000. (Cited on p. 43).

A. F. Gibson. On the depression of the filament of maximum velocity in a stream flowing through an open channel. *Proc. R. Soc. Lond. A*, 82(553):149–159, 1909. (Cited on p. 35).

R. Glowinski, T.-W. Pan, T. I. Hesla, and D. D. Joseph. A distributed Lagrange multiplier/fictitious domain method for particulate flows. *Int. J. Multiphase Flow*, 25(5):755–794, 1999. (Cited on p. 44).

D. Goldstein and T.-C. Tuan. Secondary flow induced by riblets. *J. Fluid Mech.*, 363:115–151, 1998. (Cited on pp. 34 and 139).

P. Gondret, M. Lance, and L. Petit. Bouncing motion of spherical particles in fluids. *Phys. Fluids*, 14(2): 643–652, 2002. (Cited on p. 47).

M. D. Graham and D. Floryan. Exact coherent states and the nonlinear dynamics of wall-bounded turbulent flows. *Ann. Rev. Fluid Mech.*, 53, 2021. (Cited on p. 21).

L. Grega, T. Wei, R. Leighton, and J. Neves. Turbulent mixed-boundary flow in a corner formed by a solid wall and a free surface. *J. Fluid Mech.*, 294:17–46, 1995. (Cited on pp. 31 and 182).

L. M. Grega, T. Y. Hsu, and T. Wei. Vorticity transport in a corner formed by a solid wall and a free surface. *J. Fluid Mech.*, 465:331–352, 2002. (Cited on pp. 31 and 171).

J. Guckenheimer and P. J. Holmes. *Nonlinear Oscillations, Dynamical Systems, and Bifurcations of Vector Fields*, volume 42 of *Applied Mathematical Sciences*. Springer, 1st edition, 1983. doi: 10.1007/978-1-4612-1140-2. (Cited on p. 19).

A. Gyr. Natural low-frequency riblets. *Flow, Turb. Combust.*, 60(3):235–255, 1998. (Cited on pp. 139, 142, and 149).

A. Gyr and A. Schmid. Turbulent flows over smooth erodible sand beds in flumes. *J. Hydr. Res.*, 35(4): 525–544, 1997. (Cited on pp. 37, 93, and 147).

G. H. L. Hagen. *Über den Einfluss der Temperatur auf die Bewegung des Wassers in Röhren*. Math. Abh. Akad. Wiss. Berlin, 1854. (Cited on p. 13).

P. Haldenwang, G. Labrosse, S. Abboudi, and M. Deville. Chebyshev 3-D spectral and 2-D pseudospectral solvers for the Helmholtz equation. *J. Comput. Phys.*, 55(1):115–128, 1984. (Cited on p. 49).

J. M. Hamilton, J. Kim, and F. Waleffe. Regeneration mechanisms of near-wall turbulence structures. *J. Fluid Mech.*, 287(1):317–348, 1995. (Cited on pp. 20, 34, 73, 76, 87, 145, 203, and 227).

R. Handler, T. Swean Jr, R. Leighton, and J. Swearingen. Length scales and the energy balance for turbulence near a free surface. *AIAA journal*, 31(11):1998–2007, 1993. (Cited on pp. 30 and 110).

T. Henn. *Lattice Boltzmann Methods for Particulate Flows with Medical and Technical Applications*. PhD thesis, Institute for Applied and Numerical Mathematics, Karlsruher Institut für Technologie (KIT), 2016. doi: 10.5445/IR/1000061601. (Cited on p. 5).

J. O. Hinze. *Turbulence*. McGraw-Hill book company, New York, 1959. (Cited on pp. 16 and 18).

J. O. Hinze. Secondary currents in wall turbulence. *Phys. Fluids*, 10(9):S122–S125, 1967. (Cited on p. 33).

J. O. Hinze. Experimental investigation on secondary currents in the turbulent flow through a straight conduit. *Applied Scientific Res.*, 28(1):453–465, 1973. (Cited on p. 33).

M. Hirano and T. Ohmoto. Experimental study on the interaction of between longitudinal vortices and sand ribbons. In *Proc. 6th Congr. APD-IAHR*, volume 2, pages 59–65, 1988. (Cited on pp. 140 and 141).

P. Holmes, J. L. Lumley, and G. Berkooz. *Turbulence, coherent structures, dynamical systems and symmetry*. Cambridge monographs on mechanics. Cambridge Univ. Press, 1st edition, 1996. doi: 10.1017/

CBO9780511622700. (Cited on p. 19).

E. Hopf. A mathematical example displaying features of turbulence. *Comm. Pure Appl. Math.*, 1(4): 303–322, 1948. (Cited on p. 19).

S. Hoyas and J. Jiménez. Scaling of the velocity fluctuations in turbulent channels up to $Re_{\tau} = 2003$. *Phys. Fluids*, 18(1):011702, 2006. (Cited on pp. 17, 18, 23, and 297).

S. Hoyas, M. Oberlack, F. Alcántara-Ávila, S. V. Kraheberger, and J. Laux. Wall turbulence at high friction Reynolds numbers. *Phys. Rev. Fluids*, 7:014602, 2022. (Cited on p. 22).

T.-J. Hsu. Causality between fluid motions and bathymetric features. *J. Fluid Mech.*, 936:F1, 2022. (Cited on pp. 1 and 81).

T. Y. Hsu, L. M. Grega, R. I. Leighton, and T. Wei. Turbulent kinetic energy transport in a corner formed by a solid wall and a free surface. *J. Fluid Mech.*, 410:343–366, 2000. (Cited on p. 31).

A. Huser and S. Biringen. Direct numerical simulation of turbulent flow in a square duct. *J. Fluid Mech.*, 257:65–95, 1993. (Cited on pp. 27 and 203).

N. Hutchins and I. Marusic. Evidence of very long meandering features in the logarithmic region of turbulent boundary layers. *J. Fluid Mech.*, 579:1–28, 2007a. (Cited on p. 31).

N. Hutchins and I. Marusic. Large-scale influences in near-wall turbulence. *Philos. Trans. R. Soc. A*, 365 (1852):647–664, 2007b. (Cited on pp. 115 and 196).

Y. Hwang and C. Cossu. Self-sustained process at large scales in turbulent channel flow. *Phys. Rev. Lett.*, 105:044505, 2010a. (Cited on pp. 24, 35, and 145).

Y. Hwang and C. Cossu. Linear non-normal energy amplification of harmonic and stochastic forcing in the turbulent channel flow. *J. Fluid Mech.*, 664:51–73, 2010b. (Cited on pp. 68, 69, 73, 74, 76, 77, 78, 79, and 240).

Y. Hwang and C. Cossu. Self-sustained processes in the logarithmic layer of turbulent channel flows. *Phys. Fluids*, 23(6):061702, 2011. (Cited on pp. 24, 35, 79, and 145).

S. Ikeda. Self-formed straight channels in sandy beds. *J. Hydr. Div.*, 107(4):389–406, 1981. (Cited on pp. 36, 91, 144, and 146).

S. Ikeda. Lateral bed load transport on side slopes. *J. Hydr. Div.*, 108(11):1369–1373, 1982. (Cited on pp. 54, 98, and 147).

H. Imamoto and T. Ishigaki. Visualization of longitudinal eddies in an open channel flow. In *Flow Visual. IV: Proc. 4th Int. Symp. Flow Visual.*, pages 333–337. Hemisphere Washington, DC, 1986. (Cited on p. 36).

R. Jackson. *The dynamics of fluidized particles*. Cambridge University Press, 2000. (Cited on pp. 5 and 10).

T. O. Jelly and A. Busse. Reynolds and dispersive shear stress contributions above highly skewed roughness. *J. Fluid Mech.*, 852:710–724, 2018. (Cited on p. 132).

M. Jenny and J. Dušek. Efficient numerical method for the direct numerical simulation of the flow past a single light moving spherical body in transitional regimes. *J. Comput. Phys.*, 194(1):215–232, 2004. (Cited on p. 43).

J. Jeong and F. Hussain. On the identification of a vortex. *J. Fluid Mech.*, 285:69–94, 1995. (Cited on pp. 113 and 232).

J. Jeong, F. Hussain, W. Schoppa, and J. Kim. Coherent structures near the wall in a turbulent channel flow. *J. Fluid Mech.*, 332:185–214, 1997. (Cited on pp. 20 and 34).

J. Jiménez. Coherent structures and dynamical systems. *CTR Annual Research Briefs*, pages 323–324, 1987. (Cited on p. 19).

J. Jiménez. The largest scales of turbulent wall flows. *CTR Annual Research Briefs*, pages 137–154, 1998. (Cited on p. 22).

J. Jiménez. Turbulence. In G. K. Batchelor, H. K. Moffat, and M. G. Worster, editors, *Perspectives in fluid dynamics - A collective introduction to current research*, page 231 – 288. Cambridge University Press, 2000. (Cited on p. 16).

J. Jiménez. Turbulent flows over rough walls. *Annu. Rev. Fluid Mech.*, 36:173–196, 2004a. (Cited on pp. 58, 105, 108, 109, 139, 147, 173, and 241).

J. Jiménez. Lecture notes 'Turbulence and vorticity' (unpublished). École Polytechnique, Palaiseau, France, 2004b. (Cited on p. 13).

J. Jiménez. Cascades in wall-bounded turbulence. *Ann. Rev. Fluid Mech.*, 44:27–45, 2012. (Cited on pp. 20, 206, 212, 215, 223, and 244).

J. Jiménez. Near-wall turbulence. *Phys. Fluids*, 25(10):101302, 2013a. (Cited on pp. 16, 20, 22, 23, 24, 34, 127, 130, 145, 194, 215, 225, 241, and 242).

J. Jiménez. How linear is wall-bounded turbulence? *Phys. Fluids*, 25(11):110814, 2013b. (Cited on pp. 18, 110, 112, and 113).

J. Jiménez. Coherent structures in wall-bounded turbulence. *J. Fluid Mech.*, 842:P1, 2018. (Cited on pp. 3, 16, 18, 20, 22, 24, 25, 34, 35, 67, 68, 76, 79, 81, 82, 93, 115, 126, 127, 144, 145, 147, 148, 192, 194, 197, 200, 201, 203, 204, 206, 212, 214, 223, 226, 228, 241, 242, 244, and 297).

J. Jiménez and G. Kawahara. Dynamics of wall-bounded turbulence. In P. Davidson, Y. Kaneda, and K. Sreenivasan, editors, *Ten chapters in turbulence*, pages 221–268. Cambridge University Press, 2013. (Cited on p. 20).

J. Jiménez and P. Moin. The minimal flow unit in near-wall turbulence. *J. Fluid Mech.*, 225:213–240, 1991. (Cited on pp. 20, 22, 25, 28, 87, 138, 145, 149, 157, and 242).

J. Jiménez and A. Pinelli. The autonomous cycle of near-wall turbulence. *J. Fluid Mech.*, 389:335–359, 1999. (Cited on pp. 20, 115, 142, 149, and 242).

J. Jiménez, M. Uhlmann, A. Pinelli, and G. Kawahara. Turbulent shear flow over active and passive porous surfaces. *J. Fluid Mech.*, 442:89–117, 2001. (Cited on p. 104).

J. Jiménez, J. C. Del Álamo, and O. Flores. The large-scale dynamics of near-wall turbulence. *J. Fluid Mech.*, 505:179–199, 2004. (Cited on pp. 115 and 147).

J. Jiménez, G. Kawahara, M. P. Simens, M. Nagata, and M. Shiba. Characterization of near-wall turbulence in terms of equilibrium and "bursting" solutions. *Phys. Fluids*, 17(1):015105, 2005. (Cited on p. 22).

O. C. J. Jones. An improvement in the calculation of turbulent friction in rectangular ducts. *J. Fluids Eng.*, 98(2):173–180, 1976. (Cited on pp. xxii, 174, 175, and 185).

Y. Joung and S.-U. Choi. Direct numerical simulation of low Reynolds number flows in an open-channel with sidewalls. *Int. J. Num. Meth. Fluids*, 62(8):854–874, 2010. (Cited on p. 31).

I. Karcz. Harrow marks, current-aligned sedimentary structures. *J. Geology*, 75(1):113–121, 1967. (Cited on p. 35).

G. Kawahara and E. Kamada. Vortical structures in the corner region of a turbulent square-duct flow. *Trans. Japan Soc. Mech. Eng. B*, 66:89–97, 2000. (Cited on p. 27).

G. Kawahara and S. Kida. Periodic motion embedded in plane Couette turbulence: regeneration cycle and burst. *J. Fluid Mech.*, 449:291, 2001. (Cited on p. 21).

G. Kawahara, J. Jiménez, M. Uhlmann, and A. Pinelli. Linear instability of a corrugated vortex sheet - a model for streak instability. *J. Fluid Mech.*, 483:315, 2003. (Cited on pp. 20 and 76).

G. Kawahara, T. Nakatsuji, M. Shimizu, M. Uhlmann, and A. Pinelli. Turbulence structures and secondary flow in a square duct. In *Proc. Mech. Eng. Congr. Japan*, page JO54023, 2012a. doi: 10.1299/jsmemecj.2012._J054023-1. (Cited on pp. 151, 223, 224, 229, 271, 272, and 276).

G. Kawahara, M. Uhlmann, and L. van Veen. The significance of simple invariant solutions in turbulent flows. *Ann. Rev. Fluid Mech.*, 44:203–225, 2012b. (Cited on pp. 19, 21, 29, 156, 245, and 271).

R. R. Kerswell. Recent progress in understanding the transition to turbulence in a pipe. *Nonlinearity*, 18(6):R17–R44, 2005. (Cited on pp. 21, 28, and 73).

R. R. Kerswell. Nonlinear nonmodal stability theory. *Ann. Rev. Fluid Mech.*, 50(1):319–345, 2018. (Cited on p. 67).

K. Kevin, J. P. Monty, H. Bai, G. Pathikonda, B. Nugroho, J. M. Barros, K. T. Christensen, and N. Hutchins. Cross-stream stereoscopic particle image velocimetry of a modified turbulent boundary layer over directional surface pattern. *J. Fluid Mech.*, 813:412–435, 2017. (Cited on p. 34).

K. Kevin, J. Monty, and N. Hutchins. Turbulent structures in a statistically three-dimensional boundary layer. *J. Fluid Mech.*, 859:543–565, 2019a. (Cited on pp. 34, 148, and 226).

K. Kevin, J. Monty, and N. Hutchins. The meandering behaviour of large-scale structures in turbulent boundary layers. *J. Fluid Mech.*, 865:R1, 2019b. (Cited on pp. 25, 34, 124, 125, 145, and 148).

S. Kida and H. Miura. Swirl condition in low-pressure vortices. *J. Phys. Soc. Japan*, 67(7):2166–2169, 1998. (Cited on pp. 215 and 229).

A. Kidanemariam and I. Marusic. On the turbulence-generated free-surface waves in open-channel flows. In *Proc. 22nd Austral. Fluid Mech. Conf. (AFMC2020)*. The University of Queensland, 2020. doi: 10.14264/0912f75. (Cited on p. 232).

A. G. Kidanemariam. *The formation of patterns in subaqueous sediment*. PhD thesis, Institute for Hydromechanics, Karlsruhe Institute of Technology (KIT), 2016. doi: 10.5445/KSP/1000054986. (Cited on pp. 46, 47, 88, 90, 95, 131, 236, and 259).

A. G. Kidanemariam and M. Uhlmann. Direct numerical simulation of pattern formation in subaqueous sediment. *J. Fluid Mech.*, 750:R2, 2014a. (Cited on pp. 3, 38, 83, 84, and 88).

A. G. Kidanemariam and M. Uhlmann. Interface-resolved direct numerical simulation of the erosion of a sediment bed sheared by laminar channel flow. *Int. J. Multiphase Flow*, 67:174–188, 2014b. (Cited on pp. 46 and 84).

A. G. Kidanemariam and M. Uhlmann. Formation of sediment patterns in channel flow: minimal unstable systems and their temporal evolution. *J. Fluid Mech.*, 818:716–743, 2017. (Cited on pp. 3, 38, 45, 86, 88, 90, 96, 104, 135, 148, 149, 152, 157, 164, 231, 232, 234, and 237).

A. G. Kidanemariam, C. Chan-Braun, T. Doychev, and M. Uhlmann. Direct numerical simulation of horizontal open channel flow with finite-size, heavy particles at low solid volume fraction. *New J. Phys.*, 15(2):025031, 2013. (Cited on pp. 3, 108, 117, 135, and 259).

A. G. Kidanemariam, M. Scherer, and M. Uhlmann. Open-channel flow over evolving subaqueous ripples. *J. Fluid Mech.*, 937:A26, 2022. (Cited on pp. 95, 100, 106, 131, 132, 136, 154, 155, and 231).

K. T. Kiger and C. Pan. Suspension and turbulence modification effects of solid particulates on a horizontal turbulent channel flow. *J. Turb.*, 3:N19, 2002. (Cited on p. 102).

H. Kikkawa, A. Kitagawa, and S. Ikeda. Flow and bed topography in curved open channels. *J. Hydr. Div.*, 102(9):1327–1342, 1976. (Cited on p. 99).

H. T. Kim, S. J. Kline, and W. C. Reynolds. The production of turbulence near a smooth wall in a turbulent boundary layer. *J. Fluid Mech.*, 50(1):133–160, 1971. (Cited on p. 20).

J. Kim, P. Moin, and R. Moser. Turbulence statistics in fully developed channel flow at low Reynolds number. *J. Fluid Mech.*, 177:133–166, 1987. (Cited on pp. 20, 27, 41, 48, and 82).

K. C. Kim and R. J. Adrian. Very large-scale motion in the outer layer. *Phys. Fluids*, 11(2):417–422, 1999. (Cited on pp. 23 and 31).

R. Kinoshita. An analysis of the movement of flood waters by aerial photography concerning characteristics of turbulence and surface flow. *J. Japan Soc. Photogr.*, 6(1):1–17, 1967. (Cited on pp. 35

and 110).

M. G. Kleinhans, A. W. E. Wilbers, A. De Swaaf, and J. H. Van Den Berg. Sediment supply-limited bedforms in sand-gravel bed rivers. *J. Sedimentary Res.*, 72(5):629–640, 2002. (Cited on pp. 38, 159, 160, 226, 233, and 246).

S. J. Kline, W. C. Reynolds, F. A. Schraub, and P. W. Runstadler. The structure of turbulent boundary layers. *J. Fluid Mech.*, 30(4):741–773, 1967. (Cited on pp. 20 and 195).

D. W. Knight and H. S. Patel. Boundary shear in smooth rectangular ducts. *J. Hydr. Eng.*, 111(1):29–47, 1985. (Cited on p. 26).

A. N. Kolmogorov. The local structure of turbulence in incompressible viscous fluid for very large Reynolds numbers. *Dokl. Akad. Nauk SSSR*, 30:301–305, 1941a. (Cited on p. 15).

A. N. Kolmogorov. Dissipation of energy in the locally isotropic turbulence. In *Dokl. Akad. Nauk SSSR*, volume 32, pages 16–18, 1941b. (Cited on p. 15).

Y. Kwon and J. Jiménez. An isolated logarithmic layer. *J. Fluid Mech.*, 916:A35, 2021. (Cited on pp. 145, 149, and 242).

L. D. Landau and E. M. Lifshitz. *Fluid Mechanics*, volume 6 of *Course of Theoretical Physics*. Pergamon Press, 1959. doi: https://doi.org/10.1016/C2013-0-03799-1. (Cited on pp. 5, 19, and 41).

O. E. Lanford. The strange attractor theory of turbulence. *Ann. Rev. Fluid Mech.*, 14(1):347–364, 1982. (Cited on p. 19).

V. Langlois and A. Valance. Initiation and evolution of current ripples on a flat sand bed under turbulent water flow. *Eur. Phys. J. E. Soft Matter*, 22(3):201–208, 2007. (Cited on pp. 90 and 231).

M. Lee and R. D. Moser. Direct numerical simulation of turbulent channel flow up to $Re_\tau \approx 5200$. *J. Fluid Mech.*, 774:395–415, 2015. (Cited on pp. 22 and 177).

G. Lemoult, J.-L. Aider, and J. E. Wesfreid. Turbulent spots in a channel: large-scale flow and self-sustainability. *J. Fluid Mech.*, 731:R1, 2013. (Cited on p. 28).

X. Li, S. Balachandar, H. Lee, and B. Bai. Fully resolved simulations of a stationary finite-sized particle in wall turbulence over a rough bed. *Phys. Rev. Fluids*, 4:094302, 2019. (Cited on p. 217).

A. E. Lobkovsky, A. V. Orpe, R. Molloy, A. Kudrolli, and D. H. Rothman. Erosion of a granular bed driven by laminar fluid flow. *J. Fluid Mech.*, 605:47–58, 2008. (Cited on pp. 88 and 91).

E. N. Lorenz. Deterministic nonperiodic flow. *J. Atmos. Sc.*, 20(2):130 – 141, 1963. (Cited on p. 19).

A. Lozano-Durán. *Time-resolved evolution of coherent structures in turbulent channels*. PhD thesis, Universidad Politécnica de Madrid, 2015. doi: 10.20868/UPM.thesis.36534. (Cited on p. 23).

A. Lozano-Durán and J. Jiménez. Effect of the computational domain on direct simulations of turbulent channels up to $Re_\tau = 4200$. *Phys. Fluids*, 26(1):011702, 2014a. (Cited on p. 22).

A. Lozano-Durán and J. Jiménez. Time-resolved evolution of coherent structures in turbulent channels: characterization of eddies and cascades. *J. Fluid Mech.*, 759:432–471, 2014b. (Cited on pp. 24, 25, and 245).

A. Lozano-Durán, O. Flores, and J. Jiménez. The three-dimensional structure of momentum transfer in turbulent channels. *J. Fluid Mech.*, 694:100–130, 2012. (Cited on pp. 23, 24, 30, 34, 92, 113, 126, 130, 145, 148, 208, 209, 212, 213, 215, 222, 226, 242, 245, and 297).

A. Lozano-Durán, M. Holzner, and J. Jiménez. Multiscale analysis of the topological invariants in the logarithmic region of turbulent channels at a friction Reynolds number of 932. *J. Fluid Mech.*, 803: 356–394, 2016. (Cited on pp. 114 and 197).

A. Lozano-Durán, H. J. Bae, and M. P. Encinar. Causality of energy-containing eddies in wall turbulence. *J. Fluid Mech.*, 882:A2, 2020. (Cited on p. 145).

D. Lucas and T. Yasuda. Stabilization of exact coherent structures in two-dimensional turbulence using time-delayed feedback. *Phys. Rev. Fluids*, 7:014401, 2022. (Cited on p. 21).

A. Majda and A. L. Bertozzi. *Vorticity and incompressible flow*. Cambridge texts in applied mathematics. Cambridge University Press, 1st edition, 2002. doi: 10.1017/CBO9780511613203. (Cited on pp. 5 and 9).

W. V. R. Malkus. Outline of a theory of turbulent shear flow. *J. Fluid Mech.*, 1(5):521–539, 1956. (Cited on p. 68).

F. Marqués. On boundary conditions for velocity potentials in confined flows: Application to Couette flow. *Phys. Fluids A: Fluid Dynamics*, 2(5):729–737, 1990. (Cited on pp. 9 and 271).

I. Marusic and J. P. Monty. Attached eddy model of wall turbulence. *Ann. Rev. Fluid Mech.*, 51:49–74, 2019. (Cited on pp. 23 and 24).

I. Marusic, R. Mathis, and N. Hutchins. Predictive model for wall-bounded turbulent flow. *Science*, 329 (5988):193–196, 2010a. (Cited on p. 20).

I. Marusic, B. J. McKeon, P. A. Monkewitz, H. M. Nagib, A. J. Smits, and K. R. Sreenivasan. Wall-bounded turbulent flows at high Reynolds numbers: recent advances and key issues. *Phys. Fluids*, 22(6):065103, 2010b. (Cited on pp. 16, 20, and 81).

M. Maxey. Simulation methods for particulate flows and concentrated suspensions. *Ann. Rev. Fluid Mech.*, 49(1):171–193, 2017. (Cited on pp. 42, 43, and 44).

M. R. Maxey and J. J. Riley. Equation of motion for a small rigid sphere in a nonuniform flow. *Phys. Fluids*, 26(4):883–889, 1983. (Cited on p. 42).

M. Mazzuoli and M. Uhlmann. Direct numerical simulation of open-channel flow over a fully rough wall at moderate relative submergence. *J. Fluid Mech.*, 824:722–765, 2017. (Cited on pp. 106, 108, 109, 227, and 244).

M. Mazzuoli, A. G. Kidanemariam, and M. Uhlmann. Direct numerical simulations of ripples in an oscillatory flow. *J. Fluid Mech.*, 863:572–600, 2019. (Cited on p. 97).

W. D. McComb. *The physics of fluid turbulence*. Oxford University Press, 1990. (Cited on p. 16).

B. J. McKeon. The engine behind (wall) turbulence: perspectives on scale interactions. *J. Fluid Mech.*, 817:P1, 2017. (Cited on p. 67).

S. J. McLelland, P. J. Ashworth, J. L. Best, and J. R. Livesey. Turbulence and secondary flow over sediment stripes in weakly bimodal bed material. *J. Hydr. Eng.*, 125(5):463–473, 1999. (Cited on pp. 34, 93, 140, 141, 144, 146, 158, 160, 165, 168, 225, 230, and 241).

T. Medjnoun, C. Vanderwel, and B. Ganapathisubramani. Effects of heterogeneous surface geometry on secondary flows in turbulent boundary layers. *J. Fluid Mech.*, 886:A31, 2020. (Cited on p. 34).

R. Mejia-Alvarez and K. T. Christensen. Low-order representations of irregular surface roughness and their impact on a turbulent boundary layer. *Phys. Fluids*, 22(1):015106, 2010. (Cited on p. 34).

A. Melling and J. H. Whitelaw. Turbulent flow in a rectangular duct. *J. Fluid Mech.*, 78(2):289–315, 1976. (Cited on p. 26).

E. Meyer-Peter and R. Müller. Formulas for bed-load transport. In *IAHSR 2nd meeting, Stockholm, appendix 2*, pages 39–64. IAHR, 1948. (Cited on pp. xix, 54, 95, 96, 98, 165, 166, and 251).

R. Mittal and G. Iaccarino. Immersed boundary methods. *Ann. Rev. Fluid Mech.*, 37(1):239–261, 2005. (Cited on p. 44).

Y. Mizuno and J. Jiménez. Wall turbulence without walls. *J. Fluid Mech.*, 723:429–455, 2013. (Cited on p. 145).

D. Modesti, S. Pirozzoli, P. Orlandi, and F. Grasso. On the role of secondary motions in turbulent square duct flow. *J. Fluid Mech.*, 847:R1, 2018. (Cited on pp. 27 and 229).

J. Mohd-Yusof. Combined immersed-boundary/B-spline methods for simulations of flow in complex geometries. *CTR Annual Research Briefs*, pages 317–327, 1997. (Cited on p. 44).

F. Moisy and J. Jiménez. Geometry and clustering of intense structures in isotropic turbulence. *J. Fluid Mech.*, 513:111–133, 2004. (Cited on pp. 208, 228, 244, and 266).

M. Möller. Studien über die Bewegung des Wassers in Flüssen mit Bezugname auf die Ausbildung des Flussprofiles, 1883. (Cited on p. 35).

J. Monty, N. Hutchins, H. Ng, I. Marusic, and M. Chong. A comparison of turbulent pipe, channel and boundary layer flows. *J. Fluid Mech.*, 632:431–442, 2009. (Cited on p. 31).

M. Moriche, M. Uhlmann, and J. Dušek. A single oblate spheroid settling in unbounded ambient fluid: a benchmark for simulations in steady and unsteady wake regimes. *Int. J. Multiphase Flow*, 1346: 103519, 2021. (Cited on p. 45).

Y. Motoori and S. Goto. Generation mechanism of a hierarchy of vortices in a turbulent boundary layer. *J. Fluid Mech.*, 865:1085–1109, 2019. (Cited on pp. 25 and 113).

Y. Motoori and S. Goto. Hierarchy of coherent structures and real-space energy transfer in turbulent channel flow. *J. Fluid Mech.*, 911:A27, 2021. (Cited on pp. 25, 113, 133, and 145).

T. Mullin. Experimental studies of transition to turbulence in a pipe. *Ann. Rev. Fluid Mech.*, 43:1–24, 2011. (Cited on p. 28).

R. Nagaosa. Direct numerical simulation of vortex structures and turbulent scalar transfer across a free surface in a fully developed turbulence. *Phys. Fluids*, 11(6):1581–1595, 1999. (Cited on pp. 30, 214, 219, and 245).

R. Nagaosa and R. A. Handler. Statistical analysis of coherent vortices near a free surface in a fully developed turbulence. *Phys. Fluids*, 15(2):375–394, 2003. (Cited on p. 31).

R. Nagaosa and R. A. Handler. Characteristic time scales for predicting the scalar flux at a free surface in turbulent open-channel flows. *AIChE Journal*, 58(12):3867–3877, 2012. (Cited on p. 31).

M. Nagata. Three-dimensional finite-amplitude solutions in plane Couette flow: bifurcation from infinity. *J. Fluid Mech.*, 217:519–527, 1990. (Cited on pp. 21, 242, and 271).

T. Nakatsuji. Large-scale structures and secondary flow in turbulent square-duct flow. M.Sc. thesis, Osaka University, 2012. (Cited on pp. 209, 213, 223, 224, 228, 244, 265, 266, 271, and 272).

D. Naot and W. Rodi. Calculation of secondary currents in channel flow. *J. Hydr. Div.*, 108(8):948–968, 1982. (Cited on pp. 27 and 55).

I. Nezu. Open-channel flow turbulence and its research prospect in the 21st century. *J. Hydr. Eng.*, 131 (4):229–246, 2005. (Cited on pp. 2, 33, 36, 37, 126, 179, 190, 226, and 243).

I. Nezu and H. Nakagawa. Cellular secondary currents in straight conduit. *J. Hydr. Eng.*, 110(2):173–193, 1984. (Cited on pp. 2, 35, 39, and 55).

I. Nezu and H. Nakagawa. Self forming mechanism of longitudinal sand ridges and troughs in fluvial open-channel flows. In *Proc. 23rd Congr. IAHR*, pages B65–B72. IAHR Madrid, 1989. (Cited on pp. 36, 158, 162, and 225).

I. Nezu and H. Nakagawa. *Turbulence in open channels*. IAHR/AIRH Monograph, A. A. Balkema, 1993. doi: 10.1201/9780203734902. (Cited on pp. 1, 2, 26, 30, 32, 33, 36, 51, 81, 110, 128, 144, 158, 171, 177, 182, 187, 190, 226, 227, 232, and 244).

I. Nezu and W. Rodi. Experimental study on secondary currents in open channel flow. In *Proc. 21st Congr. IAHR*, pages 115–119. IAHR Melbourne, 1985. (Cited on pp. 32, 33, 154, 158, 173, and 193).

I. Nezu and W. Rodi. Open channel flow measurements with a laser doppler anemometer. *J. Hydr. Eng.*, 112(5):335–355, 1986. (Cited on pp. 37 and 71).

I. Nezu, H. Nakagawa, and N. Kawashima. Cellular secondary currents and sand ribbons in fluvial channel flows. In *Proc. 6th Congr. APD-IAHR*, volume 2, pages 51–58, 1988. (Cited on pp. 140 and 141).

Y. Niño and M. García. Gravel saltation: 2. modeling. *Water Resour. Res.*, 30(6):1915–1924, 1994. (Cited on p. 100).

Y. Niño, M. García, and L. Ayala. Gravel saltation: 1. experiments. *Water Resour. Res.*, 30(6):1907–1914, 1994. (Cited on pp. 100 and 167).

V. Nikora and A. G. Roy. Secondary flows in rivers: Theoretical framework, recent advances, and current challenges. In M. Church, P. M. Biron, and A. G. Roy, editors, *Gravel-bed rivers: Processes, tools, environments*, pages 3–22. John Wiley & Sons, USA, 2012. (Cited on pp. 36 and 37).

V. Nikora, D. Goring, I. McEwan, and G. Griffiths. Spatially averaged open-channel flow over rough bed. *J. Hydr. Eng.*, 127(2):123–133, 2001. (Cited on p. 131).

V. I. Nikora, I. McEwan, S. McLean, S. E. Coleman, D. Pokrajac, and R. Walters. Double-averaging concept for rough-bed open-channel and overland flows: Theoretical background. *J. Hydr. Eng.*, 133 (8):873–883, 2007. (Cited on pp. 83 and 130).

J. Nikuradse. *Untersuchung über die Geschwindigkeitsverteilung in turbulenten Strömungen.* PhD thesis, Georg-August-Universität Göttingen, 1926. (Cited on p. 26).

A. M. Obukhov. On the distribution of energy in the spectrum of turbulent flow. *Bull. Acad. Sci. USSR, Geog. Geophys.*, 5:453–466, 1941. (Cited on pp. 15 and 18).

S. Okino and M. Nagata. Asymmetric travelling waves in a square duct. *J. Fluid Mech.*, 693:57–68, 2012. (Cited on p. 29).

S. Okino, M. Nagata, H. Wedin, and A. Bottaro. A new nonlinear vortex state in square-duct flow. *J. Fluid Mech.*, 657:413–429, 2010. (Cited on p. 29).

J. G. Oldroyd and A. H. Wilson. On the formulation of rheological equations of state. *Proc. R. Soc. Lond. A*, 200(1063):523–541, 1950. (Cited on p. 55).

K. W. Olesen. Alternate bars in and meandering of alluvial rivers. *Commun. on Hydr., 1983-01*, 1983. (Cited on pp. 54 and 98).

K. Onitsuka and I. Nezu. Generation mechanism of turbulence-driven secondary currents in open-channel flows. In T. Kambe, T. Nakano, and T. Miyauchi, editors, *IUTAM Symp. Geom. Stat. Turbulence*, pages 345–350. Springer, 2001. doi: 10.1007/978-94-015-9638-1_45. (Cited on pp. 36, 37, and 190).

P. Orlandi. Vortex dipole rebound from a wall. *Phys. Fluids A: Fluid Dyn.*, 2(8):1429–1436, 1990. (Cited on pp. 32, 182, 183, and 217).

S. A. Orszag. Numerical simulation of incompressible flows within simple boundaries. I. Galerkin (spectral) representations. *Stud. Appl. Math.*, 50(4):293–327, 1971a. (Cited on p. 41).

S. A. Orszag. On the elimination of aliasing in finite-difference schemes by filtering high-wavenumber components. *J. Atmos. Sc.*, 28(6):1074–1074, 1971b. (Cited on p. 49).

M. Ouriemi, P. Aussillous, and É. Guazzelli. Sediment dynamics. Part 2. dune formation in pipe flow. *J. Fluid Mech.*, 636:321–336, 2009. (Cited on p. 231).

J. Page and R. R. Kerswell. Koopman analysis of Burgers equation. *Phys. Rev. Fluids*, 3:071901, 2018. (Cited on p. 21).

J. Page and R. R. Kerswell. Koopman mode expansions between simple invariant solutions. *J. Fluid Mech.*, 879:1–27, 2019. (Cited on p. 21).

J. Page and R. R. Kerswell. Searching turbulence for periodic orbits with dynamic mode decomposition. *J. Fluid Mech.*, 886:A28, 2020. (Cited on p. 21).

J. Page, M. P. Brenner, and R. R. Kerswell. Revealing the state space of turbulence using machine learning. *Phys. Rev. Fluids*, 6:034402, 2021. (Cited on p. 21).

G. Parker. Discussion of "lateral bed load transport on side slopes" by S. Ikeda (November, 1982). *J. Hydr. Eng.*, 110(2):197–199, 1984. (Cited on pp. 54, 98, 99, and 147).

G. Parker and E. D. Andrews. Sorting of bed load sediment by flow in meander bends. *Water Resour. Res.*, 21(9):1361–1373, 1985. (Cited on pp. 99 and 147).

G. Parker, G. Seminara, and L. Solari. Bed load at low shields stress on arbitrarily sloping beds: Alternative entrainment formulation. *Water Resour. Res.*, 39(7), 2003. (Cited on p. 99).

J. P. Parker and T. M. Schneider. Variational methods for finding periodic orbits in the incompressible Navier–Stokes equations. *J. Fluid Mech.*, 941:A17, 2022. (Cited on p. 21).

A. E. Perry and M. S. Chong. On the mechanism of wall turbulence. *J. Fluid Mech.*, 119:173–217, 1982. (Cited on pp. 23 and 297).

C. Peruzzi, D. Poggi, L. Ridolfi, and C. Manes. On the scaling of large-scale structures in smooth-bed turbulent open-channel flows. *J. Fluid Mech.*, 889:A1, 2020. (Cited on p. 31).

C. S. Peskin. Flow patterns around heart valves: a numerical method. *J. Comput. Phys.*, 10(2):252–271, 1972. (Cited on p. 44).

C. S. Peskin. The immersed boundary method. *Acta numerica*, 11:479–517, 2002. (Cited on p. 44).

T. Pestana, M. Uhlmann, and G. Kawahara. Can preferential concentration of finite-size particles in plane Couette turbulence be reproduced with the aid of equilibrium solutions? *Phys. Rev. Fluids*, 5: 034305, 2020. (Cited on p. 242).

A. Pinelli, M. Uhlmann, A. Sekimoto, and G. Kawahara. Reynolds number dependence of mean flow structure in square duct turbulence. *J. Fluid Mech.*, 644:107–122, 2010. (Cited on pp. 28, 29, 49, 50, 130, 148, 152, 157, 172, 178, 179, 183, 185, 200, 201, 211, 215, 218, 228, and 229).

M. Pinelli, H. Herlina, J. Wissink, and M. Uhlmann. Direct numerical simulation of turbulent mass transfer at the surface of an open channel flow. *J. Fluid Mech.*, 933:A49, 2022. (Cited on pp. 31, 219, and 245).

S. Pirozzoli. Revisiting the mixing-length hypothesis in the outer part of turbulent wall layers: mean flow and wall friction. *J. Fluid Mech.*, 745:378–397, 2014. (Cited on pp. 71 and 72).

S. Pirozzoli, D. Modesti, P. Orlandi, and F. Grasso. Turbulence and secondary motions in square duct flow. *J. Fluid Mech.*, 840:631–655, 2018. (Cited on pp. 27, 29, 172, and 245).

S. B. Pope. *Turbulent Flows*. Cambridge University Press, 5th edition, 2000. doi: 10.1017/CBO9780511840531. (Cited on pp. 5, 7, 14, 15, 16, 18, 41, 42, 55, 58, 69, 71, 105, 106, 108, 153, and 204).

L. Prandtl. Über Flussigkeitsbewegung bei sehr kleiner Reibung. *Verhandl. III, Internat. Math.-Kong., Heidelberg, Teubner, 1904*, pages 484–491, 1904. (Cited on p. 16).

L. Prandtl. Bericht über die Entstehung der Turbulenz. *Z. Angew. Math. Mech.*, pages 136–139, 1925. (Cited on p. 18).

L. Prandtl. Über die ausgebildete Turbulenz, Verhandlungen des II.Internationalen Kongresses für Technische Mechanik 1926. In W. Tollmien, H. Schlichting, and H. Görtler, editors, *Ludwig Prandtl - Gesammelte Abhandlungen zur angewandten Mechanik, Hydro- und Aerodynamik: in 3 Teilen, (published 1961)*, pages 736–751. Springer, 1926. (Cited on p. 26).

G. Pujals, M. García-Villalba, C. Cossu, and S. Depardon. A note on optimal transient growth in turbulent channel flows. *Phys. Fluids*, 21(1):015109, 2009. (Cited on pp. 68, 76, 79, 144, and 240).

M. R. Raupach and R. H. Shaw. Averaging procedures for flow within vegetation canopies. *Boundary-Layer Meteorology*, 22(1):79–90, 1982. (Cited on p. 135).

M. R. Raupach, R. A. Antonia, and S. Rajagopalan. Rough-wall turbulent boundary layers. *Appl. Mech. Rev.*, 44(1):1–25, 1991. (Cited on p. 105).

S. C. Reddy, P. J. Schmid, and D. S. Henningson. Pseudospectra of the Orr–Sommerfeld operator. *SIAM J. Appl. Math.*, 53(1):15–47, 1993. (Cited on p. 68).

O. Reynolds. An experimental investigation of the circumstances which determine whether the motion of water shall be direct or sinuous, and of the law of resistance in parallel channels. *Philos. Trans. R. Soc. Lond.*, 174:935–982, 1883. (Cited on p. 14).

O. Reynolds. On the dynamical theory of incompressible viscous fluids and the determination of the criterion. *Proc. R. Soc. Lond.*, 56(336-339):40–45, 1894. (Cited on p. 14).

W. C. Reynolds and W. G. Tiederman. Stability of turbulent channel flow, with application to Malkus's theory. *J. Fluid Mech.*, 27(2):253–272, 1967. (Cited on pp. 68, 70, 71, 79, and 144).

L. F. Richardson. *Weather prediction by numerical process*. Cambridge University Press, 1922. doi: 10.1017/CBO9780511618291. (Cited on pp. 15 and 18).

S. K. Robinson. Coherent motions in the turbulent boundary layer. *Ann. Rev. Fluid Mech.*, 23(1):601–639, 1991. (Cited on pp. 20, 23, and 200).

J. F. Rodríguez and M. H. García. Laboratory measurements of 3-D flow patterns and turbulence in straight open channel with rough bed. *J. Hydr. Res.*, 46(4):454–465, 2008. (Cited on pp. 33, 36, 158, 187, 227, and 244).

D. Ruelle and F. Takens. On the nature of turbulence. *Comm. Math. Phys.*, 20(3):167–192, 1971. (Cited on p. 19).

Y. Sakai. *Coherent structures and secondary motions in open duct flows*. PhD thesis, Institute for Hydromechanics, Karlsruhe Institute of Technology (KIT), 2016. doi: 10.5445/KSP/1000082066. (Cited on pp. 25, 29, 31, 32, 33, 49, 50, 129, 152, 154, 157, 158, 172, 173, 174, 175, 178, 179, 182, 183, 185, 188, 189, 191, 192, 193, 201, 209, 210, 211, 213, 215, 217, 218, 219, 223, 224, 225, 228, 229, 243, 244, 245, 261, 262, 265, and 266).

M. Scherer, A. G. Kidanemariam, and M. Uhlmann. On the scaling of the instability of a flat sediment bed with respect to ripple-like patterns. *J. Fluid Mech.*, 900:A1, 2020. (Cited on pp. 38, 47, 86, 88, 136, 141, 142, 148, 155, 157, 164, 233, 236, 237, 238, and 246).

M. Scherer, M. Uhlmann, A. G. Kidanemariam, and M. Krayer. On the role of turbulent large-scale streaks in generating sediment ridges. *J. Fluid Mech.*, 930:A11, 2022. (Cited on p. 45).

P. J. Schmid. Nonmodal stability theory. *Ann. Rev. Fluid Mech.*, 39(1):129–162, 2007. (Cited on pp. 55, 56, 67, and 68).

P. J. Schmid and D. S. Henningson. *Stability and transition in shear flows*, volume 142 of *Applied Mathematical Sciences*. Springer, 2001. doi: 10.1007/978-1-4613-0185-1. (Cited on p. 67).

T. M. Schneider and B. Eckhardt. Edge of chaos in pipe flow. *Chaos*, 16(4):041103, 2006. (Cited on p. 29).

W. Schoppa and F. Hussain. Coherent structure generation in near-wall turbulence. *J. Fluid Mech.*, 453: 57–108, 2002. (Cited on pp. 20, 34, 76, 124, and 145).

A. Sekimoto. *Coherent Structures and Secondary Flow in Turbulent Square-Duct Flow.* PhD thesis, Osaka University, 2011. URL `https://hdl.handle.net/11094/58272`. (Cited on pp. 28, 29, 49, 174, and 214).

M. Sekine and H. Kikkawa. Mechanics of saltating grains. II. *J. Hydr. Eng.*, 118(4):536–558, 1992. (Cited on p. 100).

M. Sekine and G. Parker. Bed load transport on transverse slope. I. *J. Hydr. Eng.*, 118(4):513–535, 1992. (Cited on pp. 98 and 100).

G. Seminara. Fluvial sedimentary patterns. *Ann. Rev. Fluid Mech.*, 42:43–66, 2010. (Cited on pp. 53 and 98).

G. Seminara, L. Solari, and G. Parker. Bed load at low shields stress on arbitrarily sloping beds: Failure of the bagnold hypothesis. *Water Resour. Res.*, 38(11):31–1, 2002. (Cited on p. 100).

A. Shields. *Anwendung der Ähnlichkeitsmechanik und der Turbulenzforschung auf die Geschiebebewegung*. PhD thesis, Preußische Versuchsanstalt für Wasserbau und Schiffsbau, Technische Hochschule Berlin, 1936. (Cited on p. 84).

A. B. Shvidchenko and G. Pender. Macroturbulent structure of open-channel flow over gravel beds. *Water Resour. Res.*, 37(3):709–719, 2001. (Cited on pp. 2, 36, 37, and 297).

J. A. Sillero. *High Reynolds numbers turbulent boundary layers*. PhD thesis, Universidad Politécnica de Madrid, 2014. doi: 10.20868/UPM.thesis.32158. (Cited on pp. 23, 116, 203, 206, 208, 209, 211, 212, 213, 214, 228, and 244).

J. A. Sillero, J. Jiménez, and R. D. Moser. Two-point statistics for turbulent boundary layers and channels at Reynolds numbers up to $\delta^+ \approx 2000$. *Phys. Fluids*, 26(10):105109, 2014. (Cited on pp. 206 and 244).

A. J. Smits, B. J. McKeon, and I. Marusic. High-Reynolds number wall turbulence. *Ann. Rev. Fluid Mech.*, 43:353–375, 2011. (Cited on pp. 3, 16, 20, 24, 35, 81, 177, 193, 215, and 241).

R. L. Soulsby, R. J. S. Whitehouse, et al. Threshold of sediment motion in coastal environments. In *Pacific Coasts and Ports' 97: Proc. 13th Austral. Coastal and Ocean Eng. Conf. and the 6th Austral. Port and Harbour Conf.; Volume 1*, page 145. Centre for Advanced Engineering, University of Canterbury, 1997. (Cited on pp. 84, 96, and 166).

P. R. Spalart. Strategies for turbulence modelling and simulations. *Int. J. Heat Fluid Flow*, 21(3):252–263, 2000. (Cited on p. 67).

C. G. Speziale. On turbulent secondary flows in pipes of noncircular cross-section. *Int. J. Eng. Science*, 20(7):863–872, 1982. (Cited on pp. 26, 27, and 54).

C. G. Speziale. On nonlinear k-l and k-ε models of turbulence. *J. Fluid Mech.*, 178:459–475, 1987. (Cited on pp. xxiii, 27, 55, 78, and 249).

F. Stearns. On the current-meter, together with a reason why the maximum velocity of water flowing in open channels is below the surface. *Trans. Am. Soc. Civil Eng.*, 12(1):301–338, 1883. (Cited on pp. 32, 35, and 171).

A. Stroh, Y. Hasegawa, J. Kriegseis, and B. Frohnapfel. Secondary vortices over surfaces with spanwise varying drag. *J. Turb.*, 17(12):1142–1158, 2016. (Cited on p. 34).

A. Stroh, K. Schäfer, P. Forooghi, and B. Frohnapfel. Secondary flow and heat transfer in turbulent flow over streamwise ridges. *Int. J. Heat Fluid Flow*, 81:108518, 2020a. (Cited on p. 34).

A. Stroh, K. Schäfer, B. Frohnapfel, and P. Forooghi. Rearrangement of secondary flow over spanwise heterogeneous roughness. *J. Fluid Mech.*, 885:R5, 2020b. (Cited on p. 34).

K. Takeishi, G. Kawahara, H. Wakabayashi, M. Uhlmann, and A. Pinelli. Localized turbulence structures in transitional rectangular-duct flow. *J. Fluid Mech.*, 782:368–379, 2015. (Cited on p. 28).

A. Tamburrino and J. S. Gulliver. Large flow structures in a turbulent open channel flow. *J. Hydr. Res.*, 37(3):363–380, 1999. (Cited on pp. 2, 33, and 36).

A. Tamburrino and J. S. Gulliver. Free-surface visualization of streamwise vortices in a channel flow. *Water Resour. Res.*, 43(11):W11410, 2007. (Cited on pp. 2 and 36).

T. Tatsumi and T. Yoshimura. Stability of the laminar flow in a rectangular duct. *J. Fluid Mech.*, 212: 437–449, 1990. (Cited on p. 27).

H. Tennekes and J. L. Lumley. *A First Course in Turbulence*. MIT Press, 1972. doi: 10.7551/mitpress/3014.001.0001. (Cited on pp. 13 and 16).

T. Theodorsen. Mechanisms of turbulence. In *Proc. 2nd Midwestern Conf. Fluid Mech., 1952*, 1952. (Cited on pp. 22, 23, and 297).

V. Theofilis, P. W. Duck, and J. Owen. Viscous linear stability analysis of rectangular duct and cavity flows. *J. Fluid Mech.*, 505:249–286, 2004. (Cited on p. 27).

S. Toh and T. Itano. Interaction between a large-scale structure and near-wall structures in channel flow. *J. Fluid Mech.*, 524:249–262, 2005. (Cited on pp. 20, 87, and 115).

A. Tominaga, I. Nezu, K. Ezaki, and H. Nakagawa. Three-dimensional turbulent structure in straight open channel flows. *J. Hydr. Res.*, 27(1):149–173, 1989. (Cited on pp. 32, 33, 172, and 189).

A. A. Townsend. Equilibrium layers and wall turbulence. *J. Fluid Mech.*, 11(1):97–120, 1961. (Cited on pp. 24, 25, 206, and 212).

A. A. Townsend. *The structure of turbulent shear flow*. Cambridge Univ. Press, Cambridge, 2nd edition, 1976. (Cited on pp. 16, 18, 24, 25, 109, 116, 204, 212, 241, and 244).

S. Türk, G. Daschiel, A. Stroh, Y. Hasegawa, and B. Frohnapfel. Turbulent flow over superhydrophobic surfaces with streamwise grooves. *J. Fluid Mech.*, 747:186–217, 2014. (Cited on p. 34).

M. Uhlmann. The need for de-aliasing in a Chebyshev pseudo-spectral method. Technical Note No. 60, Potsdam Institute for Climate Impact Research, Potsdam, Germany, 2000. URL `https://www-turbul.ifh.uni-karlsruhe.de/uhlmann/reports/dealias.pdf`. (Cited on p. 49).

M. Uhlmann. An immersed boundary method with direct forcing for the simulation of particulate flows. *J. Comput. Phys.*, 209(2):448–476, 2005. (Cited on pp. 5 and 44).

M. Uhlmann. Interface-resolved direct numerical simulation of vertical particulate channel flow in the turbulent regime. *Phys. Fluids*, 20(5):053305, 2008. (Cited on pp. 100, 131, and 135).

M. Uhlmann. Lecture notes 'Turbulent flows' (unpublished). Institute for Hydromechanics, Karlsruhe Institute of Technology (KIT), 2014. (Cited on p. 5).

M. Uhlmann and A. Chouippe. Clustering and preferential concentration of finite-size particles in forced homogeneous-isotropic turbulence. *J. Fluid Mech.*, 812:991–1023, 2017. (Cited on p. 45).

M. Uhlmann and T. Doychev. Sedimentation of a dilute suspension of rigid spheres at intermediate Galileo numbers: the effect of clustering upon the particle motion. *J. Fluid Mech.*, 752:310–348, 2014. (Cited on p. 45).

M. Uhlmann and J. Dušek. The motion of a single heavy sphere in ambient fluid: a benchmark for interface-resolved particulate flow simulations with significant relative velocities. *Int. J. Multiphase Flow*, 59:221–243, 2014. (Cited on p. 45).

M. Uhlmann and M. Nagata. Linear stability analysis of flow in an internally heated rectangular duct. *J. Fluid Mech.*, 551:387–404, 2006. (Cited on pp. 27 and 29).

M. Uhlmann, A. Pinelli, G. Kawahara, and A. Sekimoto. Marginally turbulent flow in a square duct. *J. Fluid Mech.*, 588:153–162, 2007. (Cited on pp. 28, 29, 49, 152, 174, 200, 215, 229, and 265).

M. Uhlmann, G. Kawahara, and A. Pinelli. Travelling-waves consistent with turbulence-driven secondary flow in a square duct. *Phys. Fluids*, 22(8):084102, 2010. (Cited on pp. 29, 73, and 297).

L. C. van Rijn. Sediment transport, Part I: bed load transport. *J. Hydr. Eng.*, 110(10):1431–1456, 1984. (Cited on pp. 88 and 97).

L. van Veen. A brief history of simple invariant solutions in turbulence. In A. Gelfgat, editor, *Computational Modelling of Bifurcations and Instabilities in Fluid Dynamics*, pages 217–231. Springer International Publishing, 2019. doi: 10.1007/978-3-319-91494-7_7. (Cited on p. 21).

C. Vanderwel and B. Ganapathisubramani. Effects of spanwise spacing on large-scale secondary flows in rough-wall turbulent boundary layers. *J. Fluid Mech.*, 774:R2, 2015. (Cited on p. 34).

C. Vanderwel, A. Stroh, J. Kriegseis, B. Frohnapfel, and B. Ganapathisubramani. The instantaneous structure of secondary flows in turbulent boundary layers. *J. Fluid Mech.*, 862:845–870, 2019. (Cited on p. 34).

V. A. Vanoni. Transportation of sediment in suspension. *Trans. Am. Soc. Civil Eng*, 101:67–133, 1946. (Cited on pp. 2, 35, 38, 158, and 159).

J. G. Venditti, R. J. Hardy, M. Church, and J. L. Best. *What is a Coherent Flow Structure in Geophysical Flow?*, chapter 1, pages 1–16. John Wiley & Sons, Ltd, 2013. doi: 10.1002/9781118527221.ch1. (Cited on p. 36).

J. G. Venditti, J. A. Nittrouer, M. A. Allison, R. P. Humphries, and M. Church. Supply-limited bedform patterns and scaling downstream of a gravel-sand transition. *Sedimentology*, 66(6):2538–2556, 2019. (Cited on pp. 38 and 160).

R. Verzicco and P. Orlandi. A finite-difference scheme for three-dimensional incompressible flows in cylindrical coordinates. *J. Comput. Phys.*, 123(2):402–414, 1996. (Cited on p. 45).

R. Vinuesa, E. Bartrons, D. Chiu, K. M. Dressler, J.-D. Rüedi, Y. Suzuki, and H. M. Nagib. New insight into flow development and two dimensionality of turbulent channel flows. *Exp. Fluids*, 55(6):1–14, 2014a. (Cited on pp. 25, 27, 154, and 185).

R. Vinuesa, A. Noorani, A. Lozano-Durán, G. K. E. Khoury, P. Schlatter, P. F. Fischer, and H. M. Nagib. Aspect ratio effects in turbulent duct flows studied through direct numerical simulation. *J. Turb.*, 15 (10):677–706, 2014b. (Cited on p. 27).

R. Vinuesa, P. Schlatter, and H. M. Nagib. On minimum aspect ratio for duct flow facilities and the role of side walls in generating secondary flows. *J. Turb.*, 16(6):588–606, 2015. (Cited on pp. 27 and 33).

R. Vinuesa, C. Prus, P. Schlatter, and H. M. Nagib. Convergence of numerical simulations of turbulent wall-bounded flows and mean cross-flow structure of rectangular ducts. *Meccanica*, 51(12):3025–3042, 2016. (Cited on p. 188).

R. Vinuesa, P. Schlatter, and H. M. Nagib. Secondary flow in turbulent ducts with increasing aspect ratio. *Phys. Rev. Fluids*, 3(5):054606, 2018. (Cited on p. 27).

D. Viswanath. Recurrent motions within plane Couette turbulence. *J. Fluid Mech.*, 580:339–358, 2007. (Cited on p. 21).

D. Viswanath. The critical layer in pipe flow at high reynolds number. *Philos. Trans. R. Soc. A*, 367(1888): 561–576, 2009. (Cited on p. 21).

B. Vowinckel, V. I. Nikora, T. Kempe, and J. Fröhlich. Momentum balance in flows over mobile granular beds: application of double-averaging methodology to DNS data. *J. Hydr. Res.*, 55(2):190–207, 2017a. (Cited on p. 3).

B. Vowinckel, V. I. Nikora, T. Kempe, and J. Fröhlich. Spatially-averaged momentum fluxes and stresses in flows over mobile granular beds: a DNS-based study. *J. Hydr. Res.*, 55(2):208–223, 2017b. (Cited on pp. 3, 38, and 133).

A. Wachs. Particle-scale computational approaches to model dry and saturated granular flows of non-brownian, non-cohesive, and non-spherical rigid bodies. *Acta Mechanica*, 230(6):1919–1980, 2019. (Cited on pp. 43, 44, and 45).

F. Waleffe. On a self-sustaining process in shear flows. *Phys. Fluids*, 9(4):883–900, 1997. (Cited on pp. 20, 76, and 203).

F. Waleffe. Exact coherent structures in channel flow. *J. Fluid Mech.*, 435:93, 2001. (Cited on p. 21).

F. Waleffe. Homotopy of exact coherent structures in plane shear flows. *Phys. Fluids*, 15(6):1517–1534, 2003. (Cited on pp. 20, 21, 29, 73, 242, and 271).

J. M. Wallace, H. Eckelmann, and R. S. Brodkey. The wall region in turbulent shear flow. *J. Fluid Mech.*, 54(1):39–48, 1972. (Cited on pp. 24, 92, and 127).

Z. Wang and N. Cheng. Secondary flows over artificial bed strips. *Adv. Water Res.*, 28(5):441–450, 2005. (Cited on pp. 3, 26, and 34).

Z. Wang and N. Cheng. Time-mean structure of secondary flows in open channel with longitudinal bedforms. *Adv. Water Res.*, 29(11):1634–1649, 2006. (Cited on p. 34).

D. D. Wangsawijaya, R. Baidya, D. Chung, I. Marusic, and N. Hutchins. The effect of spanwise wavelength of surface heterogeneity on turbulent secondary flows. *J. Fluid Mech.*, 894:A7, 2020. (Cited on pp. 34, 66, and 78).

W. H. Warner. Poloidal and toroidal potentials for solenoidal fields. *Z. angew. Math. Phys.*, 23(2):221–230, 1972. (Cited on p. 271).

D. Willingham, W. Anderson, K. T. Christensen, and J. M. Barros. Turbulent boundary layer flow over transverse aerodynamic roughness transitions: induced mixing and flow characterization. *Phys.*

Fluids, 26(2):025111, 2014. (Cited on p. 34).

M. G. Wolman and L. M. Brush. Factors controlling the size and shape of stream channels in coarse noncohesive sands. *U.S., Geol. Surv., Profess. Papers*, 282-G:183–210, 1961. (Cited on pp. 35, 37, 91, 93, 140, 141, 144, 146, 149, 158, 159, 225, 226, and 241).

M. Wong and G. Parker. Reanalysis and correction of bed-load relation of Meyer-Peter and Müller using their own database. *J. Hydr. Eng.*, 132(11):1159–1168, 2006. (Cited on pp. xxvii, 84, 95, 96, 97, 98, 147, 165, and 166).

M. X. Xie, D. Chung, and N. Hutchins. Turbulent flow over spanwise-varying roughness in a minimal streamwise channel. In *J. Phys.: Conference Series*, volume 1522, page 012018. IOP Publishing, 2020. doi: 10.1088/1742-6596/1522/1/012018. (Cited on p. 186).

Y. Yamamoto and Y. Tsuji. Numerical evidence of logarithmic regions in channel flow at $\mathrm{Re}_\tau = 8000$. *Phys. Rev. Fluids*, 3:012602, 2018. (Cited on p. 22).

J. Yang and W. Anderson. Numerical study of turbulent channel flow over surfaces with variable spanwise heterogeneities: topographically-driven secondary flows affect outer-layer similarity of turbulent length scales. *Flow, Turb. Combust.*, 100(1):1–17, 2018. (Cited on p. 34).

G. Zampino, D. Lasagna, and B. Ganapathisubramani. Linearised Reynolds-averaged predictions of secondary currents in turbulent channels with topographic heterogeneity. *J. Fluid Mech.*, 944:A4, 2022. (Cited on pp. 67, 78, and 240).

A. Zampiron, S. Cameron, and V. Nikora. Secondary currents and very-large-scale motions in open-channel flow over streamwise ridges. *J. Fluid Mech.*, 887:A17, 2020a. (Cited on pp. 34, 66, 78, 133, and 136).

A. Zampiron, V. Nikora, S. Cameron, W. Patella, I. Valentini, and M. Stewart. Effects of streamwise ridges on hydraulic resistance in open-channel flows. *J. Hydr. Eng.*, 146(1):06019018, 2020b. (Cited on p. 34).

A. Zampiron, S. Cameron, and V. Nikora. Momentum and energy transfer in open-channel flow over streamwise ridges. *J. Fluid Mech.*, 915:A42, 2021. (Cited on p. 133).

L. Zeng, S. Balachandar, and F. M. Najjar. Wake response of a stationary finite-sized particle in a turbulent channel flow. *Int. J. Multiphase Flow*, 36(5):406–422, 2010. (Cited on p. 217).

N. Zgheib, J. Fedele, D. Hoyal, M. Perillo, and S. Balachandar. Direct numerical simulation of transverse ripples: 1. pattern initiation and bedform interactions. *J. Geophys. Res.: Earth Surface*, 123(3):448–477, 2018a. (Cited on p. 234).

N. Zgheib, J. Fedele, D. C. J. D. Hoyal, M. M. Perillo, and S. Balachandar. Direct numerical simulation of transverse ripples: 2. self-similarity, bedform coarsening, and effect of neighboring structures. *J. Geophysical Res.: Earth Surface*, 123(3):478–500, 2018b. (Cited on p. 234).

H. Zhang, F. X. Trias, A. Gorobets, Y. Tan, and A. Oliva. Direct numerical simulation of a fully developed turbulent square duct flow up to $Re_\tau = 1200$. *Int. J. Heat Fluid Flow*, 54:258–267, 2015. (Cited on pp. 27 and 29).

P. Zhang, Y. Duan, D. Li, J. Hu, W. Li, and S. Yang. Turbulence statistics and very-large-scale motions in decelerating open-channel flow. *Phys. Fluids*, 31(12):125106, 2019. (Cited on p. 31).

Z. Zhou, C. Xu, and J. Jiménez. Interaction between near-wall streaks and large-scale motions in turbulent channel flows. *J. Fluid Mech.*, 940:A23, 2022. (Cited on pp. 25, 115, 145, 147, and 242).

List of Figures

List of Tables

Dissertationsreihe am Institut für Hydromechanik
Karlsruher Institut für Technologie (KIT)

ISSN 1439-4111

Die Bände sind unter www.ksp.kit.edu als PDF frei verfügbar oder als Druckausgabe bestellbar.

2004/1 **Volker Weitbrecht**
Influence of dead-water zones on the dispersive mass-transport in rivers.
ISBN 3-937300-07-4

2004/2 **Christof Hinterberger**
Dreidimensionale und tiefengemittelte Large-Eddy-Simulation von Flachwasserströmungen.
ISBN 3-937300-18-X

2005/1 **Cornelia Frank**
Wirksamkeit von dünnen Windschutzstreifen auf Sockelwällen in luv- und leeseitiger Anordnung.
ISBN 3-937300-55-4

2005/2 **Carl Friedrich von Carmer**
Shallow turbulent wake flows: momentum and mass transfer due to large-scale coherent vortical structures.
ISBN 3-937300-63-5

2005/3 **Muhammad Ikhwan**
Investigation of flow and pressure characteristics around pyramidal buildings.
ISBN 3-937300-72-4

2005/4 **Herlina**
Gas transfer at the air-water interface in a turbulent flow environment.
ISBN 3-937300-74-0

2006/1 **Manual García-Villalba Navaridas**
Large eddy simulation of turbulent swirling jets.
ISBN 3-86644-015-4

2006/2 **Klaas E. Heinrich**
Hydraulische Steuerung einer Wasser-Alkohol-Zirkulation mittels Grundwasser-Zirkulations-Brunnen.
ISBN 3-86644-055-3

2006/3 **Tobias Bleninger**
Coupled 3D hydrodynamic models for submarine outfalls. Denvironmental hydraulic design and control of multiport diffusers.
ISBN 978-3-86644-160-6

2006/3 **Patrick Heneka**
Schäden durch Winterstürme – das Schadensrisiko von Wohngebäuden in Baden-Württemberg.
ISBN 978-3-86644-117-0

2007/1 **Maria Eletta Negretti**
Hydrodynamic instabilities and entrainment in two-layer stratified flows down a slope.
ISBN 978-3-86644-170-5

2007/2 **Gregor Kühn**
Untersuchungen zur Feinsedimentdynamik unter Turbulenzeinfluss.
ISBN 978-3-86644-205-4

2008/1 **Martin Detert**
Hydrodynamic processes at the water-sediment interface of streambeds.
ISBN 978-3-86644-244-3

2008/2 **Christof Gromke**
Einfluss von Bäumen auf die Durchlüftung von innerstädtischen Straßenschluchten.
ISBN 978-3-86644-339-6

2009/1 **Ekkehart Bethge**
Risikoberechnung zum Schadstoffeintrag aus Hochwasserretentionsräumen
in einen Grundwasserleiter.
ISBN 978-3-86644-426-3

2012/1 **Clemens Chan-Braun**
Turbulent open channel flow, sediment erosion and sediment transport.
ISBN 978-3-86644-900-8

2015/1 **Todor Doychev**
The dynamics of finite-size settling particles.
ISBN 978-3-7315-0307-1

2016/1 **Aman Ghebremichael Kidanemariam**
The formation of patterns in subaqueous sediment.
ISBN 978-3-7315-0532-7

2019/1 **Yoshiyuki Sakai**
Coherent structures and secondary motions in open duct flow.
ISBN 978-3-7315-0790-1

2023/1 **Michael Werner Tobias Krayer**
A hydrodynamical perspective on the turbulent transport of bacteria in rivers.
ISBN 978-3-7315-1310-0

2023/2 **Markus Maximilian Scherer**
Turbulent coherent structures, Secondary currents and Sediment ridges.
ISBN 978-3-7315-1318-6